MW00837243

An ICMA Green Book

EMERGENCY MANAGEMENT

Principles and Practice for Local Government

SECOND EDITION

An ICMA Green Book

EMERGENCY MANAGEMENT

Principles and Practice for Local Government

SECOND EDITION

Edited by
William L. Waugh Jr.
Georgia State University

Kathleen Tierney
University of Colorado at Boulder

ICMA

Leaders at the Core of Better Communities

ICMA advances professional local government worldwide. Its mission is to create excellence in local governance by developing and advancing professional management of local government. ICMA, the International City/County Management Association, provides member support; publications, data, and information; peer and results-oriented assistance; and training and professional development to more than 9,000 city, town, and county experts and other individuals and organizations throughout the world. The management decisions made by ICMA's members affect 185 million individuals living in thousands of communities, from small villages and towns to large metropolitan areas.

Library of Congress Cataloging-in-Publication Data

Emergency management : principles and practice for local government / edited by William L. Waugh Jr., Kathleen Tierney. — 2nd ed.
p. cm.
Includes bibliographical references and index.
ISBN-13: 978-0-87326-719-9
1. Emergency management. 2. Central-local government relations. 3. Hazard mitigation.
I. Waugh, William L. II. Tierney, Kathleen J.
HV551.2.E457 2007
363.34'560973--dc22
2007041959

Copyright © 2007 by the International City/County Management Association, 777 North Capitol Street, N.E., Suite 500, Washington, D.C. 20002. All rights reserved, including rights of reproduction and use in any form or by any means, including the making of copies by any photographic process, or by any electronic or mechanical device, printed, written, or oral or sound or visual reproduction, or for use in any knowledge or retrieval system or device, unless permission in writing is obtained from the copyright proprietor. This copyright does not extend to Chapter 10, which was co-written by a federal government employee as part of his official duties and therefore considered a work of the United States government.

Printed in the United States of America
2014 2013 2012 2011 2010 2009
5 4 3 2

ABOUT THE EDITORS

William L. Waugh Jr. (Editor and Chapters 1 and 16) is professor of public administration in the Andrew Young School of Policy Studies at Georgia State University in Atlanta. He also teaches in the graduate program in crisis and emergency management at the University of Nevada, Las Vegas. He serves on the Emergency Management Accreditation Program (EMAP) Commission and has also served on the Certified Emergency Manager (CEM) Commission. Recognized nationally and internationally for his scholarship and contributions to the emergency management profession, Dr. Waugh is the editor-in-chief of the *Journal of Emergency Management* and is on the editorial board of *Public Administration Review.* He attended the University of Maryland in Munich, Germany, and College Park and earned his bachelor's degree from the University of North Alabama, his master's degree from Auburn University, and his doctorate in political science from the University of Mississippi.

Kathleen Tierney (Editor and Chapter 16) is professor of sociology and director of the Natural Hazards Center at the University of Colorado at Boulder. The Hazards Center is housed in the Institute of Behavioral Science, where Dr. Tierney holds a joint appointment. Prior to her arrival at the University of Colorado in 2003, she was the director of the Disaster Research Center at the University of Delaware. Dr. Tierney is also a member of the executive committee of the National Consortium for the Study of Terrorism and Responses to Terrorism, a Department of Homeland Security social science academic center of excellence headquartered at the University of Maryland. Her research interests include the study of crisis-related collective behavior; extreme event preparedness among governmental, for-profit, and nonprofit organizations; the use of new information technologies in emergency management; and disaster and homeland security policy and institutions. Her publications include numerous articles and book chapters, among which are chapters on business disaster preparedness and on homeland security policy. Dr. Tierney teaches courses on qualitative research methods and on societal dimensions of hazards and disasters at the University of Colorado.

ICMA's "Green Books" (a designation derived from the original bright green cloth covers) have a long history as the authoritative source on local government management. They are used by local government managers in cities and counties worldwide, by university professors and students as textbooks for undergraduate and graduate courses, and by public safety professionals in preparation for promotional exams. The Green Books cover the range of local government functions, linking the latest theories and research to specific examples of day-to-day decision making and the nuts and bolts of management. Current titles in the Green Book series include

The Effective Local Government Manager

Emergency Management: Principles and Practice for Local Government

Local Government Police Management

Management Policies in Local Government Finance

Managing Fire and Rescue Services

Managing Local Government Services

The Practice of Local Government Planning

Service Contracting: A Local Government Guide

CONTENTS

Sidebars

Table

Figures

FOREWORD

Since the 1980s, ICMA has been a leader in the field of emergency management, disseminating knowledge, sharing successful practices, publishing the first comprehensive text, and supporting emergency management practitioners and organizations in efforts to set standards for the profession.

In 1991, with partial funding from the Federal Emergency Management Agency (FEMA), ICMA published the first edition of *Emergency Management: Principles and Practice for Local Government*. Edited by Thomas E. Drabek and Gerard J. Hoetmer, that groundbreaking volume was embraced by the emergency management community and by the professors and researchers who were seeking to establish emergency management in the public administration curriculum.

At the time, emergency management was a comparatively new and evolving field that often existed "under the radar"—at least as far as citizens were concerned. In the intervening years, however, a number of events have raised the profile of emergency management and changed its focus: the attacks of September 11, 2001, the devastation wrought by Katrina and other hurricanes, catastrophic tsunamis and earthquakes in Asia and elsewhere, train and subway bombings in Madrid and London. Emergency management became headline news, and awareness of vulnerability to disaster—natural or man-made—increased.

In the wake of these and other events, the context of emergency management has changed, and this completely new edition reflects those changes. On the national level, the focus has broadened from natural and technological hazards to include terrorism and threats to public health.

Another change is the increasing recognition that disaster knows no geographical or political boundaries. Collaboration among all levels of government is crucial, and local governments have learned that they need to prepare for disasters that extend beyond their borders. Many local governments now, for example, have mutual aid agreements that cross state lines and offer resources to disaster-stricken communities in distant states.

Still another change is the emphasis on intersectoral cooperation. Today's emergency managers coordinate not only with government agencies but also with medical and public health organizations, social service agencies, private firms, nonprofits, faith-based organizations, neighborhood and community groups, volunteer organizations, the media, and others. Such entities participate in the development of emergency plans, provide emergency services, and may even operate emergency operations centers.

Views of recovery have also evolved: the point is no longer simply to restore infrastructure and services, but to take restoration to a new level by creating a more resilient—and more sustainable—community. To that end, communities that have been struck by disaster must make choices during recovery that will strengthen their economic and social infrastructure and mitigate future damage from disaster. Recovery is a lengthy process, and one that necessarily involves government and nongovernmental organizations as well as the private sector.

In the background of these and other changes is the maturation of the profession, as evidenced by the proliferation of emergency management programs in colleges and universities, the rise of organizations that build the capacity of emergency managers and first responders, and the development of standards for the field.

Local governments organize the emergency management function in many ways—but whatever the size, budget, and organizational placement of the emergency services agency, it is the lead entity in all four phases of emergency management: mitigation, preparedness, response, and recovery efforts.

Collectively, local governments are the backbone of the national emergency management system. This book is designed to assist current and future emergency managers in fulfilling that crucial function by providing context, sharing experience, and offering insights for practitioners and students alike. It is ICMA's hope that this volume will become the standard text and reference for emergency management in the early years of the twenty-first century.

ICMA would like to thank the editors of this volume, William L. Waugh Jr. and Kathleen Tierney, and the members of the advisory board, whose role was to review the proposed book outline, suggest sidebars and case studies, and recommend authors. Advisory board members were William A. Anderson, associate executive director, Division on Earth and Life Studies, National Research Council, National Academies; Erik Auf der Heide, Agency for Toxic Substances and Disease Registry, U.S. Department of Health and Human Services; Emily Bentley, director and assistant professor, Homeland Security and Emergency Management Program, Savannah State University; Arrietta Chakos, former assistant city manager, Berkeley, California; Frances L. Edwards, associate professor, Department of Political Science, San José State University; David R. Godschalk, professor emeritus, Department of City and Regional Planning, University of North Carolina; Gerard J. Hoetmer, executive director, Public Entity Risk Institute; David A. McEntire, associate professor, Emergency Administration and Planning, University of North Texas; Brenda Phillips, professor, Fire and Emergency Management Program, Oklahoma State University; Richard A. Rotanz, special assistant to the president and provost, Emergency Management Program, Adelphi University; Daryl Lee Spiewak, CEM, emergency programs manager, Brazos River Authority, Waco, Texas; and Richard T. Sylves, professor, Department of Political Science and International Relations, University of Delaware.

We are particularly grateful to advisory board member Gerry Hoetmer for encouraging ICMA to develop this new volume, for sharing ideas and information, and for providing thoughtful reviews of many chapter drafts, and to Sandra F. Chizinsky, whose skillful editing lent a consistent voice to the work of the book's multiple authors. We also thank the authors for sharing their expertise and for their cooperation and patience during the review and editorial phases.

A number of ICMA staff members, former staff, and other individuals contributed as well, each playing a role in research, editing, and managing the numerous tasks that go into the development and production of a major book. Working under the leadership of Ann I. Mahoney, director of publishing, these individuals were Jane C. Cotnoir, Nedra M. James, Valerie Hepler, Will Kemp, Barbara H. Moore, and Jane E. Lewin.

Robert J. O'Neill
Executive Director
ICMA
Washington, D.C.

PART ONE

Context and Organization of Emergency Management

CHAPTER 1

Local emergency management in the post-9/11 world

William L. Waugh Jr.

This chapter provides an understanding of

- Increasing vulnerability to disaster
- The social and political context of emergency management
- Emerging forms of governance
- Emergencies, disasters, catastrophes, and incidents of national significance
- Standards for emergency management
- The profession of emergency management.

On September 11, 2001, the world of local emergency managers became much more complex. The roles and functions of emergency managers did not change dramatically, but the attention of the public and its elected representatives shifted rapidly from natural disasters to intentional man-made disasters—terrorism, in particular—and emergency managers were expected to shift their priorities accordingly.

Terrorism had been a concern of emergency managers long before the attacks on the World Trade Center and the Pentagon. Indeed, emergency managers have long viewed terrorism as one of the many threats that they had to address, although they tended to focus on the similarities between the risks posed by terrorism and those posed by other hazards, rather than on the differences. Moreover, emergency managers tended to see terrorism as a federal, rather than a state or local problem.[1] But in the years since 9/11, local emergency managers have been compelled to divide their attention between a new national priority, the "war on terror," and the "old" wars against the more common—and more likely—natural and technological disasters. For some emergency managers, particularly those in cities and urban counties, a greater emphasis on terrorism has seemed reasonable, given that terrorists have selected large urban centers and symbols of Western culture as targets. For others, continuing to focus primarily on floods, fires, tornadoes, hurricanes, earthquakes, and other hazards has seemed to be the reasonable course. Particularly in a small community, the threat of an international or domestic terrorist attack seems much less compelling than the threat of a flood, fire, or windstorm.

Nevertheless, even in large cities and urban counties, there is a disconnect between national and local priorities: federal programs focus principally on the threat to civil aviation and on the porous American borders, the two vulnerabilities made most evident by the 9/11 attacks. Officials in heavily populated areas, in contrast, are principally concerned about the vulnerability of critical infrastructure, government facilities, port facilities, mass transit systems, and other potential targets. A series of events has drawn attention to, and confirmed the seriousness of, such vulnerabilities:[2] the 1993 attack on the World Trade Center by international terrorists; the 1994 sarin attack on the Tokyo subway system; the 1995 attack on the Murrah Federal Building in Oklahoma City by domestic terrorists; the bombings in the Madrid and London transportation systems in 2004 and 2005, respectively; and the attempted bombings in London and the attack on the Glasgow airport in 2007.

The biggest lesson of the 2005 hurricanes was that the nation's emergency management system is broken, and that officials need to rebuild local, state, and regional capacities to reduce hazards and respond to disasters.

Professional emergency managers continue to debate their role in dealing with the threat of terrorism. Many feel, for example, that responsibility for prevention rests with law enforcement rather than with emergency management. However, the 2007 edition of the Emergency Management Accreditation Program (EMAP) standards, which provides benchmarks for emergency management programs, includes terrorism as one of the many hazards with which emergency management programs must deal. In addition, the 2007 edition of NFPA 1600—the National Fire Protection Association's Standard on Disaster/Emergency Management and Business Continuity Programs—integrates terrorism prevention into the comprehensive emergency management model, and there is at least some expectation that emergency management programs will share responsibility for dealing with the terrorist threat.[3] But the inclusion of prevention programs within emergency management has serious implications for how emergency management agencies function and how they relate to the public. For example, the transparency and collaborative processes that are essential to effective emergency management are difficult to maintain when information cannot be shared and all stakeholders cannot be involved in decision making.

Meanwhile, although the threat of terrorism is still viewed as important, the more time that passes without a terrorist attack on American soil, the lower the priority assigned to the threat.[4] Moreover, the busy hurricane season of 2004 and the embarrassingly poor responses to Hurricanes Katrina and Rita in 2005 forced another reevaluation of priorities (see Figure 1–1). The biggest lesson of the 2005 hurricanes was that the nation's emergency management system

Photo courtesy of FEMA/Jocelyn Augustino

Figure 1-1 Aerial photo of a flooded neighborhood in New Orleans after Hurricane Katrina

is broken, and that officials need to rebuild local, state, and regional capacities to reduce hazards and respond to disasters. The hurricanes also vividly illustrated the political costs of failing to prepare for disaster. Because the federal government has been slow to respond to the failures of 2005, state and local officials have focused on improving the Emergency Management Assistance Compact (EMAC), which is a state-level mutual aid system, and on expanding community-to-community mutual assistance compacts.

Increasing vulnerability to disaster

Unfortunately, communities in the United States are becoming more and more vulnerable to major disasters. Many settlements are already located on waterways, near mountain passes, and along coastlines—in areas prone to floods, wildfires, earthquakes, landslides, hurricanes, tsunamis, volcanic activity, and other hazards. The population is continuing to migrate to regions that are at greater risk of disaster. According to the National Oceanic and Atmospheric Administration, more than 50 percent of Americans live within fifty miles of the coast; by 2025, 70 percent will live within that zone.[5] Population density is also increasing in coastal areas that are known to be hazardous. Climate change promises rising sea levels, more frequent and more powerful coastal storms, and the spread of tropical diseases, as well as increased incidence of flooding, wildfires, drought, and heat waves. And while land use planners and emergency managers try to discourage building in the urban-woodland interface, on floodplains, and in other hazardous areas, policy makers, developers, and homeowners are not necessarily heeding the warnings. Finally, modern cities have fragile power, communications, and transportation infrastructure, and the built environment is increasingly vulnerable to structural fires, hazardous materials accidents, and other technological disasters.

In short, the potential for catastrophe is growing. Of particular concern, however, is the uneven distribution of vulnerability. As Hurricane Katrina demonstrated, the poor, the elderly, and those with chronic medical conditions need extraordinary assistance, and are often hardest hit by disaster. In New Orleans, many residents were unable to evacuate—because they lacked the necessary transportation, financial resources, and medical assistance—or chose not to evacuate because they did not understand or believe the warnings, needed to care for housebound relatives, or wanted to protect their pets or belongings. Although the regional evacuation was remarkable in its scale and effectiveness, the distressing images of evacuees

Figure 1-2 New Orleans residents lining up to get into the Superdome, which was opened as a hurricane shelter in advance of Hurricane Katrina

Photo courtesy of FEMA/Marty Bahamonde

in the New Orleans Superdome and convention center were testament to the poor execution of the city's plan to deal with residents who could not or would not evacuate (Figure 1–2). Neither the Superdome nor the convention center had adequate food or water, and neither was managed effectively as an emergency shelter.[6]

The possibility of catastrophic disaster—including events such as comet strikes and gamma-ray bursts—has always appealed to the popular imagination. Movies, documentaries, and television shows feature every conceivable disaster, from super-volcanoes and super-tsunamis to asteroid strikes and bioterrorism attacks. While the potential for such disasters is real, the threats that they pose are rarely of immediate concern to public officials. However, the combination of imagined disasters and real ones—from the Northridge, California, earthquake of 1994 to the terrorist attacks of 2001, the Asian tsunami of 2004, and the hurricanes of 2005—does focus public attention on the need to prepare for disaster. But whether greater public awareness translates into increased understanding of disaster and greater willingness to prepare for it is questionable.

Neighborhood, family, and individual preparedness efforts can greatly increase the local capacity to respond to and recover from disasters.

Recent studies have concluded that the public is not confident that government can respond to disasters effectively, and is not as prepared as it should be. A 2006 survey by Irwin Redlener and colleagues at the Mailman School of Public Health at Columbia University noted a decline in public confidence in the government's ability to protect citizens. In addition, despite the fact that an increasing percentage of respondents are concerned about a terrorist attack (82 percent, up from 78 percent in 2005 and 76 percent in 2004), relatively few respondents in their study (31 percent) had a family emergency plan.[7] In a survey undertaken in the summer of 2006, Paul C. Light of the Brookings Institution found similar results: although 44 percent of those surveyed had increased their preparedness after the September 11 attacks, 50 percent had maintained about the same level of preparedness. And while 34 percent of respondents had increased their disaster preparedness after Katrina and Rita, 57 percent did little.[8] Light recommends reforms at the local, state, and federal levels to increase public confidence and to ensure a more effective response to the next catastrophic disaster. Among other strategies, he supports local capacity building and making emergency management a more visible function of government.

Increasing public preparedness is an objective of several governmental and nongovernmental organizations. EMAP, for example, has developed guidelines for community preparedness (2006); the American Red Cross has its own public education programs; and the U.S. Department of Homeland Security's ReadyAmerica, ReadyBusiness, and ReadyKids programs are designed to increase both public awareness and public preparedness.[9] Nevertheless, emergency managers continue to be concerned by the lack of preparedness among households and public and private sector organizations.[10]

Despite the multitude of changes that have occurred since 9/11, the safety and security of our communities still depend on the capacity of local emergency management agencies, and on nongovernmental resources ranging from volunteer organizations to private firms. When a disaster strikes, the first people on the scene are typically friends and neighbors, who begin search and rescue and provide medical assistance. Thus, neighborhood, family, and individual preparedness efforts can greatly increase the local capacity to respond to and recover from disasters. The overwhelming majority of disasters are handled by local first responders, with the help of local agencies—such as the Red Cross, the Salvation Army, and other community organizations—and individual volunteers (Figure 1–3). In large-scale disasters, local agencies are responsible for managing the event until help arrives. For that reason, local capacity building has been a priority for decades, and professional emergency managers have fought to prevent cuts in Emergency Management Performance Grants and similar programs, on which they rely to maintain and expand critical functions.

The social and political context of emergency management

The social and political context within which local emergency managers work today is very different than it was in the late 1990s and has undergone significant changes since 9/11. One major shift that has affected all public officials, including emergency managers, is the contracting out of government services to private and nonprofit organizations. As a result of this change, private firms, faith-based organizations, community groups, and other organizations now participate in the

Photo courtesy of FEMA/Mark Wolfe

Figure 1-3 An American Red Cross setup in Lady Lake, Florida, a community hard-hit by tornadoes in February 2007

Building local government capacity: Findings from ICMA's Homeland Security Survey, 2005

Following the events of September 11, 2001, local governments took steps to increase their capability to deal with future emergencies, according to an ICMA survey of local government managers.

A large majority of respondents (nearly 79 percent) conducted a homeland security-related risk assessment, and nearly 90 percent reported that they consider a terrorist threat to government buildings or installations to be the greatest potential threat to their jurisdictions. More than half (60 percent) developed a comprehensive plan or amended an existing plan; 55 percent provided training for non-first responders (such as administrative staff); and 54 percent conducted an emergency drill or exercise.

Homeland security issues cut across jurisdictional lines, and 91 percent of respondents reported collaborating with other local governments, 75 percent with the state government, and 60 percent with a regional organization. Not surprisingly, local governments also took steps to improve the communication and coordination among jurisdictions. At least 61 percent adopted the National Incident Management System (NIMS), a protocol developed by the Federal Emergency Management Agency to help ensure consistency of training and approach for first responders at all governmental levels. Nearly a third of respondents hired or appointed a manager to help coordinate intergovernmental security functions.

Cities and counties used their own funds for many homeland security activities, primarily for equipment (66 percent), disaster mitigation and preparedness (64 percent), drills or training exercises (63 percent), and disaster response (60 percent). They also sought and received funding from both state and federal governments.

When asked about requests for state or federal funding for homeland security-related needs, approximately 60 percent of respondents reported requesting such funds for equipment. It appears that most requests for funding were granted, as the percentage reporting awards closely tracked the percentage requesting funds.

Nonetheless, with the exception of funds for equipment, a majority of local governments did not report requesting either federal or state funds. This may support anecdotal information suggesting that the process of requesting funding is prohibitive for local governments. The average amount awarded by the federal government was $434,000. Only 19 percent of respondents reported that the municipality had experienced budget shortfalls as a result of homeland security activities during the two fiscal years prior to 2005.

Source: ICMA, "Homeland Security, 2005" survey. The findings are based on responses from 2,786 local governments nationwide with populations over 2,500. Aggregate results are available on the ICMA Web site, icma.org/main/bc.asp?bcid=141&hsid=1&ssid1=44&ssid2=80&ssid3=215.

development of emergency plans, provide emergency services, and even operate emergency operations centers, along with providing other services and products. Response and recovery efforts following the landfall of Katrina, for example, were managed largely by contractors, raising concerns about questionable expenditures of public monies by contractors; quality control (because so many contractors and subcontractors were used); the use of out-of-state rather than in-state contractors; and the sheer size of the bill. The contracts cost billions of dollars, and the benefits were not always apparent.

During the 1990s, mitigation became the mantra of the emergency management community: the common wisdom was that a dollar spent on mitigation saved four dollars in recovery. Although it has always been easier to get funding after a disaster than before, preventing or reducing the impact of disaster is a far more effective policy than merely reacting to one. Project Impact, for example, a program created by the Federal Emergency Management Agency (FEMA) in 1997, was designed to address vulnerabilities and thereby reduce losses. Although the program was disbanded in 2001, FEMA has made hazard reduction—through effective land use regulation, zoning ordinances, and building codes, among other measures—a priority.[11] Nevertheless, progress has been slow.[12]

In the mid-1990s, the goal of disaster resistance was expanded to cover disaster resilience, which would both reduce losses and speed recovery. Increasingly, disaster recovery has come to be synonymous with sustainable development to reduce hazards. In fact, one of the lessons of both 9/11 and the Katrina disaster is that recovery plans should be linked to long-term development plans, so that communities can use redevelopment opportunities both to reduce the risk of future disasters and to address community needs. A number of measures can serve both objectives: for example, residential areas can be rebuilt outside of floodplains; rights-of-way can be acquired for mass-transit lines; schools can be relocated and redesigned; and business districts can be reconfigured to lessen vulnerability and increase economic viability.[13] Along the coasts of Louisiana and Mississippi, communities that were devastated by Hurricanes Katrina and Rita have been redesigned in ways that are intended to reduce vulnerabilities (although there are strong pressures to permit homeowners to rebuild in hazardous areas).

The traditional "cavalry" role of government–rushing in to save the day–has given way to a more collaborative and cooperative set of relationships among networks of public, private, and nonprofit organizations.

The biggest changes, however, concern individual responsibility and participation in decision making. Greater self-reliance is encouraged at the level of the community, the family, and the individual. In emergency management, that change has meant that the traditional "cavalry" role of government—rushing in to save the day—has given way to a more collaborative and cooperative set of relationships among networks of public, private, and nonprofit organizations.[14] In a related shift, communities now have greater say in setting priorities. In the past, localities relied on outside authorities to deal with disaster; increasingly, however, state and federal authorities actively support disaster operations but defer to local officials for critical decisions. For example, although state authorities generally have the power to take over in extreme circumstances, they seldom do. Similarly, although federal authorities now have the power to determine that a disaster is an incident of national significance (INS) and to assume authority for disaster operations, doing so might precipitate a political conflict with state and local authorities. More practically, preemption of state or local authority might interfere with the intergovernmental collaboration that is necessary in a large-scale disaster.

Given that decisions made during disaster operations can have profound implications for recovery, engaging local officials and their constituents in all phases of decision making is critical. Along the Gulf Coast, for example, Katrina and Rita destroyed the basic infrastructure of many communities. Rebuilding infrastructure to facilitate redevelopment has been a very slow process, in part because it is difficult to determine how to restore infrastructure without exposing it to loss or damage in future storms. The scale of the damage wrought by the 2005 hurricanes has spurred a reexamination of disaster recovery, compelling citizens and officials to examine new development in light of its sustainability.

Emergent organizations, as well as individual volunteers, have always been important actors in disasters and even in the management of environmental and other hazards, and their roles are expanding. In short, the new context for emergency management is one in which the emergency manager must work closely with social, political, and economic networks whose trust must be earned, and over which the emergency manager has little or no authority. It is also a context in which state and federal officials have limited authority and, therefore, cannot assume that they can preempt local prerogatives. The networked environment requires greater investment in relationship building and more inclusive approaches to decision making.

Emerging forms of governance

Large-scale disaster response requires multiorganizational, intergovernmental, and multisector collaboration. Preparedness and recovery involve social networks in which authority is shared, responsibility is diffused, and resources are widely distributed. The management of such systems is increasingly referred to as "governance" rather than "government,"[15] but it also goes by a number of other names: in the United States, it is called "collaborative government" or "network management"; in the United Kingdom, it is "joined up" government; Bob O'Neill, executive director of ICMA, has referred to it as the "working together" approach[16] (see sidebar on pages 10–11). However this new form of governance is described, it is not characterized by centralized decision making; instead, authority and responsibility are shared.

Local government's role in natural disaster

My challenge . . . is to give a post-Katrina perspective on emergency planning and response in the United States. First, I would like to place emergency preparedness in the context of federalism in the United States. Second, I will discuss the systematic challenges to our current approaches. Third, what we have learned over the last year, and finally, what ICMA has suggested as an approach to improve our ability to respond to large-scale disasters.

As was reported around the globe, the U.S. response to the hurricanes was fraught with problems. Undoubtedly, bureaucratic processes contributed. The United States has a tradition of federalism, which "is built on the premise that incidents are handled at the lowest level of government possible."[1] Conceptually, this sounds beneficial to local governments, but in an emergency the process can defeat the best of intentions, especially when the process isn't flexible enough to adapt to instantaneous changes in the scope and impact of the event. . . .

On paper, our response process is described in the National Response Plan (NRP), which was developed in December 2004 and provides a structure for coordinating the local, state, and federal governments' response to disasters. . . . Our federal system require[s] a complex approach to emergency response. Because there are vast resources to call upon, it can be impressive. However, who is in charge, who can access the resources and when, and who pays represent huge challenges. It has been described as a system where everyone can say no but no one can say yes.

First, a bit of background: There are four triggers that relate to the Incidents of National Significance (INS) that activate the National Response Plan, but there is lack of clarity on how and when an event becomes an INS and, perhaps more importantly, where responsibility lies for the declaration of such an event. Before any federal action is initiated, local governments must request assistance from their state government, and states must provide available relief before seeking federal assistance. States provide relief to one another through the Emergency Management Assistance Compact (EMAC), which is an interstate mutual aid agreement.

In addition to the NRP and the four triggers that initiate it, there is separate legislation, the Stafford Act, which further confuses the process. One of the obstacles is that although the president may declare that an incident is an "emergency" at his own discretion, emergency assistance is limited to $5 million without notification to Congress. A "major disaster" has no limit on financial assistance, but the president only declares a major disaster at the request of a state governor. The distinction between an emergency and a major disaster is difficult to ascertain during the event, possibly resulting in no request and no declaration.

When the president makes the declaration, the federal response is activated. There are numerous federal agencies responsible for various emergency support functions. The relatively small Federal Emergency Management Agency (FEMA), under the Department of Homeland Security (DHS), is responsible for coordinating these emergency support functions. FEMA had become a well-respected agency when knowledgeable and experienced people were appointed to key positions in the 1990s. After Hurricanes Hugo and Andrew, reflection on the emergency response at all levels of government prompted recommendations for change at all levels of government. There was anticipation that these changes would be positive. And, had some of these changes been implemented, we might have been in a less chaotic situation in August 2005. Unfortunately, when the massive new Department of Homeland Security was established in 2003 with terrorism as its focus, FEMA got lost in the bureaucracy. Many of the strong professionals left the agency, leaving FEMA with inadequate professional resources. It was reported that although FEMA typically has 2,500 employees, when Katrina hit, there were only 2,000—a loss of 20 percent of the workforce. To give you a realistic picture, DHS has approximately 180,000 employees, including thousands of employees from independent agencies, such as the Coast Guard, U.S. Secret Service, Bureau of Customs and Border Protection, and FEMA. . . .

The confusion around what qualifies as an INS and in what form the request for federal assistance must be presented explains some, but not all, of the slow response to Katrina. Another challenge arose at the state level because the EMACs were flooded with requests, which soon became backlogged due to lack of capacity to process them.

We knew we had trouble with our disaster response when the Canadian Mounties arrived in Louisiana before our U.S. government was able to get help there. Local governments across the country were ready to go, but few of them were able to get through the state and federal government's red tape to get into the region. Those that received authorization did not always have the most targeted experience. . . .

In a disaster, such as Katrina, local and state capabilities are quickly overwhelmed, especially when the personnel responsible for response are often victims. Overwhelmed state and local response capabilities, combined with an unclear process for initiating federal response, as demonstrated in Katrina, illustrated the need to take a fresh look at how disaster response is organized.

In the aftermath of Hurricanes Katrina and Rita, congressional committees and federal and state agencies examined what went wrong and what could be learned from the experience. There were several consistent themes regardless of the agency reporting.

- We must recognize that the first information received in a disaster situation is often inaccurate, incomplete, or wrong.
- We must improve communication and interoperability.
- We must restructure federal agencies and align federal processes for clearly defined roles and responsibilities.
- We must clarify the appropriate role for the Department of Defense and its relationships with other disaster response participants. . . .

The Katrina experience was a wake-up call for many of us. Our country is relearning the value of professionalism. . . . We are relearning the value of relationships and giving more thought to how we can build stronger networks that are based on training and relationships. We have opportunities to make changes community by community. We held the Restoration 2006 conference . . . in New Orleans to bring all of the national associations of local governments together with businesses, nonprofits, and academicians who have studied the difference between communities that are successful in rebuilding and those that struggle. . . .

Learning from tsunamis, floods, and terrorist events like [the 2005 London bombings], we all have struggled to organize the most effective response. . . . It's vital that we share our challenges and successes, and our resources and our expertise. . . .

The "working together" approach resonates with local government professionals. Historically, the United States has tried to manage disasters with a "command and control" approach. ICMA and other organizations representing local governments are proposing a dynamic and network-centered approach that has the flexibility to move resources and assets where they need to be, when they need to be there. Working together gives us the greatest hope of not repeating the past. We must work together to

- Restructure federal agencies and align federal processes for clearly defined roles and responsibilities.
- Establish region-based teams of federal, state, and local employees
- Involve private and nonprofit sectors in planning and response
- Improve communication/interoperability
- Provide first responder training and credentialing
- Demand accountability
- Ensure risk assessment and management
- Improve our ability to collect accurate information, analyze it, and respond accordingly. We must be able to adapt quickly to changing conditions.
- Take advantage of the transport expertise and other services the military can provide in emergencies. We need to develop collaborative relationships that allow all involved—military, federal, state, local, and nongovernmental organizations—to bring to bear the best they have to offer in ways [that] minimize bureaucratic obstacles and complement and strengthen the response.

Source: Excerpted and adapted from the keynote address delivered by Bob O'Neill, executive director of ICMA, at the 2006 National Congress of Local Government Managers, May 23, 2006, Australia, available at lgma.org.au/national/2006NationalCongress/papers/Bobs%20speech.pdf (accessed August 29, 2007).

[1]"National Preparedness—A Primer," in *The Federal Response to Hurricane Katrina: Lessons Learned* (Washington, D.C., February 2006), 14, available at whitehouse.gov/reports/katrina-lessons-learned.pdf (accessed August 29, 2007).

Figure 1-4 The only surviving home, built using many FEMA standards, in an area completely destroyed by Hurricane Katrina

Photo courtesy of FEMA/John Fleck

As a result, political leadership in the governance era requires new approaches focused on transparency, inclusiveness, and relationship building.[17]

Collaborative processes may well achieve purposes that centralized processes cannot. A study of FEMA's efforts to encourage safe construction practices, for example, concluded that the agency's power to influence such practices derives largely from its authority to regulate development on floodplains (which comes, in turn, from its management of the National Flood Insurance Program). Nevertheless, the agency also used collaborative and cooperative means, including formal partnerships, economic and regulatory incentives, and direct, informal contacts. In many cases, for example, FEMA officials personally encouraged local champions of safe construction. FEMA also worked with the entities that develop and promote building standards, including the building industry, the insurance industry, national code organizations, and the agricultural extension programs that build model homes to encourage energy conservation[18] (see Figure 1-4). Collaborative processes have proven more effective than other approaches in local government as well.[19]

Emergencies, disasters, catastrophes, and incidents of national significance

According to the National Response Plan (NRP), three documents—the Homeland Security Act of 2002, Homeland Security Presidential Directive 5, and the Stafford Act of 1988—provide a comprehensive, all-hazards approach to "domestic incident management."[20] Under new laws and presidential directives issued since 9/11, disaster declarations are now viewed as matters of domestic incident management, and emergencies and disasters are categorized as "incidents of national significance." As defined in the NRP, an INS is "an actual or potential high-impact event that requires coordination of Federal, State, local, tribal, nongovernmental and/or private sector entities in order to save lives and minimize damage."[21] In the event of an INS, money from the Presidential Disaster Trust Fund and other sources may be directed to pre-event mitigation or prevention. Although all major disasters, emergencies, and catastrophic incidents declared by the president are incidents of national significance, not all incidents of national significance "necessarily result in disaster or emergency declarations under the Stafford Act."[22]

The NRP also added a new category, "catastrophic incidents," which are

> any natural or manmade incident, including terrorism, that results in extraordinary levels of mass casualties, damage, or disruption severely affecting the population,

infrastructure, environment, economy, and national morale and/or government functions. A catastrophic event could result in sustained national impacts over a prolonged period of time; almost immediately exceeds resources normally available to State, local, tribal, and private-sector authorities in the impacted area; and significantly interrupts governmental operations and emergency services to such an extent that national security could be threatened. All catastrophic incidents are considered Incidents of National Significance."[23]

In another shift, the president now has authority to issue both pre- and post-event disaster declarations for "national special security events" (NSSEs)—"high-profile, large-scale events that present high-probability targets," such as the G-8 Summit, the Republican and Democratic national conventions, and any other event the president believes may be vulnerable to terror attack.[24] A local emergency manager would be well-advised to keep track of NSSE declarations since they can potentially demand significant mobilization, response, and spending on the part of local emergency management agencies. Annexes to the NRP indicate that the president, under the Homeland Security Act of 2002,[25] also has formal authority to issue a disaster declaration for bioterror, cyberterror, food and agricultural terror attacks; nuclear or radiological incidents; oil and hazardous materials incidents of national significance; and other large-scale emergencies.[26]

Standards for emergency management

Standards for emergency management, especially those embodied in NFPA 1600 and the Emergency Management Accreditation Program (EMAP), provide guidance that can assist states and localities in dealing with risk. Perhaps more importantly, however, the standards create an obligation: local officials may be held legally, as well as politically, liable for failure to recognize and address hazards. Thus, at a minimum, officials must (1) review the history, frequency, and severity of potential hazards; (2) assess the vulnerability of their communities; (3) identify steps that can be taken to mitigate the hazards; and (4) on the basis of risk and vulnerability assessments, determine how resources can best be allocated to reduce losses.

When it comes to standards, conflicts between national priorities and state and local priorities are all too common. In 2006, for example, in the wake of the Katrina disaster, federal officials declared that all cities should have mass evacuation plans. Officials in many large cities, however, felt that the need to evacuate an entire city was a very remote possibility; partial evacuation or sheltering in place might be needed in the event of a hazardous materials spill or a terrorist bombing, but the location of such an incident would determine the kind of protective action needed, the areas of the city that might be affected, and the number of people who would need to be evacuated or sheltered. Local response plans generally include provisions for such contingencies. Nonetheless, the federal government issued an unfunded mandate requiring mass evacuation plans, and local governments were compelled to divert resources from other purposes in order to comply.

Local officials may be held legally, as well as politically, liable for failure to recognize and address hazards.

The environment in which local emergency managers function is political (indeed, often partisan), and priorities are determined by available funding (see sidebar on page 14). Since 9/11, federal policy makers have been focused on the threat of terrorism, and federal funding has followed. Although it has been a source of frustration for many emergency managers, the shift in federal focus has not been without positive impacts. For example, although the adoption of the National Incident Management System (NIMS) may be problematic because of the dominance it assigns to the federal role, NIMS does provide a common language and a common structure to guide disaster response. Moreover, 9/11 created a "window of opportunity" for federal funding—and even though the funding might not support local priorities, it has

Funding homeland security

Because homeland security programs are often unfunded or underfunded mandates, local authorities have had to rely on their own resources–often, current budgets–or simply to decline to implement the programs. Like their state and federal counterparts, local officials do not like to raise taxes–and the risk of disaster, including terrorist attacks, is generally not compelling enough to persuade them to propose new taxes.

The two accompanying tables illustrate the limitations on funding homeland security and the principal methods used.

What are your jurisdiction's biggest homeland security problems?

	City, %	County, %
Lack of money	66	52
Personnel limitations	38	40
Technology/interoperability	24	36
Lack of clear plan/roles	13	15
Local health care capacity	11	19
Lack of external cooperation	10	15
Lack of internal cooperation	10	13
We don't have a problem	10	5

How will your jurisdiction pay for its portion of homeland security costs?

	City, %	County, %	Total, %
Dedicate a special sales tax	2	0	1
Issue bonds	3	0	2
Raise property taxes	6	11	7
Raise utility rates	7	2	5
Existing budget/revenue/funds	48	55	49
Reallocate/cut other spending	18	18	16
Asset seizure funds	16	7	12
General fund	43	37	41

Source: "Homeland Security Outlook: Local Governments Reveal Purchasing Plans for 2006," *American City & County* (January 2006): 38, 40. © 2006 Penton Media, Inc. Available at americancityandcounty.com/mag/homelandsecurity06.pdf (accessed August 27, 2007).

been used to enhance and expand local capabilities. Local emergency managers have generally pursued a "dual use" strategy, acquiring equipment, training, and other resources that are useful in natural and technological disasters, as well as in terrorism-related disasters.

The profession of emergency management

In September of 2001, the profession of emergency management was beginning to find itself. After a series of failures and scandals in the 1980s, FEMA had been reinvented with a focus on mitigation and with strong links to state and local governments. The "golden age" of FEMA had a positive impact on the agency's state and local counterparts. The Certified Emergency Manager (CEM) program was created in the early 1990s (see accompanying sidebar), NFPA 1600 was issued in 1995, the FEMA Higher Education Project was initiated in 1995, and EMAP was initiated in 1997.[27] At least at a broad level, a number of questions—What is a professional emergency manager? What should emergency management programs look like, and what should they do? What education and training are appropriate for entry-level, management, and executive positions in the field?—had clear answers. On 9/11, the political context changed. Emergency management became a subset of homeland security, and decisions were made by officials who had little or no understanding of the role or function of emergency management. Professional emergency managers were frustrated—but tried, for the most part, to negotiate the new political milieu while sustaining their focus on all the hazards that threatened their communities.

Meanwhile, because citizens and government officials continued to confuse "emergency management" with "emergency response," resources tended to be targeted to first responders,

The Certified Emergency Manager program

In the early 1990s, the International Association of Emergency Managers (IAEM) developed a certification program designed to raise and maintain professional emergency management standards. The program is designed to certify the achievements of any emergency management professional, regardless of job title, who meets the requirements.

Candidates for certification must meet a multitude of criteria and go through a rigorous peer-reviewed process. After the initial certification, a certified emergency manager (CEM) must recertify every five years or lose the use of the CEM designation. Because of these requirements, the CEM credential is one of the most difficult certifications to obtain and is highly valued by those who possess it.

Progress toward certification begins when a candidate enrolls in the program. Successful candidates must demonstrate their knowledge, skills, and abilities by documenting emergency management experience, continuing education, and contributions to the profession. Candidates must also have superior written communications skills and must pass a multiple-choice exam.

The specific program requirements are as follows:

- **Experience:** Three years of full-time comprehensive emergency management experience by date of application. The experience must include participation in a full-scale exercise or actual disaster.
- **References:** Three professional references, one of whom must be the candidate's current supervisor.
- **Education:** Any four-year baccalaureate degree. To satisfy this requirement, candidates may substitute additional years of experience at the rate of two years of experience for 30 college credits, up to the 120 credits that make up most baccalaureate degree programs. (Emergency managers who are interested in obtaining recognition without meeting the education requirement may be eligible for the Associate Emergency Manager [AEM] credential.)
- **Training:** One hundred contact hours in emergency management *and* 100 contact hours in general management, with no more than 25 hours accepted for credit in any one topic.
- **Contributions to the profession:** Six separate contributions in such areas as professional membership, speaking, teaching, publications, service on volunteer boards or committees, special assignments, awards, and other areas beyond the scope of routine emergency management job requirements.
- **Management essay:** Completion of an essay based on a real-life emergency management scenario, with the response demonstrating comprehensive emergency management knowledge, skills, and abilities.
- **Multiple-choice examination:** Candidates may sit for the 100-question exam any time after their initial application.

Source: Daryl Spiewak, CEM Commission; for more information, go to iaem.org and click on "Certification."

[1]Candidates may download an application from IAEM's Web site at iaem.com/certification/GettingStarted/intro.htm, or they may contact IAEM headquarters. The links on IAEM's Web page explain the certification process and fees. Other links provide access to application tips and recent articles from the *IAEM Bulletin* regarding the certification requirements, which a candidate can read before deciding to enroll in the program. For additional information, write to cem@iaem.com, or call 703-538-1795.

and relatively little was invested in improving local emergency management capabilities. Following the 2001 attacks, resources were also diverted to law enforcement and national security programs to protect the nation from terrorism. Many of the mistakes made in the response to Hurricane Katrina arose from confusion about the very nature of emergency management. Decision makers at all levels failed to make the best use of the available expertise.[28] Another problem was that emergency managers were having a difficult time explaining their roles to policy makers and to the public. Discussions within the profession indicated a need for greater clarity about the role and nature of emergency management.

Some of the problems that emergency managers have experienced with homeland security policies and programs reflect basic differences of opinion about the role and function of emergency management, and about public participation in decision making, transparency in decision making, and information sharing with other agencies and the public. Just as federal priorities often conflict with local, state, and regional needs, the federal approach to emergency management often conflicts with the approach of professional emergency managers.

FEMA's Emergency Management Higher Education Project

To better face the challenges confronting the nation, it is essential to ensure that emergency services personnel in both the public and private sectors are more professional, more skilled, more diverse, and better educated than they have been in the past. It is also crucial to foster greater "disaster sensitivity" among professionals in other fields. Thus, one of the goals of the Federal Emergency Management Agency is to encourage and support the expansion of emergency management education in colleges and universities across the United States so that future emergency services personnel, both in government and in the private sector, will come to the job with not just a college education but at least a course of study, if not a degree, in emergency management. Such efforts are contributing to both the professionalization of emergency management as a discipline, and the expansion of emergency management knowledge and principles into other professions and disciplines—two outcomes that will help to build the disaster-resilient communities of tomorrow.

In 1995, the Emergency Management Institute, located at the National Emergency Training Center in Emmitsburg, Maryland, developed the Higher Education Project, whose primary purpose is to encourage and support the development of bachelor's and graduate degree programs in U.S. colleges and universities.[1] In the year the project was undertaken, there were only four collegiate emergency management programs in the country, only one of which offered emergency management courses for academic credit. By 2007, there were 127 college-level programs spread throughout all but three states. These programs included

- More than seventy certificates, minors, and diplomas
- Thirty-five associate's (two-year) degrees
- Sixteen bachelor's (four-year) degrees
- Forty-seven emergency management-related master's degrees
- Seven emergency management-related doctoral degrees
- One Ph.D. in emergency management.

There were also sixty-one programs focused on homeland security, fourteen focused on public health and disasters, and nine focused on humanitarian assistance. Emergency management higher education programs have been growing at the rate of one to two per month.

Additional information concerning the Emergency Management Higher Education Project can be accessed at training.fema.gov/EMIWeb/edu/ (accessed August 29, 2007).

Source: B. Wayne Blanchard, manager, Emergency Management Higher Education Project, Emergency Management Institute, National Emergency Training Center, Federal Emergency Management Agency, U.S. Department of Homeland Security.

[1]The Higher Education Project has also funded the development of twenty-two "prototype" courses that are available to faculty.

In particular, the centralization of decision making encouraged by federal authorities conflicts with the decentralization of decision processes at the local level.[29]

In the fall of 2006, Michael Selves, president of the International Association of Emergency Managers (IAEM), suggested a review of the profession. In early 2007, with the assistance of FEMA's Emergency Management Institute, a roundtable discussion was held that included representatives from IAEM, EMAP, the National Emergency Management Association (the professional organization of state emergency management directors), the NFPA 1600 committee, the private sector, the academic community, and other stakeholder groups.[30] The objectives of the meeting were (1) to define emergency management and its mission and (2) to identify the core principles of the profession and the practice of emergency management. Although the existing literature mentioned a number of broad principles, a set of core principles had never been identified and codified.

The core principles were drawn from the literature, the CEM program, NFPA 1600, and EMAP standards. The starting point was the concept of comprehensive emergency management, a principle that is emphasized in education and training programs, and that serves as an organizing principle for emergency management agencies and programs.[31] Agreement on the next seven principles took more time, but differences tended to concern semantics rather than substance (see accompanying sidebar).

Comprehensive emergency management encompasses all hazards, all phases, all functions, and all sectors. In the years since 9/11, emergency managers have often had occasion to remind federal policy makers that the field includes all hazards, including terrorism. Moreover,

Emergency management definition, mission, vision, and principles

Definition

Emergency management is the managerial function charged with creating the framework within which communities reduce vulnerability to hazards and cope with disasters.

Mission

To protect communities by coordinating and integrating all activities necessary to build, sustain, and improve the capability to mitigate against, prepare for, respond to, and recover from threatened or actual natural disasters, acts of terrorism, or other man-made disasters.

Vision

Safer, less vulnerable communities with the capacity to cope with hazards and disasters.

The principles of emergency management

- **Comprehensive:** Emergency managers consider and take into account all hazards, all phases, all stakeholders and all impacts relevant to disasters.
- **Progressive:** Emergency managers anticipate future disasters and take preventive and preparatory measures to build disaster-resistant and disaster-resilient communities.
- **Risk-driven:** Emergency managers utilize sound risk management principles: hazard identification, risk analysis, and impact analysis. Priorities and resources are assigned on the basis of this process.
- **Collaborative:** Emergency managers create and sustain broad and sincere relationships among individuals and organizations to encourage trust, advocate a team atmosphere, build consensus, and facilitate communication.
- **Integrated:** Emergency managers are responsible for ensuring, to the highest possible degree, unity of effort among all levels of government and all elements of a community.
- **Coordinated:** Emergency managers organize all relevant stakeholders with a common purpose.
- **Flexible:** Emergency managers rely on creative and innovative approaches to solving disaster challenges. This is especially the case after disasters when predefined approaches may be inadequate to the situation at hand.
- **Professional:** Emergency managers value a science and knowledge-based approach based on education, training, experience, ethical practice, public stewardship and continuous improvement.

Source: Adapted from Michael D. Selves, "From the President: Update on the Emergency Management Roundtable Project," *IAEM Bulletin* (July 2007): 2-3, reprinted with permission of the International Association of Emergency Managers, www.iaem.com.

limiting the purview of emergency management to "consequence management" in the wake of a terrorist attack—as does, for example, the Nunn-Lugar-Domenici Act of 1996[32]—ignores the importance of measures that can prevent or reduce the impact of such attacks; it also ignores the importance of other hazards. Similarly, the focus on prevention from the perspective of law enforcement or national security, rather than in the broader sense of mitigation, ignores opportunities to prevent terrorism-related disasters by addressing precipitants, changing land use and building design, and undertaking other mitigation measures.[33]

Emergency management must be *progressive:* that is, emergency managers must anticipate and prepare for future disasters. The long-term goal is the development of disaster-resilient communities—that is, communities that address their vulnerabilities and enlarge their capacity to respond and recover—and emergency managers must be proactive in pursuing that goal. A narrow focus on the most recent category of disaster, or on a single type of disaster, limits both strategic and tactical options.

Emergency management should be *risk-driven:* that is, decisions must be based on real, measured risk, rather than on worst-case scenarios that are unaccompanied by estimations of probability. In recent years, the allocation of federal emergency management and Department of Homeland Security (DHS) funds to states and localities on the basis of criteria other than risk—using simple population formulas, for example—has generated controversy. An emergency manager's first responsibility is to protect his or her community: most often, that means focusing on the most likely natural hazards, rather than on terrorism or on other threats identified by federal or state officials. Although assessments of potential hazards may be problematic, making decisions without at least attempting to undertake a full assessment may be downright risky.

Effective emergency management requires *collaboration* rather than a command-and-control approach. Collaboration is a matter of building relationships and building consensus. It is founded on trust and open communication, and has become essential in the networked world of emergency managers.[34] In fact, the ability to work collaboratively with public, private, and nonprofit agencies, as well as with volunteers and emergent groups, has become the critical leadership skill for emergency managers.[35]

Overcoming disciplinary boundaries, professional blinders, and personal predispositions in order to ensure effective collaboration is a difficult and time-consuming task. When organizational cultures have historically encouraged parochial perspectives, an individualist orientation, and classic bureaucratic structures, collaboration has not been easy. Nonetheless, sensitivity to the values and norms of other organizations can make collaboration possible. In practical terms, such collaboration often means learning to work with individuals and entities that are less oriented toward formal structures, less responsive to outside authority, and less willing to follow rules without explanation.

In addition to being well trained and well educated in emergency management, effective emergency managers are skilled communicators, "big-picture" thinkers, and adept at administration and politics.

Emergency managers facilitate *integration:* this means bringing together all levels of government, the private sector, the nonprofit sector, and volunteers, and making them active participants in all phases of emergency management. The American emergency management system is deeply rooted in voluntarism, and communities still rely heavily on the American Red Cross, the Salvation Army, and other community organizations, both faith-based and secular. These organizations represent the surge capacity of communities, states, and the nation in the face of every type of natural or man-made disaster, including terrorist attacks.[36] It is essential, however, to cultivate links to volunteers, nongovernmental organizations, and the private sector before disaster strikes.

Emergency managers *coordinate* organizational efforts: that is, they organize disparate stakeholders behind a common purpose.[37] Importantly, the function of the emergency operations center (EOC) is to provide both a venue and mechanisms to facilitate communication and collaboration (see Figure 1–5). Effective intra- and interorganizational communication improves operational decision making. The critical importance of coordination is evident in the various functions of the EOC—in particular, in the activation of a multiagency coordination system and unified command structures.

Effective emergency management requires *flexibility.* Because emergency managers have to adapt plans to circumstances, innovate, and improvise when necessary, rigid plans and organizational structures are to be avoided. The plan is a starting place rather than an immutable guide to action. In a healthy emergency management environment, officials have the discretion to interpret plans and respond to circumstances. Craig Fugate, emergency management director for the state of Florida, has argued, for example, that although the Incident Command System (ICS) is useful for structuring response efforts, it is only a tool and may have to be adapted to circumstances.[38] Fugate has also observed that "ICS zealots" can actually hamper response operations by limiting flexibility.[39] Flexibility does not mean, however, that there are no rules governing actions. Understanding the boundaries of authority and action requires knowledge of the legal, social, and political context of emergency management, which emergency managers gain through a combination of experience and education. The IAEM has created a code of conduct to guide ethical practice.

Professional emergency management is a science- and knowledge-based profession: emergency managers take action and make decisions on the basis of evidence. Education and train-

Figure 1-5 Local officials responding to Hurricane Katrina at the Harrison County emergency operations center in Gulfport, Mississippi

Photo courtesy of FEMA/Mark Wolfe

ing programs provide them with the technical skills and knowledge base they need to increase the effectiveness of emergency management programs.

Emergency management requires good problem-solving skills, as well as expertise in planning, organizing, management, staffing, coordination, evaluation, and budgeting. Emergency managers need not be technical specialists, but they should have some knowledge of the science of natural and technological hazards, including terrorism. In addition to being well trained and well educated in emergency management, effective emergency managers are skilled communicators, "big-picture" thinkers, and adept at administration and politics. A commitment to public service and public stewardship—that is, to the long-term public interest—is also critical. Effective emergency managers are committed to continuous improvement, and to identifying and applying lessons learned. Finally, they are involved, trusted members of the community and the profession. All these qualifications are required for the CEM credential; they are also essential to working effectively in a network environment.

The organization of this book

The homeland security environment has not been a hospitable one for emergency managers, especially those at the local level. As a result of constantly shifting priorities, standards, and requirements, the authors of this volume were often in the position of attempting to hit moving targets; thus, the sixteen chapters in this volume reflect the conflicts and tensions that prevail in the field. Conflict is not new to emergency management of course: the first edition of this textbook, which came out in 1991, was being written when FEMA was still being castigated for failures in the response to Hurricane Hugo in 1989. Those problems surfaced again in 1992, with the inadequate response to Hurricane Andrew. FEMA and the national emergency management system were reinvented, reformed, and restaffed soon thereafter. Some things have not changed, however. In the introduction to the 1991 edition, Gerard Hoetmer referred to the changing environment of emergency management and suggested that greater interdepartmental collaboration was needed—and, indeed, would render day-to-day local government operations more effective; this recommendation is equally appropriate today.[40]

The chapters in this volume are organized around three themes: the context and organization of emergency management (Chapters 1 through 5), the functions and phases of emergency management (Chapters 6 through 11), and major issues in emergency management (Chapters 12 through 16).

This first chapter identifies the major trends in emergency management—in particular, the shift in national priorities from natural and technological hazards to counterterrorism in the wake of the attacks of September 11, 2001. The chapter also traces the development of standards for emergency management and the professionalization of the field. In Chapter 2, "Local Emergency Management: Origins and Evolution," Claire B. Rubin offers a brief history of emergency management in the United States, and considers the larger organizational and intergovernmental context within which emergency management operates. In Chapter 3, "Organizing for Emergency Management," Frances Edwards and Dan Goodrich examine the administrative and business continuity responsibilities of emergency management agencies, describe how the emergency management function fits into the local government organization, consider the role of emergency management in each phase of emergency management, and trace historical changes in emphasis and approach within the emergency management function. In Chapter 4, "The Intergovernmental Context," David McEntire and Gregg Dawson discuss the impact of the federalist system on emergency management; they also describe emergency management networks, the structures used to coordinate interagency and interjurisdictional operations, and approaches to financing emergency management. In Chapter 5, "Collaborative Emergency Management," Ann Patton covers the origins, context, benefits, and importance of collaborative management, and provides practical recommendations for successful collaboration.

In Chapter 6, "Mitigation," David Godschalk focuses on the value and context of mitigation and the mitigation planning process. The chapter also describes hazard-specific loss-reduction strategies and addresses the link between mitigation and sustainability. In Chapter 7, "Planning and Preparedness," Michael Lindell and Ronald Perry present the principles of emergency planning, explain the components of an emergency plan and the basic analyses that underlie it, identify conditions in the community that promote effective emergency response, and briefly discuss preparedness by households, businesses, and government agencies that do not have explicit emergency management missions. In Chapter 8, "Applied Response Strategies," Richard Rotanz focuses on the development of a response management strategy. Topics covered include the phases of disaster, agent-generated versus response-generated demands, and emergency operations centers.

In Chapter 9, "Disaster Response," Ronald Perry and Michael Lindell address the context of emergency response—from the initiation of local response operations to the use of ICS and NIMS and the request for a Presidential Disaster Declaration. In Chapter 10, "The Role of the Health Sector in Planning and Response," Erik Auf der Heide and Joseph Scanlon cover all the practical issues related to emergency medical care in disasters: both informal, on-the-spot accommodations and established formal structures, including to the National Disaster Medical System and the Medical Reserve Corps. In Chapter 11, "Recovery," Brenda Phillips and David Neal define recovery and discuss the recovery planning process, the dimensions of recovery, and recovery resources.

In Chapter 12, "Legal Issues," William C. Nicholson addresses five topics: the laws that are specific to the emergency management function; the role of tort law in emergency management; governmental immunity from liability; evolving standards that have legal implications for emergency management; and legal issues in preparedness, response, recovery, and mitigation. In Chapter 13, "Identifying and Addressing Social Vulnerabilities," Elaine Enarson addresses the needs of populations with high social vulnerability—the social groups that have the fewest defenses against disaster and are least resilient in the aftermath of disaster. The chapter discusses the changing concept of social vulnerability, the complexity of the term in practice, and structural trends in the United States that increase vulnerability. The chapter also describes tools and strategies that emergency managers can use to protect the most vulnerable members of society. In Chapter 14, "New Information Technologies in Emergency Management," Susan L. Cutter, Christopher T. Emrich, Beverley J. Adams, Charles K. Huyck, and Ronald T. Eguchi examine the role of information technology in emergency management since 1989, new tools and applications appropriate to specific phases of emergency management, and barriers to the use of information technology in emergency management practice. Among the technologies discussed are geographic information systems, remote sensing, database management, warning systems, and resource management.

In Chapter 15, "Budgeting for Emergency Management," Richard T. Sylves discusses the new prominence of homeland security as a local government responsibility. He also examines local, state, and federal funding for emergency management. Finally, Chapter 16, "Future Directions in Local Emergency Management," identifies critical issues that will shape local emergency management in the coming decade and examines their implications for local emergency management.

Conclusions

As noted at the beginning of this chapter, local emergency management is much more complex than it was in the late 1990s. A number of events—most particularly those of September 11, 2001, but also the 2004 and 2005 hurricane seasons; the Asian earthquake and tsunami in 2004; the earthquake in Pakistan in 2005; the bombing of the Madrid train station in 2004; and the bombing of the London Underground in 2005—have affected policy and funding priorities, the placement and organization of FEMA within the federal government, and, perhaps most important, the relationship between government and the public. Since the formation of DHS in 2003, emergency management at the federal level has been characterized by less transparency, less public participation in policy making, and less willingness on the part of federal officials, particularly those in DHS, to work with the networks that are the foundation of the national emergency management system.[41]

Although the attacks of 9/11 certainly had a major impact on the task environment of emergency managers, one thing remains clear: the primary responsibility of local emergency managers is to protect their communities from the hazards that pose the greatest risk. The most significant lesson of Hurricanes Katrina and Rita is that the national emergency management system has to be rebuilt. The process of rebuilding should focus on two goals: developing the capacities of local emergency managers and first responders, and increasing the disaster resilience of communities.

The emergency management agency has limited scope and limited authority, and the issues it needs to address are bigger than the agency's mission. It is therefore critical that emergency managers educate other officials, and assist them to address risks and engage in preparedness efforts. For example, emergency managers are increasingly involved in decisions about zoning, land use, and other aspects of local government that have the potential to reduce the risk of hazards. As Michel Doré, emergency management director for the province of Québec, observed in a session at the 2007 Natural Hazards Research and Applications Workshop, emergency managers have to learn to play in other "sandboxes." To succeed, emergency managers must have good interpersonal skills, good political skills, and good administrative skills, as well as skills in emergency planning, coordination, and other technical functions.[42]

Public education is essential because many officials and the public at large still see emergency management as largely a response function. The view that emergency management is reactive—specifically, that it is largely a matter of consequence management—is limiting. Professional emergency managers use a comprehensive approach, in which response is an essential component but by no means the only component. The poor response to Hurricanes Katrina and Rita demonstrated, among other things, a lack of understanding of the role and nature of emergency management among federal, state, and local officials. The work of the Emergency Management Roundtable promises to define the practice and profession of emergency management, which will ultimately help policy makers and the public better understand both—and, in particular, to understand the distinction between the emergency management community and first responders.

Since the 1980s, the field of emergency management has become rapidly professionalized. Nevertheless, many local emergency managers are still part-time staff members or unpaid volunteers. Many emergency management offices are still grossly understaffed and underfunded. Too many communities cannot even achieve basic emergency management capabilities. Clearly, lacking a qualified emergency manager, an emergency plan, or adequate communications equipment is a problem, just as is failing to take the necessary actions to prevent disaster and to reduce its effects. The question that officials struggle with, however, is how much is enough?

How much do they need to invest in emergency management programs and staff? How much capacity is needed? NFPA 1600 and EMAP provide benchmarks for emergency management programs, and it is the responsibility of public officials to follow their guidance.

Homeland security plans that do not have an all-hazards perspective and that cannot be easily adapted to nonterrorist emergencies, and federally sanctioned decision structures that fail to take into account nongovernmental organizations and volunteers—and may, in fact, discourage information sharing within and between agencies—will continue to be a source of conflict in the field. The most important task of the coming years will be to reconcile the priorities of local emergency management with those of the federal government.

Notes

1 William L. Waugh Jr., "Terrorism as Disaster," in *Handbook of Disaster Research,* ed. Havidán Rodríguez, E. L. Quarantelli, and Russell R. Dynes (New York: Springer, 2007), 388–404.

2 William L. Waugh Jr., "Mass Transit Security after the Madrid and London Bombings," in *Handbook of Transportation Policy and Administration,* 2nd ed., ed. Jeremy Plant (New York: CRC Publishers, 2007).

3 NFPA 1600, Standard on Disaster/Emergency Management and Business Continuity Programs, which is becoming a widely accepted standard in the field, has been endorsed by both Congress and the 9/11 Commission. The National Fire Protection Association has been involved in fire safety for more than one hundred years and has issued approximately three hundred codes and standards to reduce the risk of fire and other hazards. See www.nfpa.org for more information.

4 Waugh, "Terrorism as Disaster."

5 Kristen M. Crossett et al., *Population Trends along the Coastal United States: 1980–2008,* Coastal Trends Report Series (Washington, D.C.: National Oceanic and Atmospheric Administration, U.S. Department of Commerce, 2004).

6 William L. Waugh Jr., "The Political Costs of Failure in the Responses to Hurricanes Katrina and Rita," in *Shelter from the Storm: Repairing the National Emergency Management System after Hurricane Katrina,* ed. William L. Waugh Jr. (Thousand Oaks, Calif.: Sage, 2006), 10–25.

7 Irwin Redlener et al., "Where the American Public Stands on Terrorism, Security, and Disaster Preparedness: Five Years after September 11, One Year after Hurricane Katrina" (New York: Mailman School of Public Health, Columbia University, September 2006), available at ncdp.mailman.columbia.edu/files/2006_white_paper.pdf (accessed August 1, 2007).

8 Paul C. Light, "The Katrina Effect on American Preparedness: A Report on the Lessons Americans Learned in Watching the Katrina Disaster Unfold" (New York: Center for Catastrophe Preparedness and Response, New York University, 2006), available at crid.or.cr/digitalizacion/pdf/eng/doc16262/doc16262-contenido.pdf (accessed August 30, 2007).

9 See Ready.gov.

10 There is debate even within the emergency management community, however, as to how much preparedness is desirable for households: some believe that each household should have a kit that will enable it to be self-reliant during the first seventy-two hours of a disaster; others believe that kits should provide sustenance and essential gear for a week to ten days, or even a month.

11 Ironically, the president announced the termination of Project Impact on the same day that the Nisqually earthquake in Washington State demonstrated the effectiveness of the program in encouraging local mitigation efforts. The program had helped Seattle retrofit buildings to reduce seismic vulnerabilities. See Eric Holdeman, "Destroying FEMA," *Washington Post,* August 30, 2005, A17, available at washingtonpost.com/wp-dyn/content/article/2005/08/29/AR2005082901445_pf.html (accessed August 30, 2007).

12 Raymond Burby, "Hurricane Katrina and the Paradoxes of Government Disaster Policies: Bringing about Wise Governmental Decisions for Hazardous Areas," in *Shelter from the Storm* (see note 6), 171–191.

13 William L. Waugh Jr. and R. Brian Smith, "Economic Development and Reconstruction on the Gulf after Katrina," *Economic Development Quarterly* 20, no. 3 (August 2006): 211–218.

14 See, for example, Steven Goldsmith and William D. Eggers, *Governing by Network: The New Shape of the Public Sector* (Washington, D.C.: Brookings Institution Press, 2004).

15 See, for example, Steven Kelman, "The Transformation of Government in the Decade Ahead," in *Reflections on 21st Century Government Management,* ed. Donald F. Kettl and Steven Kelman (Washington, D.C.: IBM Center for the Business of Government, 2007), 33–55, available at businessofgovernment.org/pdfs/KettlKelmanReport.pdf (accessed August 30, 2007).

16 Bob O'Neill, "Local Government's Role in Natural Disaster," Keynote Address, 2006 National Congress of Local Government Managers, May 23, 2006, Australia, available at lgma.org.au/national/2006NationalCongress/papers/Bobs%20speech.pdf (accessed August 29, 2007).

17 William L. Waugh Jr. and Gregory Streib, "Collaboration and Leadership for Effective Emergency Management," *Public Administration Review* 66, Special Issue on Collaborative Management (December 2006): 131–140.

18 William L. Waugh Jr., *Using Networks to Leverage National Goals: FEMA and the Safe Construction Networks* (Washington, D.C.: PricewaterhouseCoopers Foundation for The Business of Government, 2002).

19 See, for example, Robert Agranoff and Michael McGuire, *Collaborative Public Management: New Strategies for Local Governments* (Washington, D.C.: Georgetown University Press, 2003), and Robert Agranoff, *Managing within Networks: Adding Value to Public Organizations* (Washington, D.C.: Georgetown University Press, 2007).

20 U.S. Department of Homeland Security (DHS), *National Response Plan* (Washington, D.C.: DHS, 2004), 5, available at dhs.gov/xlibrary/assets/NRP_FullText.pdf (accessed August 29, 2007). In July 2007, a draft of the *National Response Framework* (NRF) was released for comment. When approved, the NRF will supersede the NRP. The new document mentions natural disasters more frequently and terrorism less frequently than the NRP. In addition, it pays more attention to the role of the private sector and nongovernmental organizations. The document states that one objective is to ensure that the relationships among participants are more collaborative; it characterizes these relationships as "engaged partnerships." Questions remain, however, concerning the involvement of local and state officials in drafting the document. See DHS, *National Response Framework,* Pre-Decisional and Deliberative Draft

(Washington, D.C.: DHS, July 2007), available at hlswatch.com/sitedocs/20070801-1nrf-framework.pdf (accessed August 29, 2007); see also Eileen Sullivan, "DHS Achieves Target Weight for National Response Guidelines," *CQ Homeland Security,* August 1, 2007, available at public.cq.com/docs/hs/hsnews110-000002564676.html (accessed August 29, 2007).

21 DHS, *National Response Plan,* 67.

22 Ibid., 7.

23 Ibid., 43.

24 Ibid., 4.

25 Public Law 107-296.

26 *National Response Plan,* xii.

27 Keith Bea, *Emergency Management Preparedness Standards: Overview and Options for Congress,* CRS Report RL 32520 (Washington, D.C.: Congressional Research Service, Library of Congress, updated February 4, 2005), available at fas.org/sgp/crs/homesec/RL32520.pdf (accessed August 30, 2007).

28 Waugh, "Terrorism as Disaster."

29 Ibid., and William L. Waugh Jr., "Mechanisms for Collaboration: ICS, NIMS, and the Problem of Command and Control," in *The Collaborative Manager,* ed. Rosemary O'Leary (Washington, D.C.: Georgetown University Press, 2008).

30 The group has come to be known as the Emergency Management Roundtable. In the fall of 2007, the group was drafting "The Doctrine of Emergency Management" to operationalize the principles.

31 Despite the importance of comprehensive emergency management, however, it had clearly been misunderstood by policy makers when the national emergency management system was reorganized in 2003.

32 Public Law 104-201, formally known as the Defense Against Weapons of Mass Destruction Act.

33 Waugh, "Terrorism as Disaster."

34 Roger L. Kemp, "Emergency Management and Homeland Security: An Overview" (Washington, D.C.: ICMA, August 2006).

35 Waugh and Streib, "Collaboration and Leadership for Effective Emergency Management."

36 William L. Waugh Jr., "Terrorism, Homeland Security and the National Emergency Management Network," *Public Organization Review* 3, no. 4 (2003): 373–385; and William L. Waugh Jr. and Richard T. Sylves, "Organizing the War on Terrorism," *Public Administration Review* 62, suppl. 1 (September 2002): 145–153.

37 William L. Waugh Jr., "Co-ordination or Control: Organizational Design and the Emergency Management Function," *International Journal of Disaster Prevention and Management* 2 (December 1993): 17–31.

38 Waugh, "Mechanisms for Collaboration."

39 These are not new problems or even problems unique to emergency management. The U.S. military has relied upon career development programs, from tracks providing increasing responsibility to professional education, to broaden perspectives from specialist to generalist. The Certified Emergency Manager (CEM) program similarly assumes professional growth and a broadened perspective for professionals in the field.

40 Gerald J. Hoetmer, "Introduction," in *Emergency Management: Principles and Practice for Local Government* (Washington, D.C.: International City/County Management Association, 1991), xxxiv.

41 Even the Congressional Research Service and the Government Accountability Office have had trouble getting information out of DHS. Moreover, FEMA has been reorganized several times since 2003, and even congressional committees have complained that they cannot keep track of the changes within DHS.

42 Thomas E. Drabek, *The Professional Emergency Manager* (Boulder, Colo.: Institute for Behavioral Science, University of Colorado, 1987).

CHAPTER 2

Local emergency management: Origins and evolution

Claire B. Rubin

This chapter provides an understanding of

- The history of emergency management in the United States
- The organizational and intergovernmental context of emergency management.

Since 1900, many disasters—from the San Francisco Earthquake of 1906, to the Great Dust Bowl of the 1930s, to the terrorist attacks of 2001, to Hurricane Katrina in 2005—have become part of U.S. political history and cultural heritage. Interest in these "focusing events," as some scholars have called them, remains high; they have been brought to life through a number of popular nonfiction accounts,[1] and many even have Web sites devoted to them. The stories surrounding these events provide a fascinating glimpse of American history, but they also raise the questions that inevitably surround a disaster: Was it predictable? Was it preventable? Was the response timely? Adequate? Were the recovery efforts sufficient? And, most importantly, how do the systems that are in place work—in theory and in actuality?

The history of emergency management in the United States became the focus of critical interest in 2005, when three hurricanes—Katrina, Rita, and Wilma—caused record-breaking damage to the Gulf Coast area. These hurricanes did for natural-disaster response what the terrorist attacks of September 11, 2001, did for counterterrorism: by glaringly displaying the weaknesses and failures of systems, processes, and leadership, these events prompted the deepest and most sustained examination of public emergency management ever conducted in U.S. history. National attention was focused on emergency management at every level of government, with hearings, studies, and investigations undertaken by Congress and the White House; by agencies involved in the response; and by scholars, think tanks, and many others.[2] Citizens, the media, and officials at every level of government questioned the emergency management systems that were in place. These exhaustive analyses and evaluations underscore the role of past experience in determining current practice. Santayana's well-known maxim—"Those who cannot remember the past are condemned to repeat it"—is perhaps more relevant to emergency management than to any other governmental function.

This chapter provides a brief history of emergency management in the United States, with a focus on the changing role of the federal government. It also describes the context of emergency management—in particular, the extent and reach of disasters, and the larger network of governments and organizations of which local emergency management is a part.

A short history of emergency management in the United States

Human beings have always had to deal with natural and man-made disasters—from famine and flood to plagues, drought, earthquakes, and war. Relying on limited personal and communal resources, they respond as best they can. Such efforts—now categorized as personal preparedness—remain an important part of emergency management. Even today, as most studies show, the first responders to a disaster are typically survivors: family, friends, and neighbors who are sufficiently healthy and mobile to help those in need.

Before 1900

In the early years of the Republic, Americans—whether they were in cities, rural settlements, or frontier lands—believed that they were responsible for their own welfare. Victims of calamity, from barn fires to major floods, dealt with the problems themselves—or, at best, relied on assistance from neighbors. This pattern of self-reliance prevailed because farm families and ranchers rarely had formal communities or local governments to rely on. Early examples of ad hoc emergency assistance included the formation of posses to enforce the law, and the use of volunteer fire brigades to fight fires. As the population grew and settlement patterns increased risk, new organizational arrangements arose to deal with emergencies and disasters. When it became clear, for example, that well-intentioned but ill-prepared volunteer fire brigades could not provide adequate protection, professional fire departments began to emerge in major cities.

Gradually, emergency management became the responsibility of local and, to some extent, state government. During the nineteenth century, several national-level organizations were created to provide assistance when major natural disasters struck, and some existing organizations gradually expanded their basic services in order to help with disasters: for example, the American Red Cross, which was established in 1881; the National Weather Service; the U.S. Army; and two component services of what is now the U.S. Coast Guard all aided in disaster response efforts. Nevertheless, most disaster response efforts undertaken by federal (including military) agencies were ad hoc rather than part of a national strategy or system, while few, if any, emergency management systems or organizations (other than the Red Cross and faith-based organizations) existed at the local level.

Over time, it became clear that planned responses had clear advantages over ad hoc actions, and formal systems and organizations became an increasingly important feature of emergency management. In particular, the federal government's limited, ad hoc approach to disaster began to shift. As noted by hazards/disaster consultant David Butler, three factors helped to change attitudes toward natural disaster: (1) events such as the Great Chicago Fire in 1871 and the Johnstown Flood in 1889, which raised awareness of the potential level of devastation to communities subject to such disasters; (2) increasing scientific understanding of natural disasters, which was accompanied by the realization that the effects of disasters could be mitigated; and (3) the formal charter bestowed by Congress on the American Red Cross in 1905, making it the official disaster response agency.[3] This last factor represented the first, tentative step toward establishing limited, ongoing federal involvement in emergency management.

After 1900

In the early years of the twentieth century, disasters that reached catastrophic proportions in terms of deaths, injuries, and property loss focused government and public attention on the need to develop formal systems to respond to such events. In particular, the Galveston Hurricane of 1900 and the San Francisco Earthquake of 1906 demonstrated the vulnerability of large urban communities located in disaster-prone areas (see sidebars on pages 28–31). Although named for their respective cities of impact, these two disasters were of national significance even by today's standards. Since no formal plans or systems for response or assistance existed yet at the local, state, or federal levels, response and recovery were still ad hoc. Interestingly, some of the problems that arose—including conflicts over leadership and coordination, the denial of known hazards, poor management of victims, the convergence of unwanted donations, and the emergence of spontaneous volunteers—are still common today. But while the recovery efforts that were mounted in their wake were unprecedented in scale, neither the federal, state, nor local governments took significant steps to attempt to mitigate the risks posed by such hazards until many years later.

In the early years of the twentieth century, disasters that reached catastrophic proportions in terms of deaths, injuries, and property loss focused government and public attention on the need to develop formal systems to respond to such events.

During the 1920s and 1930s, continuing experience with major disasters slowly changed perceptions of government's role in relation to such events, eventually affecting the extent of federal, state, and local government involvement in emergency management. As noted in the sidebar on pages 32–33, the Great Mississippi Flood of 1927 was in some ways a major milestone in federal disaster policy. In the aftermath of the flood, President Calvin Coolidge appointed Herbert Hoover, his secretary of commerce at the time, as "Flood Czar," and put him in charge of coordinating relief efforts.[4] The flood also led to an expansion of the federal government's role in flood control. In parallel fashion, the Long Beach Earthquake of 1933 led the state of California to address earthquake risks, which included the development of building standards for structures in vulnerable areas. Finally, the ecological disaster known as the Great Dust Bowl, which spanned most of the 1930s, led to increased state and federal government involvement in land management.

Nevertheless, with the significant exception of flood-control efforts, primarily on the Mississippi River, all levels of government continued, for the most part, to limit their role to post-disaster assistance, and failed to develop formal systems to deal with pre-disaster mitigation or post-disaster recovery. Until the last four decades of the twentieth century, the prevailing notion was that mitigation and recovery were primarily local and state government responsibilities; federal responsibility was limited to providing relief in the aftermath of

The Galveston Hurricane of 1900

The Galveston Hurricane of September 8, 1900, remains the deadliest disaster in U.S. history to date. The number of people killed was never precisely determined, in part because bodies were buried on land and at sea or cremated en masse immediately after the storm. But the number is usually cited as at least 6,000, and it may have been as great as 12,000. Approximately one of every six Galveston citizens died in the disaster.

The event

The U.S. Weather Bureau in Washington, D.C., began sending its Galveston office warnings of an approaching storm as early as September 4. Although the exact evolution of the storm cannot be known, it apparently was a tropical storm when it passed over Cuba on September 4 and 5; it then exploded into a major hurricane when it passed over the warm water of the Gulf of Mexico. After receiving reports of severe storm damage along the Gulf Coast, the U.S. Weather Bureau finally ordered storm warnings from Pensacola, Florida, to Galveston, Texas. At the time, Galveston was a thriving city of more than 40,000.

Floodwaters began to creep into the city on September 8, and by midmorning the train tracks leading into Galveston were flooded. By early afternoon there was rain, and a steady northeasterly wind was blowing. By 5 PM, the weather bureau in Galveston was recording sustained hurricane force winds. The winds, however, were not the primary cause of damage. The low-lying island of Galveston was simply inundated by a storm surge of more than 15 feet. At one point, the sea actually rose a full 4 feet in four seconds. The encroaching waters acted like a riverine or flash flood, compiling and pushing debris inland, each row of structures adding more mass to the mountain of wreckage as the buildings were pushed off their foundations. More than 3,600 homes were destroyed.

Although the actual magnitude of the storm is not known, the damage and an apparent storm surge of 15½ feet has led the National Oceanic and Atmospheric Administration to estimate that it was a Category 4 hurricane (i.e., winds of 131–155 mph and a storm surge of 13–18 feet).

The storm should not have been unexpected, since "hurricanes had periodically raked the Gulf of Mexico coast—at least eleven times in the nineteenth century. Yet the inhabitants [of Galveston] denied the threat to their island community."[1] Since its founding in 1839, Galveston had weathered numerous storms and dodged many others, but in no case had it faced a direct hit. The result seems to have been complacency.

The aftermath

By all accounts the immediate aftermath of the hurricane was horrific. As in many other disasters, during and immediately after the storm, the victims had to rely on each other for medical care, shelter, food and clothing, search and rescue, and locating and burying the dead, and there were many, many bodies amid the debris.[2] The recovery operations, particularly the recovery and disposal of bodies, continued for weeks. As with many disasters, there was an outpouring of donations, which sometimes proved as much a burden as a blessing. "The Red Cross gave out food and clothing, but found much of its supply of donated clothing unusable, either too warm for the climate or too shabby."[3]

a disaster. To a large extent, this view continues to be a central tenet of emergency management; federal assistance is still considered appropriate only to supplement that of local and state governments.

During the first half of the twentieth century, industrial accidents and the explosion of munitions ships and cargo ships expanded the notion of disaster to include unintentional events related to complex industrial systems.[5] World War II further expanded the definition of potential threats to human existence, leading to the development of civil defense systems designed to prevent or to cope with the consequences of intentional harm. During the cold war era, the acquisition of nuclear weapons by the Soviet Union (and later by other countries) underscored this threat and placed new demands on emergency management.

Initially, the federal government viewed the need to develop effective civil defense systems as an opportunity to formalize response to natural disasters, industrial accidents, and intentionally harmful acts. Thus, shortly after the end of World War II, the federal government introduced the dual-use system, which was designed for response to many types of disasters. While the initial reaction among emergency managers and others to the dual-use system was favorable, many organizations still maintained separate systems for dealing with disasters: one for natural disasters, another for hazardous materials (hazmat) incidents and technological accidents, and still another for intentional attacks.

After the hurricane Galveston decided to rebuild. Although a decade earlier the city had rejected the construction of a seawall, it now adopted the idea. Considered a modern engineering marvel, the entire wall took almost sixty years to complete. It rises 15½ feet above sea level and is constructed with an advance barrier of granite boulders 27 feet wide. The city also physically elevated itself: engineers used dredged sand to raise the city—as well as 2,100 buildings—by as much as 17 feet.

Lessons learned

Galveston demonstrates what is, perhaps, the most fundamental lesson of emergency management: the danger of hubris, denial, and complacency.

There was much evidence indicating that Galveston could be struck by a severe hurricane. But for various reasons—psychological, social, economic, or some combination thereof—the people of Galveston, and particularly those responsible for the city's welfare and safety, ignored the warning signs. Reporters and researchers have been critical of the U.S. Weather Bureau; one researcher, Erik Larson, cites the bureau's territorial jealousies, centralized command and control by its Washington headquarters, inability to recognize its own shortcomings and gaps in knowledge, and reluctance to issue warnings and share information.[4]

Many others living in the city also denied the threat. Even though the 1886 Indianola hurricane had prompted a civic movement in Galveston to construct a protective seawall, that notion was quickly replaced by complacency and denial. Even after the 1900 hurricane, the editor of the *Galveston Tribune* wrote, "There prevails a belief that Galveston is subject to severe storms. . . . That is a mistake."[5]

Although some believed that the Galveston hurricane was a freak of nature, the city was stricken again by a hurricane fifteen years later. In hindsight, it is easy to see that Galveston was in danger in 1900, and even today the island remains one of the major population areas on the Gulf Coast endangered by hurricanes. Many climatologists feel that with global warming and the consequent increases in water temperatures and sea level, the risk is even greater than previously supposed. But regardless of climatological changes, there is no doubt that the risk in the Gulf region is greater simply because many more people now live there.

Source: Excerpted and adapted from David Butler, "Focusing Events in the Early Twentieth Century: A Hurricane, Two Earthquakes, and a Pandemic," in *Emergency Management: The American Experience, 1900-1950*, ed. Claire B. Rubin (Fairfax, Va.: Public Entity Risk Institute [PERI], 2007), 15-20, available from PERI at riskinstitute.org.

[1]Philip L. Fradkin, *The Great Earthquake and Firestorms of 1906: How San Francisco Nearly Destroyed Itself* (Berkeley: University of California Press, 2005), 19.

[2]Casey Edward Greene and Shelly Henley Kelly, *Through a Night of Horrors: Voices from the 1900 Galveston Storm* (College Station: Texas A&M University Press, 2000), 15-19.

[3]Erik Larson, *Isaac's Storm: A Man, a Time, and the Deadliest Hurricane in History* (New York: Random House/Vintage Books, 1999), 256.

[4]Ibid., 79.

[5]As cited in Fradkin, *The Great Earthquake*, 19.

The cold war era Direct federal involvement in the response to natural disasters was formalized in 1950 with the passage of the Federal Disaster Act (Public Law 81-875), an important first step in a series of legislative acts that enabled the federal government to provide disaster assistance on something other than an ad hoc basis. Under this act, authority for federal response resided in the office of the president, and administrative control of relief resided in the Housing and Home Finance Administration. The wording of the act emphasized that state and local governments had primary responsibility for dealing with the consequences of natural disasters and unintentional technological events, and that the federal role was supplemental; nevertheless, the legislation formalized the role of the federal government in the disaster relief process.

The year 1950 also saw the passage of the Civil Defense Act (Public Law 920), which established the Federal Civil Defense Administration (FCDA). The FCDA was given authority to coordinate civil defense efforts undertaken by local, state, and federal governments; in fact, however, local and state governments retained primary responsibility in this area. Eventually, natural disaster relief and civil defense were merged.

During the 1950s and 1960s, escalating tensions between the United States and the Soviet Union, especially with respect to nuclear arms, led to greater emphasis on the civil defense side of emergency management. Although funding for disaster relief increased, the amounts

The San Francisco Earthquake of 1906

On April 18, 1906, at 5:12 AM, a magnitude 7.8 earthquake struck northern California. The epicenter was along the coast near Daly City, just southwest of San Francisco, and the rupture extended along almost 300 miles of the northern end of the San Andreas Fault (compared with the 1989 Loma Prieta earthquake, which involved a rupture of about 25 miles). The earthquake was felt from southern Oregon to south of Los Angeles and inland as far as central Nevada. The event became legendary in American culture as the Great San Francisco Earthquake.[1]

Fires caused by ruptured gas lines, broken electrical lines, damaged and fallen chimneys, spilled lanterns, and other incendiary hazards (including arson to secure insurance payments) erupted immediately. With many burst water lines and the fire alarm system knocked out by the quake, the Great San Francisco Earthquake became the Great San Francisco Fire. Many smaller fires joined to become one great firestorm, and for three days the flames raged, obliterating about 4.7 square miles of the central city and destroying more than 28,000 buildings.

Initially the commander of the U.S. Army relief operations reported 498 deaths in San Francisco, but that number was later revised to 674. Years later, however, a San Francisco archivist concluded that at least 3,000 people had died, making it by far the deadliest earthquake in U.S. history.[2] Furthermore, of the city's population of about 400,000, some 225,000 were rendered homeless.[3]

Response and recovery

Fires and earthquakes were not new to San Francisco. In 1905, the National Board of Fire Underwriters had rated the probability of a major fire in San Francisco (even without an earthquake) as "alarmingly severe." They went on to say, "San Francisco has violated all underwriting traditions and precedent by not burning up."[4]

The response to the earthquake and ensuing three days of fire was both unique and typical of other major events. Although one was never officially declared, a de facto state of martial law governed the city. As in almost any large disaster, the ability of government and the military to improvise solutions to problems and establish mechanisms of control where none had previously existed was essential to the response. Soldiers, police, and vigilantes were instructed to brook no civil disobedience or unrest and to shoot "looters" on sight. They apparently did so, with estimates of shootings ranging from single digits to more than 500.

The earthquake disabled the city's water supply, which the National Board of Fire Underwriters considered to be inadequate anyway. Since fighting the fire with water was therefore not an option, the immediate response was to use explosives in order to blast firebreaks.

As the fires abated, city, state, and federal entities began caring for the dispossessed. Three organizations provided the bulk of the relief: (1) the Red Cross, which had the expertise; (2) the U.S. Army, which had the manpower; and (3) James Phelan, a prominent citizen who headed the local citizens committee and controlled relief funds.[5] There were issues about the equity of the distribution of aid.

Still, a vast relief effort was mounted almost immediately and actually continued for more than three years. Refugees were initially housed in army tents or makeshift shelters and then in semi-permanent camps constructed by the U.S. Army. In the end, the citizens of San Francisco—particularly the powerful and rich—moved quickly to rebuild their city. The city was essentially reconstructed by 1910; just five years later it hosted the Panama-Pacific International Exposition, which celebrated both the opening of the Panama Canal and the revival of San Francisco.

Mistakes made and lessons learned

As much as any other disaster in the twentieth century, the 1906 earthquake revealed the problems and mistakes that often surround a major catastrophe. It also offers many lessons for modern emergency managers, and the parallels to the Hurricane Katrina disaster are obvious.

Denial of risk There was ample historical evidence that San Francisco was at risk for both fire and earthquake. Similarly, the risk of hurricane and flood to New Orleans was clearly identified before 2005.[6]

Poor planning and unsafe construction At the time of the earthquake, San Francisco was still evolving from a shoddily built boomtown and industrial city to a better constructed, more stable, and organized community. Some of the newer buildings in the financial district and other wealthy areas of town were built to withstand earthquakes and endure. However, efforts to strengthen the building codes were mostly defeated—although the city did pass a new ordinance in 1909, after most of the rebuilding was complete.[7] Also, in the summer after the disaster, the city passed a law requiring that new buildings be able to withstand a wind force of thirty pounds per square foot, but in the following decades that standard was eventually cut in half. It would take additional earthquakes—Santa Barbara in 1925 and Long Beach in 1933—to prompt the state of California to enact laws mandating safe construction based on standards relating to earthquakes.[8]

Ineffective fire control San Francisco had failed to upgrade its firefighting capabilities—in particular, its water distribution system. The result was "the greatest single fire loss in U.S. history. . . . The National Board of Fire Underwriters concluded that even under normal conditions the multiple simultaneous fires would probably have overwhelmed a much larger department, such as New York's, which had three times the apparatus."[9]

Law enforcement and the military in response Immediately following the earthquake, the mayor and other local leaders made law enforcement a priority, and the regular police, the U.S. military, the national guard, and the state militia were joined by vigilantes and others in the effort. City officials, military commanders, and others had feared anarchy, especially looting, in the aftermath of the quake and fire; and although martial law was not officially imposed, most citizens assumed it to be in effect.

Relief The military was particularly effective in providing aid to disaster victims. Its organization and efficiency proved essential in establishing temporary shelter and managing supplies; this latter task has increasingly been recognized as a major problem for disaster responders.

Reconstruction Immediately after the earthquake, San Francisco's mayor established a "committee of fifty" to oversee the response, recovery, and rebuilding of the city. This extragovernmental committee, made up of prominent, wealthy citizens, ultimately usurped the power of the city's elected board of supervisors. Even though it was replaced a month later by a "committee on reconstruction," its original subcommittee on finance and its chairman, James Phelan, retained control of all relief and redevelopment funds.

Denial after the fact Notable in the aftermath of the disaster were the attempts of those in charge to deny, or at least minimize, the effects of the earthquake. For financial reasons, San Francisco's moneyed and powerful tried to ensure that the disaster was identified as a fire, not an earthquake, because insurance companies and potential investors were more likely to perceive fire as a manageable hazard, whereas an earthquake was seen as a capricious "act of God."

In the final analysis

The Great San Francisco Earthquake and Fire surpasses most disasters as a national event etched in memory, evoking images of the human struggle against catastrophe. It has since been joined in modern consciousness by the attacks of September 11, 2001, and by Hurricane Katrina.

In terms of corrective legislation, California did not act until three decades later, following the 1933 Long Beach earthquake. In the end, however, although intervening earthquakes were often the precipitating events and more proximal causes of change, researchers suggest that virtually all later programs and legislation enacted by California can be traced in part to information and insights derived from the Great San Francisco Earthquake of 1906.[10]

Analyses of the vulnerability of San Francisco today are sobering. Researchers at the U.S. Geological Survey warned in 2003 that "there is a 62 percent probability of at least one magnitude 6.7 or greater quake, capable of causing widespread damage, striking somewhere in the San Francisco Bay region before 2032. A major quake can occur in any part of this densely populated region. Therefore," they concluded, "there is an ongoing need for all communities in the Bay region to continue preparing for the quakes that will strike in the future."[11]

Source: Excerpted and adapted from David Butler, "Focusing Events in the Early Twentieth Century: A Hurricane, Two Earthquakes, and a Pandemic," in *Emergency Management: The American Experience, 1900–1950*, ed. Claire B. Rubin (Fairfax, Va.: Public Entity Risk Institute [PERI], 2007), 20–32, available from PERI at riskinstitute.org.

[1]The data in this paragraph are from the U.S. Geological Survey (USGS), "The Great 1906 San Francisco Earthquake," at earthquake.usgs.gov/regional/nca/1906/18april/index.php (accessed August 2, 2007); Philip L. Fradkin, *The Great Earthquake and Firestorms of 1906: How San Francisco Nearly Destroyed Itself* (Berkeley: University of California Press, 2005); and Simon Winchester, *A Crack in the Edge of the World: America and the Great California Earthquake of 1906* (New York: HarperCollins, 2005). See also the Web site of the Virtual Museum of the City of San Francisco at sfmuseum.org/1906/06.html (accessed August 2, 2007).

[2]Gladys Hansen, "Who Perished: A List of Persons Who Died as a Result of the Great Earthquake and Fire in San Francisco on April 18, 1906" (San Francisco: San Francisco Archives, 1980), available at sfmuseum.org/perished/index.html; and Gladys Hansen and Emmet Condon, *Denial of Disaster: The Untold Story and Photographs of the San Francisco Earthquake and Fire of 1906* (San Francisco: Robert A. Cameron, 1989).

[3]USGS, "Casualties and Damage after the 1906 Earthquake," at earthquake.usgs.gov/regional/nca/1906/18april/casualties.php (accessed August 2, 2007).

[4]As cited in Fradkin, *The Great Earthquake*, 37, 237.

[5]Ibid., 205, 207.

[6]See Shirley Laska, "What If Hurricane Ivan Had Not Missed New Orleans?" *Natural Hazards Observer* 29, no. 2 (2004).

[7]Stephen Tobriner, "An EERI Reconnaissance Report: Damage to San Francisco in the 1906 Earthquake: A Centennial Perspective," *Earthquake Spectra* 22, no. S2 (2006): S38.

[8]Charles Smith, "What San Francisco Didn't Learn from the '06 Quake," *San Francisco Chronicle*, April 15, 2006, available at sfgate.com/cgi-bin/article.cgi?f=/c/a/2006/04/15/HOGQ9I7P2T1.DTL (accessed August 3, 2007); and Ted Steinberg, *Acts of God: The Unnatural History of Natural Disaster in America* (New York: Oxford University Press, 2000), 36–37.

[9]C. Scawthorn, T. D. O'Rourke, and F. T. Blackburn, "The 1906 San Francisco Earthquake and Fire: Enduring Lessons for Fire Protection and Water Supply," *Earthquake Spectra* 22, no. S2 (2006): S135, S139.

[10]Jeanne B. Perkins et al., "A Retrospective on the 1906 Earthquake's Impact on Bay Area and California Public Policy," *Earthquake Spectra* 22, no. S2 (2006): S237–S259.

[11]USGS, "Is a Powerful Quake Likely to Strike in the Next 30 Years?" USGS Fact Sheet 039-03 (2003), at pubs.usgs.gov/fs/2003/fs039-03/ (accessed August 3, 2007).

Federal flood management and the Great Mississippi Flood of 1927

The Great Mississippi Flood of 1927 has been termed "the greatest natural disaster to befall [the United States] in terms of total human misery and suffering,"[1] although Hurricane Katrina has perhaps eclipsed it.

The event

The event was the result of unprecedented amounts of precipitation over various parts of the Mississippi drainage basin, which stretches from New York to Montana and includes tributaries in thirty-one states and southern Canada. The heavy rains began in the summer of 1926 and continued through the winter. The actual flooding, which spanned the spring and summer of 1927, began on April 16 when, thirty miles south of the confluence of the Ohio and Mississippi rivers, a 1,200-foot section of the government-financed levee collapsed and 175,000 acres were flooded. The flood affected Arkansas, Illinois, Kentucky, Louisiana, Mississippi, Missouri, and Tennessee. By May, the Mississippi River below Memphis, Tennessee, was sixty miles wide.

The Red Cross reported 246 peopled killed, but subsequent reports indicate that the actual toll may have been several times greater. More significantly, an estimated 700,000 people were rendered homeless; more than 150 Red Cross camps were quickly constructed to care for some 300,000 victims for months after the flood. About the same number of people received food and other assistance outside the camps. Some 200,000 buildings were damaged or destroyed, and at least 20,000 square miles of land were flooded.

Floods before 1927

Communities along the Mississippi had been involved in flood control—primarily levee construction—since 1726, when residents of New Orleans began building levees to protect their city. Throughout the nineteenth century, the federal government became increasingly involved in flood control efforts along the Mississippi, which ultimately led to a federal policy of using levees—and, importantly, only levees—to control flooding. By 1890, the entire lower Mississippi Valley, from St. Louis to the Gulf of Mexico, was divided into state and locally organized levee districts. In 1926, the U.S. Army Corps of Engineers publicly declared that the levee system would prevent future floods. But in 1927 the levee system along the lower Mississippi failed, thereby revealing the inherent flaws in this one-sided "levees-only" policy.

Federal floodplain management after 1927

After the Great Flood, the public and Congress began to favor a comprehensive federal program to address flooding in the Mississippi Valley, a marked change from the federal government's previously limited involvement in what had been considered a local issue. The resulting legislation was the Flood Control Act of 1928, in which Congress adopted a flood control plan that abandoned the levees-only approach. With this plan, "the government recognized that major floods, such as happened in 1927, involved drainage from far outside the lower Mississippi valley, that locals were unable to finance effective flood control measures, and that local governments were already making enormous contributions to flood control."[2]

The act provided for a partnership between the federal and local governments for the construction and maintenance of control structures. "The law set a precedent of direct, comprehensive, and vastly

were relatively small compared with the funding directed toward civil defense. Moreover, while the link between civil defense efforts and the U.S. Department of Defense was strengthened, the inclusion of natural and technological disasters with civil defense proved problematic. Essentially, the two communities were not compatible.

The advent of the Federal Emergency Management Agency During the 1960s and 1970s, the search for a balance between an effective system for dealing with natural and technological disasters and an effective civil defense system led to a series of reorganizations and reassignments at the federal level. Meanwhile, a series of disasters—including a major earthquake in Alaska in 1964, Gulf Coast hurricanes (Betsy in 1965, Camille in 1969, and Agnes in 1972), and the accidental radiological release at the Three Mile Island nuclear plant in Middletown, Pennsylvania, in 1979—demonstrated the weaknesses of the existing response systems. Once again, public and political pressure mounted on the federal government to redesign the response systems. The eventual result, which occurred in 1979 under President Jimmy expanded federal involvement in local affairs. . . . [It] reflected a major shift in what Americans considered the proper role and obligations of the national government."[3] The year following the 1927 flood, Congress spent nearly $300 million on flood control projects along the lower Mississippi. Federal involvement continued to expand, and eventually Congress passed the Flood Control Act of 1936, which essentially acknowledged that flood control was a "federal activity in the national interest."[4]

Not surprisingly, more flooding led to more legislation. Between 1936 and 1952, Congress appropriated more than $11 billion for flood control projects that helped prevent substantial flood damage.

However, some floodplain managers, engineers, and scholars realized that this reliance on engineering and construction to mitigate floods was, in its own way, a limited policy. Their concerns were underscored by studies showing that, despite structural measures to control floods, monetary losses were continuing to increase. With the country's growing urban population, the national flood damage potential was increasing faster than construction programs could control. These critics called for a broader approach that considered all possible means to mitigate flooding, including land use planning, zoning and restriction of human occupancy of flood zones, flood proofing of structures, and insurance.

This call for a broader federal approach can be traced back to the New Deal and a few staff members of President Franklin Roosevelt's administration in the 1930s. Perhaps the best known was Gilbert White, who described the prevailing national policy as "one of protecting the occupants of floodplains against floods, of aiding them when they suffer flood losses, and of encouraging more intensive use of floodplains."[5] As an alternative, White advocated "adjusting human occupancy to the floodplain environment so as to utilize most effectively the natural resources of the floodplain, and at the same time, of applying feasible and practicable measures for minimizing the detrimental impacts of floods."[6] Because of his advocacy of a broad range of human adjustments to floods, White became known as the father of modern floodplain management.

It took several decades for such ideas to be fully embraced by government, but in the end they took hold. The National Flood Insurance Program, which is administered by the Federal Emergency Management Agency (FEMA), requires local government to institute effective flood control measures, including land use planning. Moreover, to its credit, when examining local flood problems, the Army Corps of Engineers now considers a wide range of measures to reduce flood losses. The Great Mississippi Flood of 1927 was the seminal event of the twentieth century that launched these changes and elevated the control of floods to a national problem and a federal responsibility.

Source: Excerpted and adapted from David Butler, "The Expanding Role of the Federal Government, 1927-1950," in *Emergency Management: The American Experience, 1900-2005*, ed. Claire B. Rubin (Fairfax, Va.: Public Entity Risk Institute [PERI], 2007), 49-56, available from PERI at riskinstitute.org.

[1]James M. Wright, *The Nation's Responses to Flood Disasters: A Historical Account* (Madison, Wis.: Association of State Floodplain Managers, 2000), 9, available at floods.org/PDF/hist_fpm.pdf (accessed August 2, 2007).

[2]Ibid.

[3]John M. Barry, *Rising Tide: The Great Mississippi Flood of 1927 and How It Changed America* (New York: Simon & Schuster, 1997), 407.

[4]Nicole T. Carter, *Flood Risk Management: Federal Role in Infrastructure*, CRS report RL33129 (Washington, D.C.: Congressional Research Service, Library of Congress, 2000), 2, available at ncseonline.org/NLE/CRSreports/07Feb/RL33129.pdf (accessed August 2, 2007).

[5]Gilbert F. White, *Human Adjustment to Floods: A Geographic Approach to the Flood Problem in the United States* (PhD diss., University of Chicago, 1942), 32.

[6]Ibid., 2.

Carter, was the consolidation of multiple federal agencies and programs in a new, independent federal agency: the Federal Emergency Management Agency (FEMA). In the early years of FEMA's operation, the dual-use system enjoyed a resurgence.

During the 1980s and 1990s, the concept and practice of emergency management expanded because of changing expectations with regard to emergency management in general and governmental responsibilities in particular. Meanwhile, state and local officials attempted to apply federal funds earmarked for civil defense to meet all hazard-related demands.

But expectations regarding improved federal response to natural disasters were shattered by FEMA's poor performance in dealing with hurricanes in Puerto Rico and South Carolina, earthquakes in California, and Hurricane Andrew, which caused massive damage in Florida and significant, although less severe, damage in Louisiana in 1992, just weeks before the presidential election. In the wake of Hurricane Andrew, a number of studies were undertaken by the National Academy of Public Administration, the U.S. Government Accounting Office, the Congressional Research Service, and FEMA's own inspector general, among others. In

1993, when President Bill Clinton and his newly appointed FEMA director, James Lee Witt, took office, they were determined to improve emergency management at all levels. Following up on recommendations included in the aforementioned studies and in after-action analyses, Clinton and Witt altered FEMA's focus and organization in fundamental ways: the emphasis on civil defense was reduced; the natural disaster response system was given greater priority; and mitigation and preparedness were given greater emphasis. Gradually, the concept of "all-hazards" emergency management emerged.

The dismantling of the Soviet Union and the subsequent temporary reduction in external threats added support to FEMA's shift in priorities. The level of funding for disaster relief increased, the types of disasters eligible for relief expanded, and the FEMA director was invited to attend cabinet meetings. Three factors—growing political support for an all-hazards approach, more effective interactions among all levels of government, and increasing professionalism in the field of emergency management—created an environment that fostered the expansion of the federal government's role in emergency management. Both Congress and the public began to perceive FEMA, its programs, and the complementary programs of other federal departments as elements in a competent, professionally run federal emergency management program.

At the state and local levels, meanwhile, many analogous changes occurred. Florida, for example, increased the staffing and budget for its office of emergency management, and instituted new requirements for localities, including mandated recovery plans for major coastal cities and counties. Many other states reorganized their state emergency services organizations and their emergency support functions for better alignment with FEMA's organizational structure.

After 2000

The attacks of September 11, 2001, led to far-reaching changes in both the concept and conduct of emergency management, which were perhaps most evident in the heightened awareness of the threat of terrorism and the creation of the U.S. Department of Homeland Security (DHS). The attacks and their aftermath have been well-documented elsewhere—for example, in the *9/11 Commission Report*[6] and in numerous White House and congressional documents. Although these and other sources identified problems with the response—in particular, communications difficulties, which led to a high rate of death among New York City's public safety officers—studies show that, for the most part, the response was well managed by local, state, and federal agencies under existing enabling legislation and the Federal Response Plan, which had been published in 1992.

The attacks of September 11, 2001, led to far-reaching changes in both the concept and conduct of emergency management.

Nevertheless, in 2003, when DHS was created, FEMA was subsumed within it and thereby lost its independent agency status. The formation of DHS, which incorporated all or part of twenty-two agencies, marked a major shift in federal philosophy, policy, and priorities: today, emergency management has to be considered in the context of homeland security. The ramifications of this change are covered elsewhere in this book.

The context of local emergency management

The United States is a large country with widely varying topography, more than fifty state and territorial governments, and great variations in population density. Thus, the size, organization, and effectiveness of a local emergency management agency are determined by a number of factors both internal and external to the jurisdiction, including financial capacity, political will, demographic characteristics, the nature and severity of hazards, and intergovernmental requirements. Moreover, some of these factors—in particular, local hazards and mandates from other levels of government—change over time. Thus, local emergency management is a constantly evolving undertaking that requires flexibility on the part of both emergency management practitioners and local authorities.

The level and reach of disaster

The hierarchy of disasters can be thought of as a pyramid with three levels. At the base are emergencies—fires, floods, landslides, hazmat incidents, and other events that are capable of being addressed by local public agencies. At the next level are events that are dealt with primarily by local government, but with added assistance from the state government. At the apex of the pyramid are the small number of disasters that require local, state, and federal resources. Because of their size, nature, and impact, these events tend to receive sustained attention from all levels of government, as well as significant attention from the media.

Most adverse events are emergencies rather than disasters and are dealt with at the local level. But even larger-scale events can be said to start—and eventually end—locally: although the state and federal governments provide assistance for recovery, it is for a relatively short period. Long-term recovery remains a local responsibility. Indeed, as William Waugh Jr. notes, "Emergency management is the quintessential government role. It is the role for which communities were formed and governments were constituted in the first place—to provide support and assistance when the resources of individuals and families are overwhelmed."[7] Nevertheless, since local governments generally have a smaller tax base than other levels of government and face a great variety of demands, funding for emergency management is often limited. Thus, although a local jurisdiction is typically closest to a given disaster, it is also least capable of mustering the resources required to deal with the event. This is particularly the case for smaller communities, which may be unable to sustain an ongoing emergency management office and staff, let alone cope with the devastating consequences of disasters. Some small local governments opt for a minimally staffed and funded emergency management department; others contract for services from a neighboring community or from the county.

Although a local jurisdiction is typically closest to a given disaster, it is also least capable of mustering the resources required to deal with the event.

It is important to note that disasters are not always single, quick-onset, high-impact, short-term events. Some disasters, such as droughts, are slow to take effect and slow to be recognized. Others, such as floods, may continue over two or more rainy seasons. Still others may have both primary and secondary impacts: a hurricane or an earthquake, for example, can trigger hazmat releases, oil spills, or dam failures. Such secondary effects may affect a larger population and/or geographic area than does the initial event.

It is one of the axioms of emergency management that disasters do not respect political boundaries. Although the impact of an event—particularly a natural disaster—may begin locally, it can rapidly expand. Floodplains and watersheds for large rivers often span two or more states. Earthquake zones, such as the New Madrid fault (extending from Illinois, through Missouri, through Arkansas), cover several states and hundreds of municipalities. Hazmat incidents, such as major oil spills or the release of toxic chemicals, could affect the entire United States and its neighboring countries—which is why the United States, Mexico, and Canada have jointly agreed on a formal response plan. And a disaster such as a flu pandemic has the potential to affect the entire world.

Thus, emergency management is rarely the domain of a single organization. In the case of a flu pandemic, for example, numerous federal agencies—including the U.S. Department of Health and Human Services, the Centers for Disease Control and Prevention, and the Public Health Service—are responsible for tracking the onset and spread of the disease, purchasing and stockpiling vaccines, and issuing health and medical advisories. At the local level, the local medical community provides health advisories and care for victims, and distributes information and pharmaceuticals. At each level of government, both emergency management and public health officials have lead roles.

A complex network

As described in the first section of this chapter, the history of emergency management is largely a story of increasing federal involvement in disaster response—and, since 2001, increasing federal involvement with respect to the threat of terrorism. But federal assistance rarely comes without strings attached. For example, if a community wants to participate in the National Flood Insurance Program, which provides federally supported flood insurance policies for property owners in flood zones, it must comply with federal requirements that address, among other things, setbacks, elevations, and construction standards. Moreover, federal laws, regulations, policies, and programs are the source of a number of mandated activities, not all of which are funded. Finally, because federal funding for emergency management has become essential to state governments as well as local ones, and because the states establish the conditions governing the pass-through of these funds to local governments, state emergency management agencies significantly influence the planning, staffing, and training of local emergency management departments.[8]

Knowing whom to call (and knowing each contact personally), knowing what resources to ask for, and knowing how the intergovernmental system works are crucial to effective and efficient action.

The national emergency management system is a complex network that includes federal, state, and local government agencies; special districts and quasi-governmental organizations; nonprofit organizations; volunteers (both organized and spontaneous); and private sector firms that provide services and products under contract. Thus, one of the primary responsibilities of the emergency manager involves anticipating how and where to obtain additional resources in the event that the locality is overwhelmed by the demands of a disaster.

To facilitate coordination and cooperation across different levels of government, many larger cities in the United States are part of regional entities such as councils of governments, airport authorities, and port authorities. In the Washington, D.C., metropolitan area, for example, the federal government has the National Capital Planning Region, a special emergency planning entity.

Researchers have documented that the likelihood of effective response and recovery is increased if key personnel at all levels of government have established relationships before a disaster occurs. Knowing whom to call (and knowing each contact personally), knowing what resources to ask for, and knowing how the intergovernmental system works are crucial to effective and efficient action. As emergency managers often say, a disaster is not the time to be exchanging business cards.

Summary

The history of federal involvement in emergency management in the United States is a lengthy, colorful, and often sobering story.[9] For more than a hundred and fifty years after the nation was founded, natural calamities were viewed as acts of God—unavoidable events that could not be prevented or redressed by governmental action. Disaster response was the moral responsibility of neighbors, religious groups, and the community at large. Thus, the federal government rarely responded to such events—and when it did, the response often occurred well after the disaster. Most assistance came from charitable and civic organizations.

Although federal disaster assistance dates back to the early 1800s, for most of the nation's history, response to natural disasters was a local and state responsibility. The Galveston Hurricane of 1900 and the San Francisco Earthquake of 1906 heralded a gradual shift to the view that government involvement should be consistent and sustained. The first significant expansions of the federal role, however, did not occur until the 1920s. With every successive administration since then, the organizational forms and functions of emergency management have continued to evolve, culminating in the establishment of FEMA in 1979 and the creation of DHS in 2003.

In the view of many practitioners and researchers, the gradual "federalization" of emergency management has led to a change in the perspective of average citizens: where disaster victims once expected compassion—from friends, family, and local organizations—they now have a sense of entitlement to government aid. Expectations of citizens—and of state and local officials as well—have noticeably increased during the past two decades. Some of these new expectations are realistic; others are not.

Hurricane Katrina, one of the worst disasters of the past hundred years, was a low point for local, state, and federal emergency management. Because of the level of destruction, the inadequacies of response and recovery, and the consequent political controversies, Katrina is certain to become a turning point in U.S. emergency management. It is worth noting, however, that few of the problems associated with Katrina were totally unexpected; in fact, many were recurrences of problems that had been identified in disasters of the previous century. Today, most Americans regard emergency management as, in Waugh's words, the "quintessential" public service. Yet in 2005, when that public service failed at every level, the results were tragic. Public pressure to improve the public management of disaster at all levels of government is not likely to abate.

Notes

1 Some examples are Erik Larson's *Isaac's Storm: A Man, a Time, and the Deadliest Hurricane in History* (New York: Random House/Vintage Books, 1999), about the Galveston hurricane; John M. Barry, *Rising Tide: The Great Mississippi Flood of 1927 and How It Changed America* (New York: Simon & Schuster, 1997); and Simon Winchester, *A Crack in the Edge of the World: America and the Great California Earthquake of 1906* (New York: HarperCollins, 2005).

2 See "Disasters and the Law: Katrina and Beyond," an excellent collection of materials compiled by the Boalt Hall School of Law, University of California, Berkeley, California, available online at http://128.32.29.133/disasters.php?categoryID=11 (accessed August 3, 2007).

3 See David Butler, "Focusing Events in the Early Twentieth Century: A Hurricane, Two Earthquakes, and a Pandemic," 11–48, and "The Expanding Role of the Federal Government: 1927–1950," 49–80, in *Emergency Management: The American Experience, 1900–2005*, ed. Claire B. Rubin (Fairfax, Va.: Public Entity Risk Institute [PERI], 2007), available from PERI at riskinstitute.org.

4 It is interesting to note that after Hurricane Katrina in 2005, the Congressional Research Service investigated and analyzed the concept of a flood czar. President George W. Bush created a Gulf Coast Recovery Council but did not attempt to create another "czar" position.

5 For a fuller account of the post–World War II era in emergency management, see Keith Bea, "The Formative Years, 1950–1978," 81–109, in Claire B. Rubin, ed., *Emergency Management* (see note 3).

6 National Commission on Terrorist Attacks upon the United States, *The 9/11 Commission Report* (New York: Norton and Co., 2004), available at 9-11commission.gov/report/911Report.pdf (accessed August 7, 2007).

7 William L. Waugh Jr., *Living with Hazards, Dealing with Disasters: An Introduction to Emergency Management* (New York: M. E. Sharpe, 2003), 3.

8 U.S. territories, possessions, and tribal governments deal directly with the federal government on emergency management funding.

9 For a recent history of federal involvement in emergency management in the United States over the past century, see Rubin, ed., *Emergency Management*, which was used extensively for preparation of this chapter.

CHAPTER 3

Organizing for emergency management

Frances L. Edwards and Daniel C. Goodrich

This chapter provides an understanding of

- The administrative and business continuity responsibilities of the emergency management department
- How the emergency management function fits into the local government organization
- Historical changes in emphasis and approach within the emergency management function
- The role of the emergency management department in each phase of emergency management
- Resources for collaboration and cooperation.

Emergency management fulfills two roles within the structure of local government: first, it is part of local government's overall administrative functions. Specifically, the emergency management department takes the lead in mitigation, preparedness, response, and recovery. Second, the emergency management department leads organizational efforts to ensure business continuity for the local government. The development of redundant communications systems and the safe caching of vital records are examples of such activities.

This chapter describes the administrative and business continuity responsibilities of the emergency management department. It then examines a number of factors that can affect the nature and effectiveness of the emergency management function—in particular, size and structure, location within the local government organization, and staffing arrangements. Next, the chapter discusses the principal changes that have shaped emergency management in the past half century: new national perspectives, the shift in emphasis from response and recovery to mitigation and preparedness, and the change from sworn to civilian leadership. Finally, it describes the role of the emergency management function in each phase of emergency management and various approaches to strengthening that function, from partnering with local volunteer organizations to obtaining certification through the Emergency Management Accreditation Program.

Administrative functions

The emergency manager is responsible for ensuring that all resources are in place during all four phases of emergency management: mitigation, preparedness, response, and recovery. To determine what level of resources may be needed, the emergency manager works with other local government departments to develop a comprehensive risk analysis—an evaluation of the variety, scope, and frequency of the threats that the community faces. This analysis becomes the basis for emergency planning; it is also crucial to ensuring that the emergency management function is the right size for the community. (In the case of smaller local governments, "rightsizing" the emergency management function may require a countywide collaborative approach.)

The nature of the threats to a community and the resource base required to deal with them successfully together determine the size and structure of the emergency management organization.

The nature of the threats to a community and the resource base required to deal with them successfully together determine the size and structure of the emergency management organization. The role of the local emergency manager is to be familiar with the hazards, the threat levels, and the expectations of higher levels of government. Working from this knowledge base, the emergency manager advises the local government on the appropriate size of the emergency response organization, including the number of people who need to be trained and the size of the emergency response facilities required.

The National Incident Management System (NIMS), which the federal government mandates for all communities, includes six components: command and management, preparedness, resource management, communications and information management, supporting technologies, and ongoing management and maintenance. Within each component are structural elements called sections. There are typically four sections under the general heading of command and management: operations, planning (which sometimes includes intelligence), logistics, and finance and administration. However, the command section chief performs any functions that are not separately staffed. Thus, if most threats to a community are seasonal, predictable, and limited to a specific type—such as ice storms or snow emergencies—the command section chief will perform several functions, and only specific departments may be required to provide support.

Seasonal emergencies can usually be predicted at least a few hours in advance, and they recur often enough for responders to understand the need for planning, stockpiling supplies, and maintaining field-level training on the use of strategies and equipment. In the case of cold weather-related emergencies, for example, the focus might be on traffic control by the law enforcement agency, road clearance by the public works agency, and provision of alternative shelter through a partnership between the American Red Cross and the school district or parks and recreation department. Other local government entities may have no role in the emergency management structure at all.

On the other hand, in areas with the potential for high-impact events, such as hurricanes or earthquakes, most local government entities would be an integral part of the emergency plan, and nongovernmental organizations (NGOs)—such as the Red Cross, the Salvation Army, and other members of VOAD (Voluntary Organizations Active in Disaster)—would be actively involved. For example, librarians might support the care and shelter function by assisting with victim registration and helping victims obtain information on recovery resources. Staff from the city clerk's office might be included in the planning and intelligence section to assist with data collection and management. Staff from the housing department would be an integral part of the recovery process, finding available local housing and working with the Federal Emergency Management Agency (FEMA) to create temporary housing for disaster victims.[1] Bilingual local government employees could be instrumental as translators, both in shelters and at community meetings.

The emergency operations center (EOC) is the location from which a jurisdiction manages a disaster. The center should be located in a building that (1) can withstand the principal threats to the community, (2) has backup electrical power and fuel, and (3) is large enough to house the following:

- A command section, which includes a public information officer, a safety officer, an agency liaison officer, and the section chief who makes the final decisions
- An operations section, which includes branches such as law enforcement, fire and rescue, public works, communications, and care and shelter
- A logistics section, which includes branches such as purchasing and procurement, information technology, and telecommunications
- A planning and intelligence section, which conducts damage assessment, creates maps of the disaster, monitors the weather, creates the written version of the mandated incident action plan, and oversees the creation of mandated reports
- The finance/administration section, which manages accounting functions and ensures that the necessary procedures are followed to obtain reimbursement from insurance carriers and from state and federal agencies.

Planning for risks that are likely to have serious impacts should be ongoing, and plans should be continually revised on the basis of lessons learned from previous events. Comprehensive staff training should be conducted annually, and stockpiles of equipment and supplies should be reviewed annually. When an event such as a hurricane threatens, the EOC should be fully staffed from prediction through recovery. In addition to the requirements noted above, the EOC should be in a building that is outside the storm surge or flood zone and has redundant communications systems. Communities with a high likelihood of major emergencies should have space in the EOC for all five NIMS sections to be staffed by multiple staff members, around the clock, for days at a time. For each NIMS position, at least three people should be trained to work in the EOC.[2]

The 2005 hurricane season clearly demonstrated the importance of a robust emergency management program. Because the New Orleans EOC was located in a vulnerable area, the facility was destroyed; consequently, the mayor lost face-to-face contact with much of his emergency management staff and was unable to communicate with either first responders or the community at large.[3]

Business continuity

Planning for the business of local government to go forward in the event of disaster is just as important as protecting the community. Continuity-of-government plans were first developed during the era when the emphasis in emergency management was on civil defense: the plans were intended to be implemented after a nuclear disaster—in which, it was anticipated, local officials might be killed and local institutions destroyed.[4] Today, the emphasis is on ensuring that the government's business can go forward—even in the face of blocked roads, power outages, and loss of communications infrastructure. According to FEMA, "The [continuity of government] plans should describe strategies to mitigate risks, and alternative methods for ensuring the continuity of the [government's] core business functions. The plans should provide for the loss of services outside the agency's control—such as the ability of suppliers to provide products, services, or data—or the loss of critical infrastructure."[5]

A business continuity plan is more than a list of alternate EOCs and seats of government: it identifies facilities that have alternate power supplies, describes the implementation of redundant communication systems, and outlines methods for protecting vital records.[6] Alternate EOCs are housed in local government–owned facilities in secure locations that are least likely to be damaged by anticipated threats—for example, away from flood zones and liquefaction zones. Communications and power resources may be stockpiled in advance at these locations, or the business continuity plan may include provisions for the rapid acquisition of resources. Like the primary EOC, the alternate EOC will require both generators and fuel. The generators must be tested at least annually, for at least twenty-four hours, to ensure operating capability for prolonged disasters. The fuel must be recycled or used up annually to ensure that its quality is maintained.

Should severe devastation or airborne hazards prevent local elected officials from meeting as usual, an alternate seat of government would allow them to continue to conduct official business. Criteria for the location of an alternate seat of government are similar to those for an alternate EOC. Options include governmental facilities outside the community, such as a county seat or the town hall of a nearby community. As is the case with the alternate EOC, the plan must include power and communications capabilities at the alternate location. Storage of records essential to ongoing governance—such as the emergency operations plan (EOP) and key local ordinances—should also be arranged at the alternate seat of government. One option is to create a reciprocal arrangement with another local government, in which each would store both hard and electronic copies of critical documents for the other.

A business continuity plan identifies facilities that have alternate power supplies, describes the implementation of redundant communication systems, and outlines methods for protecting vital records.

Most government agencies rely heavily on electronic storage of both day-to-day work products and vital records, such as ordinances, property and permit information, and all types of plans. Storing copies of vital records off-site helps to speed recovery. For example, in the wake of a disaster, property records and birth, death, and marriage records are essential for community members to file for insurance benefits and restart their lives. These documents can generally be stored in a secure location in another community, in microfilm or microfiche form. Storing copies of software is also essential to business continuity. The local government's information technology department should maintain an annual inventory of all software programs in use and ensure that backup copies are stored safely off-site. All local government departments should have daily and weekly backup plans for work products to ensure that if computers are destroyed by physical or cyber disasters, essential data can be recovered and work can continue. Laptop computers with cigarette-lighter adapters can create a mobile business capability for the community: the city-owned vehicle fleet can be the power source used to recharge the laptop batteries.

There are two options for records storage. "Hot site" storage means that the most current records are maintained on the computers of another agency or service provider; the business continuity plan specifies how, in the event of a disaster, these records will be accessed by e-mail or the Internet. "Cold site" storage means that hard copies of records will be held in a vault for delivery to alternate business locations. Both systems can be useful in safeguarding vital records at minimal cost.

Of course, local elected officials may themselves be disaster victims, or may be stranded by a disaster. The local governing body should pass an ordinance specifying how it will be reconstituted after a disaster. There are several approaches to ensuring a functional governing body. Some communities simply lower the number of members required for a quorum; others allow each member of the body to designate one or two alternates, such as a chief of staff or a political ally: in the event that any member of the governing body is incapacitated or unavailable following a disaster, an alternate would serve. If there are not enough municipal officials to carry out the community's business, some municipal governing bodies opt to have the county board of supervisors serve in their stead.

The local governing body should pass an ordinance specifying how it will be reconstituted after a disaster.

To create the business continuity plan, the emergency manager leads a group of senior appointed officials who have collective responsibility for records and services, and knowledge of how to secure critical infrastructure. Information technology professionals are needed to design backup systems for both public safety communications and local government business communications. Although, under ordinary circumstances, land lines and cellular phones provide most communication links, disasters may take both types of phones out of service. Land-line capacity, for example, is designed on the assumption that no more than 10 percent of subscribers will be using the phones at any one time—a service level that is quickly surpassed after a disaster. Moreover, many phone companies have a policy of channeling only those calls that are going out of the area to open lines, further limiting capacity.[7] Most phone companies offer local governments, hospitals, and other critical facilities the opportunity to have their lines designated "essential services lines," which means that they will receive more rapid restoration of service. The committee creating the business continuity plan must ensure that all the proper lines have been so designated and that the agreement is up-to-date.

Cellular phones, for their part, rely on "cell sites" that cannot operate unless they have power and are properly oriented. During a disaster, the loss of cell sites—or of the power to run them—may take all cell phones out of service. Portable cell sites, backup power, and other techniques may help to keep some essential cell phones working.[8] Another option is a Government Emergency Telecommunications Service (GETS) card, which expands access to telephone service during an emergency.[9]

Public safety answering points (PSAPs) answer 911 calls from the community. After a disaster, it is essential that such calls receive rapid response. To ensure that PSAPs are not adversely affected by disasters, redundant power supplies (including battery arrays and generators) and alternate PSAP facilities need to be part of the business continuity plan. Another strategy is mobile dispatching capability, sometimes called "dispatcher-in-a-box": a self-contained, wheeled radio unit that has an antenna, runs on battery power, and can be deployed to any safe location as an alternate PSAP.

Emergency management within the local government organization

The nature and effectiveness of the emergency management function are affected by three principal factors: the location of the function within the local government organization, the size and structure of the emergency management department, and the staffing arrangements—specifically, whether the position of emergency manager is held by a political or a professional appointee.

Location

Historically, the emergency management function generally fell within the purview of the fire department, or (less often) the police department or sheriff's office. Since the end of the cold war, however, changes at the federal, state, and local levels have demonstrated that the emergency plan, like the budget, is an executive function. For example, the U.S. Department of Homeland Security (DHS), which includes FEMA, is part of the executive branch of the federal government. In California, Florida, and Texas, the state director of emergency services is part of the governor's office. Furthermore, an all-hazards EOP encompasses roles for many local government departments, and only a department head—not a subordinate member of a public safety agency—can supervise other department heads working in the EOC. Thus, the person who is responsible for making and administering the plan needs to work directly for the executive of the organization and to be on a par with other department heads.

In keeping with the state and federal model, many communities have situated the emergency management function within the office of the city's chief administrative officer (i.e., the mayor or the local government manager). This gives the emergency manager a purview that is clearly jurisdiction-wide, making it easier to obtain the cooperation of other departments in the planning, training, and exercises that lead to effective emergency management. It also ensures that the emergency manager has the opportunity for face-to-face contact with the organization's chief administrator.

Size and structure

In small to medium-sized local governments, one person performs all the emergency management functions. Larger organizations may have a small staff of specialists led by a generalist. In very large or disaster-prone cities, emergency management staff may number in the dozens. In the 1980s, FEMA's staffing guidelines for local emergency management suggested one professional for each 100,000 in population, and one support staff member for every three professionals. Few communities can afford to maintain that ratio today. For example, during the ten years following the Oakland Hills firestorm (1992–2002), Oakland's population was approximately 250,000, and the Office of Emergency Services had twelve staff members. In New York City in 2002, the population was 9 million, and the Office of Emergency Management had 100 staff members.

Emergency management is a cross-disciplinary effort, and representatives of other departments are a crucial part of the emergency management team. The committee that creates the EOP will be staffed by representatives from other departments, and department heads from across the local government will typically staff the key positions in the EOC. Nevertheless, heads of other departments will be able to devote only a limited amount of time to emergency management activities. For that reason, it is important for the emergency manager to have not only a broad background in public administration, but also a good working knowledge of the supporting disciplines—including law enforcement, fire services, emergency medical services, public health, urban planning, and civil engineering. This knowledge will enable the emergency manager to get the most benefit from the committee meetings, training, and exercises that bring various departments and disciplines together.

Staffing arrangements: Political versus professional

In some cases, the emergency manager is a politically appointed individual with ties to the governing body and a strong relationship with the mayor. In other cases, the emergency manager is a career public servant with either college- or FEMA-based training and a history of increasingly responsible appointed positions.

Emergency management today is an extremely complex responsibility. Emergency managers need basic skills in public administration (personnel management, budgeting, purchasing); the ability to write clear and comprehensive reports and to speak effectively in public; and knowledge of local ordinances, state laws and regulations, and federal codes, statutes, and regulations. They must have a basic understanding of the services delivered by first responders and of the role of first responders in disasters, as well as of the work of partner agencies, such as public health and code enforcement. To engage in mitigation activities and response planning, emergency managers must be knowledgeable about the various threats to their communities and how to address them, and they must be able to understand the work of engineers, geologists, seismologists, geographers, and chemists. To deal with chemical, biological, radiological, nuclear, and explosive threats, emergency managers also need some familiarity with military tactics, police tactics, and public health issues.[10] Given this scope of work, it is clear that a political appointee whose tenure is tied to that of the mayor or other elected official would face a significant challenge merely in attempting to master the complete acronym set, let alone the twenty-five pages of FEMA-recognized competencies, within the four-year period that an elected official typically holds office.[11]

The career public servant typically enters emergency management from a particular functional area, such as fire administration, police administration, or urban planning, or as an administrative generalist, such as a budget analyst, program analyst, or training specialist. Promotion to the position of director of emergency services usually requires five to ten years of experience in emergency management or a closely related field. To ensure that the emergency manager has strong, field-specific skills and will be active and effective in the lead emergency management role, the local government should make the director's position a senior management position.

Regardless of whether the emergency manager is a public servant or a political appointee, he or she needs the necessary knowledge and skills to effectively navigate in the political realm.

However, in communities where the principal role of the director of emergency services is to interpret the mayor's policies for the staff, a person whom the mayor trusts might be more effective than a professional emergency manager. For example, the mayor's primary concern may be to ensure that emergency management efforts are compatible with his or her overall policies: while a trained emergency manager might, for example, oppose low-income housing in an industrial area because of the presence of hazardous materials, a politically appointed emergency manager might be able to argue that the day-to-day benefits to the environment of the mayor's walk-to-work residential development would outweigh the environmental threats from a potential industrial accident.

Political advocacy of legislation and lobbying for grants and funding are part of the role of the emergency manager. Public servants understand the complex relationships required for effective and efficient service delivery, but they may have little knowledge of how to work as advocates. Many mitigation programs, for example, require significant federal funding from diverse agencies. A politically appointed emergency manager might have better access to elected officials at higher levels of government, and might therefore be in a better position to negotiate for such funding. Moreover, in some jurisdictions, public servants are forbidden by law from advocating positions before higher-level elected officials; instead, their role is to provide position papers for lobbyists or local elected officials. In these jurisdictions, a politically appointed emergency manager might have fewer limitations on efforts to advocate for legislation or grants. But regardless of whether the emergency manager is a public servant or a political appointee, he or she needs the necessary knowledge and skills to effectively navigate in the political realm.

In sum, professional emergency managers are likely to be better at comprehensive emergency management than politically well-connected people with little emergency management training or experience. On the other hand, politically appointed administrators with extensive political experience are likely to be better advocates for policy positions, with a greater chance of influencing both local elected officials and higher levels of government. Each community must carefully evaluate the scope of the emergency manager's job responsibilities and determine the staffing arrangement that will best serve those responsibilities.

The changing face of emergency management

From the 1950s to the 1970s, when emergency management was synonymous with civil defense, the emergency manager was often a politically appointed civil defense director—most often, a retired senior military officer—and the focus was on military-oriented defense planning: natural hazards were accepted as acts of God, and the emphasis was on response and recovery. Since the 1970s, emergency management has gone through a number of evolutions. The first was a shift from war planning and radiological defense to an all-hazards approach (which actually emphasized natural hazards). With this shift, the profession evolved to include civilians with professional backgrounds, who entered emergency management from a particular functional area or as administrative generalists.[12] Eventually, the ranks of emergency managers grew to include earth scientists, meteorologists, engineers, and public information and marketing specialists. In general, however, the senior emergency management position was still held by the fire chief or police chief.

In the early 1980s, when the AIDS crisis arose, public health agencies began partnering with local emergency managers to create public information campaigns and to prepare for the potential effects of contagious diseases. Emergency managers also helped train public safety personnel in personal protection and in techniques for preventing disease transmission.

In the mid-1980s, in the wake of Love Canal and Bhopal, a new emphasis on hazardous materials management came to the fore, along with a recognition of the interconnectedness between environmental issues and emergency management. Chemists were added to emergency management offices to manage the implementation of the Emergency Planning and Community Right-to-Know Act (EPCRA) and the Resource Conservation and Recovery Act (RCRA),[13] and some emergency managers obtained certificates in hazardous materials management.

After the earthquake in Northridge, California, in January 1994, many residents chose to live outdoors, in makeshift encampments, rather than go to shelters (see Figure 3–1). These residents came from areas of the world where buildings collapsed from earthquakes, so they felt safer outdoors. Teams of city building officials, social workers, and religious leaders had to circulate through the encampments, reassuring residents that it was safe to return to their homes.[14] The aftermath of the earthquake demonstrated the need for partnerships between emergency management and social service agencies; agencies serving particular cultural, religious, and ethnic groups; and agencies serving groups with special needs.[15]

Perhaps the greatest change in the face of emergency management was brought about by the terrorist attacks of September 11, 2001. For the first time, heavy-equipment operators, welders, construction safety personnel, and transit workers were recognized as first responders because of their lifesaving work at the World Trade Center. The anthrax attacks in Florida, New York City, and Washington, D.C., brought public health professionals, clinical caregivers, and environmental cleanup specialists into the emergency management realm. Homeland Security Presidential Directive 8, issued in 2003, lists a number of medical and public works–related professions not previously included in emergency management.[16]

Figure 3-1 Encampment for Northridge residents displaced by the earthquake

Photo courtesy of Robert A. Eplett/OES CA, California Governor's Office of Emergency Services

The student manual for California's Standard Emergency Management System, which was written in the early 1990s, was prescient: the manual defines emergency response agencies as "all agencies assisting with emergency response, whether at the scene, in a command post or in an emergency operations center," and emergency response personnel as "all personnel of the emergency response agencies."[17] From laboratory scientists to public health nurses to crane operators, all joined the ranks of emergency responders and became partners in local emergency management.

Changes at the national level

During the civil defense era, the organization of emergency management was top down: federal officials made policies, and local agencies complied with them. As the all-hazards approach gained favor, it became clear that top-down policy making did not suit the diversity of local needs: the four phases of emergency management look very different in a hurricane-prone area, where the onset of threats is seasonal and prediction is possible, than they do in an earthquake-prone area, where there is no "season" and no reliable short-term warning system.

It was Thomas P. (Tip) O'Neill, Speaker of the House from 1977 to 1987, who said "All politics is local."[18] James Lee Witt, director of FEMA during the Clinton administration, was often quoted as saying "All disasters are local."[19] Increasing recognition of the local nature of disasters led to an evolution in FEMA requirements, as they shifted from one-size-fits-all plans to more general guidelines that could be customized for various disaster types.

In September 2004, Tom Ridge, then secretary of DHS, announced that to be eligible for federal emergency preparedness and homeland security grants, state and local governments must comply with Homeland Security Presidential Directives, the National Response Plan, and NIMS.[20] For states (such as California and Florida) that have already adopted the Incident Command System (the field command-and-control element of NIMS), the transition should be smooth.[21] For states without a uniform emergency management system, the transition may require significant political management at the local level.[22] Moreover, the NIMS configuration for response and recovery may change the traditional relationships between state agencies and local governments. For example, funding for response and recovery may now come through state channels rather than directly from the federal government.

Evolution of emphases

As the focus in emergency management has shifted—from civil defense, to an all-hazards approach, to a renewed concentration on defense from attack—so too has the emphasis assigned to the various phases. During the civil defense era, emergency management was focused on disaster response; one result was the placement of the emergency management function within fire or police departments. In the 1980s, the primary emphasis was on planning and preparedness, including the development of EOCs and EOPs. Public outreach—which, during the civil defense era, had focused on encouraging the construction of fallout shelters—began to emphasize neighborhood self-sufficiency: for example, community residents were encouraged to keep all-hazards emergency kits on hand.

During the Clinton administration, FEMA developed a program called Project Impact: Building Disaster-Resistant Communities, which stressed the importance of mitigation. In 1997, as part of this program, the federal government provided grants to pilot communities to support the development of mitigation partnerships among local government, residents, and the business community. One goal of the program was to achieve concrete changes—for example, to create better firebreaks around homes, develop ordinances to prevent development in high-hazard areas, and install siren systems and levees.[23] Although the program no longer exists, some of the partnerships have continued.

In the preparedness phase, the focus has also shifted to self-sufficiency and neighborhood support. After the Loma Prieta earthquake in 1989, millions watched on television as Oakland residents, armed only with ladders and tools from their homes, tried to rescue strangers who had been trapped by the collapse of the Cypress Structure, a part of the Interstate 880 roadway system (Figure 3–2). Fortunately, none of the volunteers were killed, but their lack of training

Figure 3-2 The Cypress Structure, part of the Nimitz Freeway, after the 1989 Loma Prieta earthquake

Photo courtesy of Robert A. Eplett/OES CA, California Governor's Office of Emergency Services

and proper equipment put them at significant risk. Public officials, however, took notice: the volunteers' efforts demonstrated that most people will do the best they can to save a life, so public safety departments in several California communities began offering organized training in emergency medical care and light search and rescue. This training augmented courses in personal preparedness and the use of fire extinguishers, which had long been offered by California fire departments.

In the early 1990s, the Emergency Management Institute, in Emmitsburg, Maryland, began offering "train-the-trainer" courses to emergency managers and public safety professionals who were interested in bringing Community Emergency Response Training (CERT) to their communities; the program was modeled on the work of the Los Angeles fire department. After 9/11, as a means of establishing relationships between community volunteers and local government, President George W. Bush created the Citizen Corps, which incorporated three existing programs—CERT, Neighborhood Watch, and Volunteers in Police Service—and added the Fire Corps and the Medical Reserve Corps.[24] In addition, CERT was revised to include new information on disaster psychology, household hazardous materials, and terrorism awareness.

The goal of CERT is to ensure that every neighborhood has a leadership team that will be self-sufficient during the initial phases of a disaster, when professional emergency responders are overwhelmed or may be unable to access an area. The concept of neighborhood preparedness has moved beyond the three-day supply of food and water recommended during the civil defense era: CERT addresses general preparedness, firefighting techniques, basic medical care, light search and rescue, and disaster psychology. The focus is on self-help through organizational readiness.[25]

The shift from sworn to civilian emergency managers

As noted earlier, the emergency management function was historically situated within the fire department, the police department, or the sheriff's office. The work of the emergency manager focused on developing and maintaining emergency response plans, maintaining the EOC, and educating the public about the risks threatening the community. Typically, the role of emergency manager was filled by a sworn staff member, such as a battalion chief or a lieutenant, who was supported by civilian training specialists. The sworn staff member held the position for a two- to four-year rotation—in other words, just long enough to develop an understanding of the system—then moved on to another special assignment.

As emergency management developed into a more complex operation, the rotation of the senior position became counterproductive, and knowledge of field-level response management became less important than the ability to write plans, represent the city at multijurisdictional meetings, understand complex interdepartmental and intergovernmental relationships, and speak effectively at public meetings. Thus, the overall emergency management function came to focus more on the integration of resources—across the community and across departments—and less on field command. Meanwhile, field command remained under the control of the lead field response agency. The shift in emphasis from field command to broader responsibilities required emergency managers who had been trained to handle a larger role.

Thus, the role of emergency manager gradually became a full-time position for someone who had received college-level or FEMA training. The importance of both education and experience was institutionalized in the Certified Emergency Manager (CEM) program, which was developed by FEMA and is offered by the International Association of Emergency Managers.[26] Certification requires a bachelor's degree, 100 hours of education in emergency management and administration, and three years of experience.

A successful emergency manager need not come from any particular background. Emergency managers who get their start in police or fire services have a background in command and control; experience in decision making and problem solving; and an understanding of how to work with volunteers, community members, and the media. Urban planners understand plan development and the scientific bases of hazards, and are accustomed to solving problems and undertaking mitigation efforts in a multidisciplinary environment. Administrative generalists have a broad background in program management, personnel management, and budgeting and financial management, and often have public outreach experience. But whatever an emergency manager's background, the core knowledge of the field must be gained through a specific course of study, which requires both time and dedication. A CEM is the "gold standard" for professionalism in the field: CEMs have the knowledge, skills, and dedication to successfully manage all four phases of a community's emergency management work.

The role of the emergency management department

The emergency management department plays a different role during each of the four phases of emergency management: mitigation, planning, response, and recovery.

Under the Disaster Mitigation Act of 2000, local governments are required to develop local hazard mitigation plans. Mitigation involves both physical projects (such as the elimination of construction in flood zones) and public education and community outreach efforts.[27] The emergency manager takes the lead in analyzing threats to the community. Working with other local government departments, the manager develops a plan to address the most severe and frequent threats, protect critical infrastructure, and improve the overall safety of the community.

During the preparedness phase, the emergency management staff uses the risk analysis undertaken during the mitigation phase to develop a series of plans: the EOP, standard operating procedures (SOPs), and the EOC manual. The EOP, which describes how the community's resources—for example, NGOs and volunteer organizations—will be organized and deployed in the event of a disaster, must be based on NIMS. The SOPs for each department, which describe how specific departmental assets—that is, personnel and equipment—will be integrated into the community-wide response effort, are based on the EOP. Finally, the EOP drives the design and configuration of the EOC—the location from which emergencies are managed.

Working with other local government departments, the manager develops a plan to address the most severe and frequent threats, protect critical infrastructure, and improve the overall safety of the community.

The EOC must be physically designed to support all five sections of the NIMS organizational structure: command, operations, planning, logistics, and finance/administration. To ensure the effectiveness of the EOP and the supporting SOPs, the emergency management department must offer training in both NIMS and local emergency response systems and plans. Once the training is complete, the EOP and SOPs must be tested through a series of tabletop, functional, and facilitated exercises. Some communities conduct full-scale exercises every few years.

During the response phase, the emergency manager's job is to ensure that the EOP is followed. The emergency management department has overall responsibility for the smooth functioning of the EOC, ensuring that the facility is ready to support those who are responding to the disaster. NIMS mandates the use of the Incident Command System in the field, and the work of the incident commanders is coordinated and supported through the operations section of the EOC. Generally, the local fire chief or police chief serves as the operations section chief, with the support of various other elements of the local government, including fire and rescue, parks and recreation, communications, and public works.[28] Like the operations section chief, the chiefs of the logistics, planning and intelligence, and finance and administration sections are also department heads; in some cases, branches under these sections are managed by other department heads or by deputy department heads. The agency liaison officer in the management section of the EOC coordinates mutual aid if it is needed.

During the recovery phase, the emergency manager usually takes the lead in working with representatives from FEMA and from the state to obtain financial assistance under the Stafford Act and related state laws. The emergency management staff provides documentation on the disaster on the basis of EOC records, and works with departments involved in the response to obtain further documentation of reimbursable expenses. The emergency management staff also works with the federal government and with NGOs to meet the needs of local disaster victims.

Coordination and collaboration

Because emergency management departments are often small and their staffs may lack specialized skills or training, it is often necessary to form partnerships with other departments, levels of government, jurisdictions, organizations, and agencies. A local EOP must take into account all cross-governmental coordination and resources that would be required for a successful response to disaster. Partnerships with local, regional, state, and federal agencies are essential, and leaders from these other organizations must be a part of the planning process. (Chapter 5 discusses collaborative emergency management in detail.)

Local partners

As noted earlier, effective emergency management requires the active involvement of other local government departments. For example, public works staff are likely to have the knowledge of geology and hydrology that is needed for a mitigation plan. Mitigation efforts for wildland-urban interface fires require support from the fire department, the code enforcement department, the planning department, building officials, attorneys, arborists, and biologists. Such cross-departmental and cross-disciplinary ties ensure that plans are realistic and address the most salient aspects of disasters.

Every phase of emergency management also requires coordination across levels of government. Municipal resources are ordinarily sufficient for responding to normal emergencies—but by definition, a disaster typically overwhelms local resources, requiring personnel, equipment, or financial assistance from other jurisdictions, and often from state and federal partners. In California, for example, public health, mental health, social services, vector control, and environmental health all reside at the county level. Municipal EOPs must include checklists for the use of these county resources, either as local integrated resources or as external resources that would be made available through mutual aid agreements. Some states have designated substate regions for mutual aid purposes: California, for example, has six mutual aid regions.[29]

Special districts—single-purpose governmental units such as school districts, water districts, and transit districts—can also be crucial to emergency management. In most communities, for example, schools provide most of the shelter sites.[30] The water district may have a hydrologist who can analyze flooding risks, a meteorologist who can predict the weather for the watershed, and intelligence sources such as stream gauges and reservoir monitors. Transit-district vehicles can be used to evacuate people with disabilities or people who do not have cars. During rapid-onset events such as urban residential fires, buses make expedient shelters. Even cemetery districts and landscaping districts can be useful partners: they have heavy equipment and trained operators to assist with emergency response to landslides or to deal with mud left behind on streets. The EOP needs to specify both the resources that special districts will provide and the methods for contacting district staff outside business hours.

Volunteer organizations support the emergency management department in all four phases. As noted earlier, in the wake of the September 11 attacks, President Bush created the Citizen Corps; through a national network of state, local, and tribal Citizen Corps Councils, the corps works to solidify the relationship between volunteer organizations and local government.[31] These councils are to be integrated into the EOP, included in exercises, and trained to take part in the overall community response to disaster. An organization known as RACES (Radio Amateur Civil Emergency Service) has been a major source of volunteer support since the civil defense era.[32] RACES members are licensed by the Federal Communications Commission to operate amateur radios on a variety of frequencies. In 1952, FEMA's predecessor organization sanctioned the formation of RACES chapters in every local jurisdiction in order to provide auxiliary communications for communities in time of disaster or war.[33] RACES members provide their own equipment and meet regularly for training and exercises; typically, they also support other exercises conducted for the community, for local government departments, and for other volunteer groups.

Municipal resources are ordinarily sufficient for responding to normal emergencies—but by definition, a disaster typically overwhelms local resources, requiring personnel, equipment, or financial assistance from other jurisdictions, and often from state and federal partners.

So that they can work successfully with other NGOs and help support civil authority during disasters, NGOs create their own disaster plans and train their volunteers in plan implementation. For example, American Red Cross chapters have contracts with local school districts for the use of school facilities as shelters.[34] Volunteer Centers have plans to manage and deploy post-disaster volunteers for local governments.[35] Many communities have local collaborating organizations that bring local nonprofit agencies and community-based organizations into a planned response structure. In some cases, larger NGOs take the lead on a specific disaster response need and receive support from smaller NGOs. For example, the county food bank may coordinate the provision of food and receive assistance from soup kitchens run by communities or churches. At the national level, Voluntary Organizations Active in Disaster (VOAD) serves as an umbrella group for volunteer efforts.[36] Begun in 1970 with ten member NGOs, NVOAD now has twenty-eight members and two affiliates.

Citizens' groups such as community emergency response teams and Neighborhood Watch can assist with the pre-disaster education of their members. They can also become part of disaster-notification systems, participating in telephone trees or in arrangements for the use of neighborhood "runners." Members of homeowners' associations, parent-teacher associations, sports teams, and other affinity groups, who are already organized for other purposes, can apply their existing organizational structures to support emergency management. Newsletters and regular meetings sponsored by such groups offer the emergency management department opportunities to provide information or ask for assistance with disaster preparedness.

The business community is also a potential source of allies. Mitigation helps decrease insurance costs for businesses and lessens the likelihood of business interruptions.[37] A study of the business impacts of the 1995 Kobe earthquake in Japan found that about 70 percent of the manufacturing companies in Kobe had not regained their pre-earthquake sales and production levels one year later.[38] Other researchers have noted that small businesses that are closed for more than a few days never reopen.[39] Thus, emergency preparedness should be an essential feature of local economic development strategies. Partnerships may be formed with the chamber of commerce or with civic groups that include a business component, such as the Kiwanis Club or the Rotary Club. Milpitas, California, sponsors Business Partners for Emergency Preparedness, a joint government/business organization that holds quarterly workshops on current topics in mitigation and emergency preparedness.

Independent emergency preparedness groups in the business sector can also partner with government. The Business and Industry Council for Emergency Planning and Preparedness (BICEPP), in Southern California,[40] and the Business Recovery Managers Association (BRMA),[41] in Northern California, provide forums for the exchange of ideas and best practices among businesses that share the same geographically based threats. Local governments are very active in BICEPP, and BRMA actively partners with local governments in education and outreach events.

Regional partners

Many parts of the country have active councils of governments (COGs), which serve as regional planning organizations for transportation, housing, and the environment. COGs can also play a role in the development of mitigation and recovery plans. In the San Francisco Bay Area, the Association of Bay Area Governments (ABAG) has long been a leader in creating research products to support regionwide emergency management. Among the most important products are studies of potential earthquake damage to the transportation system, and "earthquake shaking maps," which help local emergency planners understand the likely effects of various earthquake scenarios on their communities.[42]

In 2004, ABAG received a FEMA grant to develop the federally mandated regional portion of the disaster mitigation plan required by the Disaster Mitigation Act of 2000. The regional portion includes maps and linked geographic representations of data on housing, employment, and transit; the linkages depicted in the regional plan provide a basis for generating similar information for each community. The partnership between Bay Area municipal governments and ABAG created a common basis for emergency plans throughout the region.

Federal resources

Federal resources are also crucial to local emergency management. Only the National Weather Service (NWS), for example, is responsible for providing warning of weather-related disasters. Hurricane and storm alerts, watches, and warnings issued by NWS provide the basis for activating local mitigation and emergency plans. Local governments obtain information on geological hazards—from volcanoes and earthquakes to landslides and groundwater conditions—from the U.S. Geological Survey, whose maps and guidance documents form the basis for risk analysis, and whose ongoing research may trigger changes in local evaluations of the probability of disasters. NWS and the U.S. Geological Survey are the only sources of official information in their respective fields.

The Posse Comitatus Act, which was passed in 1878, forbids the use of federal troops to keep order within the United States. Nevertheless, military commanders are authorized to provide "civil support to local authority." Under this doctrine, a local base commander may join forces with the local government to sandbag levees, provide drinking water, or render other immediate aid that would not compromise the ability to carry out military obligations.[43] The military also has specialized units that could be of assistance to a community subject to a terrorist attack. In the fall of 2003, when a series of sniper attacks were carried out in metropolitan Washington, D.C., Secretary of Defense Donald Rumsfeld ordered the U.S. Army's Northern Command to use its intelligence capabilities to try to find the snipers.[44]

The federal government also has specialized resources that are designed for response to weapons of mass destruction; such resources can be integrated into the terrorism response annexes of local EOPs. One example is the Marine Corps' Chemical/Biological Incident Response Force (CBIRF), which can "assist local, state, or federal agencies . . . by providing capabilities for agent detection and identification; casualty search, rescue, and personnel decontamination; and emergency medical care and stabilization of contaminated personnel."[45]

Other federal agencies also have resources that could be incorporated into EOPs. For example, the Nuclear Emergency Search Team of the Department of Energy can be sent to a community where a suspected release of radiological material (whether accidental or deliberate) has occurred.[46] DHS maintains Disaster Medical Assistance Teams (DMATs), which can provide medical care at disaster sites.[47] FEMA urban search and rescue (US&R) teams can undertake heavy-rescue operations and can also send specialized resources, such as cadaver search dogs and high-angle rescue specialists, to assist at disaster scenes.[48] After the 9/11 attacks, DMATs provided occupational health services at the World Trade Center, and US&R teams worked side by side with local first responders.

Enhancing response capacity

Mutual aid agreements provide for a rapid expansion of resources in case of need, enabling communities to avoid having to staff up for "worst case" events. Since 1996, the federal government has supported the development of the Emergency Management Assistance Compact (EMAC), which provides for state-level mutual aid in the case of major disasters. NIMS envisions the extension of EMAC to all states.[49] A related federal initiative is the Emergency Management Accreditation Program (EMAP) for state and local governments. Founded by FEMA, the International Association of Emergency Managers, and the National Emergency Management Association's EMAP is now an independent entity that relies on the National Fire Protection Association NFPA 1600 standard as the measure for accreditation.[50] The goal of the program is to enable state and local emergency management programs to be benchmarked against national best practices.[51]

The *9/11 Commission Report* recommended NFPA 1600 for the private sector's evaluation of its disaster response capabilities. The EMAP Commission, the program's nine-member governing body, is encouraging government to follow the same set of criteria: in order to obtain accreditation, a local government must (among other things) achieve a perfect score on an evaluation based on NFPA 1600, as modified by EMAC. NIMS includes a requirement for the credentialing of staff and certification of programs that will evolve over time.[52] EMAP has a set of national standards for the accreditation of local and state government emergency management programs. Florida was the first state to be accredited.

Summary

Emergency management defies simple description. It consists of layers of people and organizations at the municipal, county, regional, state, and federal levels. It encompasses residents, nongovernmental organizations, and members of the business community. It is based on the Incident Command System and NIMS but is locally controlled. It is carried out within a framework established by the federal government but must be responsive to local needs.

The local emergency manager may be full-time, part-time, or volunteer, and may be a career public servant or a political appointee. The emergency management unit may be a department in its own right or within a public safety department. Emergency management programs may include planning, training, exercising, public education, and legislative analysis. The one common goal of all emergency managers and emergency management programs is to create and sustain a disaster-resilient community, whose residents feel confident that their jurisdiction is prepared to cope with disaster, no matter what its origin.

Notes

1 For details on the rehousing efforts put forth by the Federal Emergency Management Agency (FEMA) in St. Bernard Parish in New Orleans, see CNN, "Katrina Victims: 'Living in Barns,'" December 13, 2005, available at cnn.com/2005/US/12/13/katrina.trailers/ (accessed July 2, 2007). For a similar story from across the region, see Johanna Neuman, "As Katrina's Homeless Wait, Trailers Sit Empty in an Arkansas Town," *Boston Globe*, February 12, 2006, available at boston.com/news/nation/articles/2006/02/12/as_katrinas_homeless_wait_trailers_sit_empty_in_an_arkansas_town/ (accessed July 2, 2007).

2 This presumes the use of twelve-hour shifts; five people per position are needed for eight-hour shifts.

3 Christopher Rhoads, "At Center of Crisis, City Officials Faced Struggle to Keep in Touch," *Wall Street Journal*, September 9, 2005, 1.

4 Continuity is variously referred to as "business continuity," "continuity of government" (COG), or "continuity of operations" (COOP).

5 FEMA, "Business Continuity and System Contingency Plans" (accessed October 23, 2004, from the FEMA Web site [no longer available]).

6 For requirements, see U.S. Department of Homeland Security (DHS), "2003 UASI Grant Guidance," available at dhs.gov/xlibrary/assets/grants_audit_fy03uasigrant.pdf (accessed July 2, 2007); for detailed COOP plans, see DHS, "COOP Multi-Year Strategy and Program Management Plan," available at fema.gov/government/coop/ (accessed July 2, 2007).

7 The theory behind this practice is that calls to undamaged areas are more important because that is how communities get help. The problem with this approach is that while the local government might be able to call the governor, it will be unable to reach the water service headquarters across the street.

8 For a description of emergency communications systems, see the document from the Federal Communications Commission, "Communicating during Disasters,"

available at fcc.gov/cgb/consumerfacts/emergencies.html (accessed July 2, 2007).

9 For detailed information on GETS, see the Government Emergency Telecommunications System Web site at gets.ncs.gov.

10 The debate over the knowledge, skills, and abilities required by emergency managers has gone on for some years. One comprehensive presentation is International Association of Emergency Managers, "Emergency Program Manager: Knowledges, Skills, and Abilities," which lists twenty-five pages of items for effectiveness; this is available at training.fema.gov/emiWeb/edu/EmergProgMgr.doc (accessed July 5, 2007).

11 Ibid.

12 For a more detailed discussion of the development and professionalization of emergency management, see Chapter 1 in this volume.

13 EPCRA is a stand-alone provision, Title III, of the Superfund Amendments and Reauthorization Act of 1986 (SARA). For more information on the requirements of SARA Title III, see ingham.org/hd/lepc/pamphlets/hazwaste/sara.html (accessed July 5, 2007); for information on the implementation of RCRA, see "RCRA Online," available at epa.gov/rcraonline/ (accessed July 5, 2007).

14 California Seismic Safety Commission, *Hearings on the Northridge Earthquake,* Los Angeles, California, July 1995.

15 Ibid.

16 Brian Michael Jenkins and Frances Edwards-Winslow, "Saving City Lifelines: Lessons Learned in the 9-11 Terror Attacks," MTI Report 02-06 (San José, Calif.: Mineta Transportation Institute, September 2003), 28–29, 31.

17 California Office of Emergency Services, *Standard Emergency Management System Student Manual* (Mather, Calif., 1996), 2.

18 Thomas P. O'Neill, with Gary Hymel, *All Politics Is Local, and Other Rules of the Game* (New York: Times Books, Inc., 1993).

19 Frances Edwards-Winslow, "Telling It Like It Is: The Role of the Media in Terrorism Response and Recovery," Perspectives on Preparedness #9 (Cambridge, Mass.: Harvard Executive Session on Domestic Preparedness, 2002).

20 Tom Ridge, "Letter to Governors," September 8, 2004, available at gov/doc/nims/letter_to_governors_09082004.doc (accessed July 5, 2007).

21 The National Incident Management System (NIMS) is based on the Incident Command System and on California's Standard Emergency Management System. See DHS, *National Incident Management System* (Washington, D.C.: DHS, March 2004), available at dhs.gov/xlibrary/assets/NIMS-90-web.pdf (accessed September 7, 2007).

22 Under NIMS, the local government manager is in charge of the emergency operations center and elected officials are given a "policy-making" role. In addition, the police chief and fire chief are in charge of only their departments' field response, not the overall community response. These changes may have significant political implications within an organization.

23 Details of the program are available in "Disaster Resistant Communities Initiative: Evaluation of the Pilot Phase Year 2," Project #EMW-97-CA-0519 (Newark: Disaster Research Center, University of Delaware, 2001), available at udel.edu/DRC/projectreport41.pdf (accessed July 5, 2007). A firebreak is a section of cleared or plowed land that is used as a barrier to check a forest or grass fire.

24 For details on the Citizen Corps, see its Web site at citizencorps.gov/. The Citizen Corps Council also attempted to create Operation TIPS (Terrorism Information and Prevention System) "to establish a reliable and comprehensive national system for reporting suspicious, and potentially terrorist-related, activity" (TIPS Web site, August 8, 2002 [no longer available]), but the program was withdrawn because many citizens viewed it as intrusive.

25 For details on community emergency response teams, see citizencorps.gov/cert/ (accessed July 5, 2007).

26 For details, see Chapter 1 in this volume.

27 For a complete discussion of mitigation, see Chapter 6 in this volume.

28 The coroner's office and the public health department may be located at either the municipal or county level.

29 The Southern Region, headquartered in Los Alamitos on the border of Los Angeles County and Orange County, includes Mutual Aid Regions 1 and 6; the Coastal Region is coterminous with Mutual Aid Region 2; and the Inland Region comprises Regions 3, 4, and 5 and is co-located with the State Operations Center in Sacramento.

30 The American Red Cross prefers the cooking and restroom facilities in high schools because of their size and the fact that they comply with the standards in the Americans with Disabilities Act.

31 For the work of the Citizen Corps, see its Web site (note 24); see also Citizen Corps, "Programs & Partners," available at citizencorps.gov/programs/ (accessed July 5, 2007).

32 See the Radio Amateur Civil Emergency Service (RACES) Web site at races.net.

33 The site for this information, fema.gov/library/civilpg.shtm, was accessed December 29, 2004, but is no longer available.

34 American Red Cross congressional charter of January 5, 1905, and Federal Disaster Relief Act of 1974 (Public Law 93-238 as amended by the Stafford Act).

35 For information on the Volunteer Center National Network, go to pointsoflight.org/centers/ (accessed September 7, 2007).

36 See the National Voluntary Organizations Active in Disaster (NVOAD) Web site at nvoad.org/.

37 The Institute for Business & Home Safety offers guidance through its publication "Open for Business," available at ibhs.org/business_protection/ofb.asp (accessed July 5, 2007). FEMA and state offices of emergency management also offer publications to assist businesses with mitigation and preparedness.

38 Guna Selvaduray, "Effect of the Kobe Earthquake on Manufacturing Industries," paper presented at the Business Continuity Workshop, SEMI, San José State University, March 19, 2003, available at www2.sjsu.edu/cdm/public/SEMI-Workshop.pdf (accessed July 5, 2007).

39 Daniel J. Alesch et al., "Organizations at Risk: What Happens When Small Businesses and Not-for-Profits Encounter Natural Disasters" (Fairfax, Va.: Public Entity Risk Institute [PERI], 2001), available at riskinstitute.org/PERI/PTR/Disaster + Management_PTR_1028.htm (accessed July 5, 2007).

40 See the Web site of the Business and Industry Council for Emergency Planning and Preparedness at bicepp.org/WHOWEARE.stm.

41 See the Web site of the Business Recovery Managers Association at brma.com/.

42 See "ABAG Shaking Intensity Maps and Information," available at abag.ca.gov/bayarea/eqmaps/mapsba.html (accessed July 5, 2007).

43 Barry Kellman, *Managing Terrorism's Consequences: Legal Issues* (Oklahoma City, Okla.: National Memorial Institute for the Prevention of Terrorism, 2002), available at terrorisminfo.mipt.org/pdf/managingterrorismsconsequenceskellmantoc.pdf. Because National Guard troops qualify as a state militia, they can, at the direction of the governor, be used to keep order.

44 Although Secretary Rumsfeld's decision was controversial, the fact remains that the military is a potential resource.

45 See "CBIRF Mission" at cbirf.usmc.mil (accessed September 7, 2007).

46 U.S. Department of Energy, "Order DOE 5530.2: Nuclear Energy Search Team," September 20, 1991, available at fas.org/nuke/guide/usa/doctrine/doe/o5530_2.htm (accessed July 5, 2007).

47 U.S. Department of Health and Human Services, "Emergency Response: Disaster Medical Assistance Teams (DMATs) and Disaster Mortuary Operational Response Team (DMORT)," September 11, 2001, available at hhs.gov/news/press/2001pres/20010911c.html (accessed July 5, 2007).

48 FEMA, "Urban Search and Rescue (US&R)," available at fema.gov/emergency/usr/ (accessed July 5, 2007).

49 FEMA, "Resource Management," available at fema.gov/emergency/nims/mutual_aid.shtm (accessed July 5, 2007).

50 "NFPA 1600 Standard on Disaster/Emergency Management and Business Community Programs: 2004 Edition," available at nfpa.org/PDF/nfpa1600.pdf?src=nfpa (accessed July 5, 2007).

51 See the Web site of the "Emergency Management Accreditation Program" at emaponline.org/?32.

52 National Commission on Terrorist Attacks upon the United States, *9/11 Commission Report* (Washington, D.C.: U.S. Government Printing Office, 2004), 398, available at gpoaccess.gov/911/pdf/fullreport.pdf (accessed July 5, 2007).

CHAPTER 4

The intergovernmental context

David A. McEntire and Gregg Dawson

This chapter provides an understanding of

- The management of disasters in a federalist system
- The importance of emergency management networks
- Frameworks for coordinating emergency management
- Approaches to financing local emergency management programs.

By its very nature, emergency management requires the integration of policies, programs, and operations among a variety of individuals and entities. On the one hand, the participation of diverse actors ensures access to significant knowledge, resources, and skills. On the other hand, this diversity may complicate or even hinder mitigation, preparedness, response, and recovery efforts. Thus, to improve the quality of emergency management, it is essential to understand the overall context in which it operates.

This chapter examines the management of disasters under the federalist system of government. Given the extraordinary number of actors involved in emergency management, coordination and networking are essential. Equally important, however, is adequate funding from both internal and external sources. Because public officials and citizens are often apathetic about the need to proactively mitigate against disasters, emergency managers are faced with two sets of tasks: the routine responsibilities of running a department, and the broader job of mustering the necessary political support to effect change.

Managing disasters in a federalist system

To understand emergency management in the United States, it is necessary to understand the federalist system of government. The country's founders, concerned about the potential for abuse of power, opposed the idea of a unitary government in which all authority resided with a king or other central official. Instead, they created a political system in which authority is shared among local, state, and federal levels of government; among geographical jurisdictions; and among the executive, legislative, and judicial branches. The U.S. Constitution therefore reflected dual aims: (1) to separate the functions of government and (2) to allow for some degree of sovereignty at lower levels of government.

What does federalism have to do with emergency management? Federalism permeates every governmental function, and emergency management is no exception. In practical terms, emergency management activities involve a vast number of stakeholders—from the president to Congress, federal and state agencies, and local and state officials. Although federal agencies provide substantial resources for emergency management, local governments bear most of the responsibility for the success of mitigation, preparedness, response, and recovery efforts. States often act as intermediaries, helping to implement federal policies, training communities in best practices, and funneling federal grant monies.

The federalist nature of emergency management in this country is also reflected in the law: all lower levels of government are expected to comply with national legislation and its accompanying regulations. Thus, lower levels of government must follow directives that may conflict with their own needs and desires.

More than two decades ago, sociologist Thomas Drabek examined six major disasters over a two-year period and found that the responses to these events revealed four characteristics of the American emergency management system: localism, lack of standardization, unit diversity, and fragmentation.[1] *Localism* refers to the fact that municipalities have a substantial responsibility for emergency management. In contrast to many centralized governments, the U.S. federal government relies heavily on municipalities—the lowest level of government—as the main responders in any disaster. *Lack of standardization* refers to the variation in how emergency management is organized and undertaken. For instance, some communities have independent emergency management offices; other communities locate their emergency management functions within the office of the chief executive or within the fire or police department. Because no two communities or states have identical policies or organizational arrangements, emergency management agencies operate within distinct contexts. *Unit diversity* refers to the fact that entities of many different sizes and types commonly respond to disaster. Drabek notes, for example, that at least seventy-eight organizations became involved in the aftermath of the Lake Pomona (Kansas) tornado in 1978. Unit diversity is even more extensive today: more than two hundred agencies and organizations responded to the 2001 terrorist attack on the World Trade Center.

The first three characteristics—localism, lack of standardization, and unit diversity—combine to yield the fourth: *fragmentation*. When federal, state, and local governments' approaches to emergency management diverge, the overall emergency management effort

Intergovernmental distance and integration

In a 1989 article, Llewellyn Toulmin, Charles Givans, and Deborah Steel asserted that governments and public agencies are confronted with the challenge of "intergovernmental distance."[1] This concept refers not to spatial detachment but to differing and conflicting organizational cultures (e.g., missions, budgetary priorities, bureaucratic procedures, and terminology).

Examples of intergovernmental distance in emergency management include the following:

- Emergency managers give special emphasis to emergency preparedness, whereas floodplain managers pay more attention to structural and nonstructural mitigation.
- Fire and police departments use different terminology for on-scene communication.
- Local emergency management departments view federal disaster programs as inapplicable to the jurisdictional context.
- The priorities of the state emergency management agency conflict with the goals of the state economic development department.
- The forms used for damage assessment vary dramatically among organizations that participate in this post-disaster function.
- Municipal and county finance departments are unaware of federal requirements for disaster assistance.

The theory of intergovernmental distance explains why multi-organizational responses are problematic in small disasters and especially difficult in larger events. The theory implies that local governments should make an effort, before disaster strikes, to minimize the horizontal and vertical strains that inhibit integrated policy making and discourage coordinated disaster management. Such efforts may include conflict resolution, role clarification, joint planning and training, mutual aid agreements, memorandums of understanding, and the establishment of interoperable communications systems.

[1]Llewellyn M. Toulmin, Charles J. Givans, and Deborah L. Steel, "The Impact of Intergovernmental Distance on Disaster Communications," *International Journal of Mass Emergencies and Disasters* 7, no 2 (1989): 116-132.

is subject to vertical strains. And when separate entities view emergency management from their own frames of reference—leading to conflicts across departments and between neighboring jurisdictions, for example—the result is horizontal fractures in the overall emergency management effort.

Thus, the decentralized nature of federalist government yields numerous emergency management organizations with differing policies, priorities, and programs. Decentralization has several practical effects. First, the proliferation of agencies that occurs under federalism creates uncertainty and disagreement about responsibility for essential emergency management functions, increasing the likelihood that a particular task or issue will fall through the cracks. Second, federal requirements inevitably conflict, at times, with state and local preferences—and may exceed state and local resources. However, when state or community leaders opt to ignore such requirements because they are either politically unpopular or excessively costly, those leaders risk losing vital funding for emergency management programs. It is through funding that the federal government encourages state and municipal governments to support national programs—and this means of control has only intensified since the attacks of September 11, 2001, and the creation of the U.S. Department of Homeland Security (DHS).

The federalist system is not without its advantages, however. Among the greatest strengths of federalism are flexibility, diversity, and redundancy. Because the Constitution limits the power and authority of the federal government, state and local governments may determine, to a certain extent, their own emergency management priorities. This is vitally important in a large and geographically diverse country such as the United States. Each state and community has its own hazards and vulnerabilities to deal with; no single approach will work in every situation. Diversity means that each level of government, including all its departments and agencies, brings specialized knowledge and skills to emergency management. Disasters could not be effectively addressed if only state governments existed, or if every municipal department was responsible for only firefighting or building code enforcement. Finally, certain disasters (e.g., catastrophic events), by their very nature, overwhelm local resources. Redundancy in the federalist system allows disaster-stricken communities to seek assistance from nearby jurisdictions or from state and federal agencies.

Emergency management networks

In the context of public administration, a network may be defined as a "policy making and administrative structure involving multiple nodes (agencies and organizations) with multiple linkages."[2] In the context of emergency management, such networks may include a variety of practitioners and organizations linked through formal or informal relations.

The goal of emergency management networks is to devise policy and to implement programs that will reduce vulnerability, limit the loss of life and property, protect the environment, and improve multi-organizational coordination in disasters. The degree to which the emergency management community collaborates *before* a disaster is a strong determinant not only of the success of mitigation and preparedness activities, but also of the effectiveness of management during the crisis and recovery periods.[3]

As noted earlier, a vast array of individuals and entities are involved in emergency management. The best way to harness the knowledge, skills, and resources of this eclectic group is to create, participate in, expand, and improve the effectiveness of emergency management networks. Successful collaboration depends on a clear understanding of each entity's roles and responsibilities in emergency management. Thus, emergency managers must familiarize themselves with the functions and capabilities of all departments, agencies, and organizations at various levels of government as well as of other entities in the private and nonprofit sectors. Emergency managers must also use networks to develop priorities, set goals and objectives, and agree on implementation strategies. Of course, this suggests the need for a great deal of collaboration across organizations. (Chapter 5 discusses collaborative emergency management in detail.)

In the wake of a disaster, networks form automatically. Numerous studies have shown that an impressive—and in some cases overwhelming—number of organizations will converge at the scene of an incident to perform vital disaster functions.[4] These organizations will attempt to communicate with one another, although some will interact more than others. Because of differences in expertise and level of authority, the boundaries of individual or overall responsibility may be unclear; nevertheless, the organizations generally attempt to coordinate their activities in a harmonious way. To make the best possible use of these spontaneously formed networks—referred to in the field as EMONs (emergent multi-organizational networks)—emergency managers not only must be aware of the number and diversity of agencies involved in disasters, but also should sharpen their administrative skills and their ability to improvise in cooperation with others. Most importantly, however, emergency managers must strengthen ties between organizations *before* disaster strikes. There is substantial evidence that preparedness networks enhance response: the structure of interorganizational relations augments the benefits of planning.[5]

Local emergency planning committees (LEPCs) are excellent examples of such preparedness networks. After the deadly chemical accident in 1984 in Bhopal, India, Congress passed the Emergency Planning and Community Right to Know Act (EPCRA; also known as SARA Title III because it was part of the Superfund and Amendments Reauthorization Act). Under EPCRA, each state is responsible for establishing a state emergency response commission (SERC) to plan for hazardous materials emergencies. One of the responsibilities of a SERC is to help develop LEPCs.

Each LEPC must identify hazardous materials within the applicable geographic area and develop emergency operations plans to deal with toxic chemical releases. The planning process associated with SARA Title III requires a network that may include emergency managers, fire department officials, public safety officials, public health organizations, environmental agencies, community-based organizations, industry representatives, and school administrators. Psychologist Michael Lindell has noted that "LEPCs represent a rejection of the historical role of the isolated Civil Defense Director who sat in the basement of the courthouse and prepared emergency plans that were never read, much less implemented."[6] Lindell also observes that although not every jurisdiction has created or fully used LEPCs, the existence of an LEPC does improve disaster planning. LEPCs are most likely to be successful when they are funded adequately by the community, membership in the LEPC is sufficiently broad, the burden of planning can be shared, there is political support for the LEPC's goals, and the LEPC's leaders and participants cultivate a climate of teamwork.[7]

LEPCs are not the only examples of emergency management networks, however. Other networks focus on specific hazards (such as earthquakes, hurricanes, and terrorism), different phases of emergency management, distinct levels of government, specific sectors (such as public, private, and nonprofit), or particular functional areas (such as land use planning, public information, emergency medical care, and the preservation of historic structures). Larger and more extensive networks may combine a number of different elements. Emergency managers should determine what networks exist or need to be created in their communities and seek ways to integrate themselves into what can be called "webs of enablement." Networking is an essential tool for helping emergency managers fulfill their responsibilities.

Coordinating emergency management

Coordination can be defined as "the collaborative process through which multiple organizations interact to achieve common objectives."[8] Although coordination can be—and has been—defined in a number of other ways, there is less controversy about its value. The purpose of multi-organizational, intergovernmental, and intersector coordination is to eliminate "fragmentation, gaps in service delivery, and unnecessary (as opposed to strategic) duplication of service."[9] Coordination is most likely to be achieved when organizations have (1) pre-disaster ties (which generate familiarity with others' knowledge, skills, and abilities), (2) a means of sharing disaster information easily and quickly, and (3) a willingness to work together to meet emergency management needs.[10]

In practice, coordination often takes the form of vertical and/or horizontal integration. *Vertical integration* occurs within the bureaucratic structure of a single entity or between different levels of government. A local emergency management office that works closely with state and federal agencies to minimize hazard vulnerability or to provide expeditious relief to disaster victims is an example of vertical integration. *Horizontal integration* occurs among distinct entities (e.g., different departments within a local government, diverse community organizations, or nearby jurisdictions within a region). Examples of horizontal integration include a joint effort to suppress a fire that has spread across territorial boundaries, and the creation of standard operating procedures throughout a metropolitan area. In both cases, the goal is to take advantage of synergy and to avoid working at cross-purposes.

Coordination can be strengthened through a number of means, including joint planning, memorandums of understanding, and mutual aid agreements; the Emergency Management Assistance Compact; the Incident Command System; the Standard Emergency Management System; the National Response Plan; and the National Incident Management System.

Joint planning

When joint planning is undertaken at the local level, the emergency manager meets periodically with representatives from various local government departments, community organizations, and other levels of government to identify hazards and vulnerabilities, agree on which risks have the highest priority, and find ways to mitigate disasters. Joint planning efforts may also be used to strengthen response and recovery capabilities. Because local businesses and nonprofit organizations play a much larger role in emergency management today than they did in the past, it is essential to include such entities in joint planning sessions. (Chapters 5 and 10 discuss joint planning in greater detail.)

Memorandums of understanding (MOUs) and mutual aid agreements are similar in that they are both designed to improve interagency, intersectoral, or interjurisdictional assistance and coordination. However, MOUs tend to be less formal than mutual aid agreements, which are often legally binding and must be approved by both the attorney and the governing body of the local jurisdictions involved. An official sanction is imperative because mutual aid may generate expenses and liabilities (specifically, the possibility of injury or death among responding parties).

The Urban Area Security Initiative

The goal of the Urban Area Security Initiative, a program of the Office of Grants and Training of the U.S. Department of Homeland Security (DHS), is to strengthen the ability of urban areas to prevent, deter, respond to, and recover from terrorism. The program encourages urban areas to employ regional approaches to preparedness and to adopt regional response structures where appropriate. The Office of Grants and Training is responsible for providing state and local jurisdictions with training, technical assistance, funds for equipment purchases, support for the planning and execution of exercises, and other support.

A number of intergovernmental issues arise once funds have reached state and local governments. As funds are distributed across the urban area, local governments and regional partners must work together to determine how and when the funds will be dispersed and which projects will be implemented.

The distribution of funds can take many forms. If a regional entity (such as a council of governments) covers all or most of the urban area, that entity can control the disbursement of funds according to project goals. For example, if the goal is to establish a secure wireless network across the region, the council of governments can distribute funds to each unit of local government as it implements its portion of the project. Alternatively, funds can be distributed to individual jurisdictions to use at their discretion, with the understanding that regional projects will be given priority; in this case, a local government unit may dedicate its funds to the construction of an intelligence analysis center, with the understanding that the local government will share information with the county, which maintains corrections records.

However funds are distributed, it is still necessary to decide which projects should take priority if the grant award is less than the requested amount. Setting priorities is especially important when local priorities do not mesh with either regional priorities or those put forth by DHS. For example, it may be very important to the security of a local government to purchase a backup generator for the continuity of operations in the event of an emergency. However, if the regional and national priority is to enhance interoperable communications, the generator may not be the ideal use of DHS funds. Ultimately, the projects that are given priority and implemented must strike a balance between meeting the needs of the local government and the needs of the region; they must also be realistic and attainable given the available funding.

Source: Rocky Vaz, manager, Intergovernmental Affairs, Dallas, Texas.

Mutual aid agreements are beneficial for a number of reasons. First, federal grants for emergency management endorse mutual aid agreements. Second, in the case of some preparedness grants, the federal government supplies equipment on the assumption that it will be shared among numerous municipalities and counties in a region. Third, and most important, mutual aid agreements improve the coordination of disaster response by resolving, during the calm of the planning period, challenging questions about joint assistance. Emergency managers should seriously consider participating in MOUs or mutual aid agreements.

The Emergency Management Assistance Compact

States can also work to improve coordination. In the wake of Hurricane Andrew, which occurred in 1992, Florida governor Lawton Chiles worked with the nineteen-state Southern Governors' Association to develop the Southern Regional Emergency Management Assistance Compact, an interstate disaster assistance program.[11] The goal was to provide member states with fast and flexible assistance, and to enable them to rely on each other, rather than on the federal government alone, for resources and support. In 1995, this regional compact was opened to all states that wanted to join. Congress endorsed the new organization, known as the Emergency Management Assistance Compact (EMAC), in 1996, and today all fifty states, the District of Columbia, Puerto Rico, and the Virgin Islands are members. States are encouraged but not required to donate $1,000 annually to maintain the system.

EMAC is essentially a state-level mutual aid agreement. It uses standard forms for assistance requests, helps affected states acquire resources from other states that are willing to donate them, and ensures that legal questions about employee benefits and response expenses are resolved before a disaster strikes. In the near future, EMAC's members plan to standardize some of the aid packages that will be sent to states in need. Since its founding, the compact has already proven useful in disasters ranging from wildfires and ice storms to the terrorist attacks of September 11, 2001. EMAC was also used extensively after Hurricane Katrina in 2005.

The Incident Command System

The Incident Command System (ICS) is widely recognized as a means of strengthening coordination, generally among first responders at the operational level. Conceived after a series of major fires in southern California in 1970, ICS is a widely used framework for emergency response; its purpose is to standardize operations across agencies while allowing sufficient flexibility to respond to unique circumstances.[12] The principal elements of ICS (known as "sections") are planning, operations, logistics, and finance and administration. ICS is implemented under the direction of the on-scene leadership. Because it improves communication within and across organizations, ICS also improves coordination, especially when the responding individuals and agencies focus on collaboration rather than on control. (ICS and other emergency management frameworks are discussed in detail in Chapters 7 and 9.)

The Standard Emergency Management System

Like ICS, the Standard Emergency Management System (SEMS) was first developed in California, after a major disaster: the 1991 East Bay Hills fire in Oakland. The response to this incident was inadequate, and state senator Dominique Petris introduced Senate Bill 1841, which called for the integration of emergency response operations among all agencies in California. Approved on January 1, 1993, the statute was designed to standardize emergency management training and operations, improve the flow of information, and strengthen coordination among responding agencies. Through SEMS, first responders and emergency management agencies in California were able to improve their responses to major disasters, including the Loma Prieta earthquake, the Northridge earthquake, and fires in the urban-wildfire interface.

Under SEMS, depending on the nature and scope of a disaster, five organizational levels may be activated: the field level, the local government level, the operational level, the regional level, and the state level. Because ICS is the organizing framework for disaster operations, mutual aid requests, resource management, damage assessment, and other important functions, SEMS mandates that all jurisdictions in California operate under ICS.

The National Response Plan

Federal plans offer yet another means of promoting coordination. The federal government has had a number of plans over the decades, but the most important was the Federal Response Plan (FRP), developed by the Federal Emergency Management Agency (FEMA) in 1992. The FRP was designed to coordinate the activities of all twenty-six federal agencies that were involved in disaster response operations; it also addressed the role of the American Red Cross, which was classified as a partner organization. The FRP categorized the response activities of the federal government into twelve emergency support functions, which were divided among a number of signatory agencies that had either primary or support responsibilities. Primary agencies had principal authority for a function, and support agencies provided assistance if needed. Although the FRP was implemented with increasing effectiveness during the mid- to late 1990s, the attacks of September 11, 2001, led some government officials to feel that it did not sufficiently address the threat of terrorism. In response, DHS created the National Response Plan (NRP).[13]

Like the FRP, the NRP assigns responsibilities among primary federal support agencies and support agencies, but it addresses fifteen functions instead of twelve. Although DHS maintains that the NRP represents an all-hazards approach to disaster, most scholars and practitioners of emergency management believe that it is heavily slanted toward terrorism. Moreover, under the NRP, some of the departments and agencies that were included in the FRP (FEMA among them) appear to have a less important role in managing all types of disasters than they did under the FRP. The NRP is currently under review and is likely to be revised often.

The National Incident Management System

The National Incident Management System (NIMS) was also developed by DHS in the wake of the 2001 terrorist attacks. It is designed to standardize disaster response and to address a number of problems that surfaced in the attacks—specifically, communication failures, poor information management, insufficient interorganizational coordination, failure to provide accurate and consistent public information, and flawed incident management. Developed in accordance with Presidential Homeland Security Directives 5 and 8, NIMS includes six key components: command and management, preparedness, resource management, communications and information management, supporting technologies, and ongoing management and maintenance. NIMS mandates the use of ICS and is somewhat similar to SEMS in that its goal is to integrate the activities of various governmental entities so as to achieve a holistic and well-orchestrated response.

Like ICS and SEMS, NIMS provides a set of standardized structures, but it is much broader in its intentions and implications. First, NIMS seeks to improve post-disaster operations through predisaster planning and capacity building. Second, the common processes and procedures required under NIMS are designed to improve interoperability among all types of responders, including those in the private and nonprofit sectors. Third, NIMS operates at all levels of government.

Although NIMS has been formally adopted by DHS, it is controversial. Some observers have argued that NIMS is too focused on terrorism preparedness and response, and that this focus may undermine efforts to mitigate and recover from natural disasters. Others contend that NIMS's emphasis on standards may inhibit creativity and improvisation during response. Proponents of NIMS, in contrast, believe that it is flexible enough to deal with all types of disasters, regardless of cause, size, location, or complexity. They also view NIMS as a logical approach to integrating disparate elements of response into a coherent management system. One thing is certain, however: NIMS is a new approach to disasters that will require ongoing review and modification. In the meantime, members of the emergency management community will require further training in order to implement it.

Financing local emergency management programs

Mitigation measures, emergency preparedness initiatives, and disaster response and recovery operations are expensive: among many other activities, the local government must acquire property in hazardous areas, purchase warning and communications equipment, pay overtime to emergency personnel, remove debris in the wake of a disaster, and reconstruct government buildings in affected areas. But these "routine" expenses mask the larger picture of exponential increases in disaster losses. In light of these increases, the importance of external and internal funding cannot be overstated: financial resources are the lifeblood of any emergency management program. Successful emergency management thus requires a thorough knowledge of the federal funding process, as well as of state and local government procedures for allocating funds, acquiring resources, managing grants, and developing alternative funding streams. Finally, emergency managers must recognize the contentious nature of the budgeting process and ensure that knowledgeable staff members are in charge of financial operations.

External funding

External funding for emergency management in the United States comes from the federal government. Each year, Congress appropriates money to support emergency management.[14] These funds are used to mitigate disasters and improve preparedness capabilities nationwide; to finance the day-to-day operations of DHS and FEMA; and to help implement major federal initiatives (such as earthquake mitigation, terrorism prevention, and training for emergency managers and first responders).

Congressional allocations are also directed to the Presidential Disaster Relief Fund (which is distributed by FEMA) and to numerous federal agencies that have post-disaster missions and responsibilities. These include the Small Business Administration; the Army Corps of Engineers; and the Departments of Health and Human Services, Housing and Urban Development, and Agriculture. A specific amount of money is designated for post-disaster assistance each year; if disasters are unusually frequent or severe, the result may be a shortfall in relief allocations. When this occurs, Congress will debate the merit of supplemental assistance and provide such assistance as needed.

A sizable portion of annual federal allocations is distributed to assist states and local governments in meeting their emergency management responsibilities. Most of this money is not given directly to local governments; instead, it is funneled through state emergency management offices. The state retains a portion of the funding to address its own emergency management needs; the remainder is passed on to local governments in the form of grants. In a number of instances, the regulations governing the distribution of federal funding have led to contention. Under the formula that was initially used to distribute State Homeland Security Grants, for example, some rural states, such as Alaska and Wyoming, received more money per capita than heavily populated states, such as California.

Among the federal grants for which local governments may apply are Pre-Disaster Mitigation Grants, Community Development Block Grants, Emergency Management Performance Grants, Assistance to Firefighters Grants, Community Oriented Policing Grants, State Homeland Security Grants, Community Emergency Response Team Grants, Law Enforcement Terrorism Prevention Grants, and Urban Area Security Initiative Grants. In an effort to create a "one-stop shop" for state and local governments, many of these grant programs have been consolidated under the purview of FEMA. However, these and other federal grants are administered at the state level by a state administrative agency (usually the state office of emergency management). If there is a large number of jurisdictions in the state, this task can be extremely challenging. In such cases, regional councils of government, which have well-established relationships with their member governments, play a vital role in grant administration.

Each federal grant has unique limitations and eligibility requirements.

Emergency managers can increase local government funding substantially by applying for federal grants. In doing so, however, they should be aware that each federal grant has unique limitations and eligibility requirements. For instance, Emergency Management Performance Grants cannot be used to construct emergency operations centers. Some DHS grants are available only to urban areas with large populations and vulnerable infrastructure. Other grants require the jurisdiction to have a completed emergency operations plan that meets specified criteria, or to have mutual aid agreements with neighboring communities. Many federal grants also require local matching funds.

One of the most prevalent requirements today concerns multijurisdictional projects. In fact, some federal grant programs are based on an explicitly regional approach to emergency management. In 2005, by distributing grants specifically to councils of governments in some states, DHS attempted to limit expenses and increase overall capability: for example, instead of funding a hazardous materials team for each jurisdiction, the department funded one team for the entire region. This approach was based on the assumption that jurisdictions would share resources in time of need.

Although the regional approach is an economical means of meeting needs, it may complicate efforts to apply for and manage grants. Each jurisdiction will have its own preferences, and it may be difficult for localities to agree on funding and operational priorities. Ideally, a region would have the time to develop a complete strategy and coherent logistical plan before purchasing millions of dollars worth of emergency response equipment. Fortunately, some state administrative agencies have empowered councils of governments to partially withhold funding until studies can be completed to ensure the best use and allocation of resources.

Unique rules are applicable to other grants as well. As of November 1, 2004, local governments are required to have a FEMA-approved Hazard Mitigation Action Plan (HazMAP).[15] A community that lacks an approved HazMAP is ineligible for a number of federal assistance funds, including Hazard Mitigation Grants and Pre-Disaster Mitigation Grants. Moreover, lack of a state HazMAP plan may render the entire state ineligible for FEMA's Public Assistance Grants during recovery. (Such regulations have not always been enforced, however. After

Hurricane Katrina, for example, many Louisiana communities received post-disaster assistance even though they did not have approved HazMAPs.)

Different interpretations of grant rules may lead to conflict between different levels of government. For example, post-disaster relief funds are ordinarily distributed according to a 75:25 ratio of federal to local funding. However, when an area is affected by a disaster, its senators and representatives will pressure the federal government for as much money as possible. In some cases, the president has responded by giving more than a 75 percent share to affected communities—a choice that is often made more for political reasons than for practical ones. Such decisions have led other localities, which have received only the standard amount after disasters, to claim unfair treatment. State and local governments have also objected to delays in the distribution of federal funds. In their opinion, the federal government is too slow to release money for rebuilding after a disaster.

Grants are complex in other ways. The instructions given to elected officials and chief financial officers may be ambiguous or poorly written. Application guidelines for federal grants may change over time, with little or no notice. Grants may involve many departments and agencies, complicating decisions about priorities and the application process itself. Local emergency managers must understand the importance of following grant instructions, find ways to build consensus, and ensure that applications are filled out in a clear, logical, and concise way.

If an emergency manager succeeds in obtaining grant funding, that is when the work really begins. Funds may have to be shared among as many as ten functional groups, including first responders, public health departments, and public works departments. Disagreements among these parties will almost certainly arise as funds are distributed and spent. There will also be record-keeping requirements, reporting requirements (on an annual, semiannual, or quarterly basis), and other deliverables. Some grants may require a substantial amount of paperwork for two or more years.

The management of grant funds and the associated paperwork requires extreme care. A number of recipients of DHS funds have already been audited because of concern about possible funding abuse. As the result of one audit, a municipality was required to return $300,000 to the federal government because it had used DHS funding to pay its regular personnel costs. Maintaining sound business practices in grant administration is somewhat daunting. There are, however, useful steps that local governments can take to increase the likelihood of clean audits:

- Ensure that there is a dedicated point of contact who is knowledgeable about governmental accounting and purchasing practices
- Enlist the support and involvement of the local government's finance, auditing, or purchasing department
- Develop a standardized filing and inventory system
- Identify and tag all assets purchased with DHS funding
- Create a line of succession so that disruption is minimized in the event of staff turnover
- Meet with staff from the state administrative agency or the state accounting office to learn what these agencies look for in a typical audit; use this information to improve the community's grants-management process.

Internal funding

In addition to the funds provided by the federal government, state and local governments also finance emergency management with their own funds. State governments designate tax revenues to operate state emergency management departments. Allocations vary dramatically, depending on the hazards facing the state, the strength of the state's economy, the size of its population, and the values of its constituents. For instance, California has reserve funds for disasters that affect local jurisdictions, whereas Texas does not. There are also unique state-level funding streams. For example, after Hurricane Andrew hit Florida in 1992, legislators added a tax to homeowners' insurance. This money helps support all types of emergency management activities throughout the state.

Local governments are creative when it comes to obtaining funding for emergency management. One approach is to use bond programs or special tax assessments to pay for mitigation.

Berkeley, California, for example, has passed a number of bond issues to support such efforts as the seismic retrofitting of public and private infrastructure. In parts of Oklahoma and in other jurisdictions around the country, small tax increases have been used to fund projects to prevent repeated flooding. In these cases, the immediate financial impact on citizens was small, but the long-term benefits were substantial.

Emergency managers should be prepared for political reality: when risk reduction is perceived as interfering with economic growth, both businesses and elected officials may prefer a compromise approach.

In various parts of the country, local governments have benefited from Community Development Block Grants in their efforts to buy out property in low-lying areas. Risk reduction can also be financed through special assessment districts, impact fees, zoning revisions, and changes in development rights, all of which provide either incentives or disincentives to development. Tools such as impact fees and special assessments, however, must be used with caution because they may simply drive investors to seek out communities where such fees are lower or nonexistent. Emergency managers should be prepared for political reality: when risk reduction is perceived as interfering with economic growth, both businesses and elected officials may prefer a compromise approach.

Emergency management programs can also be financed through partnerships with outside agencies in the public, private, or nonprofit sectors. For example, in exchange for limited tax abatements, Collin County, Texas, obtained permission to use the fiber-optic networks owned by the Collin County Community College District. Among other purposes, the networks were used to develop redundant communications systems for homeland security. Nearby graduate programs also offer potential partnership opportunities: when students provide low-cost technical assistance to local governments, everyone—the local government, the university, and the students—benefits from the arrangement. Because colleges and universities may have substantial investments in research projects, educational institutions can also serve as funding sources for grants that require a match. For example, the University of Texas at Dallas and the North Central Texas Council of Governments filed a joint grant application for a federal technology grant. This particular grant required a "soft" match (a soft match may include previous investments, staff time, volunteer time, and other projects or resources that are related to and/or enhance the purpose of the grant). Because the university had previously undertaken an information technology research effort, the "local" share of the required matching funds already amounted to several hundred thousand dollars.

To support emergency management programs, many localities have worked out unique memorandums of understanding with private sector organizations. In one jurisdiction, a large and well-known department store agreed to be the main distribution point for vaccinations or medications in case of a bioterrorism attack. In a different region, a large and well-known fast-food chain agreed to promote and support community emergency response teams. In yet another case, two local stores that carry hardware and home supplies have formal agreements with a city to serve as drop-off points for donations; these stores have also expressed their willingness to donate plywood, plastic, and other items from their own inventories if needed. In such arrangements, the municipality benefits by acquiring private sector resources to address local emergency management needs, and the private sector benefits from the public relations opportunity.

Although nonprofit agencies struggle to find funding at times, these organizations may nevertheless be valuable assets for preparedness. The brochures and preparedness training available through the American Red Cross, for example, can help a community prepare for any contingency. Donations and volunteer time may offer an unexpected way to meet matching requirements for federal disaster funds. For example, after the communities of Fort Worth and Arlington, Texas, were struck by a tornado on March 28, 2000, more than five thousand volunteers assisted with recovery; by applying the hourly rates established by the United Way for volunteers ($11 per hour at the time) to the number of documented hours, the local government was able to demonstrate that the community had invested more than enough funds to meet the matching requirement. (A volunteer coordinator, supported by a fiscal analyst, pulled together the documentation.)

Important considerations

Whether at the federal, state, or local level, hearings regarding appropriations for emergency management can be contentious. Most policy makers prefer to avoid spending tax revenues on disasters that may not occur, and tend instead to award funds for law enforcement, education, and roads. Governmental frugality when it comes to emergency management is nowhere more evident, however, than at the local level. Local elected officials often overlook or ignore risk, and resist allocating money for emergency management programs and activities. They may believe that disasters will not occur, or they may assume that state and federal agencies will cover all expenses if disasters do occur. If an emergency management office is located within another department, such as police or fire, funding is even more difficult to obtain because the department's mission may not correspond exactly with that of the emergency manager.

The task of obtaining internal funding is often filled with strife and turmoil. Moreover, once funding has been obtained, gaining approval for budget increases is a constant challenge. Outdoor warning systems, backup generators, upgrades to the emergency operations center, and other important projects may not be given any serious consideration, much less make it to the local government's list of capital improvements. Appearing before a local governing board to propose an annual budget or request project funds can be intimidating. Emergency managers must provide clear, succinct, articulate, and well-conceived explanations of program needs and of how funds will be spent. While denials may be common, however, persistence will eventually lead to success.

Communities often miss out on funding opportunities because they fail to provide sufficient staffing and resources to pursue or manage grants.

One of the best ways to increase emergency management funding is to select the right person to manage grants and budgets. In most states, the governing body serves, by law, as the director of emergency services and therefore has the power to remove the emergency manager for failure to acquire money for, or properly plan for, the community's emergency management needs. Local officials must also recognize that the jurisdiction may be held liable for failing to meet minimum standards for mitigation and preparedness. They need to carefully evaluate the work of the emergency manager to ensure that whoever holds this key position is proactive and keeps officials abreast of grant programs and funding opportunities. Now more than ever, emergency managers require formal training and a higher level of education to perform their duties adequately.

Moreover, because the grant management landscape changes rapidly, it will often be impossible for one person to seek out, apply for, and manage grants. Today's grants are complex, the requirements for sharing funds with other departments and jurisdictions are numerous, and time frames are often tight. Therefore, a responsive and fast-acting grant management team is virtually essential.

Communities often miss out on funding opportunities because they fail to provide sufficient staffing and resources to pursue or manage grants. For example, to be eligible for DHS funding, Texas municipalities and counties had to complete a 180-page hazard and vulnerability assessment. Such an undertaking obviously requires a large staff, but because most emergency management programs have just a limited number of employees, only 95 jurisdictions completed the assessment during the first year. In the second year, several hundred more met the requirement. As of 2004, approximately 750 of more than 1,400 potentially eligible agencies had finished their assessments. Although 750 is far from the total number of jurisdictions, the ones that did complete the study make up about 95 percent of the state's population. Nevertheless, many jurisdictions missed out on two years of DHS funding simply because they could not complete the assessment on time. Moreover, several of the assessments included incomplete or inaccurate information because the staff members who were assigned the task of filling out the form lacked essential information.

In sum, a vibrant emergency management program is impossible without adequate funding, and obtaining that funding is the emergency manager's job. Local governments finance emergency management both through the normal budgeting process and through creative ventures with other entities in the public, private, and nonprofit sectors. Funding may also include federal grants for mitigation, preparedness, response, or recovery. Although acquiring and managing grants and other intergovernmental funding can be complex, emergency managers can successfully navigate such challenges if they use political acumen and always comply with local, state, and federal regulations.

Striving for change

As noted in the preceding section, one of the most persistent problems facing emergency management is insufficient resources. And one of the principal causes of insufficient resources is apathy toward emergency management on the part of both public officials and their constituents. But it is time to rethink attitudes toward disaster. Academic research, government investigations, and insurance reports all indicate that disaster losses in the United States average about $1 billion per week.[16] Studies also reveal that disasters are becoming more frequent and intense—a trend that is expected to continue and even accelerate.[17] The growing threat of disaster results, in part, from new and more complex hazards, but it also stems from increasing vulnerability. For example, urban development continues to pack people into dangerous areas, and building codes are not always enforced. Meanwhile, the responsibilities assigned to local emergency managers are increasing, but staff and material resources are not. Emergency managers are often hard put to meet even the minimum demands placed on them.

Emergency managers and related stakeholders—meteorologists, floodplain managers, land use planners, engineers, insurance companies, business-continuity planners, workplace safety specialists, and disaster relief agencies, among others—have an obvious incentive to strive for change. These and other professionals in the disaster field must take a more active role in the policy-making process: disaster-reduction champions are urgently needed. Emergency managers, for their part, should look for opportunities to remind elected officials of their responsibility to protect people, property, and the environment. They should strive to educate civic leaders about the benefits of mitigation and preparedness so that disaster management is included on the public agenda and in debates about community priorities. The emergency manager should also ensure that other local government departments and the general public are aware of the true costs of ignoring disasters. This push for change can be accomplished over lunch, at meetings of the local governing board, in newsletters, and through public speaking. The goal is to make the case—through hard evidence and skillful argument—for increased political support, new ordinances and regulations, and additional human and material resources for emergency management. The art of the profession is at least as important as the science of the field.

Emergency managers, for their part, should look for opportunities to remind elected officials of their responsibility to protect people, property, and the environment.

Of course, those striving for change must recognize that elected officials always face constraints and are likely to make trade-offs for the public good. Budgets are tight, and emergency management needs are often politically unpopular. The seemingly uncertain benefits of mitigation and preparedness are typically overshadowed by the more obvious needs for road repairs, new schools, and additional police officers. But emergency managers must not be discouraged by these barriers: persistence and dedication will pay off in the end. Linking emergency management to wider community goals—for example, by demonstrating how disaster mitigation can strengthen the local economy—is one means of increasing support. Another strategy is to take advantage of windows of opportunity: for example, to push for change after a disaster, when it is easy to get people's attention.

Emergency managers should maintain a stack of ready-made proposals that can be given, as circumstances direct, to the local governing body. These documents may include a buyout plan for apartments located in a floodplain, bids for the installation of a new warning system, a plan for a public education program, or a draft ordinance requiring the retrofitting of buildings. But time is of the essence: both elected officials and the public are more likely to approve such ideas right after disaster strikes, and the proposals are more likely to be accepted and implemented if they have been prepared in advance.

That said, emergency managers should not wait until after a disaster has occurred to press for change. And even as policies are adopted, the emergency manager should not assume that the battle is over. Public officials may, for example, state their support for ordinances but limit compliance to the letter of the law. Or they may sponsor the creation of written plans to appease their superiors or to obtain support from funding agencies, but they may fail to follow through by developing the capability to effectively implement the plans. New Orleans, for example, had an evacuation plan, but the city does not appear to have ensured that it was capable of implementing it. Encouragement, advice, assistance, and even cajoling may be necessary to persuade public officials and citizens to support emergency management projects and programs. And once the desired programs have been established and adequately staffed, the emergency manager must ensure that they are as effective and efficient as possible. The emergency management function is not a matter of checking off items on a list but of continuous monitoring and revision.

Summary

Local emergency management is directly influenced by the federalist structure of government in the United States. Emergency managers should take advantage of the value of networking with other individuals and groups, and must employ a number of strategies to facilitate multiorganizational collaboration. The level of external and internal funding has a direct impact on local emergency management programs. Because disasters are becoming more frequent and intense, emergency managers should strive to influence disaster policies. They must make a strong and consistent effort to influence decisions that will affect the strength of emergency management institutions.

Notes

1 Thomas E. Drabek, "Managing the Emergency Response," *Public Administration Review* 45 (January 1985): 85–92.

2 Michael McGuire, "Managing Networks: Propositions on What Managers Do and Why They Do It," *Public Administration Review* 62, no. 5 (2002): 599–600.

3 See, for instance, David F. Gillespie et al., *Partnerships for Community Preparedness,* Program on Environment and Behavior Monograph #54 (Boulder: Institute of Behavioral Science, University of Colorado, 1993).

4 See, for example, Thomas E. Drabek et al., *Managing Multiorganizational Emergency Responses: Emergent Search and Rescue Networks in Natural Disaster and Remote Area Settings,* Program on Technology, Environment, and Man Monograph #33 (Boulder: Institute of Behavioral Science, University of Colorado, 1981).

5 Gillespie et al., *Partnerships for Community Preparedness.*

6 Michael K. Lindell and Resources for the Future, "Are Local Emergency Planning Committees Effective in Developing Community Disaster Preparedness?" *International Journal of Mass Emergencies and Disasters* 12 (August 1994): 163.

7 Michael K. Lindell et al., "The Local Emergency Planning Committee: A Better Way to Coordinate Disaster Planning," in *Disaster Management in the U.S. and Canada: The Politics, Policymaking, Administration and Analysis of Emergency Management,* ed. Richard T. Sylves and William L. Waugh Jr. (Springfield, Ill.: Charles C. Thomas, 1996), 234–249.

8 Thomas E. Drabek and David A. McEntire, "Emergent Phenomena and Multiorganizational Coordination in Disasters: Lessons from the Research Literature," *International Journal of Mass Emergencies and Disasters* 20 (August 2002): 199.

9 David F. Gillespie, "Coordinating Community Resources," in *Emergency Management: Principles and Practice for Local Government,* ed. Thomas E. Drabek and Gerard J. Hoetmer (Washington, D.C.: International City Management Association, 1991), 57.

10 David A. McEntire, "Towards a Theory of Coordination: Umbrella Organization and Disaster Relief in the 1997–98 Peruvian El Niño Disaster," Quick Response Report #105 (Boulder: Natural Hazards Research and Information Applications Center, University of Colorado at Boulder).

11 For more information on the Emergency Assistance Management Compact, go to emacweb.org.

12 Robert L. Irwin, "The Incident Command System (ICS)," in *Disaster Response: Principles of Preparation and Coordination,* ed. Erik Auf der Heide (St. Louis, Mo.: C. V. Mosby, 1989), 133–161.

13 National Commission on the Terrorist Attacks upon the United States, *The 9/11 Commission Report* (New York: W. W. Norton, 2004).

14 Peter J. May, "Political Influence, Electoral Benefits, and Disaster Relief," in *Recovering from Catastrophes: Federal Disaster Relief Policy and Politics,* 104–124 (Westport, Conn.: Greenwood Press, 1985).

15 Code of Federal Regulations 44, Part 201.6.

16 Dennis S. Mileti, *Disasters by Design: A Reassessment of Natural Hazards in the United States* (Washington, D.C.: Joseph Henry Press, 1999), 66.

17 See E. L. Quarantelli, "The Environmental Disasters of the Future Will Be More and Worse but the Prospect Is Not Hopeless," *Disaster Prevention and Management* 2, no. 1 (1993): 11–25.

CHAPTER 5

Collaborative emergency management

Ann Patton

This chapter provides an understanding of

- The nature of collaborative management
- The origins and context of collaborative management
- The development of emergency management partnerships
- The practice and benefits of collaborative management
- Keys to successful collaborative management.

When Frank Reddish calls a meeting on mitigation strategy, a hundred people show up—representatives from businesses, community groups, nongovernmental organizations, and local governments—all eager to lend their expertise and share in the fruits of a planning process that has yielded millions in grants in recent years. Reddish, the blunt-spoken mitigation manager for Miami–Dade County, is renowned in Florida emergency management circles for his skillful use of collaborative management: he plans and operates almost exclusively through coalitions. Why? "Because it's the way to get things done," he says.[1]

This chapter explains why and how today's emergency managers are engaging in collaborative partnerships. In an environment of increasing demands and shrinking resources, collaboration allows managers to access energy, insight, and enthusiasm that would otherwise be unavailable. In the event of disaster, collaboration fosters flexibility and improvisation—which are often key to an effective response. The chapter examines the roots of the collaborative model, as well as some of the challenges and benefits of its use.

What is collaborative emergency management?

Collaborative emergency management "gets things done" by integrating a wide range of organizations and entities into the day-to-day work of emergency management. As William Waugh and Gregory Streib note in "Collaboration and Leadership for Effective Emergency Management," the power of collaborative emergency management comes from "effective strategies and . . . a compelling vision, rather than from hierarchy, rank, or standard . . . procedures."[2] The goal of collaborative emergency management is to build the strongest possible team and to use it effectively to create a disaster-resilient community. In the words of Mike Hardin, former emergency manager for Escambia County, Florida, "All elements of the entire community . . . are stakeholders in emergency management."[3]

Collaborative management is not without its costs and challenges. Opening the planning and management process to others can complicate matters; it is often simpler—but not necessarily more effective—for one leader to make all the decisions. (As the old saw holds, dictatorship is the most efficient form of governance.) But without collaboration, it may be difficult or impossible to achieve the level of support, cooperation, and ownership that is required for success.

It is also important to note that collaborative emergency management does not undermine or dilute the emergency manager's authority or responsibility; rather, it expands the manager's reach and grasp. Ultimately, it is up to the manager to make the best decisions for the safety and welfare of the community.

Origins and context

The practice of collaborative emergency management originated in the last decade of the twentieth century with the exponential increase in demands made on emergency managers. In the years following World War II, the emphasis in emergency management began to shift from civil defense to an all-hazards approach. During the 1990s, sharp increases in disaster losses led to analogous increases in public expectations. Emergency managers were no longer expected to simply manage the response to a crisis; they were expected to manage mitigation, planning, and recovery as well. And in the face of this expansion of responsibility, community support became more and more essential; as Waugh and Streib note, "The field and profession of emergency management . . . gradually moved beyond the classic top-down bureaucratic model to become a more dynamic and flexible network model that facilitates multiorganizational, intergovernmental, and intersectoral cooperation."[4]

This fundamental shift in the leadership model was exemplified in the 1990s by the work of James Lee Witt, then director of the Federal Emergency Management Agency (FEMA). Project Impact, a prototype community-based program initiated under Witt's direction, was designed to harness local support—including that of private and public agencies—for emergency management. "Working together, we can reduce disaster losses," said Witt.[5] Over Project Impact's four-year life span, some 250 communities engaged in public-private collaboration, with the goal of building "disaster-resistant communities" (see Figure 5–1).[6] Although the national program was canceled in May 2001, the concept of developing coalitions to work

Photo courtesy of Bill Rhees

Figure 5-1 Builder Bill Rhees and a Project Impact home

toward disaster resilience had already taken firm root: many communities have continued to form partnerships and engage in collaborative projects.[7]

As demands on emergency managers continue to rise far beyond crisis response—and beyond many managers' available resources—collaboration becomes more and more crucial. The days are long gone when an emergency manager could shoulder all the work and make all the decisions alone: community support is essential to success.

Developing partnerships

Managers who are interested in collaborative management will quickly learn that they have a vast array of potential partners. Managers may wish to begin, however, by partnering with the following groups:

- *Local elected and appointed officials.* From the perspective of local officials, partnering in support of emergency management is an opportunity to improve both public safety and public relations.
- *Subject-matter experts.* Experts, including safety specialists, federal agency staff, and people with knowledge of hydrology, meteorology, sociology, urban planning, building design and construction, and other fields, often appreciate the opportunity to help educate the public and assist with crafting sound public policies.
- *Community-based organizations, social service agencies, faith-based organizations, and civic groups.* Because some organizations, such as the American Red Cross and the

United Way, are already committed to providing services to people in need, collaborating on emergency management initiatives complements and reinforces their mission. And because civic groups, such as the Lions Club and the Rotary Club, are often looking for ideas for service projects, emergency management initiatives may be a natural fit.

- *Citizens, whether affiliated with a group or independent.* Citizens often seek ways to increase their own safety as well as that of their families and their communities.
- *The private sector, including chambers of commerce and private businesses.* Many businesses, large and small, understand the importance of emergency management and have their own contingency plans in place. Ensuring that businesses can continue to function in the event of a disaster is in the best interest of the community.
- *Media organizations.* Members of the news media are potentially valuable partners; for example, some media partners have produced special reports or documentary programs about emergency management. Because of the public relations value, media executives are generally willing to support emergency management activities.

Of course, emergency managers should also reach out to officials in other local governments and in regional, state, and federal agencies. There are many established frameworks that can be used to support interjurisdictional and intergovernmental cooperation: at the local level, mutual aid agreements and memorandums of understanding offer structured opportunities to provide or receive assistance in the event of a disaster. Some localities may work closely with state emergency management agencies or with regional bodies. The Emergency Management Assistance Compact (EMAC), essentially a state-level mutual aid agreement, provides indirect support for local governments by codifying the ability and willingness of other states to provide assistance (see accompanying sidebar on EMAC).[8] Although the federal government is less likely to serve as an active partner than as a source of funding, some federally sponsored programs, such as the Citizen Corps (see sidebar on pages 76–77 on potential partners), can be integrated into collaborative management efforts.

One approach to developing partnerships is to schedule formal or informal meetings to identify priorities, goals, objectives, potential joint projects, and implementation strategies. The collaborative arrangements that emerge from this process may be reflected in written

The Emergency Management Assistance Compact

Through the Emergency Management Assistance Compact (EMAC), a national interstate mutual aid agreement, states can share resources during times of disaster—and states that provide assistance can be reimbursed by the Federal Emergency Management Agency (FEMA) for their expenses. After Hurricane Katrina, more than 24,000 people from forty-five states came to Mississippi through EMAC, and FEMA reimbursed more than $30.4 million in expenditures.[1]

In Louisiana, after Hurricane Rita stormed ashore on September 24, 2005, EMAC was crucial to getting Cameron Parish operating again. Winds clocked at 120 miles per hour and fifteen-foot storm surges nearly leveled the southwestern Louisiana parish, and almost every building along the Gulf Coast simply disappeared. Kisha Kilmer, from an adjacent parish, was brought on as acting emergency manager: "When I came in, I was alone," Kilmer said a few weeks after the storm. "All the parish leaders had been evacuated, and they were all victims themselves, so they really couldn't help." Operating from the second floor of the old courthouse, virtually the only usable building left in the town, Kilmer "ordered an overhead incident management team [from the Forest Service], and they came in from California in maybe three days.... Now we have help from all around the country—Virginia, Kansas—through the EMAC program, state to state. The people of Cameron were pretty much on their backs, and they are so proud, they won't ask for help. I sat them down and said, 'You can't do this by yourself. I can't do it by myself. We have to pull help in here from other areas. You have to focus your energies on working together.' The way everything nets together, it's very difficult to try to get everybody to pull together—but when it starts coming together, it's amazing."

[1]"FEMA Helps Mississippi Pay Florida for Disaster Aid," FEMA news release #1604-439, September 12, 2006, available at fema.gov/news/newsrelease.fema?id=29761 (accessed June 17, 2007).

[2]Kisha Kilmer, interview with author, Cameron Parish, Louisiana, November 2005.

agreements: in Corvallis, Oregon, for example, a formal citizens' advisory council assists with emergency management. Some emergency managers, however, prefer to keep things informal. Partnerships may also take advantage of existing Citizen Corps Councils or of local emergency planning committees (LEPCs).[9] Project-specific work groups function well in some jurisdictions: in Tulsa, for example, volunteers affiliated with the Disaster-Resistant Business Council encourage business continuity planning for small businesses and nonprofits, and the Language and Culture Bank provides emergency-related services to build connections and goodwill among diverse ethnic groups.

No matter how formal or informal the collaborative structure, food helps. Tommy Malmay, emergency manager of Ouachita Parish, Louisiana, says, "I always told my partners, if you can organize the meeting, I can find a way to provide the food. Breaking bread together simply helps build relationships.[10] Jeff Friedland, emergency manager for St. Clair County, Michigan, hosts an emergency management breakfast every year. The event brings together hundreds of public and private leaders from the county's thirty jurisdictions, and the cost is borne by local partners. The partnerships paid off in 2004, when severe storms roared through the county a few days after the annual breakfast. According to Friedland, "In today's world, nobody has time to waste. But when you bring together a large body of people with experience and expertise, you can only learn from them.... To make people know the plan belongs to them, they have to have ownership, and partnerships create ownership."[11]

Volunteers and the news media are among the most important partners an emergency management department can have. Both of these partners bring special advantages and special concerns. The next two sections offer guidelines on engaging and managing volunteers, and on working successfully with the news media.

Working with volunteers

Volunteers work part time, or even full time, to expand administrative capacity and help implement all phases of emergency management—from planning and executing projects to conducting training to supporting or carrying out drills and exercises. Kary Cox, emergency manager of Bartlesville, Oklahoma, relies heavily on volunteers: "We have 45 volunteers who serve the operational areas—in fact, they are our operations division, because we have only two paid employees—so we rely on volunteers for everything."[12] Through their work, many volunteers gain a deep understanding of emergency management; they also derive personal satisfaction and social connections from their experience.

Matching volunteers' skills and abilities to the tasks at hand is no less important than matching employees' skills and abilities to their job descriptions. To ensure a good fit, some emergency management departments require volunteers to go through a process of training and credentialing; for sensitive jobs, some managers even require background checks. Occasionally, it is necessary to reject a volunteer to protect the integrity of the operation. One way to build a reliable group of volunteers is to make training available to anyone, but to reserve formal volunteer status for long-term, tested workers. Kathy Hinkle, chair of the Tulsa Partners Executive Committee, which oversees the Tulsa Citizen Corps, says, "It works for us to build our volunteer cadre deliberately. We've added more than 1,500 volunteers in two years. We've only had to out-place one volunteer, because he was just too enthusiastic about anti-terror activities.... If you start with a small group of excellent... partners and volunteers, they set the team culture in a positive direction from the beginning."[13]

The use of volunteers inevitably raises liability issues; these are covered in detail in Chapter 12.[14] Regardless of legal concerns, however, the emergency manager has an ethical responsibility to protect volunteers. One way to do so is to ensure that they obtain adequate training and experience (Figure 5–2).[15] Training in the Incident Command System (ICS), which delineates widely accepted procedures for first responders, is essential; some managers require ICS training as a minimum for every volunteer.[16]

The rapid spread of information in today's society often means that in the event of an emergency, a wave of untrained volunteers can well up, washing over the operation and potentially creating a second disaster. In 1995, when a bomb destroyed the Alfred P. Murrah Federal Building in Oklahoma City, killing 168 and injuring hundreds more, the area was "overwhelmed with

Potential partners

Partners range so widely in type, size, and mission that it is impossible to list or categorize them all. The programs and organizations included in the following discussion illustrate the broad range of partnership opportunities.

Citizen Corps

Created in the wake of 9/11 as a means of establishing relationships between local government and community volunteers, the Citizen Corps operates with support from the U.S. Department of Homeland Security (DHS) and includes Neighborhood Watch, the Community Emergency Response Team, the Medical Reserve Corps, Volunteers in Police Service, and the Fire Corps. The guiding principle of the Citizen Corps is, "We all have a role in hometown security." The mission of the corps is accomplished through a national network of state, local, and tribal Citizen Corps Councils, which build on community strengths to implement the Citizen Corps programs and carry out a local strategy to have every American participate. By 2006, Citizen Corps Councils had been established in more than 2,100 communities throughout the nation.[1]

The Citizen Corps opens a wide range of possibilities for emergency managers. In Tallahassee, Florida, the Capital Area Citizen Corps regularly prepares volunteers through extensive disaster exercises—a practice that paid off in 2004 and 2005, when Florida was hit by four hurricanes.[2] In Arlington, Virginia, in 2004, Citizen Corps volunteers went house to house to deliver preparedness information to nearly 100,000 households.[3] In Washington, D.C., the Citizen Corps is mobilizing and training volunteers for the Neighborhood Corps, a program designed to empower citizens at the neighborhood level to develop neighborhood preparedness plans, obtain training, and work together to improve safety and emergency response. According to Jerome DuVal, founding director of DC Citizen Corps, which sponsors the Neighborhood Corps, the Neighborhood Corps will enable participants to cooperate not only with each other, but also with governmental first responders.[4]

Community Emergency Response Teams

The Community Emergency Response Team (CERT) program offers training through DHS and Citizen Corps. In this popular and widespread program, trainees learn to protect themselves and their neighborhoods in emergencies. Hands-on coursework includes emergency first aid, light search and rescue, and basic fire suppression. After the Whittier Narrows earthquake in 1987, Los Angeles led the nation in creating CERTs. Since the program was launched in that year, the city has trained more than 44,000 citizens. Why has Los Angeles made such a profound commitment to volunteer training? "Because of the world we live in today," says Henry Amparan, former disaster preparedness officer for the Los Angeles fire department:

> Being from California, we're always on the brink of disaster with the earthquakes. We also have floods and winter storms, and simple traffic accidents that can tie up the freeway for hours. It's important that we prepare ourselves, in our cars, work, and homes, by having an emergency plan in case of a crisis. With CERT, people can get free training so they can be self-sufficient for up to ninety-six hours after an emergency. That's our goal. The emphasis of CERT is to empower people to help themselves, their neighbors, and their communities. And after they get done doing that, they can come assist the fire department. And by the way, CERT has had an unexpected benefit for us. We find that people are more supportive of the department, and when bond issues come up, people tend to be more supportive because they understand the need to protect our community.[5]

In Portland, Oregon, neighborhood emergency teams (NETs)—based on the CERT model—train citizens in emergency first aid, fire suppression, search and rescue. "The public absolutely loves this NETs program," George Houston says. "As more and more neighborhood organizations become actively involved, there will be a growing proportional need for added training, equipment, and exercising the teams. Early on, you need to train your most dedicated and skilled volunteers to help manage the program. These citizen teams can grow into a rather nice support system for your emergency management program."[6]

The Medical Reserve Corps and Mercy Medical Airlift

Like CERT, the Medical Reserve Corps (MRC) is a Citizen Corps affiliate program. Sponsored by the Office of the U.S. Surgeon General, the MRC offers exceptional training to retired or active medical and health care professionals (and support personnel) to provide "surge" capacity during disasters, as well as ongoing health education and support services.[7]

Mercy Medical Airlift, another Citizen Corps partner, is a charitable organization whose volunteers transport patients for specialized diagnoses and medical treatments. The organization also works with the Homeland Security Emergency Air Transportation System to provide access to thousands of volunteer pilots and planes in the event of disaster.[8]

Volunteer Organizations Active in Disaster

Voluntary Organizations Active in Disaster (VOAD), an umbrella group for nongovernmental organizations, can be an invaluable source of partners. VOAD member organizations coordinate disaster-related planning, communication, and other activities among a myriad of nonprofit entities, including charities and faith-based organizations. In Monroe, Louisiana, in 2002, the emergency management department and local nonprofits organized an annual blitz: every summer since that year, the parish has hosted scores of young volunteers from all over the country, who sleep in a local gym and spend a busy week helping to repair the houses of low-income residents and mitigating hazards ranging from extreme heat to West Nile virus. The long list of co-sponsors includes the Ouachita Multi-Purpose Community Action Program, the local Girl Scouts chapter, the Retired Senior Volunteer Program, and the Group Workcamps Foundation, a faith-based volunteer organization headquartered in Colorado. "It's easier to get folks involved when they are excited about a joint project," says Tommy Malmay, who was Ouachita Parish emergency manager when the program began. "When we all work together in fair weather, we meld together a good team to pull together in the tough times."[9]

Amateur radio and the Civil Air Patrol

According to George Houston, former emergency management director for Portland, Oregon, amateur radio and the Civil Air Patrol are essential partners for the city. "An effective program demands using all available resources to staff and operate your emergency management program," Houston says. Amateur radio operators are "the quintessential communications 'go to' group. This amateur radio program really fills the need for local, regional, national, and even international communications."[10]

The Institute for Business & Home Safety

Many communities build survival capacity by cooperating with, and offering training for, local businesses. In Tulsa, the chamber of commerce, State Farm Insurance, and the Tulsa Citizen Corps have teamed up with other partners to offer training workshops for small businesses and nonprofit agencies.[11] The workshops are based on "Open for Business," a tool kit developed by the Institute for Business & Home Safety (IBHS) to help small businesses develop contingency plans for disaster.

IBHS is a nonprofit organization established by the insurance industry to help reduce deaths and damage from natural disasters.[12] As one element of its program, IBHS builds coalitions at the state and federal levels in support of building codes that will help mitigate the effects of natural disasters. After Katrina, for example, IBHS developed a strategic partnership with the International Code Council, orchestrated a state-mandated study commission, and worked to update statewide building codes.

[1]Citizen Corps Web site, at citizencorps.gov (accessed June 17, 2007).

[2]"American Red Cross: Capital Area Chapter," at tallytown.com/redcross/ (accessed June 17, 2007).

[3]Jackie Snelling, Arlington Citizen Corps, interview with author, Washington, D.C., May 2004.

[4]Jerome DuVal, interviews with author, Philadelphia, Pennsylvania, November 2002, and Washington, D.C., June 2004.

[5]Henry Amparan, interview with author, Philadelphia, Pennsylvania, November 2002.

[6]George Houston, e-mail correspondence with author, February 2005.

[7]See MedicalReserveCorps.gov (accessed June 17, 2007).

[8]"Mercy Medical Airlift/Angel Flight," available at projecttahs.org/tahsjsp/mercymed.jsp.

[9]Karina Donica, "Volunteers Repairing Houses in Monroe," *Monroe [Louisiana] News Star,* July 25, 2005; Tommy Malmay, telephone interview with author, February 2005.

[10]George Houston, personal e-mail correspondence with author, February 2005.

[11]Jeff Burton, C.B.O., IBHS building codes manager, e-mail correspondence with author, July 29, 2005; Sandy Cox, Bank of Oklahoma's business resumption coordinator, interview with author, October 2006.

[12]IBHS has an extensive Web site at www.IBHS.org.

Figure 5-2 Training and exercises to protect volunteers

Courtesy of the city of Tulsa, Oklahoma; photograph by Elaine Perkins

volunteers from the medical community." According to Dr. John Sacra, medical control director for the area's Emergency Management Services Authority, one mistake was the failure to "establish a staging area for volunteers right from the beginning":

> We had well-meaning rescuers operating with uncontrolled access and many untrained individuals in the "hot zone." One of those volunteers was Rebecca Anderson, a nurse, who was struck and killed by falling debris. Now it is clear. Rescuers, even in the early frantic moments of the disaster, must protect themselves first. If we had implemented a staging area at the time the command post was established, we might have been able to save Rebecca's life. If the bombing had involved a secondary nuclear, biological, or chemical agent, we would have suffered many, many more casualties among the first responders. One of the hard lessons we learned is the grave responsibility to train and protect your volunteers and responders.[17]

Some emergency operations plans provide guidelines for handling spontaneous volunteers. As noted by Dr. Sacra, one approach is to require volunteers to report to a staging area at some distance from the hot zone, where they can be deployed as needed. Another option is to assign a particular partner the task of fielding and deploying the volunteers. Washington, D.C., for example, contracts with Greater DC Cares, a member of the Volunteer Center National Network, to conduct skill assessments and to manage and deploy spontaneous volunteers.[18] Finally, some local governments simply discourage—or even turn away—spontaneous volunteers.[19]

Working with the news media

An emergency manager can undertake departmental communication efforts—by maintaining a Web site and publishing a newsletter, for example—but these methods will never achieve the reach and power of television, radio, and newspapers. The news media provide essential links in the chain of communication with the public, but it is essential to ensure that any collaborative arrangements protect the independence and objectivity of the media. Fortunately, most reporters and editors are eager to provide timely and accurate information to the community.

Emergency managers can take a number of steps to develop a good relationship with local news media:

- Communicate regularly and honestly to build trust well before disaster strikes.
- Make an effort to understand the media's needs.

- Ask for advice on how the media can help protect the community.
- Take every opportunity to educate reporters and editors about emergency management.
- Don't hesitate to invite members of the news media into planning and operations sessions, but remember that these individuals must remain objective and independent: don't expect information to remain off the record.

Collaborative management in practice

According to the standards established by the Emergency Management Accreditation Program, a public emergency management program is "a jurisdiction-wide system that provides for management and coordination of prevention, mitigation, preparedness, response and recovery activities for all hazards. The system encompasses all organizations, agencies, departments, entities, and individuals responsible for emergency management. . . ."[20] In other words, emergency management, as practiced today, is a team sport.[21]

Wayne Blanchard, director of FEMA's Higher Education Project, has identified ten core areas in which emergency managers must be competent, including leadership and team building; networking and coordination; and political, bureaucratic, and social contexts. Blanchard notes that because "emergency management offices are typically short staffed" or have "no staff at all—just someone with the responsibility but insufficient resources," emergency managers must "network and coordinate with a broad range of other organizations—up, down and laterally in government levels, private sector, voluntary associations and community-based organizations."[22] According to Mike Hardin, "Emergency management can only be a success if collaboration is a core value. Individual personalities and group dynamics will always prove challenging for a leader as he or she works through complex issues. However, the power of a well-conditioned team building off the individual strengths and expertise of its members is worth the effort. A collaborating team can work its way through the complex issues in all the phases of emergency management and the day-to-day operation of the organization and community."[23]

In some ways, collaborative emergency management may seem like a contradiction in terms. Professional emergency management requires a particular set of skills and abilities, including technical knowledge that other members of the team are not likely to possess. But as Waugh and Streib have noted, "the leadership strategy required for crises may well be counterintuitive": a crisis requires firm leadership and established protocols, but it also demands creative—and often ad hoc—solutions. Moreover, the energy and creativity that arise from "a style that is affiliative, open, and democratic" are most effective when the participants have already established relationships: "Part of the common wisdom of emergency management is that communication and collaboration are facilitated by personal familiarity, not just institutional contact. . . . The interpersonal contact—the working relationships—are critical."[24]

Thus, collaborative emergency management must begin long before disaster strikes. In the wake of the Oklahoma City bombing, pre-disaster collaboration among first-responder teams paid off. Responders from hundreds of communities, as well as from public, private, and non-profit organizations, converged on the scene. Dr. Sacra recalls,

> It took us all to mount the response to the Murrah bombing. . . . On the emergency medical side alone, we had fourteen ambulance companies dispatching sixty-six ambulances, and more than 140 patients were transported by ambulance in the first hour. . . . Fortunately, we had all worked and drilled together beforehand. During a disaster is no time to be exchanging business cards. Planning and practicing together beforehand allowed us to know one another and meld as a group, in the end enabling us to work far better together than would have otherwise been possible.[25]

Particularly in the case of large-scale disasters, failing to establish strong networks before a disaster risks creating a second disaster: the failure of governmental response. Hurricane Katrina exposed the weakness of rigid, hierarchical management in a catastrophe, when centralized control is least effective and when cooperative, flexible "network" management is most needed.[26] But how do emergency managers balance the need for consensus and the need for control? First, it is essential to understand that in the response phase, full consensus is seldom necessary, or even possible: "In most disaster situations, events unfold quickly and with consequences that can be life-changing, thus making it difficult to gather a consensus," says Lisa Gibney, former emergency preparedness coordinator for the Duane Arnold Energy Center, a commercial nuclear power generation station in Palo, Iowa. "Personally, I think the response phase is best suited for the Incident Command System of leadership, with ultimately the local emergency manager... in charge of coordination of the response."[27] Nonetheless, response actions will ideally be guided by an existing emergency operations plan that was developed with widespread public input.

Certainly, the heat of a crisis may require swift decisions by one individual. But the information that he or she needs is probably dispersed, and implementation may well be in the hands of scattered actors who are beyond communication or immediate control. Ultimately, success lies not in the hands of a single authority but in a complex network of others. Pearl River County, Mississippi, for example, less than an hour inland from the Gulf Coast, was right in the line of destruction on August 28, 2005, when Hurricane Katrina's 135-mile-per-hour winds passed straight over the heavily wooded county, leaving every road impassable and reducing the county's emergency operations center to toothpicks (see Figure 5-3).

But a worse disaster was yet to come. "We were absolutely on our own for four days," recalls Bobby Strahan, Pearl River County emergency manager. "Phones, cell phones, and power were all down. We lost seven people after the storm—seven people who needed oxygen who died in their homes without power. We had volunteer firefighters trying to cut their way through downed trees to reach them, passing the generators through like a bucket brigade, but we couldn't get there in time." A third disaster was the stream of evacuees running north: "They sucked us dry of food and gasoline before we even knew what was happening," Strahan said. "On the fourth day, the Florida SERT [state emergency response team] came rolling in. They got us operating with NIMS [National Incident Management System], and we started issuing mission assignments. They came in big mobile command centers, all self-contained. They brought their eighteen-wheelers of food and fed not only their own people but ours, too. They were here for eighteen days. They were tremendous."

Pearl River County survived through the collaboration of partners from without and within: Contractors helped volunteers clear the roads. The community college suffered $50 million in

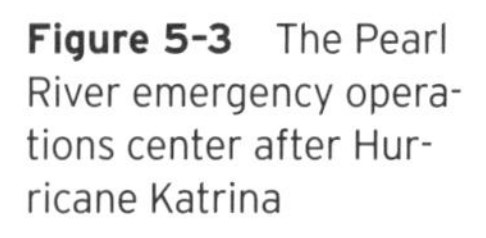
Figure 5-3 The Pearl River emergency operations center after Hurricane Katrina

Photo courtesy of Ann Patton

damages but nonetheless managed to house emergency operations and first-responder units. Teams of citizens took on the chores of allocating gasoline, dispensing food, and posting information at fire and gasoline stations. Churches, charities, and the schools operated homeless shelters and emergency feeding operations. Resurrection Life Church opened a free medical clinic in its parking lot. Several weeks after the storm, Pearl River County gave the ad hoc team that Bobby Strahan had pulled together formal status as a recovery task force, and charged the team with spearheading and coordinating rebuilding and regeneration throughout the community.[28]

The benefits of collaborative management

On its Web site, the Office of Homeland Security of Jefferson County, West Virginia, lists more than 125 partners, ranging from the AB&C Group to the West Virginia University Extension Service.[29] Why would so many different entities want to become partners in homeland security? "For openers, Jefferson County is less than 100 miles from D.C.," says Jefferson County Homeland Security director Barbara Miller. "I think our partners recognize that we all benefit when our community is safe. By partnering to create a disaster-resistant community, . . . these people, businesses, and organizations . . . can make a real difference in protecting lives and property. And, by the way, being a homeland security partner can also be excellent public relations for the company or group."[30]

Collaboration can provide an emergency manager with resources—insight, energy, ideas, talent, funding, and time. It can foster buy-in and enlarge understanding. And as collaborators work together to achieve a common goal, turf battles subside. Ultimately, collaboration yields a stronger, more resilient community. The village of Freeport, New York—with an area of 4.5 square miles and a population of 55,000—lay in the shadow of the World Trade Center. Some of the firefighters who were killed in the attack of September 11, 2001, came from the village, which has a long tradition of partnerships. "We involve our residents because they can contribute significantly to our planning process," says Freeport emergency manager Joe Madigan.

> It is part of our "community-oriented" process. The residents . . . are educated in disaster procedures and mitigation endeavors. Through this process and through CERTs [community emergency response teams], they become more self-reliant and supportive of what emergency services can provide. We have mutual aid agreements with other towns and villages in our area, as well as with the county and state emergency management offices. Partnerships with businesses and civic groups are important for us, also. And we have a cadre of volunteers who provide essential services to us. They come in and help when the emergency operations center is activated, for example. As we worked on our all-hazards plan, residents were part of the committee all the way through. They were also part of our process to formulate our flood mitigation plan.[31]

Partnerships strengthen all four phases of emergency management, and they can be of particular help in the areas of public awareness and education. For example, partners may sponsor special events or develop flyers informing residents about risks. Mike Hardin, of Escambia County, says that his department "incorporates the public into . . . planning and operations several ways, including . . . mitigation strategy development [and] operational plans."[32] Hardin partners with "private utilities, private educational institutions, . . . health care groups [and] local chambers of commerce," as well as with "more traditional partners such as amateur radio." He notes, "It's the entire community, really. All these stakeholders have a vested interest in reducing the impact . . . as well as ensuring an orchestrated response to any emergency. Private partnerships also bring a reality check to any government operation by balancing 'good ideas' with the community's comfort level or standard of care as it relates to providing services."[33]

Jim Mathie, division chief of Deerfield Beach Fire Rescue, is one of the leaders of the Beach Blowout, the annual hurricane-education expo for this Florida town of 75,000. This event, held on Deerfield's broad, blond-sand beach, draws some 15,000 every May. The Beach Blowout includes competitive events for police officers and firefighters, a lifeguard competition, a triathlon, food and drink, music, games, and vendors hawking products; but it also has

Collaborative management in action

- In 1998, when Hurricane Georges threatened Cape Coral, Florida, trained volunteers staffed an information hotline. According to Carolyn Nelson, emergency management planner for Cape Coral, "These members were invaluable in assisting with nonemergency and information-only calls, and alleviated the overloading of our 911 system."[1]
- When Seattle Emergency Management decided, as part of its Project Impact program, to offer earthquake preparedness and survival training to the public, private businesses mobilized to help. By 2001, when the Nisqually earthquake (magnitude 6.8) hit, a thousand people had taken the training, and at least three hundred had retrofitted their homes.[2]
- When tornadoes hit Polk County, Missouri, in 2003, a project that had accustomed local citizens to working with emergency management proved to be a lifesaver. In the fall of 2002, emergency manager Kermit Hargis mobilized 248 people from forty-seven organizations, including churches, schools, the Future Farmers of America, the Civil Air Patrol, Habitat for Humanity, and the local Ministerial Alliance, to clear out drains and ditches in order to help curb flooding. Hargis notes that when the tornadoes hit less than a year later, they "were so massive and...affected so many communities across Missouri and the Midwest, that all the [state and regional] organizations I had intended to use weren't available to me.... Fortunately we had trained and organized people here who could step in."[3]
- Since 2003, Kary Cox, emergency manager of Bartlesville, Oklahoma, in the heart of Tornado Alley, has helped spearhead McReady OK!, a multipartner program that offers free storm-survival information in every McDonald's restaurant in Oklahoma; for an entire month each year, the program reaches more than 150,000 customers a day.

[1]See citizencorps.gov/cert/downloads/training/use_em.doc.

[2]Robert Freitag, "The Impact of Project Impact on the Nisqually Earthquake," *Natural Hazards Observer* 35 (May 2001).

[3]Kermit Hargis, telephone interview with author, February 2005.

a serious purpose: to prepare the community for the hurricane season that begins June 1. And it's all done through partnerships. Mathie notes that

> the Deerfield Beach Blowout has become so big...that we could never pull it off without our partners. We rely on sponsorships 100 percent. It's a free event, and sponsors...[make] it possible. Last year we had more than 100 sponsors.
>
> It's a natural mix, and we all benefit. We get to understand our partners better, and they get to understand us better, and we are all going for a common cause—to help the community and to be prepared for a hurricane.[34]

Kary Cox, of Bartlesville, Oklahoma, says that public support and understanding increased after his department began to involve citizens in planning and exercises. "The community became more supportive of the needs of first responders and the equipment purchases that were made with homeland security grants," Cox says.

> It is important for the community to trust what we do and have faith in the plans; if not, they won't follow the plans, and the plans will fail. Additionally, how can one person, or one group, expect to be able to plan effectively for an entire community? Partnerships are the key to a successful program.... Progress may be slow in the beginning, but with diligence and time, the partnerships will become natural.[35]

John Sacra notes that at the time of the attack on the Murrah Building, "we were astounded at the number of different institutions within the community that were essential to the task." The Oklahoma Restaurant Association stepped in to provide an endless supply of food, sustaining rescuers throughout the sixteen-day ordeal. Architects and builders were consulted continuously, as the shattered building was taken apart in the search for survivors and bodies. Nearly all communications systems were downed by the blast, and Cellular One and Southwestern Bell Telephone donated hundreds of cell phones for responders. Seven hundred members of the clergy supported rescue workers and the families of victims. Seventy-five towns sent firefighters. Operations could not cease when night fell, so the local airport provided generators and bright, temporary lighting. Local television stations broadcast public service messages, and local businesses provided emergency supplies. When calls went out for boots, batteries, or duct tape, donations filled a warehouse—and local businesses provided the warehouse space, too.

Keys to success

Successful collaborative endeavors have a number of elements in common:

- Energy, creativity, and an openness to change
- Cooperation on the part of disparate stakeholders to achieve common goals
- A shared vision
- Shared ownership and leadership
- A willingness to expand the scope of activity and to build on success.

Ultimately, the goal of collaborative emergency management is for the community, not the emergency manager, to own the process. An observer of the collaborative process may find it difficult to identify one leader; often, different leaders emerge as different projects and programs evolve.

Put yourself in your partner's place

Dale Carnegie was right: "You can make more friends in two months by becoming interested in other people than you can in two years by trying to get other people interested in you." One key to successful partnerships is to put yourself in the place of the potential partner: how will the partner benefit from working with you? What are the partner's needs, and how can you help meet them?

However the partnership is structured, it is a good idea to follow the 60–40 rule: each partner should put in 60 percent of the effort and should be content to derive 40 percent of the rewards. When each partner gives more than his or her share, the results are synergistic.

Honor diversity

Successful collaborative managers draw as many groups as possible into the emergency management process and ensure that their needs are considered. For Ellis M. Stanley, general manager of the Los Angeles Emergency Preparedness Department, harnessing the capabilities of diverse community groups is a daily challenge and delight. Los Angeles is vulnerable to a smorgasbord of natural and man-made hazards—and, like many major cities, it is demographically diverse.

Making collaboration work

What are the secrets to making it all work? A successful collaborative emergency manager works strategically and intuitively, keeping in mind a number of proven principles. Below are some actions to avoid and others to embrace.

- Focus on how to do the most good for the most people; don't make decisions based purely on self-interest.
- Always take responsibility—but take credit only rarely.
- Don't monopolize attention or hog the spotlight.
- Celebrate when good things happen to your partners.
- Work every day to understand your constituencies: listen and learn.
- Learn to effectively mobilize, motivate, and manage staff and volunteers.
- Operate inclusively.
- Foster innovation and creativity.
- Remember that achieving consensus takes work: it's not supposed to be easy.
- Trust your partners and the process.

If Los Angeles is to prepare for and survive disaster, the city needs the broadest possible base of support. "Good emergency preparedness is all about good relationships," Stanley says.[36]

A diversity program begins with an analysis, formal or informal, of community characteristics. What is the numerical and geographic distribution of the population on the basis of age, income, language, ethnic background, and other characteristics? How can the manager incorporate the needs of this diverse population into emergency management initiatives? First, the emergency manager should strive to ensure that the composition of the staff, volunteers, and boards reflects the diversity of the community. Second, the manager should identify organizations that serve or represent particular cultural, religious, or ethnic groups, and establish relationships with them. Third, the manager must address the needs of vulnerable populations: members of the community who are poor or elderly, who live in high-hazard areas, or who have physical conditions or characteristics that may put them at risk. In Tulsa, for example, emergency planners work with the Tulsa Speech and Hearing Association to offer emergency-warning equipment to members of the deaf community who cannot hear sirens. (Chapter 13 considers the issue of social vulnerability in detail.)

Summary

As Waugh and Streib have noted, "The emergency management profession has moved beyond a focus on emergency response to a focus on all hazards and comprehensive emergency management.... Collaboration is the way professional emergency managers get the job done."[37] Since the 1990s, more and more managers are being asked to lead comprehensive programs that protect the community through all phases of emergency management: preparedness, mitigation, response, and recovery. Troy Feltman, administrator of St. Clair County, Michigan, says that because money is scarce and needs are mounting, the county leans on partnerships to help create effective systems: "We don't have as many resources as we used to have to address these needs.... We have to band together outside our individual townships. We need to learn how to work together as a matter of first resort rather than as a last resort."[38]

For a growing number of emergency managers, the ultimate goal is not merely to respond to individual events, but to create disaster-resilient communities; community support and involvement are crucial to such initiatives. Managers who work collaboratively consider themselves facilitators. Theirs is a more indirect style of leadership, and they are quick to give credit to other participants. A manager must identify appropriate team members and draw them into a joint planning process that yields a common vision and mutually beneficial goals. The inclusiveness of collaborative management may complicate planning and management, but it can yield benefits that far outweigh the costs.

In the wake of the 9/11 attacks and Hurricane Katrina, initial federal reaction favored greater centralization of authority for emergencies. But subsequent analysis suggests that centralization may not work well in a catastrophe. Network management—in which trained and coordinated team members work together, and decision making is dispersed as appropriate—is more apt to succeed. Moreover, effectiveness is increased when the members of the network have already become familiar and comfortable with each other through frequent pre-disaster contact, planning, and exercising. Good relationships make all the difference.

Collaboration can help build a sustainable, disaster-resistant community of people who care about and take care of each other. Ultimately, that kind of community may well be the best defense against disaster.

Notes

1 Frank Reddish, interview with author, Miami, Florida, March 2005.

2 William L. Waugh Jr. and Gregory Streib, "Collaboration and Leadership for Effective Emergency Management," *Public Administration Review* 66 (December 2006): 131.

3 Mike Hardin, correspondence with author, February 8, 2005.

4 Waugh and Streib, "Collaboration and Leadership," 131.

5 James Lee Witt, interview with author, Washington, D.C., December 1999.

6 In the years since Project Impact, there has been a shift in terminology and emphasis among some emergency management scholars and practitioners: "disaster resistance" emphasized the reduction of losses; "disaster resilience" emphasizes the ability to bounce back from any losses that do occur.

7 Some of these programs are affiliated with emergency management departments; others are independent.
8 "FEMA Helps Mississippi Pay Florida for Disaster Aid," FEMA news release #1604-439, September 12, 2006, available at fema.gov/news/newsrelease .fema?id=29761 (accessed June 17, 2007).
9 Local emergency planning committees, which undertake planning for toxic chemical hazards, are required under the Emergency Planning and Community Right-to-Know Act.
10 Tommy Malmay, interview with author, Monroe, Louisiana, May 2002.
11 Jeff Friedland, interview with author, St. Clair, Michigan, May 13, 2004.
12 Kary Cox, personal correspondence with author, February and March 2005.
13 Kathy Hinkle, personal interview with author.
14 Excellent information about liability issues is also available on the Web site of the Public Entity Risk Institute (PERI) at PERI.org. PERI is a nonprofit clearinghouse and resource for information about risk management issues.
15 One of the leading options for training in emergency management is the Emergency Management Institute (EMI), a FEMA-operated training academy in Emmitsburg, Maryland. EMI offers many on-site and online courses in all aspects of emergency management. For more information, see EMI Web site at training.fema .gov/emiweb/ (accessed June 18, 2007). Because a well-trained citizenry enhances the general safety of the community, many managers offer ongoing training opportunities for the general public as well as for volunteers.
16 Ellis M. Stanley, "Effective Crisis Response Begins with Preparedness," available at disaster-resource.com/ articles/effective_crisis_prep_stanley.shtml (accessed June 18, 2007). For information about FEMA's Incident Command System training program, see training.fema .gov/EMIWeb/IS/is100.asp (accessed June 18, 2007).
17 Dr. John Sacra, interview with author, Tulsa, Oklahoma, February 2005.
18 Jerome DuVal, interview with author, Philadelphia, Pennsylvania, November 13, 2002.
19 Excellent advice and examples of management plans for spontaneous volunteers can be found at the Points of Light Foundation Web site, pointsoflight.org (accessed June 18, 2007).
20 Emergency Management Accreditation Program, EMAP Standard 3.3.3, adapted from the National Fire Protection Association's Standard 1600 (NFPA 1600), quoted by Waugh and Streib, "Collaborative Leadership," 137.
21 For examples of successful collaborative efforts, see the following Web sites: CitizenCorps.gov; FEMA.gov; LLIS.gov (Lessons Learned Information Sharing, a Web site of the U.S. Department of Homeland Security that emphasizes antiterror activities); and ProjectTAHS.org (Technical Assistance in Homeland Security, a Web site of the Corporation for National and Community Services that emphasizes grassroots homeland security efforts).
22 Wayne Blanchard, telephone and e-mail interviews with author, May 2006.
23 Hardin, correspondence with author.
24 Waugh and Streib, "Collaboration and Leadership," 136–137.
25 Dr. John Sacra, speech to the American Trauma Society, 1999.
26 Waugh and Streib, "Collaboration and Leadership," 136.
27 Lisa Gibney, e-mail correspondence with author, January 7, 2005. The Incident Command System (ICS) that Gibney refers to was created in the 1970s to coordinate the work of firefighters combating large forest fires. The ICS structure has now been widely adopted by first responders, and an expanded version, Unified Command (UC), operates in multijurisdictional disasters such as Hurricane Katrina. ICS and UC offer universally accepted operational and decision structures for command-and-control response; however, they are not designed to foster the collaboration that may be needed in complex crisis situations.
28 Bobby Strahan, interview with author, Pearl River County, Mississippi, December 2005.
29 "Jefferson County Office of Homeland Security Partners," available at jeffersoncountywv.org/Partners.htm (June 18, 2007).
30 Barbara Miller, telephone and e-mail interviews with author, February 2004.
31 Joe Madigan, e-mail correspondence with author, February 2005.
32 Although community involvement in mitigation, preparedness, and response is desirable, in the case of antiterror planning, emergency managers must exercise caution: it is unwise to reveal vulnerabilities of which the general public would otherwise be unaware. Similarly, in the aftermath of an event resulting from a deliberate criminal act, the emergency manager's first responsibility is to secure the site, both to protect the community and to ensure that evidence required by law enforcement will be preserved.
33 Hardin, correspondence with author.
34 Jim Mathie, telephone interview with author, January 2005.
35 Cox, personal correspondence with author, February and March 2005.
36 Ellis M. Stanley, telephone communication with author, January 2005.
37 Waugh and Streib, "Collaboration and Leadership," 137–138.
38 Troy Feltman, interview with author, St. Clair, Michigan, May 14, 2004.

PART TWO

Functions and Phases of Emergency Management

CHAPTER 6

Mitigation

David R. Godschalk

This chapter provides an understanding of

- The value of hazard mitigation
- The context in which mitigation occurs
- Federal mitigation programs
- Hazard-specific loss-reduction techniques
- The mitigation planning process
- The link between mitigation and sustainability.

Hazard mitigation is sustained action taken to reduce or eliminate the risk to human life and property from hazards.[1] Long-term mitigation is related to, but different from, the immediate actions taken to prepare for, respond to, and recover from a disaster that is impending or has just occurred. Mitigation can take the form of physical, bricks-and-mortar projects or of planning and community education. Regardless of the form it takes, however, mitigation is a highly cost-effective investment for local governments, states, and the nation. By protecting a community against disaster before disaster strikes, mitigation saves lives, property, time, money, and resources. The mitigation process can also unite disparate stakeholders in an effort to achieve shared public safety goals. Ultimately, mitigation works to create a more resilient, more sustainable community.[2]

A resilient community is one that can withstand an extreme event without suffering devastating losses and without requiring a great deal of outside assistance.[3] The community's physical, social, economic, and institutional networks survive and continue to function even under extreme conditions; they might bend from the effects of a disaster, but they do not break. Instead of suffering repeated damage and continually demanding disaster assistance from the state and federal governments, the resilient community proactively protects itself against hazards and builds self-sufficiency. Mitigation plays a key role in breaking the cycle of damage, reconstruction, and further damage.

A sustainable community is one that seeks to meet the needs of current generations without compromising the ability of future generations to meet their own needs. Sustainable development balances competing environmental, economic, and social objectives—sometimes referred to as "the three E's": environment, economy, and equity.[4] Sustainable development cannot succeed without taking into account the risks of natural hazards. In practical terms, this means that communities must develop disaster reduction policies that foster resilience and that ensure that future development does not increase vulnerability. Thus, achieving resilience is an important part of achieving sustainability.

This chapter describes a tested and practical approach to hazard mitigation. Drawing on examples from successful mitigation efforts, the chapter explains the importance of mitigation, shows how to reduce losses for all types of hazards, and walks through the steps of the mitigation planning process. The emphasis throughout is on the connection between mitigation and sustainability.

The value of mitigation

In 2005, Hurricane Katrina sounded a national wake-up call for improved preparedness and mitigation.[5] An unparalleled catastrophe, Katrina overwhelmed the response capacities of federal, state, and local agencies. Katrina's destructiveness demonstrated that the nation has yet to develop an effective plan for coping with a major catastrophe—despite calls for such a plan dating back to Hurricane Andrew in 1992, and despite repeated and specific warnings about the vulnerability of New Orleans and the Gulf Coast to hurricane damage. With proper mitigation, much of the human suffering, physical destruction, and financial devastation associated with Katrina could have been avoided. Decisions made about mitigation will determine the safety of the region in the face of inevitable future storms.

The damage caused by Hurricane Katrina was far greater than that typically associated with a hurricane. But the costs of even noncatastrophic disasters are very high: according to the National Oceanic and Atmospheric Administration, the average annual U.S. damage from hurricanes, floods, and tornadoes is $11.4 billion.[6] An in-depth national study of potential savings from mitigation found that, on average, each dollar spent on mitigation avoids $4 in future losses.[7] Figures from communities that have undertaken mitigation programs confirm the cost savings. Kinston, North Carolina, for example, elected to move 100 houses out of the floodplain. While the cost of the effort was $2.1 million, the move avoided some $6 million in losses when Hurricane Floyd hit in 1999.[8] And by elevating 182 residences above the base flood elevation and removing its elementary school from the floodplain, the small town of Belhaven, North Carolina, avoided more than $3 million in losses from Hurricane Isabel in 2003.[9]

By reducing vulnerability, mitigation decreases disaster costs and increases sustainability. However, as the Katrina experience dramatically illustrated, those who fail to invest in mitigation cannot reap its benefits.

Context of mitigation

Like every other aspect of emergency management, mitigation takes place in an intergovernmental context. Although the major players are typically the local government, the state emergency management agency, and the Federal Emergency Management Agency (FEMA) and its regional offices, a successful mitigation process engages multiple individuals and entities that have a stake in reducing or eliminating hazard risk.

Natural-hazard mitigation plans are developed at the state and local levels, but planning requirements, technical assistance, and funding are provided by FEMA. Additional disaster assistance is provided by the U.S. Department of Housing and Urban Development, the U.S. Department of Transportation, the U.S. Environmental Protection Agency, the U.S. Army Corps of Engineers, and the Small Business Administration.

At the state, regional, and local levels, the agencies typically involved in mitigation include emergency management, planning, housing and community development, commerce, economic development, transportation, and environmental protection. It is particularly important to coordinate mitigation efforts with the development of the local comprehensive plan and the accompanying regulations.

With the rise of global terrorism and the increase in the number and severity of technological disasters, the institutional context of mitigation has become more complex. Entities that are new or are playing new roles in efforts to address man-made hazards, either intentional or accidental, include the U.S. Department of Homeland Security and public health agencies, law enforcement agencies, and agricultural agencies at both the federal and state levels.

Federal mitigation programs

Mitigation is largely shaped by federal legislation—in particular, the Disaster Mitigation Act of 2000 (DMA 2000), which amended the Stafford Act, the basic federal disaster statute. Although previous federal legislation focused on post-disaster relief and recovery, mitigation is the centerpiece of DMA 2000. Section 322 of DMA 2000 lays out the requirements for state and local mitigation planning. Through the Hazard Mitigation Grant Program (HMGP), the act provides funding for plan making; a state that develops an enhanced mitigation plan—demonstrating that it has made an extra effort to reduce losses, protect resources, and create safer communities—receives additional funding. To receive post-disaster HMGP funds, state and local governments must have FEMA-approved mitigation plans in place, including action plans that describe how proposed mitigation projects will be assigned priority, implemented, and administered. The emphasis in the action plans is on maximizing benefits; to that end, all proposed projects must be subjected to a cost-benefit review. Multijurisdictional mitigation plans seeking FEMA approval must include identifiable action items for each participating jurisdiction.[10]

To obtain FEMA approval, local and tribal plans must demonstrate that their proposed mitigation measures adequately account for the risks faced by the locality and accurately reflect the ability of the locality to deal with these risks. State governments are responsible for providing technical assistance and training to local governments and for reviewing and approving local plans. Thus, DMA 2000 not only rewards pre-disaster planning but also encourages cooperation between state and local governments.

In addition to developing the mitigation plans required under DMA 2000, many communities also participate in the Flood Mitigation Assistance (FMA) grant program and the Community Rating System (CRS), both of which are under FEMA. The purpose of FMA is to reduce claims under the National Flood Insurance Program (NFIP).[11] To that end, the program provides communities with grants for the preparation of flood mitigation plans and with separate project grants for the implementation of measures to reduce flood losses.[12] The CRS provides incentives—in the form of reduced NFIP premiums—to policyholders in communities that undertake specified measures to reduce flood hazard risk. As of 2006, 1,049 communities were participating in the CRS and receiving discounts on flood insurance premiums.[13]

To enable state and local governments to meet DMA 2000 requirements, FEMA established planning and funding criteria for state and local governments.[14] The agency's guides to state and

Figure 6-1 The phases of hazard mitigation planning

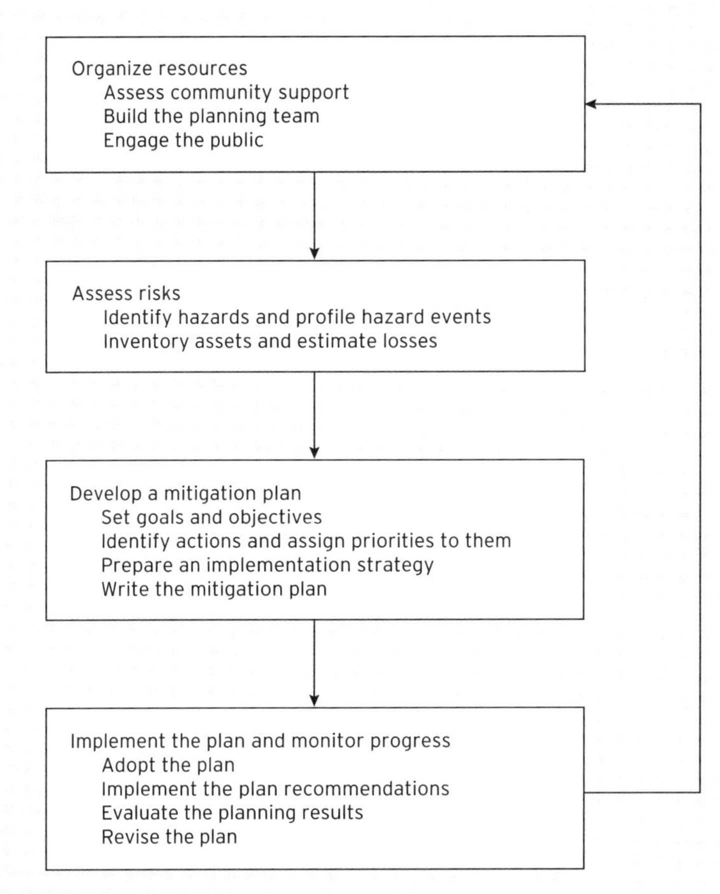

Source: Adapted from the following Federal Emergency Management Agency (FEMA) publications (see notes 1, 16, 17, and 18): *Understanding Your Risks: Identifying Hazards and Estimating Losses*, FEMA 386-2 (2001); *Getting Started: Building Support for Mitigation Planning*, FEMA 386-1 (2002); *Developing the Mitigation Plan: Identifying Mitigation Actions and Implementation Strategies*, FEMA 386-3 (2003); and *Bringing the Plan to Life: Implementing the Hazard Mitigation Plan*, FEMA 386-4 (2003).

local mitigation planning break the process into four phases: (1) organizing resources,[15] (2) assessing risks,[16] (3) developing the mitigation plan,[17] and (4) implementing the plan and monitoring progress[18] (see Figure 6–1). The last phase and the first are connected by a feedback loop.

As represented in the figure, the phases follow a linear path; in practice, however, the process is more fluid because the plan must respond to changing economic and political circumstances. Figure 6–1 represents a framework that can be adapted to the unique conditions of each community—and, in particular, to the community's growth patterns and disaster history.

The mitigation planning process must engage multiple individuals and entities that have a stake in mitigation. As has already been noted, however, the major players are typically the local government, the state emergency management agency, and FEMA and its regional offices.

Hazard-specific loss reduction

Although community mitigation plans have traditionally focused on the single most prominent local hazard, today's mitigation plans are required to address all hazards to which a community could be exposed.[19] To ensure that all relevant hazards are covered, planners can consult multihazard maps on the Web.[20] In some cases, the same mitigation techniques will apply to more than one hazard; in other cases, specific techniques will apply only to particular hazards.

Floods

Flooding typically results from three causes: (1) inappropriate development within the floodplain, which puts buildings at risk and, by obstructing the flow of water, decreases the capacity of stream channels to carry floodwaters; (2) the filling of wetlands, which decreases the natural floodwater storage capacity; and (3) extensive impervious surfaces, which decrease the ability of the soil to absorb stormwater. Flood damage results from five principal sources: (1) water forces and moisture, (2) objects carried by floodwaters, (3) erosion and undercutting of stream banks, (4) deposition of debris, and (5) the release of untreated sewage into streams (which occurs when sewage treatment plants are flooded).

Floodplain information is published on flood insurance rate maps (FIRMs), which depict areas subject to flooding and are used to calculate the cost of flood insurance. According to engineering determinations of the expected limits of the 100-year flood (also known as the base flood), the areas that are depicted on FIRMs as being within the A zone have a 1 percent chance, on average, of flooding in any given year. In addition to still-water flooding, coastal areas face the added hazard of wave action; these areas are mapped as V (or velocity) zones.[21] Because changes in upstream land use can increase flood heights, floodplain maps must be updated to keep pace with development. Accurate floodplain maps should be widely promulgated to citizens, decision makers, and developers. FEMA is gradually issuing new floodplain maps for the country, but this is a slow and expensive process.

For structures that already exist in the floodplain, two basic mitigation techniques have proven effective: elevation and relocation. Elevation physically raises the lowest habitable floor of a structure above the base flood elevation—the expected height of the 100-year flood (see Figure 6–2). Relocation removes the structure from the floodplain, either by moving it or demolishing it; the local government then acquires the land and converts it to a use that is not vulnerable to flood damage—transforming it, for example, into a greenway, a park, or a recreation area. Structural mitigation techniques, such as levees, floodwalls, and shoreline protection, can also be effective in protecting existing vulnerable areas. However, structural devices are subject to failure, as in the case of the New Orleans' levees, and may provide a false sense of security.

Nonstructural means of minimizing flood damage include (1) limitations on development within floodplains, (2) the adoption and enforcement of flood mitigation standards in zoning and subdivision regulations, and (3) the adoption of a stormwater management ordinance.

Figure 6-2 Diagram of a floodplain

The special flood hazard area–designated as the A zone on the flood insurance rate map–is subject to a 1 percent or greater chance of flooding in any given year. The special flood hazard area is divided into two sections: the floodway is the channel of a river or other water course, where development must be restricted to ensure that there are no increases in upstream flood elevations; the flood fringe is the area adjacent to the floodway, where development that does not significantly increase the base flood elevation may be permitted.

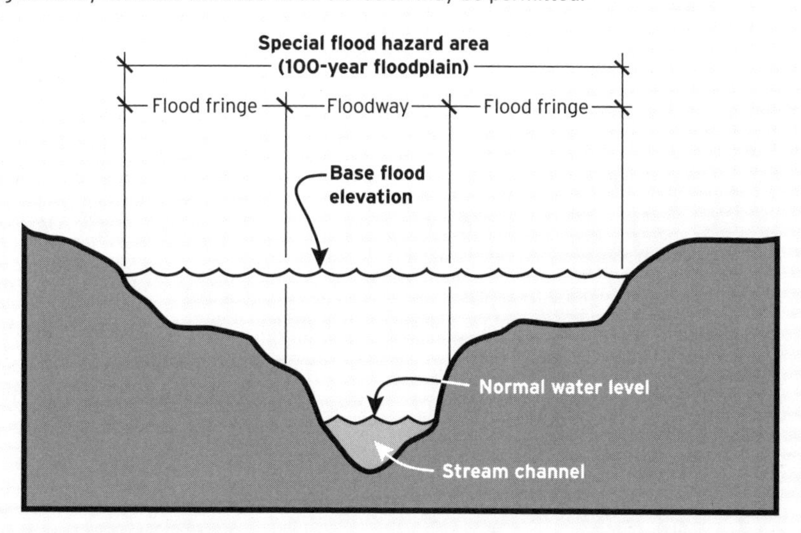

Source: FEMA, *Understanding Your Risks: Identifying Hazards and Estimating Losses*, FEMA 386-2 (Washington, D.C.: FEMA, 2001), 2-12, available at fema.gov/library/viewRecord.do?id=1880 (accessed June 7, 2007).

Earthquakes

Earthquakes are abrupt movements of the Earth's crust caused by the collision or separation of subsurface tectonic plates. Earthquakes can generate ground motion, surface faulting, and ground failure; liquefaction may also occur.[22] Earthquake damage includes settlement, structural weakening, and breakage or collapse of buildings and infrastructure (including underground pipelines, such as water, sewer, and gas lines). Secondary hazards resulting from earthquakes include landslides, fires, flooding, tsunamis, and the release of hazardous materials.

Earthquakes are measured on two scales. Magnitude—the amount of seismic energy released—is measured on the nine-level Richter Scale (see Figure 6–3). Intensity—the effect on various sites—is measured on the twelve-level Modified Mercalli Intensity Scale (see Figure 6–4). An earthquake can have only one Richter Scale magnitude but can generate a number of different Mercalli intensities, depending on the location of the site under consideration.[23]

Mitigation planners can use HAZUS (Hazards US), FEMA's loss estimation software, to develop loss estimates for earthquakes. HAZUS, which can be run on a personal computer, includes data on building stock and on economic, geologic, and other characteristics. To identify areas of severe seismic risk, mitigation planners can use maps of earthquake faults, data on soils and geologic conditions, and histories of past earthquakes. Future development should be restricted in areas that are adjacent to active faults, that contain steep slopes, or that are susceptible to liquefaction. However, mitigation planners may propose alternative land uses, such as open space or recreation, that do not place permanently occupied buildings permanently at risk. Incentive programs, such as transfers of development rights, density bonuses, and tax credits, can be used to encourage developers and building owners to undertake mitigation efforts, including building upgrades designed to increase protection against seismic events.

The adoption and enforcement of up-to-date building codes is critical to protecting new and existing buildings from earthquake forces. California, Oregon, and Washington have adopted seismic safety codes to protect schools, hospitals, and other public buildings.[24] Codes to protect private buildings—homes and businesses—should meet the current National Earthquake Hazard Reduction Program standards. Building inspectors should be trained to screen buildings for potential seismic hazards and to recommend seismic protection measures. Homeowners and businesses should also be encouraged to use nonstructural earthquake mitigation techniques, such as anchoring water heaters and bookcases.

Figure 6–3 The Richter Magnitude Scale

The severity of an earthquake is generally proportional to the amount of seismic energy it releases.

Magnitude	Effects
Less than 3.5	Recorded on local seismographs, but generally not felt.
3.5-5.4	Often felt, but rarely cause damage.
Under 6.0	At most slight damage to well-designed buildings. Can cause major damage to poorly constructed buildings over small regions.
6.1-6.9	Can cause damage to poorly constructed buildings and other structures in areas up to about 100 kilometers across where people live.
7.0-7.9	"Major" earthquake. Can cause serious damage over larger areas.
8.0-8.9	"Great" earthquake. Can cause serious damage and loss of life in areas several hundred kilometers across.
9	Rare great earthquake. Can cause major damage over a large region over 1000 km across.

Source: Natural Resources Canada, "Earthquake Magnitude Scales," available at earthquakescanada.nrcan.gc.ca/gen_info/scales/magnitude_e.php (accessed June 7, 2007). Reproduced with the permission of the Minister of Public Works and Government Services Canada, 2007. Courtesy of Natural Resources Canada, Geological Survey of Canada.

Figure 6-4 The Modified Mercalli Intensity Scale

The Modified Mercalli Intensity Scale is designed to describe the effects of an earthquake in terms of the intensity of its effects, at a given place, on natural features, on industrial installations, and on human beings. The version shown here is condensed and adapted from the longer original.

Intensity	Effects
MM I	Not felt except rarely, under especially favorable circumstances.
MM II	Felt indoors by few, especially on upper floors, or by sensitive, or nervous persons; sometimes hanging objects may swing.
MM III	Felt indoors by several, motion usually rapid vibration. Sometimes not recognized to be an earthquake. Vibration like that from passing vehicles. Movement may be appreciable on upper levels of tall structures.
MM IV	Felt indoors by many, outdoors by few. Vibration like that from passing of heavy, or heavily loaded trucks. Sensation like heavy body striking building, or falling of heavy objects to inside. Rattling of dishes, windows, doors.
MM V	Felt indoors by practically all, outdoors by many or most. Outdoors direction estimated. Frightened few—slight excitement, a few ran outdoors. Buildings trembled throughout. Broke dishes, glassware, to some extent. Cracked windows in some cases. Overturned small or unstable objects.
MM VI	Felt by all, indoors and outdoors. Frightened many, excitement general, some alarm, many ran outdoors. Persons made to move unsteadily. Trees, bushes, shaken slightly to moderately. Liquid set in strong motion. Small bells rang—church, chapel, school etc. Damage slight in poorly built buildings. Fall of plaster in small amount. Cracked plaster somewhat.
MM VII	Frightened all—general alarm, all ran outdoors. Some, or many, found it difficult to stand. Trees and bushes shaken moderately to strongly. Waves on ponds, lakes, and running water. Rang large church bells, etc. Suspended objects made to quiver. Damage negligible in buildings of good design and construction, slight to moderate in well-built ordinary buildings, considerable in poorly built or badly designed buildings.
MM VIII	Fright general—alarm approaches panic. Trees shaken strongly—branches, trunks, broken off, especially palm trees. Ejected sand and mud in small amounts. Changes: temporary, permanent; in flow of springs and wells; dry wells renewed flow; in temperature of spring and well waters. Damage slight in structures (brick) built especially to withstand earthquakes. Considerable in ordinary substantial buildings, partial collapse.
MM IX	Panic general. Cracked ground conspicuously. Damage considerable in (masonry) structure built especially to withstand earthquakes: threw out of plumb some wood-frame houses built especially to withstand earthquakes; great in substantial (masonry) buildings, some collapse in large part; or wholly shifted frame buildings off foundations, racked frames; underground pipes sometimes broken.
MM X	Cracked ground, especially when loose and wet. Landslides considerable from river banks and steep coasts. Shifted sand and mud horizontally on beaches and flat land. Changed level of water in wells. Damage serious to dams, dikes, embankments. Severe to well-built wooden structures and bridges, some destroyed. Developed dangerous cracks in excellent brick walls. Destroyed most masonry and frame structures, also their foundations.
MM XI	Disturbances in ground many and widespread, varying with ground material. Broad fissures, earth slumps, and land slips in soft, wet ground. Ejected water in large amounts charged with sand and mud. Caused sea-waves ("tidal" waves) of significant magnitude. Damage severe to wood-frame structures, especially near shock centers. Great to dams, dikes, embankments, often for long distances. Few, if any (masonry), structures remained standing.
MM XII	Damage total—practically all works of construction damaged greatly or destroyed. Disturbances in ground great and varied, numerous shearing cracks. Landslides, falls of rock of significant character, slumping of river banks, etc. numerous and extensive. Wrenched loose, tore off, large rock masses.

Source: Adapted from Natural Resources Canada, "The Modified Mercalli (MM) Intensity Scale," available at earthquakescanada.nrcan.gc.ca/gen_info/scales/magnitude_e.php (accessed June 7, 2007). Reproduced with the permission of the Minister of Public Works and Government Services Canada, 2007. Courtesy of Natural Resources Canada, Geological Survey of Canada.

Coastal storms

Several types of coastal storms—hurricanes and tropical storms, northeasters, and tsunamis—can damage development adjacent to shorelines. Hurricanes are severe cyclonic tropical storms that bring high winds ranging from 74 to as much as 155 miles per hour, intense rainfall, storm surges (ocean water rushing ashore at heights of from four to twenty feet above sea level), and flooding and erosion over wide areas. (Storms with winds below 74 miles per hour are classified simply as tropical storms.) Hurricanes occur primarily in the Atlantic and Gulf Coast states during a season that extends from June to November. Northeasters are coastal low-pressure systems that occur along the Atlantic seaboard during the winter months; they bring heavy rains, high surf, and coastal erosion.

Tsunamis ("harbor waves")—fast-moving tidal waves that build up to significant heights as they approach land—are triggered by underwater earthquakes or landslides. Because of seismic activity within the Pacific Rim, U.S. tsunamis are most common along the Pacific coast. In 1964, an earthquake with a magnitude of 8.4 generated a tsunami that struck southeastern Alaska, killing 137 and causing $350 to $400 million in damage.[25] In December 2004, an earthquake in the Indian Ocean, measuring over 9 on the Richter Scale, generated the Asian Tsunami; one of the deadliest disasters in modern history, the tsunami left some 230,000 people dead or missing.[26]

Winds from coastal storms can demolish structures, uproot trees, and send debris flying. Winds that lift off the roofs of buildings or break through walls can allow rainwater to enter and damage interiors. Storm surges and flooding can topple buildings, erode shorelines, and destroy infrastructure.

Hurricane intensities are measured on the Saffir-Simpson Scale (see Figure 6–5). Two types of maps delineate hurricane risk: floodplain maps and SLOSH (Sea, Lake, and Overland Surge from Hurricanes) maps. Floodplain maps show the expected areas of flooding; SLOSH maps indicate likely areas of surge penetration from hurricanes of varying intensity.

Hazard mitigation measures for hurricanes are similar to those for floods, but they also include protective measures to address risk from wind and storm surges. For new development, the most effective way to protect against hurricanes is to adopt comprehensive plans and zoning regulations that steer development away from hazardous coastal areas. Other options are to acquire vulnerable areas for open space or recreation (in order to prevent permanently occupied buildings from being constructed in hazard zones) and to prohibit

Figure 6-5 The Saffir-Simpson Scale

Category	Wind speed	Storm surge (feet above normal sea level)	Expected damage
1	74-95 mph	4-5 ft.	Minimal: Damage is done primarily to shrubbery and trees, unanchored mobile homes are damaged, some signs are damaged, no real damage is done to structures.
2	96-110 mph	6-8 ft.	Moderate: Some trees are toppled, some roof coverings are damaged, major damage is done to mobile homes.
3	111-130 mph	9-12 ft	Extensive: Large trees are toppled, some structural damage is done to roofs, mobile homes are destroyed, structural damage is done to small homes and utility buildings.
4	131-155 mph	13-18 ft.	Extreme: Extensive damage is done to roofs, windows, and doors; roof systems on small buildings completely fail; some curtain walls fail.
5	>155 mph	>18 ft.	Catastrophic: Roof damage is considerable and widespread, window and door damage is severe, there are extensive glass failures, and entire buildings could fail.

Source: FEMA, *Understanding Your Risks: Identifying Hazards and Estimating Losses*, FEMA 386-2 (Washington, D.C.: FEMA, 2001), 2-23, available at fema.gov/library/viewRecord.do?id=1880 (accessed June 7, 2007).

the construction of infrastructure, such as bridges and waste-treatment facilities, in high-risk areas. Such approaches, which come under the general heading of growth management, also help to ensure long-term sustainability.

Mitigation measures can be incorporated into building codes through standards for flood proofing, wind proofing, and elevation above the surge level, and through requirements for storm shutters for windows and other openings. Shoreline protection measures include seawalls, groins, jetties, and other engineering techniques that can help mitigate erosion and wave action. However, because such structures may actually increase erosion on the downstream side, some coastal states, including North Carolina, prohibit them. Every coastal area should also have an evacuation plan to move residents away from threatened locations to shelters or safe inland locations.

Tornadoes

Tornadoes—rapidly rotating funnels of air that extend from a thundercloud toward the ground—can create wind speeds of 250 miles per hour and are usually accompanied by heavy rain, hail, and lightning. When tornadoes touch down, they can cause tremendous destruction. Tornadoes can occur at any time and in any part of the country, but they are more frequent during tornado season (between March and August) in the Midwest, Southeast, and Southwest. The impact of a tornado depends on its intensity, the duration of touchdown, and the structural strength of the buildings in its path. Tornado damage includes flying debris, uprooted trees, and the destruction of vulnerable buildings, such as mobile homes. Tornado intensity is measured on the Fujita Tornado Measurement Scale (see Figure 6-6) and is based on the degree of damage; intensities range from less severe (F1) to most severe (F5).

There are two principal approaches to mitigating potential damage from tornadoes: strengthening buildings and constructing safe rooms. The adoption and enforcement of up-to-date building codes can help reduce the vulnerability of new and existing structures. Anchoring and reinforcing walls, roofs, and foundations, for example, can make buildings more resistant to wind. Safe rooms are small enclosures that are specially armored and anchored; residents can also seek protection in community tornado shelters.

Figure 6-6 The Fujita Tornado Measurement Scale

Category	**Wind speed**	**Expected damage**
F0	Gale tornado (40-72 mph)	Light damage. Some damage to chimneys, branches broken off trees, shallow-rooted trees pushed over, sign boards damaged.
F1	Moderate tornado (73-112 mph)	Moderate damage. The lower limit is the beginning of hurricane wind speed. Surface peeled off roofs, mobile homes pushed off foundations or overturned, moving autos pushed off roads.
F2	Significant tornado (113-157 mph)	Considerable damage. Roofs torn off frame houses, mobile homes demolished, boxcars pushed over, large trees snapped or uprooted, light-object missiles generated.
F3	Severe tornado (158-206 mph)	Severe damage. Roofs and some walls torn off well-constructed houses, trains overturned, most trees in forest uprooted, cars lifted off ground and thrown.
F4	Devastating tornado (207-260 mph)	Devastating damage. Well-constructed houses leveled, structure with weak foundation blown off some distance, cars thrown, large missiles generated.
F5	Incredible tornado (261-318 mph)	Incredible damage. Strong frame houses lifted off foundations and carried considerable distance to disintegrate, automobile-sized missiles flying through the air in excess of 100 yards, trees debarked, incredible phenomena occurring.

Note: The accuracy of expected damage at particular wind speeds has never been scientifically proven.

Source: FEMA, *Understanding Your Risks: Identifying Hazards and Estimating Losses*, FEMA 386-2 (Washington, D.C.: FEMA, 2001), 2-21, available at fema.gov/library/viewRecord.do?id=1880 (accessed June 7, 2007).

In May 2003, a strong (F3) tornado struck Moore, Oklahoma—near Oklahoma City, in the heart of Tornado Alley.[27] The National Weather Service issued a tornado warning fifteen minutes before the tornado hit, but Moore residents, who had experienced three tornadoes in the previous five years, were prepared. The emergency manager activated the town's new warning sirens, and residents took shelter, according to plan, in some 750 safe rooms, cellars, and other shelters. Thanks to an effective mitigation strategy, there were no deaths, and injuries were scattered. However, more than 300 buildings were destroyed, and at least 1,200 more were damaged.

Wildfires

Wildfires are intense, uncontrolled, rapidly spreading fires that sweep through forested areas or chaparral. Wildfires can damage or destroy homes; public infrastructure; and environmental, economic, and historic resources. They can also disrupt community services and devastate forestry, agriculture, and fishing resources.

The principal goal of mitigation planning for wildfires is to manage conditions in the urban-wildfire interface—the boundary zone between developed areas and surrounding forests or chaparral. Wildfire mitigation focuses on development management, fire-resistant construction, fuel management, and public outreach. Zoning regulations can be used to reduce residential density in the most vulnerable urban-wildfire interfaces. Wildfire risk can be mitigated for individual buildings through the use of nonflammable building and roofing materials, and the planting of fire-resistant trees and shrubs. Fuel management involves clearing brush to reduce available fuel, planting fire-resistant vegetation in at-risk areas, and constructing firebreaks and safety zones around homes and public lands in the urban-wildfire interface. Outreach programs encourage residents to minimize the storage of flammable materials, prune trees and vegetation to create a fire mitigation zone, ensure access routes for fire crews (by, for example, not planting trees and shrubs that would block the path between the house and the road), and maintain an outside water source that can be used to fight a fire. Finally, to ensure their availability in time of need, evacuation routes must be identified and protected.

In May of 2000, the Cerro Grande wildfire damaged or destroyed 235 homes and other structures, and led to the evacuation of 18,000 residents from the towns of Los Alamos and White Rock, New Mexico.[28] In one neighborhood, all but four homes were completely destroyed. Of those four, however, one escaped with only a moderately singed side. The house was protected by its design and by the owners' mitigation efforts: double-paned windows and solid-core doors prevented flames from entering the structure, and by pruning limbs and raking up pine needles, the owners had ensured that the property was free of ground fuel. The design and the mitigation measures were enough to protect the structure until the fire department could arrive.

Landslides

A landslide—the movement of rocks, earth, or debris down a slope—can be triggered by heavy rains, floods, earthquakes, lightning strikes, or volcanic activity. Landslides are influenced by human activities, such as land clearing and earth moving, and by natural factors, such as geology, topography, and precipitation. Landslides can occur in any state but are most common in California, Colorado, and the Appalachian Mountains. A special type of landslide, debris flow, is a rapidly moving slurry of soil, rock, water, and vegetation that takes on the consistency of mud as it flows downhill. Landslides can inundate and wipe out structures in their paths; they also pose dangers to people and natural resources.

An evaluation of landslide risk involves (1) mapping areas that are susceptible to landslides; (2) tracing the history of past landslides; and (3) analyzing a number of geological, topographical, and hydrologic features, such as bedrock type, degree of slope, presence of sandy soils, and the level of the water table. Mitigation measures include the installation of catchment basins to control runoff, the construction of retaining walls along evacuation routes, and the use of ground cover and riprap to hold soil in place on hillsides.[29]

In high-risk areas, local governments should require geotechnical investigations prior to development, and should adopt grading ordinances to regulate slope cutting and leveling; zoning ordinances to control density and site plan features (such as management of stormwater runoff and the removal of natural vegetation); and soil conservation or slope stabilization ordinances to minimize clearing and earth moving. Risk communication strategies include signage to indicate areas subject to landslides, homeowner information packets, and ordinances requiring the disclosure of landslide hazards when property is transferred.

Man-made hazards

State and local mitigation plans must address two categories of man-made hazard: technological hazards and terrorism-related hazards.[30] Technological hazards arise from human activities such as the manufacture, transport, storage, and use of hazardous materials; such hazards are assumed to be accidental. Terrorism is an intentional criminal act; it may involve biological, chemical, nuclear, or radiological weapons; arson; incendiary, explosive, or armed attacks; industrial sabotage or the intentional release of hazardous materials; or "cyber-terrorism." Whether intentional or accidental, man-made disasters involve the application of harmful force to the built environment; the result may take one or more forms: contamination (from chemical, biological, radiological, or nuclear hazards), the release of energy (from explosives, arson, or electromagnetic waves), or service disruption.

There are two types of terrorism mitigation: counterterrorism and antiterrorism. The first is designed to offensively manage the threat of terrorism. The second, which consists of defensive efforts to protect people and property, is the type most likely to involve local emergency management departments.

In addressing man-made hazards, FEMA mitigation guidelines focus on the physical aspects of vulnerability reduction. Thus, FEMA-funded mitigation plans addressing man-made hazards would focus on specific opportunities to reduce exposure to and impacts from terrorism and technological hazards. For example, if the community is planning to build a new city hall or hospital, defensive architecture, site planning, and design might be used to reduce the risk from man-made hazards. The FEMA guide to mitigation planning for man-made hazards recommends the same four-phase planning process as the FEMA guides to natural hazard mitigation—organizing resources, assessing risks, developing the mitigation plan, and implementing the plan and monitoring progress—but the nature of the activities undertaken in each phase differs.

FEMA-funded mitigation plans addressing man-made hazards would focus on specific opportunities to reduce exposure to and impacts from terrorism and technological hazards.

Mitigation strategies for man-made hazards focus on creating a safe, or "hardened," environment that resists accidents or attacks, is resilient in the face of an accident or attack, and is protective of human life in the event that an accident or attack occurs. Many strategies for mitigating other hazards will protect against man-made hazards as well; for example, strengthening buildings to resist earthquakes and hurricanes will also help to protect against explosions; fire resistance and fireproofing will help to protect against bombs and incendiary attacks.

Safe areas within buildings and safe evacuation routes from buildings are important considerations in building design. Target hardening, the primary terrorism mitigation technique, includes a number of approaches: strengthening buildings, altering land use patterns to create buffer zones around high-risk buildings, and installing security fencing and surveillance cameras. Site planning and landscape design can also reduce the impact of attacks on buildings and critical facilities.

Because of the complex underlying human factors, it is more difficult to assess the level of risk from man-made hazards than from natural causes. Nevertheless, mitigation planners can develop lists of potential man-made hazards in their jurisdictions, and can review public and private emergency and contingency plans and reports, including radiological emergency plans, the emergency plans of facilities that use hazardous materials, emergency plans for chemical stockpiles, toxic release inventory reports, and statewide domestic preparedness strategies. Under the Emergency Planning and Community Right-to-Know Act, each state must establish a state emergency response commission (SERC) to oversee planning for hazardous materials emergencies. The SERC establishes planning districts within the state, each with a local emergency planning committee (LEPC), which must include state and local elected officials as well as representatives from (1) local government departments and agencies (including public safety, emergency management, public health, environmental protection, and transportation); (2) hospitals; (3) broadcast and print media; (4) community groups; and (5) owners and operators of hazardous materials facilities.

Critical infrastructure systems are vital to the defense and economic security of the nation. Categories of critical infrastructure range from food, water, and public health to telecommunications, transportation, and banking and finance. To plan for the protection of such systems, mitigation planners can develop maps showing the community's critical facilities. To create the initial inventories, planners can use the baseline data available through HAZUS, FEMA's loss estimation software. Such maps and databases can be used to develop a protective strategy.

The mitigation planning process

The purpose of mitigation planning is to identify and implement policies that will reduce risk and future disaster losses. Making and implementing mitigation plans is very similar to making and implementing land use and other types of community plans. The four-phase FEMA mitigation planning process forms the planning framework: (1) organizing resources, (2) assessing risks, (3) developing a plan, and (4) implementing the plan.[31]

Organize resources

The goal of organizing resources is to generate community involvement in, and support for, mitigation plans. To be effective, hazard mitigation planning must bring together community stakeholders to achieve a common vision, identify desired mitigation projects, and build political support for mitigation programs.[32] The first phase of mitigation planning involves assessing the level of community support, building the planning team, and engaging the public. These activities go on simultaneously as information is gathered, stakeholders are identified, and participation is initiated. Typically, these tasks are coordinated by the local emergency management department, with the assistance of the planning department and other appropriate local government departments.

Although community support is essential to any mitigation effort, it may not be immediately forthcoming. Resistance may stem from the low priority that decision makers typically afford to mitigation, from citizens' unwillingness to acknowledge the potential for disaster, and from the view that hazard mitigation is a "technical" activity carried out by engineers and other specialists. Thus, part of the task of mitigation planners is to educate community leaders and the general public. An effective community involvement program demonstrates that mitigation is vital to the future of the community, helps officials and residents understand that disasters may well occur, and makes it clear that mitigation planning is too important to be left to technicians alone. Successfully educating the community about the importance of mitigation may well turn skeptics and opponents into supporters. For example, mapping the extent of the community's tax base that is vulnerable to a 100-year flood may help decision makers realize just how serious the threat of flooding is.

To assess community support, planners need to ask themselves some preliminary questions:

- How much do elected officials and citizens know and understand about local hazards? Has their understanding been sharpened by a recent disaster?
- Do community leaders understand the local, state, and federal context of hazard mitigation?
- How likely is it that an individual, an organization, or a business will emerge as a champion for mitigation?
- How adequate is the community's planning capacity? Is there sufficient staff? Does the community have existing disaster-related plans, a hazards database, a geographic information system (GIS), hazard maps, and plan-making and implementation procedures?
- Can current planning initiatives (such as the comprehensive plan, stormwater management plans, open space and recreation plans, transportation plans, or redevelopment and housing plans) be expanded to include mitigation policies?
- Does the community have a solid working relationship with state emergency management staff and with the FEMA regional office?

Community planning and emergency response: The need for collaboration

When engaged in community planning, emergency managers need to recognize that communities have multiple goals. A classic example concerns the physical design and layout of streets in new subdivisions.[1] In an effort to meet various community objectives—including providing for pedestrian safety and comfort, and meeting the demand for more compact, walkable neighborhoods—many jurisdictions have embraced traditional neighborhood developments (TNDs). Among other characteristics, TNDs feature narrow, tree-lined streets and allow on-street parking.

Although fire departments and other emergency responders in older towns and central cities are accustomed to working in neighborhoods with narrower (26- to 28-foot-wide) streets, some fire departments have responded to proposals for TNDs by requesting wider (36- to 40-foot-wide) streets; the fire departments claim that the wider streets are necessary to ensure good access for emergency vehicles. However, the majority of calls to which fire departments respond are medical emergencies and car crashes; home fires make up a dwindling percentage of calls. Moreover, the streets in TNDs—which are relatively short and offer multiple entry and exit points—actually provide better access for emergency vehicles than the long, winding blocks and disconnected street networks of typical suburban developments. In addition, TNDs typically have less traffic congestion than other types of development; since congestion is often a major source of delay in responding to emergency calls, emergency responders may find that TNDs actually allow improved response times. Finally, because traffic speeds are slower on TND streets, these streets are associated with a lower rate of serious injuries and fatalities from car crashes.

To address both the needs of emergency responders and the broader goals of the community, some local governments have engaged in collaborative efforts involving city planners, emergency response officials, developers, citizens, and other stakeholders. As long as all of the relevant stakeholders are in the room when decisions are being made and everyone involved is willing to see the value in each other's objectives, collaborative solutions can often be found that meet all of the community's needs.

Source: Dan Emerine, project manager, ICMA, and Steve Tracy, senior research analyst, Local Government Commission.

[1]The Local Government Commission has more information about dealing with street design and emergency response challenges available on its Web site at lgc.org.

- Is there an existing emergency operations plan or mitigation plan that needs to be revised and updated? If so, does this plan conform to state enabling legislation and to federal mitigation planning mandates?

Answers to these questions will help in selecting the mitigation planning team. This team should be broadly representative, carefully selected, and formally appointed by the governing body. Members should include interested citizens; community leaders (both officials and other recognized leaders); businesspeople; representatives of local nonprofit organizations; and representatives of local government agencies, such as public safety, engineering, planning, and utility departments. The mitigation planning team will be the workhorse, sounding board, and community link for the planning effort.

To broaden participation in the planning process, the general public should be engaged. Involving the public ensures that the planning team hears many different points of view, has the opportunity to inform and educate the community about hazard risks and mitigation, and generates support for the adoption and implementation of the final plan.

In designing the community involvement program, choices must be made about six key elements: (1) program administration, (2) objectives, (3) citizen involvement at various stages, (4) targeted groups, (5) community involvement techniques, and (6) necessary information:[33]

- Who will administer the participatory process and prepare the written participation plan?
- What are the objectives of participation: to comply with state requirements, to learn about citizens' preferences, to educate citizens about policy issues, to foster citizens' influence on decision making, and/or to mobilize an active constituency to support proposed plans?
- At what stage are citizens to be involved: during "preplanning," as the vision is framed; during planning, when alternative proposals are evaluated; and/or during the postplanning phase, when citizens will comment on the plan at a public hearing?

- What groups are to be targeted: businesspeople, elected officials, development groups, local government departments, neighborhood groups, environmental groups, and/or disadvantaged or vulnerable populations located in hazard areas?
- Which community involvement techniques will be used: public hearings, open meetings, information sessions, facilitated workshops, and/or citizen advisory committees?
- What information do citizens need in order to participate effectively: maps of hazardous areas and vulnerability, summaries of mitigation plans and policies, vision statements, federal and state requirements, and/or emergency plan elements?[34]

Assess risks

To project the effects of specific disasters, mitigation planners can use FEMA's loss-estimation guides, which cover seven natural hazards—floods, earthquakes, tsunamis, tornadoes, coastal storms, landslides, and wildfires[35]—as well as various types of man-made hazards.[36] Risk assessment can be a highly technical process, requiring specialized expertise. In this section, however, risk assessment is considered more broadly as a process in which mitigation planners identify hazards to which the community may be subject, profile hazard events, inventory assets, and estimate potential losses.

Identify hazards and profile hazard events To find out which hazards a community faces, mitigation planners can search newspapers and historical records, talk with experts (including staff at the state emergency management office), and gather information from FEMA Web sites.[37] Once the hazards have been identified, the next step is to profile the hazard events in terms of their potential impact. The primary source of such information is HAZUS, which uses GIS software to estimate losses from various hazards. Originally designed to estimate losses from earthquakes, HAZUS now includes models for estimating potential losses from wind (hurricanes, thunderstorms, tornadoes) and flood (coastal and riverine) hazards. It can also be used as a mapping and inventory tool.

Inventory assets and estimate losses An asset inventory reveals the extent of the community's vulnerability to hazards. The inventory lists the number of people and the value of the building stock located within various hazard areas, such as the 100-year floodplain. It identifies at-risk critical facilities, such as hospitals, emergency shelters, and transportation systems; facilities that face high potential losses, such as schools; and facilities that use hazardous materials. The baseline data in HAZUS will be sufficient for an initial inventory; planners can develop a more refined inventory by supplementing HAZUS with property tax and land use records.

Assessing impacts from different hazard events allows potential losses to be estimated. First to be estimated is the level of damage from a hazard event, which is calculated in one of two ways: (1) on the basis of the replacement value of a structure and its contents, and (2) on the basis of the use and function of the structure. For example, if the replacement value of a building's structure is $100,000, the replacement value of its contents is $50,000, and expected damage from a 100-year flood is 40 percent of the structure and 60 percent of the contents, then the total loss would be $70,000. Loss of function for a business would be computed as the number of non-operational days multiplied by the average daily revenue. Estimates of potential human losses, such as deaths, injuries, and disaster-related homelessness, can be drawn from HAZUS or from records of similar disasters. HAZUS can also provide estimates of indirect economic losses caused by building damage, business interruption, and rupture of lifelines (i.e., critical systems such as water, power, and telephone services). Composite loss maps can be assembled from loss estimates for individual hazards.

No jurisdiction can mitigate its hazards to such an extent that all risk is eliminated. Communities should review their composite loss maps and determine the degree of risk they are willing to accept. Decisions about acceptable risk are not merely technical; community stakeholders need to be involved.

A risk assessment should be forward looking. This means planning for buildout—the time when all projected community growth is complete.[38] For example, mitigation planners should take into account the effect of an increase in impervious surfaces on stormwater runoff—which will, in turn, affect the boundaries of the 100-year floodplain. If necessary, the community's floodplain limits and development regulations should be revised to ensure that future development is not at risk.

During the 1990s, the city of Charlotte and Mecklenburg County used buildout analysis to manage growth, achieve sustainable development, and mitigate flood hazards.[39] Their strategy has two main components: reducing flood risks and managing water quality. To reduce flood risks, the city and county revised their floodplain maps to reflect future buildout; then, on the basis of the new maps, they revised their zoning regulations, acquired flood-prone properties, and conducted flood audits for at-risk commercial structures. To manage water quality, Charlotte and Mecklenburg County adopted a surface water management program, which calls for buffers and other measures to protect and restore streams. Consensus for the plan was built through a participatory modeling and planning process. To make the new requirements palatable to the development community, the plan includes incentives and flexible mitigation options.

Under the new regulations, vegetated buffers are required adjacent to streams; the greater the amount of acreage drained by the stream, the larger the buffer. Planners used the new flood-risk maps, which showed the extent of the 100-year floodplain under buildout conditions, to determine the size of the buffers. After acquiring and demolishing 116 flood-prone properties, the city and county created open space in their stead. This component of the plan was funded by a $10.3 million Hazard Mitigation Grant and a $2.2 million contribution from the Mecklenburg County stormwater fee. A $1 million grant from the North Carolina Clean Water Management Trust Fund was used to create a nearby wetland, which filters sediment and absorbs pollutants from surface waters.

Develop a mitigation plan

The mitigation plan lays out a community's strategy for ensuring long-term sustainability by reducing its vulnerability to hazards. Plan making involves four steps: (1) setting goals and objectives, (2) identifying mitigation actions and assigning priority to them, (3) preparing an

The Holden Beach mitigation plan

The mission of the Holden Beach, North Carolina (2000 pop. 787), mitigation plan is to substantially and permanently reduce the community's vulnerability to natural and technological hazards.[1] The plan is designed to protect citizens, critical facilities, infrastructure, private property, and the natural environment. This goal is to be achieved by increasing public awareness, providing resources for risk reduction and loss prevention, and undertaking activities that will foster a safer, more sustainable community.

Holden Beach involved the community through stakeholder interviews, workshops, and public hearings. The plan's vulnerability analysis covers seventeen natural hazards and three technological hazards. An implementation matrix places each mitigation strategy into one of five categories: (1) new policy, (2) amended policy, (3) continued policy, (4) new project, and (5) continued project. For each strategy, the matrix identifies the type of strategy, the target completion date, the responsible party or organization, potential funding sources, monitoring and evaluation indicators, and the hazard addressed.

The Holden Beach plan specifies five types of mitigation strategies:

- **Preventative:** To keep vulnerability from increasing
- **Property protection:** To protect existing structures by modifying buildings, removing structures from hazardous locations, or adopting policies that address risks
- **Natural resource protection:** To preserve or restore natural areas and their mitigation functions
- **Structural projects:** To lessen the potential impact of a hazard by modifying the environment or the natural progression of the hazard event
- **Public information:** To advise citizens, business owners, potential property buyers, and visitors about hazards, hazardous areas, and mitigation techniques they can use to protect themselves and their property.

[1]See "Executive Summary," *Town of Holden Beach, North Carolina, Community-Based Hazard Mitigation Plan*, 5, available at ncem.org/mitigation/holden_beach_plan.htm (accessed June 5, 2007).

implementation strategy, and (4) writing the mitigation plan.[40] A vigorous debate about which goals are to be adopted and how they are to be achieved is likely to accompany this rational series of steps.

Set goals and objectives The first step in the development of goals and objectives is the creation of problem statements: nontechnical expressions of the findings from the risk assessment. An example of a problem statement for a coastal community might be, "The community has been damaged by winds and flooding from four hurricanes in the past decade, and 25 percent of the property tax base is located within the 100-year floodplain." Such statements, which can be easily understood by citizens and public officials, facilitate the development of goals and objectives.[41]

A goal is a broad statement of a desired long-term outcome—a problem to be alleviated, a need to be met, or an aspiration to be achieved. For example, one goal for a coastal community might be to ensure that future hurricanes do not threaten the community's economic vitality. Goals may reflect community values (e.g., the desire to protect open space), community expectations (e.g., a projected expansion of tourist-oriented businesses), or mandates from higher levels of government (e.g., state requirements prohibiting future development from infringing on sensitive coastal environments).

An objective is a tangible, intermediate step toward attaining a goal. It is specific, measurable, and attainable. For example, objectives for a coastal community might be, "The minimum setback for all future shoreline development will be twenty times the annual coastal erosion rate," or "Within ten years, all existing structures in the 100-year floodplain will be elevated above the base flood height."

Moving from problem statements to goals and objectives requires careful review and study by community stakeholders and the mitigation planning team. Different constituencies will advocate different approaches, and consensus on goals and objectives will not be easy to achieve. Nevertheless, to ensure support for implementation, it is essential for elected officials and others to understand the basis for, and agree on the content of, the plan's goals and objectives.

Identify actions and assign priorities to them Mitigation actions support the goals and objectives of the plan. These actions can range from adopting new zoning and building codes to constructing tornado shelters and retrofitting buildings. The mitigation planning team selects mitigation actions by (1) reviewing the hazards faced by the community, (2) assessing the feasibility of proposed actions in light of state and local capability, and (3) evaluating proposed actions in relation to broader community goals and to other actions that are already under way.

Mitigation capability is a function of available staff and resources. For example, a coastal community might discover that it will need outside funding in order to elevate structures in the 100-year floodplain, and that a study is needed to complete an application for funding. If a consultant is needed to prepare the study, this may extend the time horizon for achieving the objective of elevating the structures. Thus, the community might make adoption of a coastal setback its first priority and make completion of an elevation study its second priority.

In setting priorities, a number of considerations come into play, including community values, feasibility, timing, and the opportunity to undertake measures that serve multiple community goals. With respect to values, for example, some communities may be reluctant to adopt measures such as relocation, which involves the acquisition of property rights for structures in the hazard zone. Timing may come into play if stakeholders believe that actions are more likely to be implemented during the aftermath of the next disaster, when funding and political will may be more abundant. The community that has assigned priorities to actions in advance of a disaster will be better prepared to take advantage of post-disaster mitigation opportunities: as noted earlier, to be eligible for a FEMA post-disaster Hazard Mitigation Grant, a community must demonstrate that the benefits of a proposed project exceed its costs. Such cost-benefit considerations should be taken into account in setting priorities.

Prepare an implementation strategy The implementation strategy is a management tool to ensure that efforts to carry out the plan's goals and objectives are coordinated, efficient, and widely understood. The implementation strategy identifies who is responsible for mitigation actions, what resources are to be used or sought to undertake the actions, and what the projected timing is for the actions. The strategy allows progress to be tracked, responsible parties to be held accountable, and funding to be obtained as needed.

Preparing an implementation strategy involves three basic tasks. First, those responsible for actions must be identified and their roles defined. For example, the local planning department may be designated to prepare a new coastal development setback ordinance for consideration by the governing body. Second, a budget should be prepared that reflects the funding that will be necessary to support the actions. Finally, the time frame for carrying out each action should be specified to ensure that deadlines are met.

Write the mitigation plan The mitigation plan is both a guide to future actions and a record of the analysis, principles, and concepts underlying the planning process. The plan should be accurate, complete, and understandable, and should clearly link the problem statements, goals, objectives, and actions (see sidebar on pages 106–107).

Implement the plan and monitor progress

The final stage in the mitigation planning process is to put the plan into action. FEMA envisions this stage as a four-step process: (1) adopt the mitigation plan, (2) implement the plan recommendation, (3) evaluate the planning results, and (4) revise the plan.[42]

Adopt the plan Under DMA 2000 requirements, the local jurisdiction's governing body must formally adopt the plan. However, before the plan is submitted to the governing body, a draft must be reviewed by the state emergency planning office to ensure that the plan meets the criteria established in DMA 2000. Formal adoption by the local governing body prevents the plan from becoming simply another document on the shelf: it invests the plan with the authority to guide local government decisions, gives it legal status, certifies to program and grant administrators that the plan's recommendations have been officially approved, and helps ensure the continuity of mitigation efforts over time.

To facilitate adoption, it is useful to brief elected officials and community leaders throughout the planning process. The mitigation planning team could hold periodic public workshops and presentations on the progress of the plan preparation process. It is also helpful to present the governing body with letters of support from stakeholder groups.

Implement the plan recommendations In order to bring the plan to life, it needs to become part of the community's administrative, budgetary, and regulatory processes. Implementation also includes the establishment of performance indicators that will make it possible to measure the effectiveness of mitigation actions.

One way to clarify and communicate the responsibilities of all parties involved in the implementation is to draw up a memorandum of agreement outlining goals and objectives, organizational structure, responsibilities, and resources. Mitigation plans are not self-implementing: implementation is significantly affected by events during the post-plan period. To strengthen and support implementation efforts, it is important to make the mitigation concept a part of everyday local government operations—that is, to ensure that budgets, decisions, and regular activities take account of hazard mitigation needs and opportunities. Another way to obtain support for implementation is by building relationships and partnerships with businesses, volunteers, community groups, and state agencies. Mitigation can also be linked to related programs, such as open space, recreation, or community development. For example, Community Development Block Grants, which are available through the U.S. Department of Housing and Urban Development, can be used to fund the local match required for hazard mitigation projects.

Finally, mitigation can benefit from a champion—a community leader or an elected or appointed official who can make a strong and convincing case for mitigation. The success that Tulsa, Oklahoma, has enjoyed in implementing its mitigation plans can be attributed, in large

Principles and criteria for preparing and evaluating mitigation plans

Clarity of purpose

The plan should identify and explain the desired mitigation goals.
Does the plan

- Clearly state the mitigation goals?
- Link the mitigation goals to broader environmental, social, and economic goals?
- Explain clearly how mitigation goals will be achieved?
- Explain state and federal legislation that enables or requires the plan?
- Explain how legislative provisions are to be implemented?

Citizen participation

The plan should be based on explicit procedures for involving stakeholders.
Does the plan

- List the organizations and individuals that were involved in planning?
- Explain why they were involved?
- Ensure that all groups affected by the plan are represented?
- Identify the participation techniques that were used?
- Distinguish between passive involvement and active involvement?
- Give a historical account of how stakeholder involvement is related to prior planning activities?

Problem identification

The plan should discuss significant hazard mitigation problems and assign priorities to them.
Does the plan

- Identify problems related to natural hazards?
- Provide background information on each problem, including causes and effects?
- Assign priorities to problems?
- Explain the rationale for the priorities?

Specification of objectives

The plan should provide specific objectives to guide decision making and planning.
Does the plan

- Link mitigation objectives to specific actions?
- Distinguish clearly between recommended mitigation objectives (with words like *consider, should, may*) and mandatory objectives (with words like *shall, will, require*)?

Fact base

The plan should be built on a foundation of solid facts and careful analysis.
Does the plan

- Support explanations of problems, current conditions, trends, and likely future conditions with high-quality data?
- Relate mitigation objectives to scientific data and common knowledge?
- Cite the methods and models used to derive analyses?
- Include maps of hazardous areas and vulnerable structures?
- Present information in a way that is relevant and comprehensible?
- Include relevant and comprehensible tables delineating vulnerable structures and populations?
- Give sources for background information and data?

Policy integration

The actions described in the plan should be coordinated with those of other relevant agencies.
Does the plan

- Refer to other related plans, both internal and external to the community (e.g., the community's comprehensive plan or the state hazard mitigation plan)?
- Explain the relationship between the goals and objectives outlined in the plan and those in related plans?

Connection to community development

The plan should link mitigation and community development objectives.
Does the plan

- Tie mitigation goals to other publicly supported development goals?
- Use action instruments (e.g., zoning, building codes) that are also used for other community development aims?

Multihazard scope

The plan should deal with all hazards affecting the community.
Does the plan

- Include mitigation goals and policies that apply to all the hazards faced by the community?
- Require risk assessments for multiple hazards?
- Compare the risks to life and property posed by different hazards, and assign priority to the various hazards that the community faces?

Organization and presentation

The plan should be understandable to a wide range of readers.
Does the plan

- Include a table of contents, an executive summary, and a glossary?
- Cross-reference problems, goals, objectives, and actions?
- Include clear illustrations and maps?
- Include supporting documents (e.g., a video, a compact disk, geographic information system data, a Web page)?

Internal consistency

The plan should maintain consistency among its goals, objectives, and actions.
Does the plan

- Make mitigation goals comprehensive enough to accommodate strategic issues?
- Link mitigation actions to specific goals and objectives?

Performance monitoring

The plan should include indicators for assessing goal achievement.
Does the plan

- State clear objectives?
- Include performance indicators for each objective?
- Identify the organizations and entities that are responsible for monitoring and/or providing data for indicators?

Implementation

The plan should commit the community to carrying out the actions outlined in the plan.
Does the plan

- Include timelines for implementation?
- Identify the organizations and entities that are responsible for implementation actions?
- Identify funding sources for implementation?

Source: Adapted from David R. Godschalk, Edward J. Kaiser, and Philip R. Berke, "Integrating Hazard Mitigation and Local Land Use Planning," in *Cooperating with Nature: Confronting Natural Hazards with Land-Use Planning for Sustainable Communities*, ed. Raymond J. Burby (Brookfield, Conn.: Rothstein Associates, Inc., 1998), 115-117.

part, to a local champion.[43] Tulsa faces multiple hazards, including tornadoes, violent thunderstorms, and repeated flooding from the Arkansas River. Under the leadership of a persuasive staff member who enlisted the support of elected officials and community groups, Tulsa transformed itself. Before it began its mitigation efforts, Tulsa had been the site of more federally declared disasters than any other city in the country; it is now among the best examples of resiliency in the nation.

By 1993, Tulsa had implemented a program to clear 875 buildings from the floodplain. To pay for the program, the city adopted a stormwater utility fee that brings in $8 million per year. It also initiated watershed-wide development regulations, an aggressive public awareness campaign, a master drainage plan supported through a capital funding program, and a plan to transform floodplains into open space and recreation areas. Tulsa's mitigation program reduced losses from repeated flooding, improved the quality of life by expanding open space and recreation areas, returned floodplains to wetlands, and reclaimed wildlife habitat (see Figure 6–7).

Figure 6-7 Aerial view of the Robert S. Kerr Lock and Dam, a reservoir, hydroelectric facility, and recreational area on the Arkansas River Navigation System in Tulsa, Oklahoma

Photo courtesy of the U.S. Army Corps of Engineers

Evaluate the planning results Regular evaluation of mitigation outcomes keeps the community informed of progress and helps to manage implementation. The performance indicators developed during the implementation phase will reveal which actions are working well and which may need to be rethought.

The purpose of evaluation is to determine whether the mitigation actions helped to achieve the objectives outlined in the plan. For example, was there a reduction in the number of households living in unsafe areas or vulnerable structures? Was there a reduction in the level of expected losses? Did the benefits exceed the costs?

Evaluation also asks why actions did or did not work. A number of factors may account for suc-cess or failure: the presence or absence of political support; the availability of funds, staff, and technical support; and the appropriateness of the original expectations, which may have been unrealistic. Analyzing the underlying reasons for success or failure makes it possible to take the necessary corrective actions.

Revise the plan To remain eligible for disaster assistance under DMA 2000 regulations, communities must revise and update their mitigation plans at least every five years. However, changing conditions may require more frequent revisions—for example, to replace objectives and actions that turn out to be less relevant or effective than was expected, to cope with damage from a disaster, or to account for reductions in vulnerability brought about by the completion of mitigation projects. It would also be wise to revise the mitigation plan to reflect the adoption of a new comprehensive plan or set of development regulations. Following a disaster, the plan should be revised to account for lessons learned from the disaster and to make use of the window of opportunity: immediately after a disaster, long-term sustainability rises to the top of the public agenda, often allowing mitigation goals to be accomplished that might not otherwise have been achievable.

Linking mitigation to community sustainability

The community that fails to address the threat of hazards may find itself in an unsustainable position, particularly if it is highly vulnerable to risks. Because of the threat of hazards, every community needs to develop a vision of future sustainability.

Mitigation and smart growth in Kinston, North Carolina

The goal of smart growth is to shape development in ways that strengthen communities, protect the environment, and make the best use of existing infrastructure. Smart growth enjoys wide support among planners, developers, and community groups. Among the desired outcomes of smart growth are compact urban patterns, infill development, lessened automobile dependence, walkable neighborhoods, and safe environments.

Smart growth offers an excellent opportunity to integrate hazard mitigation and other community goals. In Kinston, North Carolina (2000 pop. 23,688), for example, families formerly at risk from repeated flooding were moved to safe neighborhoods, and the floodplain was converted to community open space.

Kinston is situated on the banks of the Neuse River, and much of the city is located in the 50-year floodplain. In 1996, Hurricane Fran caused major flooding: dozens of businesses and more than four hundred homes were damaged, along with a considerable amount of public infrastructure—including the sewage treatment plant, which spilled raw and partially treated sewage into the Neuse River. The city suffered serious economic disruption, and losses were estimated in the tens of millions of dollars. Fran was a wake-up call to local officials, who took on the task of developing a recovery and mitigation strategy.

Just three years later, in September 1999, Kinston was struck by Hurricane Floyd. The extent of the flooding was even greater than it had been during Fran: more than four hundred homes and two hundred businesses sustained major flood damage, the central business district was submerged under several feet of water, and the town's two wastewater treatment plants were flooded. Nevertheless, in the three years between Hurricanes Fran and Floyd, approximately one hundred houses had been acquired and vacated, and their occupants had been relocated—actions that averted an estimated $6 million in damage. (The acquisition and relocation costs were $2.1 million.) In the wake of Hurricane Floyd, Kinston moved to incorporate mitigation into its long-range comprehensive plan.

Lenoir County and the city of Kinston developed a plan to eliminate or substantially reduce the risk of repeated flooding. In addition to reducing flood hazards, the Kinston–Lenoir County plan will meet other objectives related to long-term sustainability: it will revitalize existing residential neighborhoods and businesses, empower citizens, foster self-sufficiency, and improve the quality of life. Federal, state, and community resources will be integrated not only to move families out of the floodplains, but also to address affordable housing, employment, job training, and the use of renewable resources. In a multiyear effort that will cost about $31 million, the city and county will acquire more than four hundred residences, three mobile home parks, and sixty-eight vacant lots. Of the total funding needed, $15 million will come from the Hazard Mitigation Grant Program; $12 million from U.S. Department of Housing and Urban Development, Community Development Block Grant, and Disaster Recovery Initiative funds; and $4 million from the state of North Carolina.

The Greater Kinston Urban Growth Plan was designed to guide the city's development for twenty years. The goals of the plan are to

- Expand the flood buyout program
- Prohibit, on an interim basis, new residential construction or replacement of substantially damaged housing in the 100-year floodplain
- Prepare a redevelopment plan for the flood area
- Amend the floodplain regulations, including elevation standards
- Investigate new floodplain mapping
- Prepare a master plan for new investments in housing and economic development
- Enhance the tax base by supporting development within the city boundaries
- Provide conservation areas within the floodplain that can be used for stormwater management and recreation.

Smart growth offers the opportunity to link hazard mitigation goals and broader community objectives; when mitigation and community development are coordinated, the whole is greater than the sum of the parts. The sustainable community ties its hazard mitigation plan, comprehensive plan, open space plan, and economic development plan into a unified planning and implementation strategy based on the principles of smart growth.

The sustainable community vision

A sustainable community vision—in which environment, economy, and equity are balanced—should be the centerpiece of a hazard mitigation plan. Sustainable development means two things: first, acknowledging and attempting to reduce the community's vulnerability to hazards; second, planning to ensure that future development does not increase that vulnerability.

In a sustainable community, the natural environment is conserved and enhanced, both because it is a valuable community resource that contributes to the quality of life, and because it is a valuable component of efforts to reduce the impact of hazards. A river corridor, for example, not only protects water quality and provides wildlife habitat, but also absorbs floodwaters and reduces the erosion caused by stormwater runoff.

In a sustainable community, the economy is nurtured and developed as a critical source of employment, goods, and services, and as a source of support for future generations. Sustainable economic structures and practices contribute to the community's ability to plan ahead and to ensure that future development mitigates the risk. For example, a sustainable manufacturing facility uses renewable energy sources, minimizes waste, and employs processes that protect public safety.

In a sustainable community, social equity is both a central value in resource allocation and a touchstone for mitigation planning and implementation. Those at greatest risk from hazards, especially citizens who are poor or disadvantaged or who live in high-risk areas, should be involved in mitigation planning and implementation, and their needs should be given a high priority in mitigation programs. The sustainable community seeks social as well as physical resilience.

Disaster resilience

A disaster-resilient community is designed to anticipate, weather, and recover from the impact of natural or man-made hazards.[44] A resilient community is a sustainable network of physical and social systems that are redundant, diverse, adaptable, and interdependent. The resilient community builds a broad base of commitment to mitigation and ensures that it has sufficient capability to make that commitment a reality. Mitigation objectives are incorporated not only into the emergency management program, but also into other programs, such as community development, land use, and growth management.

Berkeley, California, exemplifies a resilient community.[45] Following the 1989 Loma Prieta earthquake and the 1991 East Hills wildfire, the city of Berkeley set out to achieve sustainability within a high-risk environment. The city's multipronged mitigation strategy included the following coordinated policies and actions:

- Berkeley voters approved five local ballot measures (totaling $390 million) to fund the seismic retrofitting of municipal facilities and schools.
- The city has regularly invested in safety programs and seismic retrofitting subsidies.
- The city adopted programs that award rebates on transfer taxes and permit fees to homeowners who increase the seismic safety of their homes.
- The city initiated a loan program and a free home repair and retrofitting program for low-income seniors and disabled residents.

As a result of these actions, Berkeley's rate of seismic retrofitting is the highest in the San Francisco Bay Area.

Summary

Mitigation is central to the creation of resilient, sustainable communities. This chapter describes a proven, reliable process for successful mitigation. A strong mitigation program includes the following components:[46]

- *Preparation of hazard mitigation plans.* In collaboration with community stakeholders, the mitigation team must identify hazards and vulnerabilities and adopt strategies for dealing with them.
- *Management of development.* To avoid putting people at risk in known hazard areas, the community should use land use plans and development regulations to direct new growth

away from hazard areas. Existing buildings should be relocated from threatened areas to safe locations.

- *Protection of buildings and public facilities.* Existing structures should be flood proofed, wind proofed, and retrofitted to resist damage from earthquakes. New structures should be required to meet the standards set in building and design codes. Defensive design should be incorporated into public facilities and buildings.
- *Conservation of natural areas.* Wetlands, dunes, and forests that reduce hazard impacts should be maintained. To preserve open space and limit development in hazard areas, the community should acquire at-risk property.
- *Control of hazard risks.* To reduce risks from hazards such as flooding, landslides, and erosion, the community should use techniques such as flood control structures, slope stabilization, and shoreline setbacks.
- *Limitation of public expenditures.* To discourage development in high-risk areas such as floodplains and coastal barrier islands, the community should withhold subsidies for roads, bridges, sewage treatment systems, and other public facilities.
- *Risk communication and public education.* The community must ensure that residents, decision makers, and prospective property buyers are notified about the location of hazard areas. Residents must be informed about evacuation and shelter provisions, and builders and developers should be educated about safe construction and mitigation techniques.

Notes

1 Federal Emergency Management Agency (FEMA), *Getting Started: Building Support for Mitigation Planning,* FEMA 386-1 (Washington, D.C.: FEMA, 2002), available at fema.gov/library/viewRecord.do?id=1867 (accessed June 4, 2007).

2 FEMA, *Planning for a Sustainable Future: The Link between Hazard Mitigation and Livability,* FEMA 364 (Washington, D.C.: FEMA, 2000), available at fema.gov/library/viewRecord.do?id=1541 (accessed June 4, 2007).

3 Dennis Mileti, *Disasters by Design: A Reassessment of Natural Hazards in the United States* (Washington, D.C.: Joseph Henry Press, 1999), 32–33.

4 Conflicting goals about sustainability can make it difficult to find a balance between economic, environmental, and social equity considerations: see David R. Godschalk, "Land Use Planning Challenges: Coping with Conflicts in Visions of Sustainable Development and Livable Communities," *Journal of the American Planning Association* 70 (Winter 2004): 6.

5 U.S. House of Representatives, *A Failure of Initiative: The Final Report of the Select Bipartisan Committee to Investigate the Preparation for and Response to Hurricane Katrina* (Washington, D.C.: Government Printing Office, 2006), available at katrina.house.gov/full_katrina_report.htm (accessed June 6, 2007).

6 National Oceanic and Atmospheric Administration (NOAA), *Economic Statistics for NOAA,* 3rd ed. (Washington, D.C.: U.S. Department of Commerce, April 2004), 5, available at nws.noaa.gov/com/2004_economic_statistics1.pdf (accessed June 6, 2007).

7 Multihazard Mitigation Council, *Natural Hazard Mitigation Saves: An Independent Study to Assess the Future Savings from Mitigation Activities,* vol. 1, *Findings, Conclusions, and Recommendations* (Washington, D.C.: National Institute of Building Sciences, 2005), 5, available at floods.org/PDF/MMC_Volume1_FindingsConclusionsRecommendations.pdf (accessed June 6, 2007).

8 See "Kinston-Lenoir County Acquisition Project: Sustainable Redevelopment," at ncem.org/mitigation/case_kinston.htm (accessed June 6, 2007).

9 See "Belhaven, North Carolina—A Case Study," in *Mitigation Preliminary Performance Assessment: Losses Avoided during Hurricane Isabel in North Carolina* (Raleigh, N.C.: Flood Insurance and Mitigation Division, Division of Emergency Management, 2003), available at dem.dcc.state.nc.us/Mitigation/Library/Success_Stories/Perf%20Assessment%20NC%20Print.pdf (accessed June 4, 2007). The base flood elevation (BFE) is the computed elevation to which floodwater is anticipated to rise during the base flood—that is, the 100-year flood. BFEs are shown on flood insurance rate maps (FIRMs) and are the basis for regulatory requirements for elevation or flood proofing of structures.

10 See "Mitigation Planning Interim Final Rule," *Federal Register* 67, no. 38 (February 26, 2002), at 44 CFR Parts 201 and 206, available at fema.gov/library/viewRecord.do?id=1933 (accessed June 6, 2007).

11 The National Flood Insurance Program has three components: (1) federally backed flood insurance for homeowners, renters, and business owners located in floodplains; (2) floodplain management requirements; and (3) flood hazard mapping.

12 For more about FEMA's Flood Mitigation Assistance Program, see fema.gov/government/grant/fma/index.shtm (accessed June 6, 2007).

13 For more about FEMA's Community Rating System, see fema.gov/business/nfip/crs.shtm (accessed June 6, 2007).

14 "Mitigation Planning Interim Final Rule" (see note 10); other Mitigation Planning Interim Final Rules include October 1, 2002 (deadline extension), October 28, 2003 (local plan requirement clarification), and September 13, 2004 (state and tribal extension option), all available at fema.gov/plan/mitplanning/interim_final_rules.shtm#0 (accessed June 6, 2007).

15 FEMA, *Getting Started.*

16 FEMA, *Understanding Your Risks: Identifying Hazards and Estimating Losses,* FEMA 386-2 (Washington, D.C.: FEMA, 2001), available at fema.gov/library/viewRecord.do?id=1880 (accessed June 7, 2007).

17 FEMA, *Developing the Mitigation Plan: Identifying Mitigation Actions and Implementation Strategies,*

FEMA 386-3 (Washington, D.C.: FEMA, 2003), available at fema.gov/library/viewRecord.do?id=1886 (accessed June 4, 2007).

18 FEMA, *Bringing the Plan to Life: Implementing the Hazard Mitigation Plan,* FEMA 386-4 (Washington, D.C.: FEMA, 2003), available at fema.gov/library/viewRecord.do?id = 1887 (accessed June 4, 2007).

19 FEMA, *Rebuilding for a More Sustainable Future: An Operational Framework,* FEMA 365 (Washington, D.C.: FEMA, 2000), available at fema.gov/library/viewRecord.do?id = 1429 (accessed June 4, 2007).

20 See FEMA Multi-Hazard Mapping Initiative (MMI) Hazard Maps at gcmd.nasa.gov/records/FEMA-HazardMaps.html; FEMA's Web site for disaster and hazard information and maps at fema.gov/hazard/index.shtm; and the USGS Natural Hazard Support System at nhss.cr.usgs.gov (all accessed June 5, 2007).

21 There are six types of A zones, although all are based on the 100-year flood. Along with V zones depicting a velocity hazard from wave action, other types of zones include B zones showing moderate flood hazard; C zones showing areas outside the 500-year flood hazard; and D zones indicating undetermined but possible flood hazards. Flood maps can be obtained on the FEMA Web site (see note 20).

22 Liquefaction occurs when ground shaking causes hydric (water-bearing) or sandy soils to behave like fluids; it is prevalent in areas that were developed on filled wetlands or drained hydric soils.

23 Another common earthquake measurement is the Peak Ground Acceleration (PGA), which measures how hard the ground shakes in a given geographic area (relative to the acceleration due to gravity). The Mercalli scale is measured by personal reports; the PGA is measured by instruments, but the two generally correlate well.

24 Robert B. Olshansky, "Examples of Successful Seismic Safety Advocacy," in *Promoting Seismic Safety: Guidance for Advocates,* by Daniel Alesch et al. (Buffalo, N.Y.: Multidisciplinary Center for Earthquake Engineering Research [MCEER], 2005), 43–115; see also D. J. Alesch, L. A. Arendt, and W. J. Petak, *Seismic Safety in California Hospitals: Assessing an Attempt to Accelerate the Replacement of Seismic Retrofit of Older Hospital Facilities,* MCEER-05-0006 (Buffalo, N.Y.: MCEER, 2005).

25 See the Web site of NOAA's National Geophysical Data Center at ngdc.noaa.gov/nndc/struts/results?eq_1=4&t=101634&s=0&d=3&d=33 (accessed June 5, 2007).

26 T. Lay et al., "The Great Sumatra-Andaman Earthquake of December 26, 2004," *Science,* May 20, 2005, 1127–1133, available at sciencemag.org/cgi/content/abstract/308/5725/1127 (accessed June 5, 2007).

27 A. June Patton, *Surviving the Storm: Sheltering in the May 2003 Tornadoes, Moore, Oklahoma,* Quick Response Report 163 (Boulder: Natural Hazards Research and Applications Center, University of Colorado at Boulder, 2003).

28 See FEMA, *Rebuilding for a More Sustainable Future,* 4-21 to 4-23.

29 Robert B. Olshansky, "Land Use Planning for Seismic Safety: The Los Angeles County Experience, 1971–1994," *Journal of the American Planning Association* 67, no. 2 (2001): 173.

30 FEMA, *Integrating Manmade Hazards into Mitigation Planning,* FEMA 386-7 (Washington, D.C.: FEMA, 2003), available at fema.gov/library/viewRecord.do?id=1915 (accessed June 7, 2007).

31 FEMA, *Developing the Mitigation Plan.*

32 FEMA, *Getting Started.*

33 For an overview of the techniques and challenges of involving citizens in hazard mitigation planning, see Samuel D. Brody, David R. Godschalk, and Raymond J. Burby, "Mandating Citizen Participation in Plan Making: Six Strategic Planning Choices," *Journal of the American Planning Association* 69 (Summer 2003): 245–264; and David R. Godschalk, Samuel D. Brody, and Raymond J. Burby, "Public Participation in Natural Hazard Mitigation Policy Formation: Challenges for Comprehensive Planning," *Journal of Environmental Planning and Management* 46 (September 2003): 733–754.

34 For examples of the types of information citizens need, see Olshansky, "Examples of Successful Seismic Safety Advocacy."

35 FEMA, *Understanding Your Risks;* see also Roxanna McDonald, *Introduction to Natural and Man-Made Disasters and Their Effects on Buildings* (Oxford, UK: Architectural Press, 2003), and the FEMA HAZUS-MH Web site, fema.gov/plan/prevent/hazus (accessed June 7, 2007).

36 FEMA, *Integrating Manmade Hazards into Mitigation Planning.*

37 A useful site for identifying likely local hazards is FEMA Multi-Hazard Mapping Initiative (MMI) Hazard Maps (see note 20).

38 For a buildout assessment methodology, see David R. Godschalk, "Buildout Analysis: A Valuable Planning and Hazard Mitigation Tool," *Zoning Practice* 23 (March 2006): 2–7.

39 For information about the management of the Charlotte-Mecklenburg floodplain in North Carolina, see North Carolina Division of Emergency Management, "Integrating Water Quality into Floodplain Management—Charlotte–Mecklenburg County's Approach," available at dem.dcc.state.nc.us/mitigation/case_Mecklenburg1.htm (accessed June 4, 2007); see also the case of Charlotte–Mecklenburg County, North Carolina, as described in Association of State Floodplain Managers, *No Adverse Impact Floodplain Management: Community Case Studies 2004,* available at floods.org/PDF/NAI_Case_Studies.pdf (accessed June 4, 2007); and North Carolina Department of Crime Control and Public Safety, "Mecklenburg County Flood Audits for At-Risk Commercial Structures."

40 FEMA, *Developing the Mitigation Plan.*

41 Ibid.

42 FEMA, *Bringing the Plan to Life.*

43 Ann Patton, *From Harm's Way: Flood Hazard Mitigation in Tulsa, Oklahoma* (Tulsa: City of Tulsa Public Works Department, 1993); and D. R. Conrad, B. McNitt, and M. Stout, *Higher Ground: A Report on Voluntary Property Buyouts in the Nation's Floodplains* (Washington, D.C.: National Wildlife Federation, 1998).

44 David R. Godschalk, "Urban Hazard Mitigation: Creating Resilient Cities," *Natural Hazards Review* 4, (August 2003): 136–143.

45 Arrietta Chakos, Paula Schulz, and L. Thomas Tobin, "Making It Work in Berkeley: Investing in Community Sustainability," *Natural Hazards Review* 3 (May 2002): 55–67.

46 Godschalk, "Urban Hazard Mitigation."

CHAPTER 7

Planning and preparedness

Michael K. Lindell and Ronald W. Perry

This chapter provides an understanding of

- The guiding principles of emergency planning
- The development of an emergency operations plan
- Community preparedness analyses
- The development of emergency response resources
- A model that shows how to make the planning process happen.

During the terrorist attacks of September 11, 2001, and the flooding in New Orleans from Hurricane Katrina in 2005, responders had to deal with confusing and conflicting cues about what was happening and what to do next. The complexity of the situation, the time pressure, and the severity of the potential consequences created conditions that were unforgiving of error and thus highly stressful. To prepare for such situations—to increase organizational effectiveness when there is enough time to respond but not enough time to improvise a coordinated response—communities must engage in emergency preparedness. That is, they must engage in pre-impact activities that establish a state of readiness to respond to extreme events.

A major component of emergency preparedness is the development of emergency operations plans (EOPs), which guide emergency responders when prompt and effective response actions are needed. The foundation for any EOP is the four basic functions that emergency responders must perform in responding to disaster demands: emergency assessment, hazard operations, population protection, and incident management.[1] These four functions serve as a relatively simple key to what emergency response organizations must do regardless of the organizational structure they adopt—whether it be the Incident Command System (ICS), the Incident Management System, or the National Incident Management System's version of the ICS (see below).

The purpose of *emergency assessment* is to define the nature and magnitude of the event by evaluating conditions in the physical environment. *Hazard operations* consist of preventive and corrective actions that limit either the likelihood or the magnitude of hazard impacts, and that do so by either controlling the source (e.g., plugging leaks in hazardous materials [hazmat] containers) or mitigating the effect of the hazard (e.g., placing sandbags around structures threatened by floods). *Population protection* includes both the use of protective gear by responders and actions (such as evacuations) to protect community residents. *Incident management* ensures that emergency assessment, hazard operations, and population protection are undertaken in a timely and effective manner and that responders have sufficient resources—including support staff, equipment, and facilities—to do their jobs.

EOPs generally assign emergency response to government agencies. However, some components of the population protection function—specifically, mass care and evacuation support—are assigned to other organizations. Mass care is usually assigned to the local American Red Cross chapter or the Salvation Army, and evacuation support is often assigned to the local transit authority or school district.

Writing an EOP is an important step in establishing emergency preparedness, but it is only the beginning. To ensure that there will be an adequate emergency response, emergency planners must also develop the necessary resources. These resources include

- Equipment for hazard operations (ranging from tank-patching equipment to bulldozers and sandbags), personnel protection (ranging from respirators to traffic barricades), environmental monitoring (wind gauges, toxic gas monitoring devices), data processing (calculators and computers), and communications (telephone, fax, and radio).
- Facilities such as emergency operations centers (EOCs) for emergency managers, assembly/staging areas for emergency responders, and "safe havens" in which emergency personnel or community residents can shelter. (These safe havens, which are used during emergency response for protection from extreme environmental conditions ranging from tornadoes to hazmat releases, are different from the public shelters—known more technically as mass care centers—which are used to house evacuees. Safe havens are needed only for periods lasting from a few minutes to a few hours, whereas public shelters must provide for eating, sleeping, personal hygiene, and medical care for periods lasting from several days to several weeks.)
- Staffing to include sufficient numbers of personnel in each job title, and their organization into response teams.

For the most part, EOPs draw emergency responders from organizations (such as police, fire, and emergency medical services) that perform their normal day-to-day duties during disasters

Preparation of this chapter was supported in part by the National Science Foundation under Grants CMS 0219155 and SES 0527699. None of the conclusions expressed here necessarily reflects views other than those of the authors.

and catastrophes. But in addition, disasters and catastrophes often require normal emergency responders to perform unfamiliar tasks; and personnel from other agencies (e.g., public works) and organizations (e.g., transit districts and schools) are required to participate in the emergency response as well. Thus, to develop emergency response resources, certain activities are needed:

- Developing standard operating procedures (SOPs) and other job aids to guide emergency responders in performing tasks that are critical, infrequent, and hard to perform
- Conducting training at the individual and team levels
- Conducting periodic drills and exercises to verify that emergency responders can perform their assigned tasks and that equipment and facilities function properly
- Conducting oral and written critiques after drills, exercises, and actual incidents to identify any changes that need to be made to plans, staffing, training, procedures, equipment, and facilities in order to improve emergency preparedness
- Auditing the emergency preparedness program to ensure that the necessary emergency response resources have been acquired and are being maintained
- Conducting risk communication programs to promote emergency preparedness by households, businesses, and government agencies (e.g., tax assessors) that lack emergency missions.

This chapter examines the emergency preparedness process, emphasizing the importance of plans but also the fact that plans are only a part of preparedness. The chapter's five major sections discuss the guiding principles of emergency planning, the development of an emergency plan, the basic analyses that underlie the plan and are necessary for adapting it to local conditions, the acquisition and maintenance of emergency response resources, and a model for strengthening local emergency preparedness (i.e., for understanding the conditions in the community that promote effective performance by local emergency management agencies and formal networks, or committees).

The guiding principles of emergency planning

As noted above, planning is an important avenue to community emergency preparedness. The practice of emergency response planning is best thought of as a *process*—a continuing sequence of analyses, plan development, and the acquisition by individuals and teams of performance skills achieved through training, drills, exercises, and critiques. The process varies considerably among communities. In some communities, planning is formalized by a specific assignment of responsibility to an office having an identifiable budget. In other communities, planning is informal: responsibility is poorly defined, and a limited budget is dispersed among many agencies. Similarly, response plans and procedures may be mostly written or mostly unwritten. Such variability exists despite federal and state requirements for community emergency planning because local governments vary in their capacity (especially funding) and their commitment to emergency management. Thus, for many years, higher levels of government described their standards for emergency preparedness as "guidance."

Over the years, researchers have identified eight fundamental principles of community emergency planning that can be used to increase a community's level of preparedness regardless of the amount of funding available:[2]

1. Anticipate both active and passive resistance to the planning process, and develop strategies to manage these obstacles.
2. Address all hazards to which the community is exposed.
3. Include all response organizations, seeking their participation, commitment, and clearly defined agreement.
4. Base pre-impact planning on accurate assumptions about the threat, about typical human behavior in disasters, and about likely support from external sources such as state and federal agencies.
5. Identify the types of emergency response actions that are most likely to be appropriate, but encourage improvisation based on continuing emergency assessment.

6. Address the linkage of emergency response to disaster recovery.
7. Provide for training and evaluation of the emergency response organization at all levels—individual, team, department, and community.
8. Recognize that emergency planning is a continuing process.

Manage resistance to the planning process

Emergency planning is conducted in the face of apathy on the part of some and resistance on the part of others. People are apathetic because they don't like to think about their vulnerability to disasters. Alternatively, people resist disaster planning because it consumes resources that could be allocated to more immediate community needs—police patrols, road repairs, and the like. Thus, disaster planning requires strong support from one of the following: the jurisdiction's chief administrative officer, an issue champion (also known as a "policy entrepreneur") who has the expertise and organizational legitimacy to promote emergency management, or a disaster planning committee that can mobilize a constituency in support of emergency management.

Adopt an all-hazards approach

The plans for each hazard agent (e.g., flood, tornado, hazmat release) should be integrated into a comprehensive plan for multihazard emergency management. Emergency planners should conduct a community hazard/vulnerability analysis (discussed in a later section) to identify the types of environmental extremes (e.g., floods, tornadoes, hurricanes, earthquakes), technological accidents (e.g., toxic chemical releases, nuclear power plant accidents), and deliberate incidents (e.g., sabotage or terrorist attack involving chemical, biological, radiological/nuclear, or explosive/flammable materials) to which the community has exposure. After identifying these hazards, emergency planners should examine the extent to which different hazard agents make

Resistance to emergency management

In *Disaster Response: Principles of Preparedness and Coordination*, Erik Auf der Heide explains a number of the reasons that emergency management receives limited attention from policy makers; he also explains why emergency management efforts are actively resisted by some individuals and entities.[1] According to Auf der Heide, most people, including local officials, are unaware of the hazards that threaten their communities, and of the potential impact of disasters on their physical, social, and economic well-being. Moreover, disasters are low-probability events–that is, they occur much less frequently than, for example, petty crime, and are therefore less salient within the public imagination. Because disasters may or may not occur, emergency managers may sometimes find it difficult to substantiate the benefits of mitigation and preparedness. This may be particularly true for smaller local jurisdictions, which are less likely to experience a disaster than larger ones.

To the extent that citizens and officials *are* aware of risk, many are likely to underestimate it and assume that disasters will not affect them. Other citizens are fatalistic: they recognize risk, but believe that disasters are acts of God or nature and that they are powerless to do anything about them. Americans' historic attachment to private property rights and resistance to government intervention lead some people to believe that government should not determine whether they can build on certain lands–such as scenic but dangerous locations on mountains, near rivers, or along the coast. Corporations, for their part, may have an interest in denying that their business practices may make others vulnerable to disaster. Developers, landowners, builders, and real estate agents may resist the expense and additional regulations that accompany disaster mitigation.

Community strategies for dealing with disaster tend to focus on technological solutions, such as dams or communications equipment, which can create a false sense of security. Communities may also overestimate their ability to deal with emergencies, or may assume that state and federal agencies will pay for all expenses. Finally, local governments are chronically short on funds. When there are so many competing priorities, attracting the necessary political attention–and the accompanying funding–is often an uphill battle.

[1]Erik Auf der Heide, *Disaster Response: Principles of Preparation and Coordination* (St. Louis, Mo.: C. V. Mosby, 1989), available at orgmail2.coe-dmha.org/dr/Images/Main.swf.

similar demands on the emergency response organization; if two hazard agents have similar characteristics, they probably will require similar emergency response functions. Commonality of emergency response functions provides multiple-use opportunities for personnel, procedures, facilities, and equipment. In turn, multiple use simplifies the EOP by reducing the number of functional annexes (discussed further on); it also simplifies training and enhances performance reliability during emergencies. Only when hazard agents have very different characteristics and therefore require distinctly different responses will hazard-specific appendixes (also discussed further on) be required for any particular functional annex.

Promote multiorganizational participation

To be effective, emergency planning should promote interorganizational coordination. Mechanisms should be developed to elicit participation, commitment, and clearly defined agreement from all response organizations (see Chapter 5). These organizations would obviously include public safety agencies such as emergency management, fire, police, and emergency medical services. However, they should also include potential hazard sources, such as hazmat facilities and transporters (pipeline, rail, truck, and barge), and organizations that must protect sensitive populations, such as schools, hospitals, and nursing homes. The reason coordination is required is that emergency response organizations of differing capabilities must nonetheless work in concert to perform the four major functions of responders (emergency assessment, hazard operations, population protection, and incident management).

Rely on accurate assumptions

Emergency planning should be based on accurate knowledge of the threat, of likely human responses, and of likely aid from external sources. Accurate knowledge of the threat comes from thorough hazard/vulnerability analyses. Accordingly, emergency managers must identify hazards to which their communities are vulnerable, determine which geographical areas are exposed to those hazards (e.g., 100-year floodplains and toxic chemical facility vulnerable zones), and identify the facilities and population segments located in those risk areas. Part of knowing the threat means understanding the basic characteristics of these hazards, such as speed of onset, scope and duration of impact, and potential for producing casualties and property damage.

Planners and public officials also need accurate knowledge about likely human behavior in a disaster. Contrary to widespread belief (and common depictions in the media), people do *not* flee in panic, wander aimlessly in shock, or comply docilely with the recommendations of authorities. Instead, disaster victims typically act rationally (in terms of the limited information they have about the situation). Following impact, they are the first to search for survivors, care for the injured, and assist others in protecting property from further damage. When they seek assistance, victims are more likely to contact informal sources such as friends, relatives, and local groups than governmental agencies or even such quasi-official sources as the Red Cross. Moreover, looting in evacuated areas is extremely rare, and crime rates tend to decline following disaster impact. Finally, concerned citizens believe they can best help the victims by entering the impact area to donate blood, food, and clothing, even though doing so creates major problems of convergence.

Emergency planning should be based on accurate knowledge of the threat, of likely human responses, and of likely aid from external sources.

Disaster myths are not inconsequential, for when emergency managers believe them, emergency planning becomes less effective as resources are misallocated and information is ineffectively disseminated. An example of how belief in a myth can do harm is when officials cite expectations of panic as a justification either for giving the public incomplete information about an environmental threat or for withholding information altogether; in fact, this response to the expectation of panic is actually counterproductive: when warning messages are vague or incomplete, people are more reluctant to comply with the suggested emergency measures. Thus, the misconception that accurate information will cause panic can lead officials to take actions that frustrate their own attempts to protect the public. In other words, the planning process must be firmly grounded on (1) the physical or biological science literature about the effects of hazard agents on human safety, health, and property, and (2) the behavioral literature describing the response patterns of affected populations and emergency organizations.

Finally, emergency managers must be sure to transmit to other government agencies, businesses, and households accurate information about the amount of aid they should expect to find available from external sources. In major disasters, hospitals might be damaged or overloaded; telecommunication and transportation systems (highways, railroads, airports, and seaports) could be damaged so badly that outside assistance would be prevented from arriving for days; and water, sewer, electric power, and natural gas pipeline systems could be disrupted to the point that restoration might take much longer than a few days. Consequently, all social units must be prepared to be self-reliant for at least three to five days.

Identify appropriate actions while encouraging improvisation

Sometimes the response that is usually the most appropriate one might not in fact be most suitable, given the circumstances that arise in a specific event. Thus, emergency responders should be trained to implement the most likely responses to disaster demands, but they should also be encouraged to improvise on the basis of a continuing emergency assessment that identifies the appropriate response actions to the particular disaster well before those actions need to be implemented. In the highly charged atmosphere of imminent disaster, it is hard for an emergency manager to appear to be "doing nothing." However, it is important to recognize that the best action might be to mobilize emergency personnel and *actively* monitor the situation for further information rather than initiate unnecessary hazard operations, population protection, or incident management actions. Thus, planning and training should focus on *principles of response* rather than trying to define overly specific procedures that contain a multitude of details.

Link emergency response to disaster recovery

It is increasingly recognized that there is no clear line between emergency response and disaster recovery.[3] At any point after impact, some portions of the community will be engaged in emergency response tasks whereas others will have moved on to disaster recovery. Moreover, senior elected and appointed officials are likely to be inundated with policy decisions that need to be made to implement the emergency response at the very time they must plan for the disaster recovery. Consequently, pre-impact emergency response planning should be linked to pre-impact disaster recovery planning. Coordination between the two plans will speed the process of disaster recovery by ensuring that the priorities for disaster recovery have been clearly established so that recovery actions can be initiated while the emergency response is still under way.

Provide for training and evaluation

Emergency preparedness also has a training and evaluation component. The first part of the training process involves explaining the provisions of the plan to the administrators and personnel of the departments that will be involved in the emergency response. Second, all those who have emergency response roles must be trained to perform their duties. Of course, this includes fire, police, and emergency medical services personnel, but there also should be training for personnel in hospitals, schools, nursing homes, and other facilities that might need to take protective action. Finally, the populations at risk must be involved in the planning process so that they can become aware that planning for community threats is under way and be knowledgeable about what is expected of them under those plans. These populations need to know what is likely to happen in a disaster and what emergency organizations can *and cannot* do for them. (For a discussion about involving the most vulnerable populations in disaster planning, see Chapter 13.)

It is also essential that training be followed by evaluation in the form of tests and exercises to determine whether it has been effective. Emergency drills and exercises provide a setting in which the adequacy of the EOP, SOPs, staffing, facilities, and equipment can all be tested as well. Further, multifunctional exercises (exercises that test a jurisdiction's ability to perform all four emergency response functions—emergency assessment, hazard operations, population protection, and incident management) facilitate interorganizational contact, allowing members of different organizations to better understand each other's professional capabilities and personal characteristics. And multifunctional exercises also produce publicity for the broader emergency management process, which informs community leaders and the public that disaster planning is under way and preparedness is being enhanced.

Adopt a process of continuous planning and auditing

Preparedness is a continuing process because conditions within the community change over time, conditions outside the community can change as well, and the products of planning itself change.

Conditions inside the community include hazard vulnerability, organizational staffing and structure, and emergency facilities and equipment. Conditions outside the community include federal regulations (witness the requirements for communities to adopt the National Incident Management System [NIMS]). The preparedness process results in some products that are tangible (and have already been mentioned, such as equipment and facilities), and others that are intangible—hard to document on paper and not realized in hardware. An example of such intangible products is the development of emergency responders' knowledge about disaster demands, about their own emergency response roles, and about other agencies' capabilities.

The potential for change in all three areas dictates that the emergency planning process detect and respond to these three kinds of changes and that all elements of emergency preparedness be audited periodically—and at least annually.

Development of the emergency operations plan

As noted above, the EOP provides emergency responders with the guidance they need to take prompt and effective response action. To be useful in an emergency, the EOP must be realistic and must also have the commitment of the participating organizations. To be realistic, the EOP should be consistent with available information about the community's hazards, vulnerable populations, and organizational response capabilities. (The various analyses that present this information are addressed further on under "Community Preparedness Analyses.") In fact, many of the supporting analyses can be conducted *concurrently with* the development of the EOP. As the supporting analyses are completed, their results will form the basis for revisions to the EOP. Moreover, work on the EOP often identifies areas in which more refined supporting analyses need to be conducted. Thus, the relationship between the EOP and supporting analyses is one of continuous iteration.

The way to obtain the commitment of participating organizations is to have the EOP development process include all the organizations that will participate in the plan's implementation. In many communities, the organization responsible for a particular function (or functional component) writes the corresponding section of the plan.

This section begins with an overview of the three components of the EOP. It then discusses the three components in turn: the basic plan itself, the plan's functional annexes, and the hazard-specific appendixes to the annexes. The section concludes with a discussion of the linkage between the EOP and the disaster recovery plan.

Components of the emergency operations plan

There are many views on what elements make up an EOP. One framework, developed by the Federal Emergency Management Agency (FEMA), can be found in FEMA's *Guide for All-Hazard Emergency Operations Planning*, SLG-101,[4] but it is not mandatory, so local jurisdictions can use other frameworks in their EOPs if they choose. For example, a jurisdiction could organize its EOP along the lines of the ICS or the National Response Plan's emergency support functions.[5]

Although historically EOP organization has been left to the discretion of local jurisdictions, the federal government has been exerting increasing pressure for standardization. Requiring jurisdictions to adopt NIMS is a step in this direction, but achieving full standardization will be difficult because each jurisdiction has its own distinctive combination of hazards, vulnerable populations, resources, structures, and approaches to managing people and the environment. Consequently, the local emergency planning process must identify ways in which responses to disasters will have to accommodate the jurisdiction's individual character. Achieving standardization under these circumstances is likely to take a very long time.

The EOP has three components: the basic plan itself, functional annexes to the basic plan, and hazard-specific appendixes that may accompany annexes. The basic plan addresses issues that are common to the entire emergency response organization. Each functional annex addresses issues that are specific to a single emergency response function. Finally, hazard-specific appendixes address tasks that are performed only in response to certain types of disasters or are performed in a different way during some types of disasters. Extensive background material about hazard agents should be referenced in the EOP but located in training materials. Similarly, detailed instructions for performing specific tasks should be referenced in the EOP but located in SOPs appended to the plans of the agencies that will implement them.

The basic plan

The basic plan contains provisions identifying the legal and technical bases for the EOP and provisions describing the administrative aspects of the plan. These various provisions are often captured under eight headings: (1) authority for emergency planning; (2) EOP aim and scope; (3) statement of purpose; (4) situation and assumptions underlying the EOP; (5) overall concept of operations; (6) documentation of agreements; (7) provisions for emergency response training, drills, exercises, and critiques; and (8) procedures for administering the EOP. In addition, ideally the basic plan should address the linkage between the EOP and the pre-impact disaster recovery plan (if the recovery plan is not one of the annexes).

The authority for the plan is derived from a variety of sources, including laws, general police powers, and special statutes. The EOP's section on aim and scope establishes the conditions under which the plan is activated and deactivated and the person (or persons) who is (are) authorized to initiate this process. The statement of purpose provides a concise identification of the plan's major goals and the jurisdictional unit that is the focal organization (e.g., a municipality or county emergency management agency), and it lists any obligations under the plan for jurisdictions other than the focal organization. The situation and assumptions section summarizes community threat assessments that are produced in the hazard/vulnerability analysis (described at length in the section "Community Preparedness Analyses"); this information should include disaster incidence, average severity of impact, probability of occurrence, and the presence of hazardous materials (and their quantities) that are located at fixed sites within the community or on road, rail, water, or air transportation routes. It is especially important to identify geographical areas and special facilities within the community that are most vulnerable to these hazards.

The concept of operations section should identify the emergency response functions that must be performed and the government agencies or other community organizations that are responsible for performing each of them. It should also describe how hazard operations at the incident scene will be linked to the emergency operations center (EOC). Documentation of agreements should ensure that potential interorganizational ambiguities have been resolved to the satisfaction of all parties likely to be involved. Interorganizational problems to be addressed include the exchange of aid or support—during the period of emergency operations—among neighboring local governments, private sector parties, and state and federal government agencies. The documentation of agreements should also recount all reciprocal agreements, such as memorandums of understanding and mutual aid pacts, and should list the following items:

- Conditions under which the agreement is activated, accompanied by the organizational titles of, and instructions for contacting, the persons who have the authority to activate the assisting organization
- The specific nature of the assistance to be rendered

- The resources (both equipment and personnel) available for loan and their locations
- The way in which personnel from the assisting organization will be deployed and the person to whom they will report
- The party that will pay for the responding personnel and equipment and that bears liability for injury or damage incurred during the emergency response.

The EOP's section addressing emergency response training, drills, exercises, and critiques should briefly describe the organization of the training program, especially the agencies responsible for different types of training. Frequently, training on the basic plan is conducted by the emergency manager, whereas training on each functional annex or hazard-specific appendix is performed by the department having primary responsibility for that annex and its accompanying appendixes. This section of the basic plan should also address the methods for assessing training needs, sources of training content, and frequency of refresher training. However, it is very important to avoid including a detailed discussion of training content or actual training material because such voluminous materials will make the EOP too long and hard to use.

In addition, this section should identify procedures for verifying the competence of individuals and teams through drills and functional exercises as well as through tests of the performance of the overall emergency response organization—tests that are graded (i.e., externally evaluated) full-scale exercises. This section should indicate how often exercises will be conducted, how scenarios will be developed, how evaluators will be selected, and how oral or written feedback about performance will be obtained. Evaluations should also be conducted after actual incidents in which the emergency response organization is activated. Finally, the training provisions should describe procedures for ensuring that lessons learned from exercises and incidents are reflected in updates to the EOP.

The section describing the administration of the plan typically specifies four procedural matters:

- The process by which suggestions for revisions will be solicited and incorporated into the EOP. The change process should allow for requests by any organizations from which support is to be received or to which support will be provided. The change process should also define the frequency with which subsequent EOP revisions will be made. The EOP and SOPs should be reviewed at least annually, whereas the personnel and equipment lists in each annex (which identify the resources that will be used to implement that annex) should be reviewed quarterly.
- A record of plan distribution, which is needed to ensure that all participating agencies have current copies of the plan.
- A record of all previous plan amendments, with the dates they became effective.
- The mechanisms for coordinating pre-impact emergency response planning and pre-impact disaster recovery planning (e.g., establishment of organizational contacts and perhaps overlapping membership between the committees responsible for these two activities).

Functional annexes

Functional annexes to the basic plan should address the principal functions that must be performed in responding to disaster demands. FEMA's SLG-101 lists eight functions that should be addressed in the EOPs of local jurisdictions. Four of them (warning, evacuation, mass care, and health and medical) are components of the population protection function, and the other four (direction and control, communications, emergency public information, and resource management) are components of the incident management function. Surprisingly, none of the components of the emergency assessment or hazard operations functions is represented on this list.

In practice, the number and nature of the annexes must reflect a compromise among three considerations:

- Minimizing the number of annexes to simplify the structure of the EOP (the fewer the annexes, the clearer the overall structure of the EOP)
- Ensuring that the interrelationships among tasks are clearly identified, and that the allocation of resources and the performance of tasks are appropriately coordinated

(which may require the number of annexes to be increased or decreased, depending on circumstances)

- Assigning to each agency with emergency response duties responsibility for writing an annex that defines its own responsibilities in the emergency response (which makes it easier for agencies to develop and maintain the annexes but may increase the total number of annexes).

These considerations will not produce the same annex titles, or even the same number of annexes, in all jurisdictions. In any case, the number of annexes that clarify task interrelationships is usually smaller than the number of responding agencies but is usually larger than the number that best clarifies the basic structure of the emergency response organization.

Whatever list of functions is selected, it must be downwardly compatible with the organization of fire departments operating under the Incident Command System (ICS) or Incident Management System (IMS), upwardly compatible with state and federal emergency response organizations, and responsive to incident demands for the timely and effective performance of specific emergency response functions.

For an example of problems in reconciling these potentially competing objectives, an emergency response organization operating under NIMS ICS typically consists of the command section and the four sections reporting to it: planning, operations, logistics, and finance/administration.[6] The sections are staffed as appropriate to the incident size and conditions. Section chiefs in the incident command post located at the incident scene work with the command staff to formulate an overall emergency response strategy. The section chiefs then direct and monitor tactical operations that are implemented by specialized branches within each of the four sections that contain even more specialized sectors. This structure easily accommodates three of the four basic functions that emergency responders must perform—emergency assessment, hazard operations, and incident management—but pays less attention to the function of population protection. Consequently, emergency responders sometimes fail to respond in a timely and effective manner to the need for population protective actions such as evacuation and in-place protection in potential risk areas. Yet the IMS structure is logical because it was originally designed to provide a rational organizational structure for operations at an incident scene. Consequently, most population protection activities—which take place away from the incident scene—are usually coordinated through the jurisdictional EOC that is located in one of the principal administrative buildings. When population protection is addressed within ICS/IMS, it is usually buried deep within the operations section—and is therefore easy to overlook during an emergency response.

Hazard-specific appendixes

Hazard-specific appendixes provide information about the ways in which the response to a particular hazard agent is particular to the agent. For example, evacuations from floods would be directed away from the river and toward the hills. By contrast, evacuations from landslides or wildfires might take place in the opposite direction. The appropriate response is a function of the hazard, and these differences in response should be defined in the hazard-specific appendixes.

Linkage to the recovery operations plan

As noted above, emergency response and disaster recovery often overlap because some sectors of the community are in emergency response mode while others are moving into disaster recovery, and some organizations might be carrying on both types of activity at the same time. Disaster recovery, which has both physical and social aspects, entails (among other things) taking actions to provide temporary accommodations for displaced households, businesses, and government agencies; facilitating the repair and reconstruction of damaged property, public and private; and seeing to the restoration of disrupted community social routines and economic activities.

Because many tasks need to be accomplished very quickly and virtually simultaneously after a disaster, pre-impact planning for disaster recovery is as critical as planning for disaster response. And coordinating the two types of planning has three advantages: the allocation of resources is more effective and efficient, the delays that can occur while decisions are being made about recovery priorities may be avoided, and the likelihood of conflicts over scarce resources is lower. However, such coordinated planning involves some significant challenges because the agencies that are most often involved with the development of the EOP (e.g., police, fire, emergency medical services) and those that need to be involved in the development of the disaster recovery plan (e.g., land use planning, economic development, public works) have significantly different organizations and organizational cultures. Specifically, public safety personnel operate under paramilitary command structures that emphasize immediate response to dangerous situations, whereas planning and development personnel typically are involved in community development decisions having protracted deliberations with a wide variety of community stakeholders. Thus, in most jurisdictions it will take a determined effort to achieve the needed coordination.

Community preparedness analyses

As Figure 7–1 indicates, four analyses should be used to guide the development of the EOP: hazard/vulnerability analysis, hazard operations analysis, population protection analysis, and incident management analysis. All four of these analyses consider the demands of the different types of disasters that could strike the community and the capabilities that the community can call upon in responding. As noted above, these analyses might not be official documents produced before the EOP development process begins. Indeed, as the figure indicates, the analysis process is usually iterative. An EOP is often first developed on the basis of rather broad conceptions of the disaster demands and community capabilities, but the EOP development process might identify the need for more specific analyses. The process can continue to cycle as further planning or an actual emergency makes clear the need for further analysis.

Hazard/vulnerability analysis

The hazard/vulnerability analysis should address three major components of vulnerability: hazard exposure, physical vulnerability, and social vulnerability. Communities can obtain data for most natural hazards by accessing Web sites maintained by federal agencies such as FEMA, the U.S. Geological Survey, and the National Weather Service. Data for man-made hazards should be obtained from local industry (for fixed-site hazards) and pipeline operators, railroads, or truck carriers (for transportation hazards). In addition, some organizations

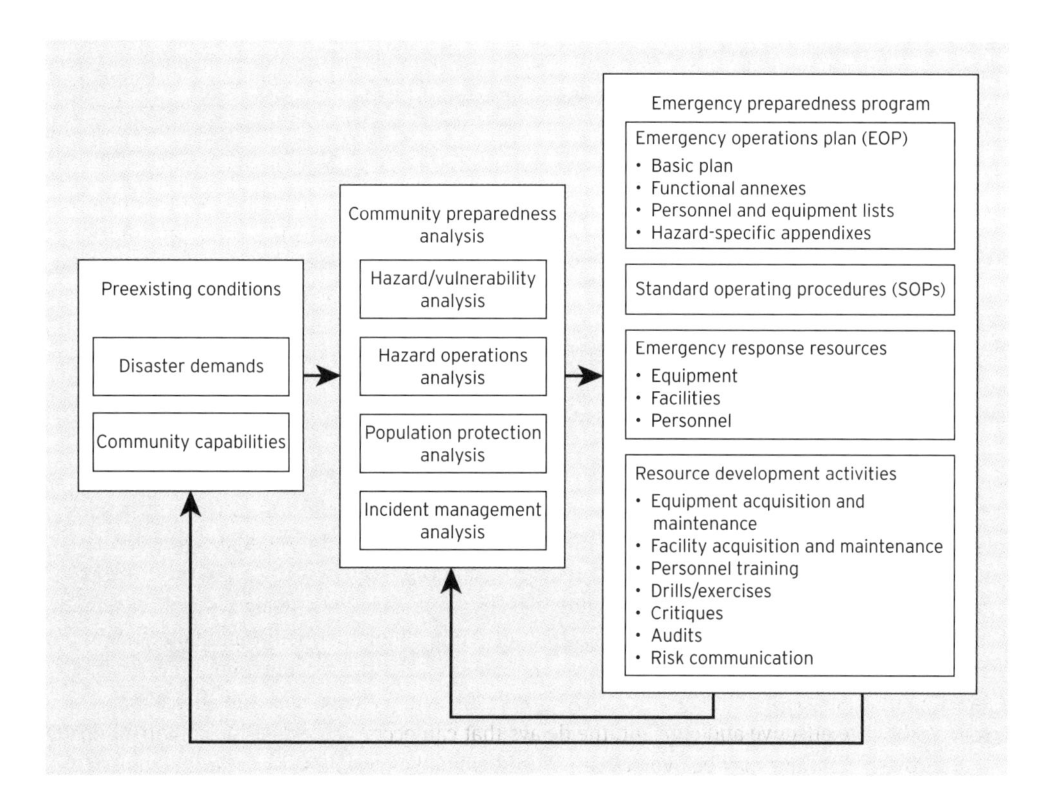

Figure 7-1 Elements of the community emergency preparedness process

provide computer software, planning guidance manuals, and training courses that explain how to assess community vulnerability.[7]

Hazard exposure Hazard exposure arises from people's occupancy of geographic areas where they could be affected by specific types of events that would threaten their lives or property. In principle, hazard exposure can be measured by the probability that an event of a given magnitude will occur, but the probability of events exceeding that magnitude can be hard to estimate reliably when the available historical data are too sparse. For example, many areas of the United States have limited meteorological and hydrological data, so estimation of 100-year floods requires extrapolation from a limited data series. Moreover, watershed urbanization causes the boundaries of floodplains to change in ways that might be hard for local emergency managers to assess. In recognition of these limitations, the hazard analysis should make use of the best data available from federal and state agencies to map the specific locations that would be affected by different intensities of impact (e.g., 50- and 100-year floodplains, areas prone to liquefaction from earthquakes, and areas within toxic chemical vulnerable zones). Local emergency planners can work with land use planners and other agencies, such as metropolitan planning organizations (MPOs) or councils of governments (COGs), to store digital maps of these risk areas in geographic information systems so that they can be manipulated, displayed, and used in the development of EOPs.

Physical vulnerability Physical vulnerability has three components: human, agricultural, and structural. Physical vulnerability analysis should be used to identify the hazard operations actions (e.g., sandbagging for floods, patching and plugging for hazmat releases) and population protection actions (e.g., evacuation, sheltering in place) that might be needed in an emergency. Identification of the likely hazard operations actions and population protection actions will, in turn, indicate how the corresponding EOP annexes and SOPs should be written and what resources will be needed to implement them.

Human beings are vulnerable to environmental extremes of temperature and pressure and to chemical exposures that can cause illness, injury, or death. Thus, the human vulnerability analysis should list critical levels of exposure to those community hazards for which such data are available (e.g., heat and cold index tables). For any hazard agent (e.g., water, wind, ionizing radiation, toxic chemicals, infectious agents), the physiological response of the affected population is often characterized by variability: that is, given the same level of exposure, some people might die, others would be severely injured, still others slightly injured, and the rest unharmed. Typically, the demographic segments of the population that are most susceptible to any environmental stressor will be the very young and the very old, as well as those with weakened immune systems.

Agricultural vulnerability refers to the susceptibility of animals and crops to the community's environmental hazards. As is the case with people, animals and crops vary in their susceptibility to environmental hazards as a function of the stage in their life cycle. For example, many annual crops such as cotton are most vulnerable at harvest time. Other analyses might identify distinctive patterns of hazard agents (e.g., nuclear power plant accident), exposure pathways (e.g., airborne release), and receptors (e.g., milk cows) that are likely to produce adverse effects (e.g., strontium in milk) that would require protective action (e.g., placing cows on stored feed). Of course, an analysis of agricultural vulnerability is especially important in a community whose economic base relies heavily on the agricultural sector.

Structural vulnerability arises when buildings are constructed using designs and materials that are incapable of resisting extreme stresses (e.g., pressure from high wind, water, or seismic shaking) or that allow hazardous materials to infiltrate the building. The construction of most buildings is governed by codes intended to protect the life safety of building occupants from structural collapse, but these codes do not necessarily prevent severe damage and loss of building function from extreme wind, seismic, or hydraulic (water pressure) loads. Nor do they require an impermeable barrier to the infiltration of toxic air pollutants. Consequently, structural vulnerability analysis must assess the community's structures (residential, commercial, and industrial buildings) and lifeline and infrastructure systems (fuel, electric power, water, sewer, telecommunications, and transportation) in terms of their ability to withstand the environmental forces predicted by hazard analyses.

Social vulnerability The third component of a hazard/vulnerability analysis, after hazard exposure and physical vulnerability, is social vulnerability, defined as "characteristics of a person or group and their situation that influence their capacity to anticipate, cope with, resist and recover from the impact of a natural hazard."[8] Whereas human vulnerability refers partly to a person's susceptibility to biological changes (i.e., the possibility of anatomical and physiological responses to disaster impacts), social vulnerability refers to a person's susceptibility to behavioral changes (i.e., the possibility of behavioral changes in response to impacts). These behavioral changes can be classified, in terms of their effects, as

- Psychological (changes in cognitive and affective states that alter patterns of task and social behavior)
- Demographic (victims emigrating out of the impact area to search for housing, and construction workers immigrating to the impact area to search for employment)
- Economic (altered patterns of savings and consumption)
- Political (mobilization by victim groups to seek policy change).

Just as hazard exposure, human vulnerability, and structural vulnerability are neither uniform (there is variability across the population) nor uniformly distributed (rates of vulnerability are higher in some areas than in others), so too is social vulnerability neither uniform nor uniformly distributed—across geographic areas, population segments (i.e., demographic categories such as income, education, age, and ethnicity), or economic sectors such as construction, manufacturing, and retail trade.[9] Social vulnerability varies systematically across communities and, moreover, across households within communities, so emergency planners can use census data to identify neighborhoods having large percentages of residents with demographic characteristics that are strongly correlated with social vulnerability. (Commonly used demographic measures include the percentage of households that are headed by women, have incomes below the poverty line, are dependent on public transit, are elderly, or are ethnic minorities.) Areas that are high on multiple predictors of social vulnerability are known as "vulnerability hot spots" and therefore should receive special attention during the emergency planning process. This is especially important if areas with high social vulnerability coincide with either high hazard exposure (e.g., location in a 100-year floodplain or in the vulnerable zone of a hazmat facility) or high physical vulnerability (e.g., residence in very old, poorly maintained houses) or, as is often the case, both.

Hazard operations analysis

The second analysis that underlies the basic emergency plan is the hazard operations analysis. For some hazard agents, especially technological hazards, it is possible to intervene in the hazard-generating process to prevent a disaster from occurring. The four types of intervention are

- *Hazard source control:* In the case of hazmat releases, for example, taking actions such as patching holes or replacing leaking valves in tank cars
- *Community protection works augmentation:* Adding sandbags to increase the height of levees or cutting fire breaks to isolate wildfires, for example
- *Building construction augmentation:* Strengthening building soft spots, as when storm shutters are installed to protect against high wind
- *Building contents protection:* Moving furniture and appliances to higher floors when flooding is forecast, for example.

The purpose of the hazard operations analysis is to identify which interventions will be needed and what resources must be acquired and maintained so that these interventions can be implemented during an emergency response.

Population protection analysis

The third analysis that underlies the EOP is the population protection analysis. Many major emergencies require local officials to initiate protective actions for the population at risk. First, all of those in the risk area must be warned about the hazard. Second, the feasibility of protective actions must be assessed.

Warning those in the risk area can be easy to do in some situations (in a small area where emergency responders can go door-to-door) but hard to do in others (in large areas at night when people are asleep). There are two types of warning mechanisms: local mechanisms (face-to-face warnings, mobile loudspeakers, sirens, and telephones) and broadcast media (commercial radio, commercial television, and National Weather Service Weather Radio [tone-alert radios]). During incidents with substantial forewarning, newspapers and the Internet can also provide useful hazard information.

These warning mechanisms differ from each other with respect to a number of important characteristics. Variability is found in

- Precision of dissemination (ability to reach all those at risk and only those at risk)
- Penetration of normal activities (e.g., the ability of tone-alert radios to wake people who are sleeping, or of sirens to be heard indoors by people who have air conditioners running and music playing)
- Message specificity (e.g., the ability of a police officer at the door to identify the hazard, describe the personal risk of staying, recommend a protective action, and indicate how to receive additional information or assistance)
- Susceptibility to message distortion (the tendency for errors to creep into messages as they are relayed from one person to another in a communications network)
- Rate of dissemination over time (the number of people warned per hour)
- Sender and receiver requirements (the amount of money, effort, specialized knowledge, and specialized equipment needed to install, maintain, and operate a warning system)
- Feedback (the ability of a warning system to provide verification that those at risk have received the warning).[10]

Communities can select the warning mechanisms that are most appropriate in terms of their own needs (e.g., the size of the population at risk), their own resources (e.g., the amount of funding available for warning systems), and the characteristics of the hazards to which they are exposed. In particular, technologies such as tone-alert radios and autodial telephone systems are desirable when a community faces hazards with a high speed of onset and a large scope of impact (e.g., tsunamis, tornadoes, and toxic chemical and radiological materials releases). If a jurisdiction does not have a warning system that is adequate for the hazards it faces, local emergency planners should request funds to eliminate the gap in warning capability.

In addition to assessing warning systems, emergency planners should assess the feasibility of protective actions. The most common protective action for environmental hazards is evacuation, and it appears to be deceptively simple: just warn everyone in the risk area to leave. Indeed, a rapid evacuation of the risk area is relatively easy to achieve when the area's population density is low, all evacuating households are united at the same physical location (or are able to contact each other to account for the whereabouts of missing members) and have their own vehicles, and the capacity of the evacuation route system is high.[11] However, when the population density is high in relation to the capacity of the evacuation route system, clearing the risk area can take many hours. In fact, estimated times for evacuating many major urban areas along the Atlantic and Gulf Coasts in anticipation of a hurricane exceed thirty hours, and some even exceed forty-eight hours.

In some circumstances, evacuation is not the protective action of choice. When exposure to the hazard conditions while in an evacuating vehicle would be more dangerous than remaining in a substantial structure, sheltering in place is preferable. (However, for many hazards, remaining in a mobile home is more dangerous than leaving.) For some hazards (e.g., tornadoes), sheltering in place is the most common recommendation for protective action, but for other hazards, choosing between evacuation and sheltering in place can be complex.[12] For example, the choice of a protective action for a toxic chemical release should depend on the chemical's toxicity, release rate, and release duration; the wind speed and direction during the entire time of the release; the air infiltration rates in local structures; and the evacuation time estimates for different types of occupancy (e.g., homes, businesses, schools, nursing homes, and hospitals) around the release point.

Incident management analysis

The fourth analysis underlying the basic plan is the incident management analysis. This analysis examines the ability of community organizations to perform the emergency response functions during a disaster. That is, emergency planners must determine whether government agencies, nongovernmental organizations (e.g., American Red Cross), and local businesses have the trained personnel, facilities, and equipment that are needed.

There is relatively little guidance on performing incident management analyses, but the basic procedure is to seek out the best available subject matter experts (SMEs) in the jurisdiction for information about each of the emergency response functions.[13] Taking the functional component "emergency public information" as an example, the best available SMEs in a jurisdiction are likely to be public information officers from public safety or public health agencies who have experience with the activity. Once the SMEs have been identified, they can be asked what specific tasks they perform under different emergency scenarios (e.g., floods, tornadoes, or toxic chemical releases). For each task, the SMEs can then be asked to identify the personnel, facilities, equipment, and materials that are needed to perform the task.

Emergency planners should be cautious when asking SMEs about the performance of tasks under conditions they have not previously experienced. Transportation personnel, for example, might assume that their experience in managing small-scale evacuations will generalize to a large-scale evacuation for a hurricane or nuclear power plant accident, but in fact large-scale evacuations are much more complex than small-scale evacuations. Thus, when the hazard/vulnerability analysis indicates that there are some disaster conditions that could be significantly different from the local SMEs' previous experiences, the emergency manager should seek expertise from outside the jurisdiction.

Emergency response resource development

Figure 7–1 indicates that, in addition to using the four types of community preparedness analyses to develop an EOP and SOPs, emergency managers must develop the emergency response resources needed to ensure an adequate emergency preparedness program. The necessary physical resources can be categorized as equipment, facilities, and personnel. Some of these resources will already be available within the jurisdiction because they have previously been acquired for response to routine emergencies. However, other resources must be acquired and specially maintained because they are needed only for response during major disasters.

Thus, emergency managers must engage in seven emergency response resource development activities. The first three are (1) acquisition and maintenance of disaster response equipment; (2) acquisition and maintenance of disaster response facilities; and (3) training to provide the skills that local personnel have not learned in their routine job duties but will need in emergency response. In addition, emergency managers must conduct (4) drills and exercises to provide personnel with opportunities to perform their emergency response tasks; (5) critiques to identify deficiencies in the emergency preparedness program; (6) audits to verify that the EOP, SOPs, and emergency response resources comply with standards, including NFPA 1600 and the Emergency Management Accreditation Program (EMAP) (both of which are discussed further on under "Audits"); and (7) risk communication programs to motivate households and businesses to increase their resources for emergency response.

Inevitably, developing the resources needed for an adequate level of emergency preparedness will take more time and money than are available in a single year. Therefore, emergency managers need to develop a strategic plan for their emergency management program that systematically directs their efforts, over the course of the years, toward closing the gap between the resources that are needed and the resources that are currently available. Specifically, each emergency manager should devise a multiyear (typically a five-year) strategic plan to reduce the capability shortfall by setting annual goals in each of the major programmatic areas for which he or she is responsible: hazard/vulnerability analysis, hazard mitigation, emergency preparedness, recovery preparedness, and risk communication, among others. Once these goals have been set, the emergency manager should establish specific annual milestones (objective indicators) for determining whether progress is being made at a satisfactory rate.

Equipment

Each agency should identify the equipment it needs to perform its assigned tasks, paying special attention to those tasks that are performed only during emergencies. Special-purpose equipment that is not used routinely will require that personnel be trained and periodically tested in its proper use. In addition, such equipment might need periodic preventive maintenance, battery checks, and recalibration. An emergency manager should maintain a computer database of emergency-relevant equipment that the jurisdiction owns. To provide a capability for rapid search during an emergency, this database should contain fields listing the equipment's name, model, and manufacturer; names and contact numbers for personnel who can authorize the release of the equipment; names and contact numbers for qualified operators; contact numbers for repairs; and the dates for critical activities such as preventive maintenance, battery checks, and recalibration.

Facilities

Two of the most important emergency response facilities are the primary EOC and an alternate EOC (in case the primary one is nonfunctional). These facilities are permanently located in areas expected to be safe from hazard exposure so that they can provide continuous technical assistance to responders at the incident scene. Such assistance includes obtaining weather forecasts, coordinating warning and evacuation, supervising mass care facilities, and disseminating public information. An EOC is also needed to request additional resources and to direct their disposition as they arrive from other local jurisdictions and higher levels of government. Moreover, because many organizations participate in the incident response and each must have a capability for obtaining and processing timely information about the incident, essential personnel with telecommunications and information-processing equipment should be located in an EOC that will provide an effective division of labor while maintaining coordination of action. Close coordination among organizations at all levels is critical because previous events have shown that while considerable decision-making authority should be allocated to organizations close to the incident site (because their knowledge of local conditions is superior), greater technical knowledge and resources generally are available at higher levels.

A jurisdiction's EOC should be sited not only, as indicated above, in an area expected to be safe from hazard exposures but also at a location that is readily accessible to those who are essential to a timely and effective emergency response. Included in this group are individuals who have technical knowledge as well as those with policy-making responsibilities. In the case of a community that is operating under the ICS and is responding to, for example, a transportation incident, an incident commander would establish an incident command post from which to direct on-scene operations while maintaining regular communication with the EOC. In addition, the incident commander would coordinate the activities of the field teams from the shipper or carrier (who are performing preventive, corrective, and hazard/environmental assessment actions) with local government response teams, such as firefighters (who are attempting to terminate the emergency and minimize population exposures). Other types of incidents (e.g., pandemic flu) probably would not have a specific scene.

Physically, an EOC must be designed with enough space to support the emergency response functions that take place within it. Moreover, it must provide a layout that places its staff in proximity to the equipment, information, and materials they need. Previous guidance and practice indicate that emergency planners should establish an EOC design team that performs the following tasks:

1. Analyzes the organization that will operate within the EOC and, especially, the specific roles that will be performed by each person within that organization
2. Assesses the flow of resources associated with each personnel position
3. Determines the workstation requirements for each personnel position
4. Assesses the environmental conditions such as lighting, heating, and quiet needed to support each personnel position
5. Determines the space needs for each personnel position

6. Summarizes the results of the analyses in tasks 1–5 in a design basis document that summarizes the design requirements for the EOC and archives them for reference in case there are later modifications to the operations center
7. Develops a design for the EOC that meets the requirements defined in the design basis document.

Personnel training

Training provides the means by which emergency personnel develop the knowledge, skills, and abilities (KSAs) needed to respond to the expected disaster demands. Many KSAs needed for emergency response differ from those used in routine job performance, so initial training is needed. Moreover, infrequently used KSAs decay over time, so refresher training is also needed. Consequently, the EOP should describe the content and frequency of response training that will be provided for all emergency personnel, including members of emergency response teams who implement the previously approved EOP and SOPs and members of the emergency management team (agency heads and elected officials who approve improvised actions that accommodate the unique demands of a specific emergency).

Classroom instruction should include discussion of each individual's assigned tasks, the individuals assigned to each role on a team and the responsibilities of those individuals, and a review of the procedures that each individual will be responsible for following. To promote flexibility in the organizational response, all members of the emergency management team and the emergency response teams should receive an explanation of the nature of the community hazards and the measures they should take to ensure their own personal protection, as well as an overview of the emergency response plan and the rationale for its various components.[14] By learning the rationale for the EOP and SOPs as well as how the entire emergency response organization has been designed to implement the EOP and SOPs, emergency response personnel will be better able to improvise effectively in response to situational needs.

Drills and exercises

Walk-through drills should be conducted to ensure that procedures and information are current (e.g., notification procedures should be tested to verify that telephone numbers are current) and to assess the capabilities of individuals or small teams to perform their designated emergency duties. They can be used, for example, to assess an emergency responder's ability to operate complex computer and communications systems, to staff and operate traffic control points, and to activate mass care facilities. These drills should focus on tasks that are hard to perform, infrequently practiced (i.e., performed only during emergencies), and critical to protecting life and property.

In tabletop exercises, department heads and their assistants gather informally to discuss solutions to very general emergency situations presented by the local emergency manager. Response actions are described but not actually implemented.

In functional exercises, staff from a single department perform a limited number of specific functions in response to a more detailed scenario that a training committee might construct. By contrast, full-scale exercises involve staff from many departments performing a wide range of emergency response functions. The scenario might include initial classification of the emergency, activation of emergency response organizations, continuing emergency assessment, communication and coordination with other responding organizations, protective response (including protective action decision making and implementation, and the use of medical support), and a declaration of the end of the emergency. The scenario for a full-scale exercise is often very detailed and is usually constructed by a training committee or external consultant. Functional exercises are typically conducted once a year, and full-scale exercises once every four years. Both types of exercises can be announced in advance but should be based on emergency scenarios whose contents are withheld from all responding personnel.

Critiques

The response to a functional or full-scale exercise should be evaluated by independent outside observers who are qualified by virtue of their knowledge of both the emergency plan and the proper use of emergency equipment and procedures. Following each exercise, a debriefing session—an oral critique—should be held to identify any problems that arose and to determine what revisions are required for the plan, procedures, or training. It is extremely important for critiques to be conducted impersonally and dispassionately. Critiques are performance appraisals, and it is a well-known principle of personnel management that one should "praise in public and criticize in private." The reason for observing this rule is to create a climate of trust and support that enhances performance motivation. Of course, it is impossible to follow this rule precisely because a critique is inherently conducted in public. However, one can minimize individuals' defensiveness by deemphasizing the assignment of personal or organizational blame for performance errors. Instead, an effective critique directs attention toward identifying the situational impediments to effective performance and devising improvements to the EOP, the SOPs, the facilities, the equipment, or the personnel training program. After the debriefing, the outside observers submit written critiques of the performance.

Drills, exercises, and critiques are important not only because they help the organization respond as it needs to but also because they provide an opportunity for personnel to test themselves as both individuals and teams.[15] That is, emergency responders can rehearse their own emergency response roles, improve their ability to coordinate their activities with those of others with whom their tasks are interdependent, and see how their roles fit into the broader functions of the emergency response organization. Exercises also make it possible to test interagency and interorganizational agreements and to obtain positive (credibility-building) publicity about community emergency management efforts.

In some jurisdictions, portions of the EOP will be tested many times a year in routine responses to actual environmental threats. This partial testing may give grounds for confidence in the ability of the agencies involved to respond effectively, but confidence should not become overconfidence. Testing portions of the EOP does not demonstrate the ability of the plan as a whole to produce effective response in functions that have not been tested or in emergencies whose scope of impact is significantly broader.

Audits

One way to audit the emergency preparedness program is by having the staff of the local emergency management agency compare the EOP, SOPs, and emergency response resources (equipment, facilities, and personnel) with the requirements identified in the community preparedness analyses. In past decades, this was the most common emergency preparedness audit procedure, although the audit was sometimes conducted by outside consultants from private industry or from state or federal agencies. Then in 1995 the National Fire Protection Association (NFPA) developed Standard 1600, which, in turn, led to the development of EMAP. In addition to providing more systematic methods for auditing the local emergency preparedness program, NFPA 1600 and EMAP serve as a basis for evaluating the entire emergency management program—including hazard mitigation and disaster recovery preparedness.

Compliance with National Fire Protection Association Standard 1600 In 2004, NFPA defined an updated set of criteria for disaster and emergency management programs (and business continuity programs).[16] The Standard on Disaster/Emergency Management and Business Continuity Programs, known as NFPA 1600, provides a means of assessing and improving existing programs and a set of guidelines for creating new programs. Although local emergency management agencies are not required to comply with NFPA 1600, the standard is the basis for EMAP, discussed in the next section.

NFPA 1600 defines an *entity* as a public or private sector organization that is responsible for emergency/disaster management or continuity of operations. The standard requires the entity to have a documented emergency management program with an adequate administrative structure, an identified coordinator, an advisory committee, and procedures for evaluation. The program must address all fourteen of the following elements:

- *Laws and authorities:* The entity will identify the legislation, regulations, directives, and industry standards that authorize the emergency management program.

- *Hazard identification, risk assessment, and impact analysis:* The entity will identify the hazards to which it is exposed, the probability of extreme events that would adversely affect it, and the potential physical and social consequences of those events.
- *Hazard mitigation:* The entity will develop a strategy to eliminate the hazards or to limit their consequences.
- *Resource management:* The entity will identify resources (personnel, facilities, equipment, materials and supplies, money) that are needed by the emergency management program, inventory those resources that are currently available, and define the resulting shortfall; specify the role of volunteers; create and participate in mutual aid agreements; calculate resource needs in terms of quantity, response times, capabilities, and other factors related to deployment; establish performance objectives for personnel, equipment, apparatus, facilities, and the like; and develop a separate set of performance objectives for each threat identified in the vulnerability analysis.
- *Mutual aid:* The entity will identify the need for resources from other entities, establish formal agreements for requesting those resources, and reference these agreements in relevant sections of its emergency management plans—plural because the *planning* element (see next bullet) requires the entity to develop an overall strategic plan as well as distinct plans for mitigation, operations, recovery, and continuity, and to specify the generic content of each type of plan.
- *Planning:* The entity will specify functional roles and responsibilities, not only within its own organization but also for external organizations or agencies that are expected to participate in mitigation, preparedness, response, or recovery.
- *Direction, control, and coordination:* The entity will establish authority for response and recovery operations by adopting an incident management system and clearly assigning functional responsibilities to specific organizations within the entity.
- *Communications and warning:* The entity will develop and periodically test equipment and procedures needed to activate the emergency response organization.
- *Operations and procedures:* The entity will develop procedures needed to respond to the hazards that have been identified (see the second bullet above); procedures must include situation assessment and resource assessment, transition from response to recovery operations, and continuity of operations.
- *Logistics and facilities:* The entity will develop systems for (1) identifying, obtaining, and delivering needed resources to response and recovery personnel and (2) handling unsolicited donations; and will fulfill the requirement for a primary EOC and an alternate EOC.
- *Training:* The entity will identify training needs (including training scope and frequency), implement a training program, and document the delivery of training to all emergency response personnel.
- *Exercises, evaluations, and corrective actions:* The entity will periodically evaluate plans and procedures, including tabletop, functional, and full-scale exercises, and will develop procedures for ensuring corrective action regarding any identified deficiencies.
- *Crisis communications and public information:* The entity will establish procedures for disseminating relevant information to the news media and the public about the operation before, during, and after the disaster.
- *Finance and administration:* The entity will develop fiscal procedures to ensure that decisions can meet the time constraints of emergencies while also conforming to accepted accounting standards. (Normally, purchasing outside resources requires multiple levels of time-consuming approval, but in emergencies, resources may have to be obtained within hours rather than days or weeks. Thus, emergency procedures are needed to ensure that purchases can be made in a timely manner without compromising accountability.)

The Emergency Management Accreditation Program EMAP is based on NFPA 1600. However, EMAP differs from NFPA 1600 in that the former contains language that is specifically geared to state emergency management agencies (SEMAs) and local emergency management agencies (LEMAs), whereas the scope of NFPA 1600 extends to private sector organizations.

EMAP has an elaborate accreditation process.[17] When an emergency management agency submits an application for accreditation, it has eighteen months to conduct a self-assessment of its compliance with EMAP's fifty-four standards. The accreditation process begins with a written expression of executive-level (e.g., mayor, city manager, county supervisor) commitment to receiving EMAP accreditation. The next step in the process is the agency's selection of an accreditation manager, who will be responsible for coordinating both the self-assessment and the accreditation application. The accreditation manager is most likely to be the local emergency manager.

The self-assessment requires the agency to examine its emergency management program and to provide a proof-of-compliance record for each of the EMAP standards. In the record, the accreditation manager identifies the documentation, interview sources, or observations on which the claim of compliance is based and explains how the documentation, interviews, or observations support the claim. In the case of written documentation, the proof-of-compliance record must be accompanied by a copy of the document or by information on how to locate a copy of it. Equivalent documentation is required for interviews or observations.

After completing the self-assessment documents, the applicant submits them to the EMAP Commission for review. If the commission judges the applicant's materials to be satisfactory, it will schedule an on-site assessment. During this assessment, the assessor team receives an orientation to the jurisdiction (including a tour of major facilities); verifies compliance claims by examining the written plans, procedures, and memorandums that were listed as proof of compliance; and—to obtain independent verification of the information listed in the application—contacts interviewees listed on the proof-of-compliance record. Finally, the assessor team inspects facilities, equipment, materials, and supplies to verify their adequacy.

The assessor team then reports its findings to the EMAP Program Review Committee at a meeting that the accreditation applicant may attend. On the basis of the report of findings, committee recommends that the applicant be accredited, conditionally accredited, or denied. This recommendation is then transmitted to the EMAP Commission for final action. An accredited applicant is issued a certificate valid for five years. During that period, the applicant must maintain accreditation by continuing to comply with the EMAP Standard and to document compliance, and by filing an annual report with the EMAP Commission. If full accreditation is not granted, the applicant is informed of the reasons for the action.

Risk communication

The emergency manager should be sure there is a risk communication program to encourage households, businesses, and other government agencies to adopt preparedness measures of their own. Communication with these social units is an important component of community emergency preparedness because when these units are prepared for disasters, they will need less help from emergency responders, who will be able to focus more of their efforts on special facilities (e.g., hospitals and nursing homes) and on dispersed population segments with special needs (e.g., people with limited mobility).

Preparedness by households Research on household emergency preparedness shows that many people are unaware of the hazards to which they are exposed, and even those who are aware of the hazards have inaccurate beliefs about how those hazards could affect them. Moreover, people are generally unaware of all the actions they could take to protect themselves, their families, and their property. And even when they are aware, they have inaccurate beliefs about those hazard adjustments. As a result, most households engage in only a few of the hazard adjustments recommended by authorities such as the American Red Cross.

One of the principal determinants of households' protective actions is personal experience. Those who have sustained property damage or injuries to members of their households adopt more hazard adjustments than those who lack such experience. Another important influence on hazard adjustment is information obtained from authorities, peers (friends, relatives, neighbors, and co-workers), and the mass media. The most important aspects of this information are the sources, the channels, and the messages:

- *Sources* can be characterized in terms of their expertise (knowledge about hazards and protective actions) and their trustworthiness (willingness to share accurate information).
- *Channels* of information can be characterized by number and type (radio, television, newspapers, magazines, brochures, face-to-face conversations).
- *Messages* providing information about hazards and protective measures can be characterized by number, specificity, and comprehensibility.

Messages about hazards should describe the threat, recommend specific protective actions, and identify whom to contact for further information.[18] Information about the threat should describe the likelihood and severity of personal consequences, such as death or injury; personal property damage; and disruption to work, school, and other daily activities. Personalizing the risk in this way arouses people's self-protective motivation. Recommending specific protective actions makes it more likely that the actions people take will be effective in protecting lives and property. People are most likely to adopt the recommended protective actions if those actions are suitable for multiple hazards; are low in cost, time, and effort; require little by way of specialized knowledge, skill, tools, and equipment; and do not depend on cooperation from other people. Risk messages should also emphasize people's personal responsibility for self-protection. In this regard, it is important to note that a major catastrophe is likely to overwhelm local government agencies, and that state and federal resources might not be able to arrive for three days or more. Thus, households must assume responsibility for protecting themselves as much as possible.

The effectiveness of risk communication is also determined by the characteristics of the *receiver*—the kind or amount of information the receiver needs, and the resources available to him or her. Indeed, the most effective risk communicators divide the population into segments on the basis of each segment's information needs and material resources. The most effective risk communicators also seek opportunities for household members to provide feedback about how well they understood the messages and how well the messages met their information needs.

Preparedness by businesses and government agencies Emergency preparedness by businesses and government agencies suffers from many of the same limitations as preparedness by households. Hazards have low salience until an imminent threat arises, so emergency preparedness must compete with routine demands for space on the organizational agenda. This tendency is especially pronounced in organizations with limited financial assets.

Businesses display limited levels of emergency preparedness, but business disruption can devastate a community's economy, so local governments should make a special effort to contact small businesses that lack previous disaster experience. The risk communication program for these businesses should adopt the same principles as the risk communication program for households. That is, the agency in charge of the program should work to develop the perception, among target businesses, of its expertise and trustworthiness, and should use multiple communication channels. It should disseminate messages describing hazards, giving guidance on protective actions, identifying sources of additional information, and stressing the importance of personal responsibility for taking protective action.

FEMA's *Emergency Management Guide for Business & Industry* outlines a planning process, identifies critical corporate emergency management functions, provides information about a variety of hazards, and lists sources to contact for further information. In addition, the Institute for Business & Home Safety has developed a program called "Open for Business," and the Public Entity Risk Institute posts disaster-relevant information for small businesses on its Web site. Finally, books have been published on emergency planning for businesses.[19]

Among government agencies without emergency missions, three factors have been consistently identified as correlates of organizational preparedness. The first is *organizational size:* larger organizations not only have more resources but also are more likely to perceive a need for strategic planning. The second is the level of *perceived risk* among organizational and department managers. The third factor that is positively correlated with emergency preparedness is the extent to which managers report *seeking information about environmental hazards.*

A model for strengthening local emergency preparedness

After reading a description of the components of an effective emergency preparedness program, one will logically ask how emergency managers can perform these activities and produce an effective emergency management program in the face of the apathy and resistance described at the outset of the chapter. Research undertaken since the mid-1970s has identified local emergency management agencies (LEMAs) and local emergency management committees (LEMCs)—the agents specifically formed and exclusively assigned to undertake activities whose ultimate goal is to reduce casualties and damage from disasters.[20]

A LEMA is a jurisdictional organization headed by the emergency manager. It might be an independent agency but is often a division of the police or fire department, and it usually has a small budget and staff. Whether an independent agency or a division of the police, fire, or other department, the LEMA is responsible for coordinating emergency planning (and is often responsible for operating the EOC during disasters). An LEMC is a generic name for a permanent task force—a coordinating committee with a very limited (or no) budget and no operational responsibility—that has representatives from the LEMA and from other emergency-relevant organizations (e.g., police, fire, emergency medical services, public works, and the American Red Cross). The members of an LEMC might be, but are not necessarily, the same as the members of the local emergency planning committee (LEPC) that performs emergency planning for toxic chemical hazards under Title III the Superfund Amendments and Reauthorization Act of 1986 (SARA Title III). When LEMCs are established at the county level, they often include representatives from municipal LEMAs and other agencies within the county.

The research findings are summarized in Figure 7–2, which diagrams the main factors that influence community emergency preparedness. As the figure indicates, the best organizational outcomes (measured by indicators such as the quality, timeliness, and cost of preparedness measures) are the direct result of the quality of individual outcomes and the quality of the planning process. The quality of the planning process, in turn, is determined by the level of hazard exposure/vulnerability, community support, community resources, extra-community resources, and staffing/organization. This section discusses all the factors, beginning with the ones that determine the quality of the planning process and ending with organizational outcomes.

The model depicted in Figure 7–2 is static: that is, the arrows begin on the left and end on the right. However, the actual process is dynamic because success tends to be a self-amplifying process in which high levels of individual and organizational outcomes produce increased

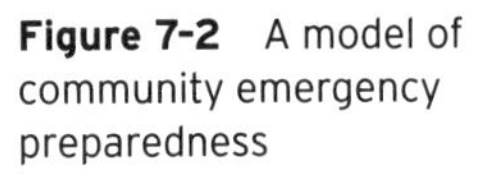
Figure 7-2 A model of community emergency preparedness

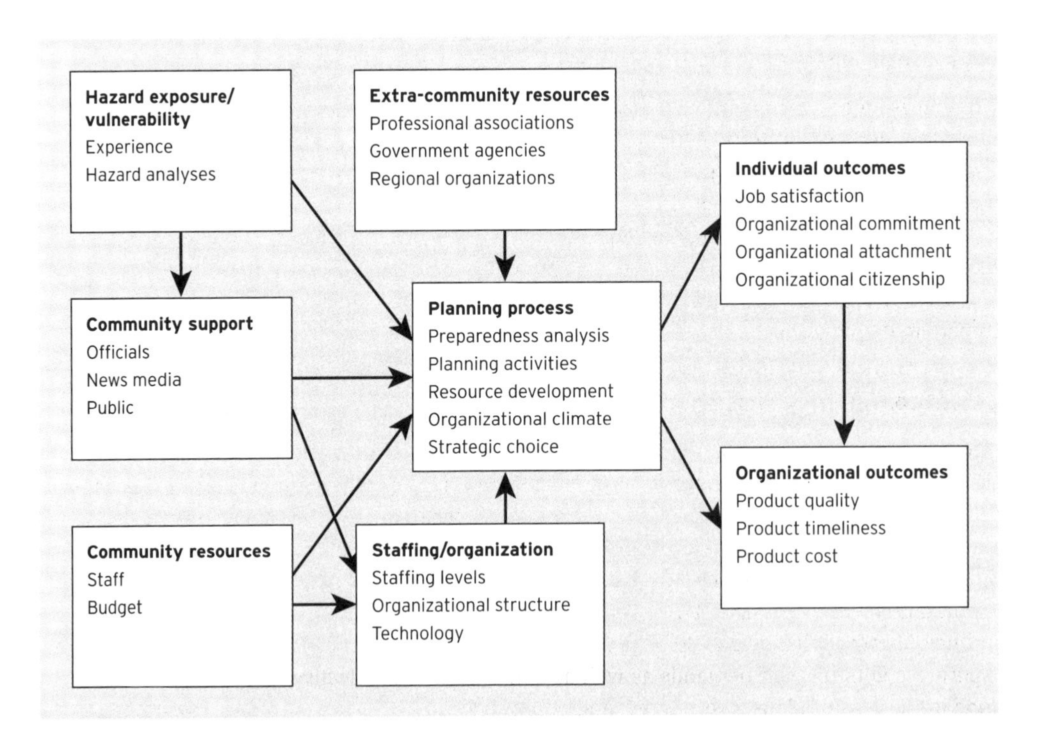

levels of *vicarious* experience with disaster demands (by increasing the frequency and comprehensiveness of emergency training, drills, and exercises), stronger community support, better staffing and organization, and more emergency planning resources.

Hazard exposure/community vulnerability

Although the evidence is mixed, many studies have found that the level of community preparedness for emergency response and disaster recovery is greater if the community has experienced a disaster or, even worse, a catastrophy.[21] Frequent, recent, and severe impacts make the community's vulnerability to hazards easier to remember, and the memories are more likely to stimulate action. But when disasters are infrequent or their impacts have been minimally disruptive, people can be made more aware of their exposure to hazards if they read or hear about the findings of the community's hazard/vulnerability analysis. And this pallid statistical information can be supplemented with vivid accounts of other communities' disasters that are transmitted through newspaper articles or television accounts or, most powerfully, through first-person accounts—especially from peers. For example, a local fire chief is most likely to be influenced by other fire chiefs' accounts of their experiences, the city manager is most likely to be influenced by another city manager's experiences, and so on.

Community support

Support from the community's senior elected and appointed officials, the news media, and the public is important because it affects the resources allocated to emergency management. As noted above, many researchers have systematically documented what numerous emergency managers have personally experienced—that emergency management is a low priority for the local elected and appointed officials who control budgets and staffing allocations. According to one police chief, "My number one priority is getting the uniforms out in response to calls. The public judges me on that performance, not [on] whether I'm planning for an earthquake that may never happen. If left alone, disaster planning would get even less attention from my office. It requires that the executive clearly make this a priority."[22]

One important way of generating community support is by developing and implementing an effective risk communication program.[23] Methods of improving community risk communication programs have already been discussed.

Community resources

Differences among communities in their level of emergency preparedness can be attributed partly to differences in those communities' human and economic resources. For example, researchers have found a significant correlation between emergency planning effectiveness and a jurisdiction's population size, percentage of urban population, police staffing, fire staffing, median household income, and budget size. However, community resources have less influence on the effectiveness of LEMAs and LEMCs than does community support. That is, even communities with limited financial resources can still develop effective emergency preparedness programs.

Extra-community resources

LEMAs and LEMCs can obtain many emergency planning resources from outside the community to increase their emergency planning effectiveness. Such resources include computer hardware and software, federal and state agency technical reports, training courses, and industry association materials. In addition, planning effectiveness is significantly related to *vertical links* to federal agencies (the U.S. Departments of Transportation and Homeland Security, and the U.S. Environmental Protection Agency) and *horizontal links* to private industry and neighboring jurisdictions. Private industry and neighboring jurisdictions can convey their experience with disaster demands as well as demonstrate the effectiveness of specific innovations in the area of plans, procedures, and equipment.

Staffing and organization

The last factor that has a significant influence on the quality of the planning process and the effectiveness of LEMAs and LEMCs is the staffing and organization of those two bodies. The most effective LEMCs have

- Defined roles for elected officials
- Clear internal hierarchy
- Good interpersonal relationships
- Commitment to planning as a continuing activity
- Member and citizen motivation for involvement
- Coordination among participating agencies
- Public-private cooperation.

Moreover, LEMCs are more effective when they have a larger number of members, more full-time paid staff, and a larger number of agencies represented, and when they are organized into subcommittees. Having representatives of elected officials and from citizens' groups is very important, and establishing an organizational structure through subcommittees is significant, probably because members can then focus on specific tasks and avoid feeling overwhelmed by all the work that needs to be done. Assigning a full-time coordinator to an LEMC is also important because this position serves as a focal point for LEMC activities. In many cases, the jurisdiction's emergency manager assumes this role.

Planning process

The planning process brings together preparedness analysis, planning activities, resource development, organizational climate, and strategic choice.

Preparedness analysis As indicated previously, the components of the community preparedness analysis are hazard/vulnerability analysis (including analyses of hazard exposure, physical vulnerability, and social vulnerability), hazard operations analysis, population protection analysis, and incident management analysis. Hazard/vulnerability analysis identifies the natural and technological hazards to which the community is exposed, the locations that would be affected, and the amount of damage associated with events of various intensities. It also assesses the community's structures (residential, commercial, and industrial buildings) and lifeline and infrastructure systems (fuel, electric power, water, sewer, telecommunications, and transportation) in terms of their ability to withstand the events identified in the hazard analysis. Finally, it assesses the community's susceptibility to psychological, demographic, economic, and political effects of a disaster, and examines both the physical and the social effects to determine differences among demographic segments and among economic sectors in susceptibility to these effects.

Hazard operations analysis and population protection analysis identify alternative responses to disaster demands and evaluate those responses in terms of both their effectiveness in protecting persons and property and their resource requirements (the amount of time, effort, money, and organizational cooperation needed to adopt and implement them). The final component of community preparedness analysis—incident management analysis—determines whether households, businesses, government agencies, and nongovernmental organizations (NGOs) have the capacity (i.e., resources) and commitment (i.e., motivation) needed to implement the hazard operations and population protection actions.

Planning activities Superior planning activities involve key personnel from diverse agencies in a participative and consensus-oriented process. As Figure 7–3 indicates, interdepartmental training—including reviews with senior officials and the establishment of interdepartmental task forces—is one of the most important activities influencing the adoption of good emergency preparedness practices. By contrast, the more routine activities, such as updates of procedures, are the least important.

Most important	Least important
Interdepartmental training	Standard operating procedure updates
Progress reviews with senior officials	Emergency operations plan updates
Interdepartmental task force	Reviews of mutual aid agreements
Community disaster assistance council	
After-action critiques	
Exercises	
Vulnerability analyses	
Meetings with managers of local television and radio stations	

Source: Adapted from Jack D. Kartez and Michael K. Lindell, "Planning for Uncertainty: The Case of Local Disaster Planning," *Journal of the American Planning Association* 53, no. 4 (1987): 494.

Figure 7-3 Most and least important planning activities vis-à-vis the adoption of good emergency preparedness practices

LEMCs are more effective when they

- Formally orient new members by providing them with written descriptions of the committee's overall function and the functions of the subcommittees
- Meet frequently (between once a month and once a quarter) in the same place, on the same day of the week, and at a time of day that is convenient for all members (whether they are full time, part time, or volunteer)
- Distribute written agendas in advance and keep written minutes of the meetings
- Set milestones and monitor progress toward annual goals.

LEMC meetings can facilitate interorganizational linkages by promoting additional informal contacts, verbal agreements, and written agreements among government agencies, NGOs, and local businesses. These contacts are especially important because the low priority given to local emergency management often makes it impossible for LEMAs to purchase needed resources outright. Consequently, local emergency managers must build capacity by collaborating with other organizations (such as police and fire departments) that do have the needed resources or the influence to obtain the funding that will allow them to make those purchases.

Of course, other organizations are more likely to collaborate with a LEMA if there are compelling reasons for them to do so. Thus, local emergency managers need to promote an awareness of potential disaster demands and of the need to avoid gaps in services or duplication of effort. They can also promote collaboration by ensuring that LEMC members have timely access to information, services, or resources; by helping LEMC member organizations develop internal organizational response capability; and by enhancing these organizations' autonomy, security, and prestige. However, interorganizational linkages can be impeded by geographical distance, lack of funds, lack of staff, incompatible professional perspectives and terminology, lack of trust in an organization or its representative, overconfidence in one's own capability, and inequality of rewards and costs of participation for those in different organizations.

Resource development As noted earlier, resource development involves obtaining emergency personnel, equipment, and facilities of many different types from a variety of sources. In addition to developing the resources needed for emergency response, it is also necessary to develop the resources needed for emergency planning. The major type of emergency planning equipment—the microcomputer—is usually available at the LEMA, but the types of high-speed/high–storage capacity computers needed for conducting hazard and vulnerability analyses are more often found in the planning department, where geographic information systems are routinely used. (As noted above, important types of information include data about geographic areas, buildings, infrastructure, and people at risk.)

Organizational climate Interagency cooperation in the work of planning committees, and thus in the effectiveness of emergency planning, has been found to be greatest in LEMCs that have positive organizational climates, which can be defined as "distinctive patterns of collective beliefs that are communicated to new group members through the socialization process and are further developed through members' interaction with their physical and social environments."[24] That is, as people work with each other in organizational settings, they often develop similar perceptions of the climate within those organizations. The five major facets of organizational climate that are relevant to LEMC effectiveness pertain to leadership, team, role, job, and reward.

LEMC leaders can establish a positive *leadership climate* by being clear about what tasks are to be performed, recognizing members' strengths and weaknesses, and being responsive to members' needs.

In a positive *team climate,* members focus on the tasks to be performed, share information, and coordinate effort. Little time is spent socializing. Members trust each other, feel that they are included in all activities, and believe that their LEMC is one of the best there is.

When team members understand what tasks are to be performed, who is to perform them, and how they are to be performed—and have enough time to perform the tasks for which they are responsible—the stress associated with role ambiguity, role conflict, and role overload is averted, and the organization may be said to have a positive *role climate.*

A positive *job climate* is one in which members have enough independence to do their work in the way they choose, as long as they deliver a quality product on time and within the resources available. Members also should be allowed both to perform a "whole" piece of work that provides a meaningful contribution to the group product and to exercise a variety of significant skills.

LEMCs will function more effectively when they have a positive *reward climate,* which is characterized by members' having opportunities to perform new and challenging tasks and to work with other people, and members' being told that other people appreciate their work.

When the leadership, team, role, job, and reward components of organizational climate are positive, outcomes at the individual and organizational levels are positive. Specifically, members have higher job satisfaction, organizational commitment, attendance, effort, and "citizenship behavior" (going beyond the minimum requirements of organizational expectations), and lower turnover intentions and actual turnover. These positive outcomes at the individual level also produce positive consequences at the organizational level by increasing the organization's stability (through decreased turnover) and productivity (through greater effort), raising product quality, improving timeliness, and lowering cost.

Strategic choice Organizational scientists generally agree that there is no single best way to organize, and, indeed, there is significant variation in the strategies and structures that individual emergency managers use to develop their communities' emergency preparedness. Some successful emergency managers enthusiastically endorse strategies that are explicitly rejected by other equally successful managers. The adoption of a strategy should certainly depend on contextual conditions in the community, but some structures and strategies are likely to significantly improve the success of all LEMAs and LEMCs regardless of context—and especially without significant expense.

There are seven such strategies, and although presented as distinct, they are not mutually exclusive. That is, emergency managers can pursue multiple strategies simultaneously.

1. A *resource building* strategy emphasizes acquisition of the human, technical, and capital resources needed for effective agency performance.
2. An *emergency resource* strategy is defined as securing the participation of emergency-relevant organizations in preparedness—particularly those that control critical emergency preparedness and response resources.
3. The *elite representation* strategy involves placing members of a focal organization (in this case, the LEMA) in positions or situations where they can interact with influential members of other emergency-relevant organizations.
4. The *constituency* strategy consists of establishing a symbiotic relationship between two organizations, with both of them benefiting from cooperation.
5. The *co-optation* strategy consists of absorbing key personnel, especially those from other organizations, into the focal organization's formal structure as directors or advisers.

6. The *audience* strategy focuses on educating community organizations and the public at large about the importance of community emergency preparedness.
7. With the *organizational development* strategy, local emergency managers actively try to increase the resource base of all local agencies, not just their own, in order to foster pre-disaster relationships among organizations that must respond to a disaster.

The seventh strategy needs more elaboration. The reason for engaging in it is that disaster planning requires emergency response organizations to recognize that community-wide disasters, unlike routine emergencies, cannot be handled by a single agency. Emergency managers can pursue the organizational development strategy by relying on committees and joint ventures to involve other community organizations. To implement this strategy, emergency managers should manage conflict over controversial issues before it gets out of control. In particular, they should work on achieving consensus about all agencies' missions. They should also have more frequent contacts and more formalized interagency agreements (e.g., memorandums of understanding) with other emergency-relevant agencies.[25]

Individual outcomes

As noted above, individual outcomes include job satisfaction, organizational commitment, organizational attachment, and organizational citizenship.[26] The importance of positive affective states such as job satisfaction and organizational commitment derives from the fact that the time people must contribute to the activities of the LEMC is often unpaid. In the absence of extrinsic rewards such as pay, intrinsic (i.e., internal) rewards must serve as motivators. Intrinsic motivation to contribute to community emergency preparedness is likely to be high if LEMC members

- Perceive social and environmental problems within their community
- Are committed to the success of these communities
- Expect the LEMC to be successful in solving these problems.

Organizational outcomes

As stated above, the ultimate goal of LEMAs and LEMCs is to reduce casualties and damage from disasters. However, disasters are rare events, so emergency managers cannot measure the effectiveness of the LEMAs and LEMCs in terms of a reduction in the number of casualties and the amount of damage from disasters. Instead, they must use other indicators of effectiveness, such as the quality of plans, procedures, training, equipment, and facilities; timeliness; and cost. The way to evaluate quality is by conducting internal audits to see if the jurisdiction meets the standards of NFPA 1600. Later, drills and exercises can be used to see if there are any deficiencies in the emergency preparedness program, and finally a jurisdiction can seek accreditation from EMAP. The way to evaluate timeliness and cost is by asking if LEMAs and LEMCs achieve progress in developing emergency preparedness within prevailing schedule and cost constraints. If they fail to meet their deadlines or overrun their budgets, they risk losing support from senior elected and appointed officials, as well as from community residents.

Summary

The two best-known aspects of a disaster planning and preparedness program are the emergency operations plan—the formal document that governs a community's preparedness program—and the standard operating procedures. Together they form the core of the program.

The EOP has a three-part structure: basic plan, functional annexes, and hazard-specific appendixes. Adoption of a standardized model plan will not provide adequate emergency preparedness because EOPs must be individualized—that is, adapted to the community's own expected disaster demands and capabilities. The vehicle for the adaptation is a community preparedness analysis, which comprises four basic analyses: hazard/vulnerability, hazard operations, population protection, and incident management.

In addition to the EOP and the SOPs, emergency preparedness requires equipment, facilities, and staffing, as well as the activities needed to develop them: training, drills and exercises, and critiques, along with audits and a risk communication program.

A model of the entire local emergency preparedness process illustrates how to make the planning process happen. The model focuses on the factors that interact with one another to produce eventual success (defined as effective performance by the local emergency management agencies and committees). To achieve the desired outcome, what does the emergency manager have to attend to? The preparedness planning process is pivotal, for individual and organizational outcomes depend on it. The several factors that feed into the preparedness planning process are hazard exposure and vulnerability, community support, community resources, extra-community support, and staffing and organization. The planning process itself is composed of preparedness analysis, planning activities, resource development, organizational climate, and strategic choice. All these elements are crucial to the goal of the planning process: to produce successful individual and organizational outcomes within the emergency management agencies and committees—and, by extension, within the community when a disaster strikes.

When disaster strikes a community, the quality of a local government's preparation will determine the quality of its response.

Notes

1 For further discussion of these basic emergency response functions and their subcomponents, see Michael K. Lindell, Carla S. Prater, and Ronald W. Perry, *Fundamentals of Emergency Management* (Emmitsburg, Md.: Emergency Management Institute, Federal Emergency Management Agency [FEMA], July 2006), available at training.fema.gov/EMIWeb/edu/fem.asp (accessed September 3, 2007).

2 Enrico L. Quarantelli, "Ten Research-Derived Principles of Disaster Planning," *Disaster Management* 2 (January–March 1982): 23–25; David A. McEntire, "Disaster Preparedness," *IQ Report* 35, no. 11 (ICMA, 2003); Michael K. Lindell et al., "The Local Emergency Planning Committee: A Better Way to Coordinate Disaster Planning," in *Disaster Management in the U.S. and Canada: The Politics, Policymaking, Administration and Analysis of Emergency Management,* ed. Richard T. Sylves and William L. Waugh Jr. (Springfield, Ill.: Charles C. Thomas, 1996), 234–249; Carla S. Prater and Michael K. Lindell, "The Politics of Hazard Mitigation," *Natural Hazards Review* 1, no. 2 (2000): 73–82.

3 James Schwab et al., *Planning for Post-Disaster Recovery and Reconstruction,* Planning Advisory Service Report Number 483/484 (Chicago: American Planning Association, 1998).

4 FEMA, *Guide for All-Hazard Emergency Operations Planning,* State and Local Guide (SLG) 101 (Washington, D.C.: FEMA, 1996), available at fema.gov/plan/gaheop.shtm (accessed September 3, 2007).

5 Alan V. Brunacini, *Fire Command: The Essentials of IMS* (Quincy, Mass.: National Fire Protection Association [NFPA], 2002); U.S. Department of Homeland Security (DHS), *National Response Plan* (Washington, D.C.: DHS, 2004), available at dhs.gov/dhspublic/interapp/editorial/editorial_0566.xml (accessed September 3, 2007).

6 See DHS, *National Incident Management System* (Washington, D.C.: DHS, March 1, 2004), 12–26, available at dhs.gov/xlibrary/assets/NIMS-90-web.pdf (accessed September 6, 2007).

7 FEMA's HAZUS-MH software is used to analyze potential loss from earthquakes, floods, and wind. For more information, see fema.gov/plan/prevent/hazus (accessed September 3, 2007). For guidance on conducting a local hazard/vulnerability analysis, see the National Oceanic and Atmospheric Association's Coastal Services Center's Community Vulnerability Assessment Tool at csc.noaa.gov/products/nchaz/startup.htm (accessed September 3, 2007).

8 Ben Wisner et al., *At Risk: Natural Hazards, People's Vulnerability, and Disasters,* 2nd ed. (London: Routledge, 2004), 11.

9 The North American Industry Classification System (NAICS) categorizes all businesses into twenty broad industrial classes; see census.gov/epcd/naics02/.

10 Michael K. Lindell and Ronald W. Perry, *Behavioral Foundations of Community Emergency Planning* (Washington, D.C.: Hemisphere Press, 1992), chapter 4.

11 Thomas Urbanik II, "Evacuation Time Estimates for Nuclear Power Plants," *Journal of Hazardous Materials* 75, nos. 2–3 (2000): 165–180.

12 Lindell and Perry, *Behavioral Foundations;* John H. Sorensen, Barry L. Shumpert, and Barbara M. Vogt, "Planning for Protective Action Decision Making: Evacuate or Shelter-In-Place," *Journal of Hazardous Materials* 109, nos. 1–3 (2004): 1–11.

13 National Response Team, *Hazardous Materials Emergency Planning Guide,* NRT-1 (Washington, D.C.: National Response Team, March 1987), 28–34, available at nrt.org/production/NRT/NRTWeb.nsf/AllAttachmentsByTitle/A-22nrt1/$File/nrt1.pdf?OpenElement (accessed September 3, 2007).

14 Irving L. Goldstein and J. Kevin Ford, *Training in Organizations: Needs Assessment, Development, and Evaluation,* 4th ed. (Pacific Grove, Calif.: Brooks/Cole, 2002); J. Kevin Ford and Aaron Schmidt, "Emergency Preparedness Training: Strategies for Enhancing Real-World Performance," *Journal of Hazardous Materials* 75, nos. 2–3 (2000): 195–215.

15 FEMA, *Exercise Design Course,* IS-139 (Emmitsburg, Md.: Emergency Management Institute, FEMA, 2003), available at training.fema.gov/EMIWeb/IS/is139.asp.

16 NFPA 1600: Standard on Disaster/Emergency Management and Business Continuity Programs (Quincy, Mass.: NFPA, 2004).

17 Emergency Management Accreditation Program (EMAP), *Candidate's Guide to Accreditation* (Lexington, Ky.: EMAP, 2004); EMAP, *EMAP Standard* (Lexington, Ky.: EMAP, 2004). Further information about EMAP can be found at emaponline.org/.

18 See Lindell and Perry, *Behavioral Foundations,* chapter 5, for detailed recommendations about the development and implementation of risk communication programs.

19 FEMA, *Emergency Management Guide for Business and Industry,* available at fema.gov/business/guide/index.shtm (accessed September 3, 2007); Paul A. Erickson, *Emergency Response Planning for Corporate and Municipal Managers* (San Diego, Calif.: Academic Press, 1999); and Robert Heath, *Crisis Management for Managers and Executives* (London: *Financial Times*/Pittman, 1998). See also the resources on the

Web sites of the Institute for Business & Home Safety at ibhs.org, and the Public Entity Research Institute (PERI) at riskinstitute.org.

20 Michael K. Lindell and Marna J. Meier, "Effectiveness of Community Planning for Toxic Chemical Emergencies," *Journal of the American Planning Association* 60, no. 2 (1994) 222–234; Michael K. Lindell, "Are Local Emergency Planning Committees Effective in Developing Community Disaster Preparedness?" *International Journal of Mass Emergencies and Disasters* 12, no. 2 (1994): 159–182; Michael K. Lindell and David J. Whitney, "Effects of Organizational Environment, Internal Structure and Team Climate on the Effectiveness of Local Emergency Planning Committees," *Risk Analysis* 15, no. 4 (1995) 439–447; Michael K. Lindell et al., "Multi-Method Assessment of Organizational Effectiveness in a Local Emergency Planning Committee," *International Journal of Mass Emergencies and Disasters* 14, no. 2 (1996) 195–220, available at ijmed.org/PDF_Files/August_1996.pdf (accessed September 3, 2007); and Lindell et al., "The Local Emergency Planning Committee."

21 Earl J. Baker, "Hurricane Evacuation Behavior," *International Journal of Mass Emergencies and Disasters* 9, no. 2 (1991): 287–310; Michael K. Lindell and Ronald W. Perry, "Household Adjustment to Earthquake Hazard: A Review of Research," *Environment and Behavior* 32, no. 4 (2000): 590–630.

22 Jack D. Kartez and Michael K. Lindell, "Adaptive Planning for Community Disaster Response," in Richard T. Sylves and William L. Waugh Jr., *Cities and Disaster: North American Studies in Emergency Management* (Springfield, Ill.: Charles C. Thomas, 1990), 13.

23 Michael K. Lindell and Ronald W. Perry, *Communicating Environmental Risk in Multiethnic Communities* (Thousand Oaks, Calif.: Sage, 2004); Daniel Alesch et al., *Promoting Seismic Safety: Guidance for Advocates*, FEMA-474 (Buffalo, N.Y.: Multidisciplinary Center for Earthquake Engineering Research, 2005), available at mceer.buffalo.edu/publications/Tricenter/04-sp02/ (accessed September 4, 2007).

24 Michael K. Lindell and Christina J. Brandt, "Climate Quality and Climate Consensus as Mediators of the Relationship between Organizational Antecedents and Outcomes," *Journal of Applied Psychology* 85, no. 3 (2000): 331.

25 David F. Gillespie et al., *Partnerships for Community Preparedness* (Boulder: Natural Hazards Research and Applications Information Center [National Hazards Center], University of Colorado, 1993); Thomas E. Drabek, *Emergency Management: Strategies for Maintaining Organizational Integrity* (New York: Springer-Verlag, 1990); Thomas E. Drabek, *Emergency Management: Strategies for Coordinating Disaster Responses* (Boulder: Natural Hazards Center, University of Colorado, 2003).

26 Consistent with contemporary theories of motivation in organizations, Whitney and Lindell define attachment behaviors as effort, attendance, and continued membership and organizational citizenship behaviors as performance that exceeds an organization's minimum requirements; see David J. Whitney and Michael K. Lindell, "Member Commitment and Participation in Local Emergency Planning Committees," *Policy Studies Journal* 28, no. 3 (2000): 467–484.

CHAPTER 8

Applied response strategies

Richard A. Rotanz

This chapter provides an understanding of

- How emergencies and disasters are defined and distinguished
- The phases of disaster
- Agent-generated versus response-generated demands
- Emergency operations centers.

Recent events—the use of sarin nerve gas in the Tokyo subway in 1995, the terrorist attacks on the World Trade Center and the Pentagon in 2001, the Northeast power outage of 2003, the Asian tsunami in 2004, the hurricanes in the Gulf states in 2004 and 2005, and the collapse of Interstate 35W in Minneapolis, Minnesota, on August 1, 2007—highlight the variety and extent of natural and man-made threats, and the importance of reducing vulnerabilities and developing effective response mechanisms. Drawing on the World Trade Center attacks as an example, this chapter examines the transition from preparedness to response in the event of disaster.

An effective response management strategy can provide flexibility in dealing with a wide range of events—from emergencies that require a single-agency response, to complex disasters that require the activation of an emergency operations center (EOC) and the establishment of an emergency response organization network (ERON). To develop such a strategy, the emergency manager must identify the characteristics of the disaster, assess the need for multiorganizational coordination, and determine which form of management is needed: incident command, unified command, or a multiagency coordination system. The nature and scale of the disaster will dictate whether mutual assistance agreements should be activated, whether state and federal assistance should be requested early on, and whether volunteer and other nongovernmental organizations (NGOs) should be mobilized.

A response strategy facilitates overall management and helps ensure a smooth and timely transition from one phase to another. This chapter identifies critical constructs and functions necessary for an effective response to complex events. The principal areas of coverage are (1) how emergency managers define events in order to guide response; (2) the phases of disaster; (3) the demands to which emergency managers must respond during disasters; and (4) how emergency managers use EOCs to coordinate multiagency, multijurisdictional response.

Of course, not all local governments have the economic resources to establish a cutting-edge EOC. Smaller jurisdictions have to make do with the resources provided by taxpayers, state and federal agencies, and other stakeholders to design and build an EOC that will fit their needs. In the simplest terms, smaller emergency management departments have less capacity to undertake specialized tasks and have to rely more heavily on outside resources, including volunteers and NGOs. Nonetheless, the same functions are necessary in complex events, regardless of the size of the jurisdiction.

Defining the event

The scale of an incident provides the context for disaster response. As noted in Chapter 1, smaller events are generally referred to as emergencies. An emergency is a relatively routine incident that does not have community-wide impact, does not require the extraordinary use of resources or procedures to bring conditions back to normal, can be successfully handled by local emergency responders, is manageable and clearly defined, is brought under control quickly, and generally involves responders who know one another. In addition, roles and responsibilities in an emergency are clear-cut, and there is a recognizable authority structure.[1]

Once an emergency has reached a certain threshold in terms of size or severity, however, it is categorized as a disaster. In New York State, a disaster is defined as an event that involves the occurrence or imminent threat of widespread or severe damage, injury, or loss of life or property resulting from any natural or man-made causes–including, but not limited to, fire, flood, earthquake, hurricane, tornado, high water, landslide, mudslide, wind, storm, wave action, volcanic activity, epidemic, air contamination, blight, drought, infestation, explosion, radiological accident, water contamination, bridge failure, or bridge collapse (see Figure 8–1).

Local governments should identify thresholds to distinguish, for example, between a routine emergency, a jurisdiction-wide emergency, and an event that might or will overwhelm local resources. For a given event, a jurisdiction can determine, on the basis of those thresholds, which agencies will respond, which will lead, and which will provide support in dealing with the event. For example, the fire department will take the lead in fire-suppression operations, while police and emergency medical services will take the lead for traffic and

Photo courtesy of Brian Peterson/*Minneapolis Star Tribune*/ZUMA Press

Figure 8-1 The collapse of Interstate 35W in Minneapolis, Minnesota

medical operations, respectively. The thresholds will also indicate when to activate mutual aid agreements; when to establish unified command, area commands, or both; what levels of notification should be initiated; and when to activate the EOC. It is also essential to develop thresholds for notifying the local chief executive or the governor so that they can fulfill their legal and political responsibilities in the event of an emergency. The chief executive and the governor may also have designated staff who monitor such events and determine when they should be informed.

The phases of disaster

To develop thresholds that trigger particular actions, it is necessary to understand how disasters typically develop. Figure 8–2 shows the phases of disasters.[2] During the *pre-disaster phase,* the emergency management department undertakes preparedness activities, including performing hazard analyses, evaluating human and material resources, developing and exercising emergency operations plans (EOPs), and conducting training and education.

Warning of an impending disaster occurs during the *pre-impact phase.* This phase is critical because the initial evaluation of the incident determines how quickly resources will be mobilized and whether shortfalls may occur. Businesses, private citizens, and governments at all levels need to understand the importance of having swift access to adequate financial, material, and skilled resources—and how quickly resources can be depleted. Rapid-onset events with little forewarning present particular difficulties in anticipating resource needs, although a well-conceived EOP can go a long way toward facilitating quick response.

The onset of the disaster—the *impact phase*—may be gradual or sudden. It is at this point that officials need to reevaluate the scale of the event and deploy resources accordingly. The effectiveness of the response operation, which is undertaken during the *emergency phase,* reflects the adequacy of planning and preparedness efforts. In the case of a rapid-onset disaster

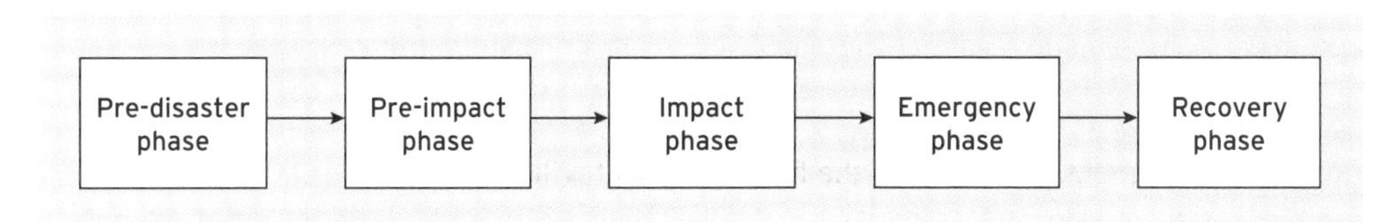

Figure 8-2 The phases of disaster

such as a terrorist attack, landslide, or industrial accident, the pre-impact, impact, and emergency phases may be compressed, and it may be difficult to distinguish between one phase and another. Nevertheless, the response strategy should help anticipate response needs and facilitate the initiation of disaster operations.

The response management strategy should incorporate guidelines for when and how to transition from one phase of an emergency to the next.

The first priority in response is always to save lives; thus, the first operations typically include search and rescue, emergency medical care, evacuation, and firefighting. Efforts to mitigate exposure to hazardous materials or to restore critical infrastructure may occur simultaneously and are usually the next priority. Eventually, the emergency phase transitions into the *recovery phase,* in which efforts are directed toward returning to normal and identifying mitigation measures that may help to avoid future disasters.

After the 9/11 attacks, the move from search and rescue to recovery at ground zero meant acknowledging that no more survivors were likely to be found; as difficult as this decision was for officials, rescue workers who had lost friends or relatives in the collapse of the towers found it traumatic. The response management strategy should incorporate guidelines for when and how to transition from one phase of an emergency to the next. Such guidelines would indicate, for example, how evacuation, sheltering, mass feeding, donation management, debris management, and other response and relief programs will be activated, managed, and deactivated.

Agent-generated versus response-generated demands

During a disaster, some demands are generated by the event, and others are generated by the response operations themselves; planning efforts should account for both.

Agent-generated demands

Warning, typically the first event-generated demand, varies with the nature of the agent involved. Local authorities must agree as to which warning methods will be used for various events, and must educate citizens about the warnings so that they can respond appropriately. It would be unwise, for example, to use a one-source warning device, such as a siren, to alert the community to more than one type of event, because different threats—a tornado versus a flash flood, for instance—require different protective actions. It is critical that citizens hear or see, understand, believe, and act upon warnings; multiple *types* of warnings, as well as *repeated* warnings, may be necessary to ensure that the public is alerted to the disaster and recognizes its seriousness.

Pre-impact preparation, another agent-generated demand, refers to response strategies that are designed to mitigate impact; such strategies include evacuation, sheltering in place, and assessing stockpiles of antibiotics in anticipation of a disease outbreak. Search and rescue is similarly unique to the disaster agent; an overturned ferry, a mass hostage situation, a wilderness search and rescue operation, and urban search and rescue following an earthquake all require different skill sets and specialized equipment. As part of pre-impact preparation, it is also necessary to identify organizations that have similar, if not identical, functions and add them to the list of response agencies. In New York City, for example, numerous agencies have the ability to respond to hazardous materials events, including the fire department (FDNY), the police department (NYPD), the sanitation department, the environmental protection department, and the New York–New Jersey Port Authority.

The management of injuries and fatalities is a critical aspect of both planning and response. Emergency services officials must determine what types of injuries or illnesses can best be managed by each hospital; for example, burn and trauma victims, and those who have been exposed to toxic agents, will need specialized care. Officials must also identify both (1) medical facilities that are capable of handling large surges and (2) alternate (nonhospital) medical facilities that can assist with the treatment of minor injuries and the distribution of emergency pharmaceutical supplies and other resources. If quarantine orders are issued, municipal officials must not only

A model emergency management program

Because the economy of Tarpon Springs (pop. 21,000), a coastal Florida community, relies so heavily on tourism, it is crucial that city operations return to normal as quickly as possible in the wake of a disaster. Thus, the city's Model Emergency Management Program is designed to provide efficient and effective emergency operations with nationally recognized facilities and a quality employee/family shelter system.

The program relies primarily on its state-of-the-art facilities. The emergency operations center (EOC) is located in a multipurpose public safety facility that houses police, fire and rescue personnel, and equipment. Opened in April 2001, the facility was built to withstand a Category 3 hurricane and to serve as an EOC, with tilt-wall construction, roll-down garage doors, hurricane-force deadbolts, impact-resistant glass, and an 800-kW generator with a 6,000-gallon fuel tank to operate the facility in the event of a power outage. Other features include a special hurricane storage area that converts into a staging area within the building; an advanced electronics system; closed circuit television; reciprocal communications systems for radio, microwave, cell, and satellite phone throughout the building; separate facilities for media and policy-making officials; interview rooms; card key-access security; and sleeping/dining facilities. (Food for other emergency workers is coordinated through the Elks Club. The city provides for a generator and food delivery.)

A second facility is located adjacent to the EOC. During nonemergency times, it houses the city's recreation department and gymnasium. During emergency planning and activation, this building, which is also hardened to withstand a Category 3 hurricane, serves as a shelter for employees, including first responders, and their family members. A separate area houses a pet shelter. The family/pet shelter is staffed by the city's human resource staff. The parking area is used as a staging area for post-emergency vehicles. This building also acts as a backup facility for the EOC.

Critical planning aspects of the city's emergency management program include

- Standing orders for generators for critical facilities and shelters
- Standing food service orders
- Maintenance of a debris management contract for post-storm activities
- Extensive coordination with utility emergency response personnel, and the provision of staging areas, shelter, and food for these workers.

The city has been recognized locally for several of these features, as well as for an all-hazards Hurricane Implementation Guide that delineates duties for all city personnel, and a strike force team approach that ensures canvassing before the event in addition to timely damage, rescue, and security needs assessments after the event.

The comprehensive emergency management program ensures the most efficient and effective emergency operations for city residents while providing for the safety and security of program personnel, their families, and their pets. Should a disaster occur, Tarpon Springs is prepared with a high level of emergency services.

The Tarpon Springs Model Emergency Management Program and city manager Ellen C. Posivach received the Public Safety Program Excellence Award in the 10,000-to-49,999 population category from ICMA in 2005.

Source: "2005 ICMA Annual Awards" supplement, *Public Management (PM)* (September 2005): 14.

ensure that citizens under quarantine receive appropriate medical care, but also make provisions for hygiene, feeding and sheltering, economic, and mental health needs. (Medical response to emergencies and disasters is discussed in detail in Chapter 10.)

Emergency managers must also plan for the possibility of mass fatalities, taking into account how the causes of death, criminal investigations, and religious requirements for burial may affect operations. Mass fatalities also create legal issues, particularly in relation to child protection; authorities must be prepared to deal with children who have been orphaned. Terrorist attacks, in particular, create demands that may complicate disaster response. For example, efforts to preserve evidence can conflict with other response operations.

Recovery, the longest part of a disaster time line, presents a multitude of challenges. Restoration of essential services is a demand typically dictated by the agent. A large hurricane will damage a large geographical area, affecting roadways, bridges, tunnels, and medical facilities, as well as police, fire, and emergency medical services stations. The rapid outbreak of a disease will have an entirely different array of effects. Measures to protect against continuing threats and to restore order are similarly unique to the agent, and are usually undertaken at the same time as efforts to restore essential services. In the wake of a disease outbreak, for example, authorities may be simultaneously trying to ensure that medical facilities have the necessary resources to deal with the illness, trying to prevent further contamination, and trying to maintain community order and stability through public outreach.

Response-generated demands

Unlike agent-generated demands, which are unique to the event, response-generated demands originate in the efforts to deal with disaster. E. L. Quarantelli, Russell Dynes, and Gary Kreps have identified five response-generated demands, first and foremost of which is *communications,* the most important function in any response operation.[3] Good communication processes are essential for emergency managers to effectively assess the situation and to *mobilize and dispatch human and material resources* where they are needed. *Continuous assessments* are required to ensure that communication processes are working and that information is being funneled to the appropriate places. Damage assessments, threat analyses, and estimations of the size and scope of impact, for example, provide essential information that enables agencies and officials to respond effectively.

Coordination is a response-generated demand that is crucial for dealing effectively with disaster. Simply put, coordination means organizing entities behind a common purpose. Generally, this involves identifying lead and support agencies for the numerous missions that must be undertaken during disasters. The need for coordination comes from the fact that disaster response is typically a multiagency, multijurisdictional endeavor; NGOs, including emergent groups that did not exist prior to the disaster, are also likely to be involved. Some of the responding entities may have no prior experience in response; others may have response experience but may not have worked together before.

Good communication processes are essential for emergency managers to effectively assess the situation and to mobilize and dispatch human and material resources where they are needed.

Integrating NGOs and emergent organizations into response is a good strategy, but it does require effort. The goal is to prevent what Dr. Erik Auf der Heide refers to as the "Robinson Crusoe syndrome," in which each organization believes that it is the only one on the island.[4] When disparate entities work together, they have to develop techniques to coordinate efforts in order to maximize the human and material resources available to cope with disaster.[5]

The last response-generated demand, but certainly not the least important, is *control and authority,* which is required for the implementation of an effective, coordinated response. The person who bears ultimate responsibility for the outcome of disaster response is the leading elected official: the mayor or county executive at the local level, and the governor at the state level. But it is also imperative that others in leadership positions—regardless of whether they are in charge of a lead or a supporting agency—send a strong message about the critical importance of coordination and cooperation across all organizations and between all governmental levels.

Emergency operations centers

Local, state, and federal governments; hospitals; large utility companies; universities; and school districts should have a designated center for the coordination of both daily and emergency operations. An EOC facilitates management of the agency's responsibilities, committed and available resources, and level of involvement in an event. Some operations centers are designed to accommodate other agencies or organizations with similar responsibilities, either at the same or at different levels of government. The FDNY and the NYPD both have operations centers that are open twenty-four hours a day, seven days a week; New York City's health department has

Emergency response in Minnesota: The collapse of Interstate 35W

Minnesota officials said the emergency response to the deadly failure of a major bridge in Minneapolis went smoothly with the exception of some communications glitches, in an event that is being viewed as a good test of a large regional city's ability to respond to terrorist attack or natural disaster.

The Minneapolis police and fire chiefs and the Hennepin County sheriff jointly led rescue and recovery efforts after the forty-year-old Interstate 35W bridge plunged into the Mississippi River, killing at least four and injuring seventy-nine. The sheriff's office organized operations in the water while the fire department managed those on the ground and police officials secured the scene, officials said. It was all part of a unified command set up according to the principles of the National Incident Management System, a federally devised plan to help governmental entities work together after terrorist attacks or natural catastrophes.

"It keeps us all on the same sheet of music," said David Berrisford, state incident manager for Minnesota's Homeland Security and Emergency Management division.

The response involved at least seventy-five state, local, and federal agencies linked through a radio system that has been enhanced with some of the $170 million in federal homeland security grants the state has received since 2002, officials said.

Although the system—rated among the nation's best this year by the Department of Homeland Security—generally worked well, officials said, there were reports of it jamming at moments of peak use.

In one case, twenty-second delays hampered the lone channel set aside for emergency medical services supervisors, said John Hick, emergency physician for Hennepin County Medical Center and an architect of the mass-casualty response plan for twenty-nine hospitals in the region. Responders sometimes turned to cell phones, which worked only intermittently, he said, but even so, the system was a "miracle compared to what we used to have."

James Franklin, executive director of the Minnesota Sheriffs' Association, said cellular access is used only to augment shared radio frequencies. "This is a problem we run into in every single disaster," he said. "Everybody begins to use cell phones and the system jams up."

Kristi Rollwagen, deputy chief of emergency preparedness for Minneapolis, called cellular problems "just an inconvenience," adding: "It did not affect us operationally."

The state has spent about $10 million to develop rescue teams for collapsed structures. At least twenty members of the teams helped with rescue and recovery efforts using listening devices, trucks, and other equipment, Rollwagen said.

The Minnesota National Guard launched a UH-60 medevac helicopter Wednesday night, said Maj. Patricia Baker, a spokeswoman. As many as 10,000 Guard members are ready to help, she said, adding that the deployment of 3,000 Guard soldiers to Iraq and elsewhere won't impede the efforts.

Area hospitals and ambulance services coped well with what turned out to be a relatively small number of victims compared with mass-casualty events they have trained for, officials said. Six ambulance services transported fifty-five victims within 1 hour 53 minutes of the first call.

In disasters large and small, it often takes days before problems with emergency response begin to emerge clearly. But so far, Berrisford said, "there haven't been any significant issues. Our first responders and our state agencies know how to respond—and, quite frankly, we just do it," he said.

Source: Christopher Lee and Paul Lewis, "With Minor Exceptions, System Worked," *Washington Post*, August 3, 2007, A09. © 2007, *The Washington Post*. Reprinted with permission.

an operations center that is open during normal business hours and during emergencies. New York State's Emergency Management Office has an operations center—referred to as an emergency coordination center—that helps coordinate nonemergency activities and, during emergencies, coordinates the resources and operations of all (and only) New York State agencies. At the federal level, the Federal Emergency Management Agency (FEMA) and the U.S. Department of Homeland Security (DHS) both have operations centers.

An EOC provides an environment for the work of an emergency response organization network (ERON)—an assemblage of representatives from all levels of government, the private sector, nonprofit organizations, and, at times, emergent groups; this network manages critical response functions, maintains continuity of critical social services, and expedites recovery.[6] The organizational representatives, while remaining autonomous, work cooperatively and synergistically, in accordance with emergency management plans and agreements, as a unified decision-making force.[7] One role of the ERON is to organize teams drawn from different entities, creating dedicated units that handle specific response functions. During the anthrax attacks of 2001, for example, temporary teams consisting of staff members from the NYPD, FDNY, and the New York City Office of Emergency Management (OEM) were organized to respond to the barrage of alarms.

Given its position as the locus for coordinating the multiagency, multijurisdictional response to a disaster, an EOC should be designed to accommodate the following ten functions: coordination, surveillance management, levels of activation, information management, planning, operations and missions management, policy making, public information, physical environment, and hosting visitors.

Coordination

The coordination of all disaster-related activities is the single most important function of an EOC. Coordination involves the interaction of multiple entities from various levels of government and from the private sector. Because it is the location where plans are activated and where personnel will be on duty during emergencies, the EOC is typically used in the design and testing of EOPs, the education and training of personnel, and the implementation of EOPs during response and recovery efforts. Thus, familiarity with the EOC is essential.

Surveillance management

Surveillance management is the monitoring of local, regional, and national conditions to maintain constant situational awareness. Items of interest might include current and forecasted weather; seismic activity; national and local terrorist threats; incidence of crime, fires, and use of emergency medical services; air-, rail-, marine-, and road-traffic conditions; the effects of construction or repair on communications and traffic flow; the supply of electrical power; disease outbreaks; and hospital surge capacities.

Surveillance is not simply a matter of being vigilant about changes that may signal the onset of an event; it means monitoring activities that may affect current operations. In the weeks and months after the attack on the World Trade Center, for example, New York City's OEM also responded to hundreds of anthrax alarms; labor unrest; the threat of Hurricane Michelle, which was traveling up the East Coast; the crash of Flight 587 in Queens; and two minor earthquakes in Queens and Manhattan. Surveillance allows the emergency manager to continuously reevaluate strengths, weaknesses, and vulnerabilities; current resources; and expected disaster demands.

Levels of activation

Levels of activation are used to determine staffing needs for the ERON. It is recommended that adjoining municipalities, neighboring counties, and nearby school districts, hospitals, and businesses use identical or compatible terms to distinguish among activation levels for their EOCs. The use of a common terminology, which may include color coding, can help neighboring jurisdictions understand the severity of the emergency or disaster, assess which level of activation might be appropriate, and determine whether requests for mutual assistance may be forthcoming.

The accompanying sidebar contains a description of activation levels from the policy section of the Nassau County, New York, EOP.

Information management

Information management is crucial for an effective response to large-scale events. Today, dedicated software programs provide the capability to manage the massive influx of data passing through the EOC. The Office of Justice Programs and the National Institute of Justice, both of which are within the U.S Department of Justice, periodically evaluate software programs made

Activation levels in the Nassau County emergency operations plan

Alert status: Office of Emergency Management (OEM) staff ensures the readiness of the emergency operations center (EOC) as well as notification of EOC representatives for a potential higher level of activation. This normally occurs during a U.S. Department of Homeland Security threat level of orange and/or the anticipation of worsening weather conditions, potential labor unrests, or any prodromal issue that may warrant an activation of levels 1 through 3 (see below):

- **Level 1:** Significant risk of civil, natural, or technical threat indicates a level of heightened awareness. Such activation will require selected OEM staff and some selected county, town, and state representatives.
- **Level 2:** Conditions posing a credible threat to public safety have been identified, or an event has occurred that has the potential to threaten public safety and stability. An activation requires all OEM staff and a more broad attendance of county, town, and possibly state representatives. New York City and Suffolk County representatives will also be requested to send representatives.
- **Level 3:** Condition of highest alert or an event has occurred or will most likely occur that will have significant impact on Nassau County. An activation requires all OEM and reserve staff, and a full range of county, town, New York State, and federal representatives. New York City and Suffolk County representatives will be requested to participate. Selected representation will be called from Nassau's villages, our two cities, hospitals, fire and school districts.

The EOC will remain in a prepared, inactivated status during our national and local threat levels GREEN, BLUE, and YELLOW. The activation of the EOC will be based on national and/or local threat levels from natural, technological, and civil hazards.

During a level ORANGE, OEM will place the EOC on alert status. Based on intelligence gathering from local, state, and federal daily situation reports, the OEM commissioner will advise the county executive to elevate the condition to level RED, remain at level ORANGE, or deescalate to level YELLOW or lower.

Once at level RED, the EOC will be activated, and county staff will be detailed to the EOC. Additionally, a joint information center will be established at the EOC, and a specific emergency plan will be reviewed and placed into effect.

Source: Adapted from the policy section of the emergency operations plan for Nassau County, New York.

specifically for EOCs; the resulting reports allow emergency managers to compare products and select those that are most appropriate for their needs.

Critical features of a software product for an EOC include

- An Oracle-based (or similar) function for encryption capabilities
- Real-time, dynamic-impact modeling of traffic, gas, or vapor plumes and high winds
- Storage of critical documents, including EOPs
- Archiving capability for audio, visual, and text data
- Current demographic data
- Conformity with the National Incident Management System (NIMS)
- Separate user log-in for individual emergency support functions
- A prompt system to remind users of situational reports, notifications, and plan activations that are due
- Databases for skills inventory, material resources, dynamic damage assessments, fatality and injury tallies, debris estimates, and similar data
- Archives for financial data, including overtime, emergency contracts, and purchases eligible for financial recovery through the Stafford Act
- Geographic information systems (GIS)
- Global positioning systems
- Capacity to import video surveillance from transportation, government, and businesses facilities
- Situational reporting.

The shuttle *Columbia* disaster: Lessons learned

At 8:00 AM on Saturday, February 1, 2003, the space shuttle *Columbia*, on her way back to the Kennedy Space Center in Florida after a successful mission, reentered the Earth's atmosphere with unknown damage to her left wing. *Columbia* broke up over the western United States at 200,000 feet and 12,000 miles per hour. The disintegration was a tragedy because the lives of seven heroic astronauts and the life's work of countless engineers were lost, and a disaster because more than 87,000 pounds of debris were strewn over 2,000 square miles of East Texas and western Louisiana.

The response

What followed was one of the largest search and recovery efforts in history. Ultimately, 25,000 people from 450 organizations worked for 100 days to recover the shattered space shuttle and her crew. At the heart of this intergovernmental, interagency, cross-sector response were the myriad local-level agencies and organizations in whose backyards this event unfolded. Public officials and emergency responders activated their command centers and marshaled their resources, responded to hundreds of calls from citizens, deployed police officers and firefighters to safeguard shuttle material, and coordinated with the media to provide guidance to citizens. Thanks to existing plans, relationships, training, and experience, communities were able to protect their citizens in the immediate term and position themselves to sustain a major national response over the coming weeks.

Meanwhile, President George W. Bush declared emergencies in Texas and Louisiana, and U.S. Department of Homeland Security secretary Tom Ridge organized recovery efforts under the Federal Response Plan, then in effect. The operation was directed jointly by the Federal Emergency Management Agency (FEMA), as the lead federal agency for consequence management, and the National Aeronautics and Space Administration, as the lead agency for the investigation of the accident, with significant leadership roles assumed by the U.S. Environmental Protection Agency (for recovery of the debris and mitigation of hazardous materials), the Federal Bureau of Investigation (for recovery of *Columbia*'s crew), the U.S. Forest Service and Texas Forest Service (for the ground-based search for debris), and the U.S. Navy (for the underwater search).

Disaster field offices were established at a Louisiana air force base and near Dallas, but most federal agencies and many state and local assets converged on the emergency operations center in Lufkin, Texas, which became the nexus of operations.

For the next two weeks, thousands of responders, many of whom were volunteers, worked to remove hazardous material that had fallen in public areas and to search for the remains of the crew. Police officers, firefighters, Salvation Army and Red Cross volunteers, National Guard troops, urban search and rescue teams, SCUBA (self-contained underwater breathing apparatus) divers, horsemen, farmers, ranchers, members of the Veterans of Foreign Wars, forest rangers, private businesses, and many others came together with representatives from thirty federal agencies to find *Columbia* and her crew.

One hundred days of hard work brought the operation to a close. The total ground, air, and water search had covered over 2.28 million acres. More than 84,000 pieces of material had been recovered, amounting to about 38 percent of the landing-configured weight of the orbiter—an astounding figure, given that the National Transportation Standards Board had predicted that only 10 to 20 percent would be found.

Ultimately, FEMA projected that it would reimburse local governments $10 million for costs associated with the effort. For the recovery operation itself, FEMA had spent about $245 million when the Lufkin Disaster Field Office was finally closed.

Lessons learned and reinforced

As communities work to refine their emergency management systems to meet the demands of today's social, political, physical, and technological environment, the lessons of *Columbia* are especially pertinent.

Lesson 10: Planning makes all the difference. Public officials in Nacogdoches, Texas, attest to the payback they reaped from having recently completed a hazard mitigation planning effort. Said the county sheriff, "The whole time we were drafting these plans, you're thinking tornadoes, hurricanes.... Never in our wildest dreams did we envision something like a space shuttle crashing upon our county. But [planning] played a crucial part. When this incident happened, we didn't spend the first two or three hours scrambling around."

Lesson 9: Focus on problem solving. Large incidents are idiosyncratic; much about them doesn't "fit the rules." Thus, these events are better addressed with flexible guidelines than with rigid rules and fixed processes. The *Columbia* response succeeded in part because the state and local governments involved gave their people enough discretion to determine what needed to be done in the specific circumstances they faced and to make decisions in real time.

Lesson 8: Identify capabilities and resource needs ahead of time. Local officials were not immediately sure what they needed and how it would be used, and were reluctant to activate resources unnecessarily. At the same time, they did not want to be caught short if the incident turned out to be larger than they initially guessed. On top of these concerns, they had to decide what to do with offers of help coming in from all over the country. And once the managers had obtained the resources they needed, they had to deploy them, direct them, track them, support them—and pay for them.

Lesson 7: Overcome jurisdictional conflicts. In the *Columbia* recovery effort, the interagency turf battles that so often plague emergency responses were almost absent. As one local official noted, "The remarkable thing is that out of between seventy and eighty agencies that were there, there wasn't the first unkind word. There was no power play, no power struggle. Everybody there just knew that the job was there to be done, and all of us had the same purpose." The lead agencies explicitly worked to develop a common understanding of the mission, universal buy-in to the division of responsibility, and clearly articulated priorities.

Lesson 6: Don't compromise on training. The *Columbia* response reaped the benefit of the commitment to good training that distinguishes many organizations in Texas and Louisiana. One state police captain explained: "We practice and practice and practice. We drill our emergency preparedness, and it's clear that because of that, we were able to respond and respond well. We knew who could do what and maybe even where some of the pitfalls would be and where the shortfalls would come."

Lesson 5: Understand the federal agencies, and help them understand you. Large-scale incidents almost certainly involve federal agencies and may be run by them. In reality, many in the federal government do not understand local capabilities very well, and federal agencies tend to make one-size-fits-all decisions meant to apply across a diverse array of localities. Thus, local officials need to educate federal responders, probably multiple times over the course of a long-duration incident.

Lesson 4: Information is hard to obtain, validate, and share. One of the biggest frustrations expressed by local officials was how hard it was to get the information they needed to direct their operations appropriately. In part, this was because good information didn't exist early on, but it was also because federal agencies were not always forthcoming—sometimes because they legitimately needed to protect sensitive information, but in other cases because it simply didn't occur to them. And when there was information to share, it was hard to distribute, especially given the need for security and the lack of infrastructure in rural East Texas. Managers had to actively seek information, figuring out who had it and convincing others to keep them "in the loop."

Lesson 3: Complex incidents require sophisticated management. To the frustration of the local governments, who were pressed by immediate response demands, it took time for a coherent incident command structure to emerge. When it did, decision authority was shared by a group of agencies that worked closely together. The recovery operated—in spirit, if not formally—as a unified command. It is important, though, for all the agencies involved to be aware of the incident command structure. On the *Columbia* response, although a command structure did exist in the disaster field office, many local governments still couldn't tell who was in charge.

Lesson 2: Be proactive about communication. One of the hardest things to do well during a large-scale incident is to communicate. On *Columbia,* the initial leadership structure was diffuse, with federal, state, and local field offices, operations centers, and command posts all directing parts of the operation. The need to coordinate well did prompt the initiation of a series of regular meetings and teleconferences, which helped facilitate information sharing and decision making. Nonetheless, local governments had to be proactive about engaging federal agencies, which often did not think to introduce themselves to the leaders of the communities in which they were working.

Lesson 1: Relationships matter. All these lessons hinge on one common element: relationships. Agencies and jurisdictions that know each other work together better. In fact, perhaps the most important benefits of joint planning and training are not the plans or skills themselves but the familiarity and understanding that accrue among the agencies involved. Building habitual relationships in times of relative peace and calm with those who will be partners during a crisis eases and enables communication, coordination, and cooperation. Mutual understanding reduces the chaos and confusion that arise simply from lack of familiarity and allows organizations to operate more seamlessly. Perhaps a silver lining to the *Columbia* incident is the host of new relationships that were forged over the course of that event and that may support future disaster responses.

Source: Adapted from Amy Donahue, "The Day the Sky Fell: The Space Shuttle *Columbia* Disaster," *Public Management (PM)* (September 2004): 8-12.

Situational reports, the most important products created through the use of computer software, should be able to compile and display the following:

- An event identifier (often, the location and the type of disaster) and information on the nature, scale, and other characteristics of the event
- An organizational chart showing current EOC staffing
- Current incident objectives
- Project objectives and missions
- Complications and impediments
- Surveillance information
- Reports from individual ERON representatives.

Periodic training in the use of dedicated emergency management software should be conducted as a shadow exercise, in conjunction with field exercises, in order to provide refresher training for current EOC staff; provide training for new ERON members; update databases; and ensure that computers, printers, plotters, faxes, screens, and other hardware are in good working order.

Planning

The EOC should be the repository of current plans from mutual aid partners and all levels of government. Stored plans should include the National Response Plan, NIMS guidance, and other relevant federal plans, as well as state and regional plans that augment local EOPs. Plans and mutual aid agreements from NGOs, special districts, schools, universities, businesses, nonprofit organizations, and Voluntary Organizations Active in Disaster (VOAD) should also be part of the EOC library of documents. Plans should be updated periodically and maintained within the EOC software so that the emergency manager can easily refer to them during response.

In addition to providing ready access to plans, the EOC should facilitate incident action planning. Such planning is often necessary during an emergency to cope with emerging issues that require a multiagency response under uncertain conditions. Once an incident action plan (IAP) has been developed, the agreed-upon course of action must be communicated to all appropriate entities to prevent confusion and conflict; this is where emergent multiorganizational networks (EMONs) come into play during response. For example, EMONs may be tasked with an emergency support function, such as providing mass feeding or emergency shelter, and the IAP should indicate when and how the appropriate EMONs will be mobilized. An IAP might also result in the formation of special task forces, and in the creation of mission statements that identify lead and supporting agencies for each mission. As noted earlier, to respond to the hundreds of calls about potential anthrax sites, New York City's OEM managed dozens of special teams.

Operations and missions management

Operations and missions management is the component of the ERON that, in effect, performs the unified command role from the EOC: this includes determining which agencies will perform specific tasks and activities, and monitoring the deployment of critical resources. The result is a level of coordination and cooperation that ensures that each organization has the necessary authority to fulfill its responsibilities. As an event progresses, both agent- and response-generated demands change, requiring the activation of planning annexes and/or the creation of IAPs that establish new missions.[8] The goal of operations and missions management is to ensure that as demands change, the requisite shifts in mission, tasks, and activities—including, for example, the transformation of lead agencies into support agencies—are seamless. For operations and missions management to fulfill its function, good communication between the emergency operations center and the field is essential.

Policy making

All disasters require high-level decisions that are based on information that the emergency manager collects, analyzes, and presents to the chief executive, chief administrative officer, and/or governing body. Such decisions must, of course, take into account ordinances and regulations at both the local and state levels. In New York State, for example, Article 2-B of the state's executive law addresses various legal issues that arise during a "declaration of state of emergency," including quarantine, criminal and terrorist investigations, mandatory evacuation, the commandeering of facilities, and the relaxation of regulations.[9] Moreover, emergency managers and legal experts must be vigilant in ensuring that response actions comply with constitutional law.

Public information

Public information is managed through a joint information center (JIC) within the EOC, where public information staff from all the agencies involved in the response effort are gathered. The purpose of the JIC is to ensure that a single, credible, and consistent message is disseminated to the media and to the public. Typically, the JIC is the source for information on personal protective measures, traffic conditions, weather forecasts, evacuation recommendations, the location of shelters, hotline numbers, and Web addresses for additional resources.

Physical environment

FEMA has a checklist of the physical features of an EOC, which focuses on survivability, security, sustainability, interoperability, and flexibility. According to the checklist, the following issues should be considered:

- Square footage, which will be affected by the following factors:
  - Anticipated staff per shift
  - Equipment
  - Security personnel
  - Parking, berthing, and landing zones
  - Press, visitors, and political leaders
- Communications systems, including
  - Phone lines
  - Computer servers
  - Satellite equipment
  - Radio and television broadcasting equipment
- Redundancies in power generation and water supply
- Conference rooms, including a policy conference room
- A secure communications room
- Multiuse spaces
- Sleeping accommodations
- Kitchen and/or cafeteria
- An operations floor large enough to accommodate the ERON members.

Pier 92, on Manhattan's West Side Highway, was home to the temporary EOC for twelve weeks after the World Trade Center attack. This space was approximately 1,000 × 250 feet, or 250,000 square feet. The operations floor housed 150 representatives from federal, state, and city government and from private and nonprofit organizations—the largest ERON ever assembled. When visitors, press, and security personnel were included, the population of the EOC reached well over a thousand.

As shown in Figure 8–3, ERON representatives who were responsible for similar emergency support functions were grouped in the same cell (some local governments refer to cells as communities, sections, or beehives). These cells (not all of which are included in the figure) were the Podium, Utilities and Infrastructure, Law Enforcement, Fire and EMS (emergency medical service), Military, Debris Management and Transportation, Joint Information Center, Human Services/Special Needs, Health and Medical, Government, Administrative, Communications, and GIS. Cells were left open for emergent entities; for example, New York Waterway, Nextel, Motorola, and IBM provided crucial support in the wake of the 9/11 attacks.

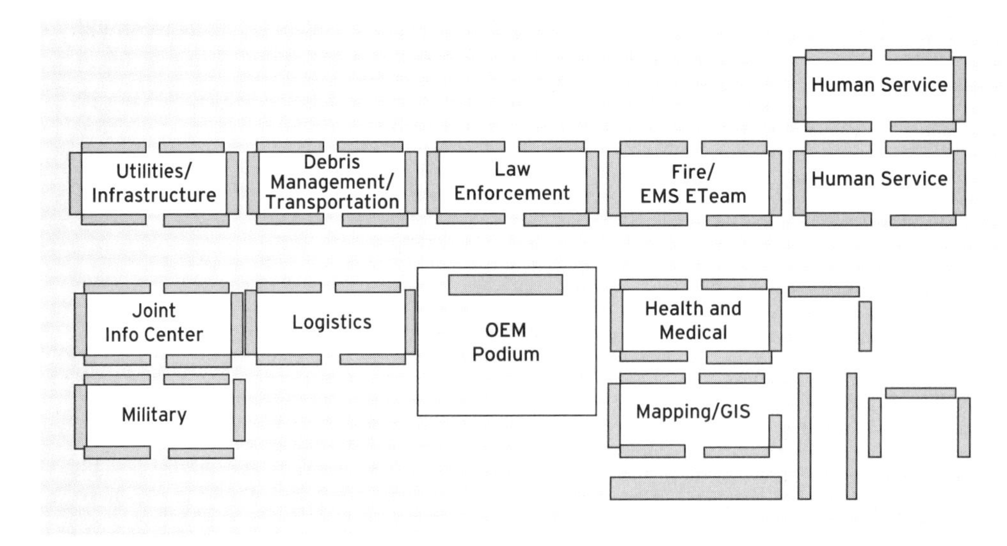

Figure 8-3 New York City's emergency operations center, January 2002

The Podium housed the emergency management staff, along with support representatives from FEMA and the state emergency management office. Among the numerous functions of the Podium staff were coordination of all EOC, ERON, and EMON functions; management of the EOC; storage and dissemination of critical information; monitoring and management of event logistics; monitoring and management of field operations and missions; generation of situation reports; establishment of time lines for operations, both in the field and within the EOC; recording and monitoring legal activities such as emergency declarations; maintaining continuity of EOC operations, including the functioning of communications and information technology systems; and credentialing.[10]

The ERON that occupied the Human Services/Special Needs cell was made up of representatives from the Red Cross, the Salvation Army, City Harvest (a nonprofit food rescue organization), the New York City Community Assistance Unit, the New York City Board of Education, New York University, New York Cares, Catholic Charities, and World Church Services—to name a few. Among the functions undertaken by that cell were volunteer and donations management, the establishment of respite centers for relief workers, and mental health assistance.

Hosting visitors

One of the purposes of the EOC is to provide an environment for learning, so that mistakes will not be repeated and exemplary practices can be integrated into operations. Areas should be set aside in the EOC for out-of-town emergency managers, who can provide backup staffing and expertise; for researchers and other potential advisers; and for VIPs, including state and federal officials, members of Congress, foreign dignitaries, and public figures.

Establishing relationships with other emergency managers—at the local, state, and federal levels—is crucial for the exchange of best practices and lessons learned; emergency managers should also make a special effort to form connections with colleges or universities that offer programs in emergency management, conduct disaster research, or both. During the first days of response to the attacks on the World Trade Center, for example, New York City's OEM gave researchers from the Disaster Research Center at the University of Delaware and from the Henry Stimson Center, a Washington, D.C., think tank, access to the EOC and other operations, so that critical issues could be analyzed firsthand. Valuable studies are still being generated as a result.

Summary

Emergency managers must constantly evaluate threats, anticipate demands, and determine how best to address those demands. The quality of the preparation largely determines the effectiveness of the response. The EOC should provide an environment that fosters cooperation, coordination, and collaboration. The design should facilitate communication among all organizations involved in response, and should be open to outsiders who can assist with operations and help identify lessons learned. Information technologies can help, but the key to communication is having close working relationships with those who are likely to be involved in disaster operations and with others who might assist.

A crucial component of preparation is the development of a response strategy; such a strategy should focus on anticipated policy decisions—not how to run a shelter, but how to determine, for example, what constitutes an acceptable shelter and how long shelters will be open. A response strategy is typically more flexible than a response plan: it provides general guidance that leaves room for improvisation as circumstances change.

The emergency manager's principal job is the development of a broad strategy that includes long-term recovery and the reduction of vulnerabilities. In fact, Lucien Canton, director of San Francisco's Offices of Emergency Services, suggests that the Achilles heel of emergency managers may be their tendency to focus on EOPs rather than on the long term.[11] When the response operation is viewed within the broader context of community resilience, the emergency manager's role is to anticipate how the disaster will develop, ensure that essential resources are available, and ensure that the community's long-term goals guide the response.

Notes

1 Gerard J. Hoetmer, "Introduction," in *Emergency Management: Principles and Practice for Local Government* (Washington, D.C.: ICMA, 1991), xvii–xxxiv; Patrick Lagadec, *Preventing Chaos in a Crisis,* trans. Jocelyn M. Phelps (New York: McGraw-Hill, 1993).

2 E. L. Quarantelli, Russell Dynes, and Gary Kreps, *A Perspective on Disaster Planning,* Report Series 11 (Newark: Disaster Research Center, University of Delaware, 1981), 8, available at dspace.udel.edu:8080/dspace/bitstream/19716/1259/1/RS11.pdf (accessed August 24, 2007).

3 Ibid, 11–12.

4 Erik Auf der Heide, *Disaster Response: Principles in Preparation and Coordination* (St. Louis, Mo.: C. V. Mosby, 1989), 57, available at orgmail2.coe-dmha.org/dr/PDF/DisasterResponse.pdf (accessed August 24, 2007).

5 See, for example, Thomas E. Drabek and David A. McEntire, "Emergent Phenomena and Multiorganizational Coordination in Disasters: Lessons from the Research Literature," *International Journal of Mass Emergencies and Disasters* 20, no. 2 (August 2002): 207–213, for a discussion of techniques to overcome coordination problems.

6 Karl E. Weick, "The Collapse of Sense-making in Organizations: The Mann Gulch Disaster," *Administrative Science Quarterly* 38, no. 4 (1993): 628–652.

7 Tricia Wachtendorf, *Improvising 9/11: Organizational Improvisation following the World Trade Center Disaster* (Newark: Disaster Research Center, University of Delaware, 2004).

8 All-hazards plans typically include annexes that cover specific kinds of disasters, such as terrorism or hazardous materials accidents, that may require specialized operations. There may also be an annex to cover particular sites (e.g., zoos, museums, and oil refineries) that may require specialized response operations. Annexes may also include additional detail about emergency support functions, such as the roles and responsibilities of lead and supporting agencies.

9 See New York State Executive Law, Article 2-B, "State and Local Natural and Man-Made Disaster Preparedness," available at semo.state.ny.us/uploads/Article%202-B.pdf (accessed August 27, 2007).

10 Participants are typically issued credentials that authorize access to one or more sites, such as the disaster site and the emergency operations center. The credentialing process has become much more sophisticated in recent years: identification badges can now be issued that include photographs, biometric information, and color codes detailing levels of access. Jurisdictions can choose among a variety of technologies for badges and identification cards.

11 Lucien G. Canton, *Emergency Management: Concepts and Strategies for Effective Programs* (Hoboken, N.J.: Wiley Interscience, 2007), 335–336.

CHAPTER 9

Disaster response

Ronald W. Perry and Michael K. Lindell

This chapter provides an understanding of

- Five disaster-response operational challenges
- The federal context of disaster response after 9/11
- The federal bureaucratic framework
- The Urban Areas Security Initiative
- The Metropolitan Medical Response System
- The National Incident Management System.

Local governments deal with emergencies, disasters, and catastrophes.[1] Briefly, emergencies, which are unforeseen but predictable and narrow in scope, occur frequently and are dealt with by means of standard operating procedures. Emergencies are house fires, vehicle accidents, medical crises, tunnel collapses, small hazardous materials (hazmat) releases, and the like, and although they may be terrible for those involved, they usually do not significantly impede the operation of the community as a whole or threaten government's ability to deliver services, nor does managing them demand many (if any) extra-community resources.

Disasters, in contrast, disrupt social interaction and interrupt the ability of major community systems to afford reasonable conditions of life; moreover, they require resources from outside the community. However, they are confined to a sufficiently narrow geographic area that resources can come from nearby "external" sources.

A catastrophe differs from a disaster in at least three ways. First, it embraces multiple communities so that support from outside the community is necessarily quite limited or curtailed. Second, the levels of damage and interruption of social intercourse are greater in a catastrophe than in most disasters: most of the built environment is damaged or destroyed; places of employment, recreation, worship, and education are gone; and community response mechanisms are interrupted. Third, most community functions are sharply and concurrently interrupted; with critical infrastructure systems failing simultaneously, emergency responders must prioritize or triage the restoration of critical services. (For more detail on the differences among emergencies, disasters, and catastrophes, see Chapter 1.)

Communities prepare for all three occurrences, but this chapter focuses on planning for response to disasters and catastrophes. The locus of planning and response for disasters is jurisdiction-wide and usually requires resources from outside the jurisdiction, sometimes from state government but often from adjacent local governments. For catastrophes, the locus is more regional, and the involvement of state and national government is required, as is regional assistance well beyond the scope of normal mutual aid agreements.

Research on disasters reveals consistent patterns of organizational response. Some aspects of these patterns are adaptive. For example, in the event of a disaster, outside individuals and organizations will converge on the impact area without a formal request for assistance. And as those outside the emergency response organization identify unmet needs, they organize to form emergent groups to fill the gaps. Where procedures for accomplishing tasks or coordinating organizations are lacking, people improvise responses to protect lives and property.

Eliminating maladaptive patterns effectively requires that the structure and resources of all organizations engaged in the local emergency response system be clarified pre-impact.

Other aspects of these same patterns, however, are maladaptive. The arrival of people and organizations with unknown capabilities, together with poor communication among organizations, creates ambiguity about who is in charge. Convergence often produces duplication of some tasks, neglect of others, underuse of resources, and tardy response to disaster demands. When maladaptive patterns are recognized during emergency response, participating organizations may attempt to strengthen interorganizational coordination by creating emergent multiorganizational networks (EMONs).

Eliminating maladaptive patterns effectively, however, requires that the structure and resources of all organizations engaged in the local emergency response system be clarified pre-impact. The comprehensive emergency operations plan (EOP) (see Chapter 7) should document the roles of all the organizations involved in response. This higher level of organizing and planning will minimize the need for improvisation in the field and will reduce reliance on EMONs.

If the organizational structure established by an EOP cannot meet the demands of a given incident, the EOP will be inadequate. In practice, no EOP is likely to be completely appropriate to the demands of every type of incident or every magnitude of event (emergency, disaster, or catastrophe). To limit the inherent potential for an event to outmaneuver the plan, so to speak, emergency managers attend to two broad areas that are critical in promoting preparedness for response operations: operational challenges and the availability of federal resources. Managers attempt to routinely incorporate into their thinking and response planning the five principal operational challenges that have been documented in disasters for decades; and they draw on resources from the national government, not only funding to develop local emergency planning but also national initiatives—which, with the associated guidance and requirements, have become more prominent since September 11, 2001.

Disaster response challenges

The five response-related principal challenges that are critical to the effectiveness of an emergency response plan are (1) understanding the responses of citizens to disasters, (2) managing the convergence of resources and people on the disaster scene, (3) limiting the likelihood of role abandonment by emergency workers who are responding to the demands of disaster operations, (4) obtaining disaster declarations, and (5) addressing communications among responders during operations. Since it is known that these issues arise in almost every disaster and are even more critical in catastrophes, emergency managers can plan for them in advance by using the process of defining the challenge and developing a protocol that can be quickly implemented. Advance planning reduces the load on personnel and emergency responders during the response phase itself.

Citizen behavior during response operations

Many laypersons—and even some emergency managers and jurisdiction officials—believe that people respond to disasters in a socially disorganized and even individually disoriented manner. Conventional wisdom holds that typically citizens respond to disasters with panic, shock, or passivity. However, social scientific studies have demonstrated that the majority of disaster victims respond constructively during disasters.[2] Most victims do not develop shock reactions; panic flight is rare; and people act in what they believe is their best interest, *given their understanding of the situation.* Behavior in the disaster response period is generally prosocial (as opposed to antisocial) as well as rational. Uninjured victims are often the first to search for survivors, care for those who are injured, and assist others in protecting property. Antisocial behaviors such as looting are relatively rare and take place most often in conflict-based situations. For example, although significant looting was reported after Hurricane Katrina struck New Orleans in 2005, it seemed to be peculiar to conditions created in that city; other cities affected by Katrina did not report significant looting.

In New Orleans, some circumstances were unique. Local government functions nearly collapsed for a long period, citizens were stranded without government contact (and without food and water), and established gangs that had not been able to evacuate were present in the impact area. To be sure, at least some of the apparent "lawlessness" in New Orleans seems to have been a product of the normal level of "incorrect" information associated with the early period following disasters: in any disaster, the situation is fluid, communications are difficult, and the priority is saving lives. Accordingly, exact information is hard to obtain; but as time passes and emergency assessment systems are implemented, initial misinformation is corrected. In New Orleans, however, what normally occurs was exacerbated inasmuch as some of the information given out by local authorities about violence, looting, and lawless behavior was simply wrong and was then injected into an environment where much unsubstantiated information was already circulating. Two reporters for the *New York Times* conducted extensive interviews and concluded that although there was much misery in the shelters and on the streets, the fear of crime and looting "far exceeded the reality."[3]

New Orleans aside, therefore, the picture that emerges from research about disaster victims is one of responsible activism, with victims attempting self-care, supporting neighbors, and ameliorating the situation as best they understand it, using whatever resources are available. Citizens confronted with disaster are thinking, acting beings who tend *not* to be frozen in fear, *not* to engage in panic flight, and *not* to engage in irrational behavior. The typical picture of people confronted with a disaster, including the certainty that they will die, is one of people who behave as the passengers on United Airlines flight 93 did. Once they understood the threat, they organized and attacked their hijackers, choosing to die in Pennsylvania rather than allow others to die with them in Washington, D.C. Understanding that this is the case makes possible several important observations about managing terrorist threats or other disasters.

Observation 1: Expect anxiety The first observation is to expect anxiety. Anxiety is a normal reaction to extreme conditions. It rarely results in the inability to act, but it does lower one's ability to effectively reason through complex problems. Anxiety is made worse by the unfamiliar, and terrorist events that involve chemical, biological, radiological, nuclear, or explosive (CBRNE) agents by definition introduce the unfamiliar. (Many of these agents are undetectable by human senses and produce both immediate and delayed negative outcomes, and citizen knowledge about them is highly limited.) Emergency managers should therefore plan to address anxiety, and should do so by providing—not withholding—information. They should use a strategy of information dissemination. The information should identify the threat, explain its human consequences (personalizing the risk), and explain what can be done to minimize negative consequences (see observation 2). If protective actions cannot be undertaken by individuals alone, the emergency manager should explain what the authorities are doing.

Observation 2: Expect citizen action A second observation is to expect citizens to act. Once citizens are informed of a danger, they will undertake actions believed to reduce that danger. If authorities do not provide protective action recommendations (PARs), citizens will take action anyway, adopting what they believe are "reasonable" protections. Thus, official messages must specify recommended protective actions, and these messages should be frequent. Messages without PARs simply enhance anxiety, which itself cannot be salved without information and action.

Observation 3: Expect citizen compliance with recommendations for action A third observation is that particularly in the case of terrorist attacks—and particularly when there is an identifiable incident scene and onset is quick—it is appropriate for authorities to expect citizens to comply with PARs.[4] In times of extreme stress, citizens look to government for guidance. When the agent is unfamiliar or intangible or the consequences are overwhelming, citizen expectations of protection and help are especially pronounced. National opinion polling following the 9/11 attacks indicated substantial increases in levels of "trust in government." Even if people later return to skeptical attitudes toward government, emergency managers have a policy window of opportunity. The expectation that citizens will comply underscores the need for authorities to have response plans in place that they are capable of executing. When such plans are lacking, citizens will subsequently hold authorities responsible through political and legal processes.

Observation 4: Expect some long-term consequences The fourth observation—one that authorities also must remember—is that the experience of any disaster may carry some long-term consequences for some of the victims. In other words, although cooperation and action can be expected in the short run, the disaster syndrome may persist for weeks or months in the form of anxiety accompanied by psychosomatic and physiological disruptions (e.g., bedwetting in children). Similarly, the disaster experience may produce psychological trauma, manifest over time in the form of persistent depression, post-traumatic stress reactions, and "survivor syndrome" (feelings of guilt and depression because one has survived while many others did not). Terrorist attacks commonly produce long-term consequences, and the extended effects of Hurricane Katrina on the victims—particularly children—are now surfacing. Long-term consequences are more likely to arise among three groups of people: (1) those who have witnessed death or handled the dead; (2) those who have been exposed to large-scale destruction of property; and (3) those whose relatives, neighbors, or friends have been seriously injured or have lost their lives. Recovery plans should include provisions for facilitating crisis and other short-term therapeutic contact as a means of reducing long-term psychological effects. Attention can also be given to citizens' need for economic support, and their need to develop a sense of closure and fit the disaster experience into a worldview that allows a transition to a stable life.[5]

Convergence and volunteers in disasters

A second response-related challenge that commonly arises comes about when a disaster-stricken community becomes the focus of aid contributed by nearby households, communities, larger political entities, and private organizations. The result is an influx of telephoned offers of help, volunteers arriving unannounced, and equipment and material being delivered. On the one hand, this convergence has a positive effect on the local authorities' resource base and on victims' morale. Victims may interpret the presence of such help as evidence that the disaster is not totally overwhelming and can be overcome. They feel less alone and more hopeful. But on the other hand, convergence can produce unprecedented communication difficulties and response-generated demands. When convergence inundates emergency responders with unanticipated people and materials, the asset becomes a liability.

Disaster plans must allow for the appropriate integration of volunteers into the response force; the management and care of volunteer labor; and the logistics of receiving, storing, and deploying material and equipment. Locally trained and equipped citizen emergency response teams can alleviate some of the challenges of volunteers. These teams can be deployed to effectively accomplish a variety of emergency functions, including managing untrained but well-meaning volunteers who appear at the scene. (For more details about managing an influx of donations and volunteers, see Chapter 11.)

Especially in the context of incidents involving CBRNE agents, there is an acute concern about controlling the exposure of volunteers. The way to handle this is to use some strategy of selective exclusion from the impact area and to provide personal protective equipment for volunteers who are operating with potential exposure. Volunteers should be overseen by professionals, protected by frequent hazard monitoring, and assured of ready access to communication with authorities.

Role abandonment by emergency professionals

A third response-related challenge has to do with the disaster myth that periodically resurfaces about emergency response personnel: that they will abandon their roles to take care of other responsibilities, particularly family obligations. One researcher who interpreted the studies on role abandonment drew three conclusions:[6] First, people who have no officially defined emergency response role and who have not themselves become victims will render aid at the disaster scene. Second, people who have no officially defined disaster role and who have been victims themselves will render aid first to family and then work outward in relationships to friends, neighbors, and strangers. Finally, people with officially defined disaster roles will execute those roles but will do so with psychological discomfort until they obtain information about the safety of their primary group.

All emergency response personnel, particularly first responders, are given formal training and made aware that their work may require separation from their families during a disaster. Almost all agencies encourage their response personnel to develop family disaster plans, including appropriate protective actions, arrangements for mutual support with neighbors and friends, and designation of procedures for reestablishing contact. Police and fire departments have institutionalized such concerns, devising protocols for agency contact with duty personnel families as well as for welfare reports to deployed employees. By supplying the safety accounting that people need if they are to be comfortable in continuing to work, practices such as these limit the probability of role abandonment. Two researchers reviewed studies compiled at the University of Delaware's Disaster Research Center and found that "in over 100 disasters studied and in the course of interviewing over 2,500 organizational officials, . . . role conflict was not a serious problem [that] creates a significant loss of manpower."[7] Indeed, in connection with the responses to the World Trade Center attack, the Pentagon attack, and the anthrax incidents that followed, no role abandonment was reported. The role abandonment by police officers in New Orleans after Hurricane Katrina in 2005 took place in an almost unique context characterized by massive destruction, little command oversight, the absence of communications, and no evidence of government intervention. In other cities during the same disaster, there is evidence that police and firefighters dutifully remained at their posts.

Disaster declarations

The fourth response-related challenge that commonly arises concerns disaster declarations. To the victims and communities involved, every disaster is catastrophic and merits a declaration, but procedures exist and must be followed.

There are a variety of governmental disaster declarations that are issued by different governments below the federal level: mayors, county executives, and governors all issue formal declarations that may or may not involve special emergency powers (for the issuing government) or open special-assistance funds. For disaster response and recovery, the primary concern is with the declaration made by the president of the United States under the Stafford Act of 1988 (the amended Federal Disaster Relief Act of 1974). The president may declare two types of Stafford Act events: a major (presidential) disaster or an emergency. The principal difference between the two relates to the types of federal resources made available and the matching funds required of the state government. While major disaster declarations can vary in resources provided, they generally activate long-term assistance programs that require cost sharing by the state at up to 25 percent. An emergency declaration provides largely short-term assistance (often aimed at preserving lives) and requires a 10 percent funding match by the state. The Federal Emergency Management Agency (FEMA), under Presidential Executive Order 12673, evaluates requests for Presidential Disaster Declarations and advises the president regarding the type of declaration and the resources appropriate.

At the municipal level, the disaster declaration process begins with damage assessment. As the process continues, the emergency director assembles data on the consequences of impact and advises the mayor on the appropriateness of a Presidential Disaster Declaration. Although a variety of issues may be raised, the scope and magnitude of damages are important determinants of declarations. Damages must outstrip the resources of both the local and state governments.

Once the mayor has decided to request a Presidential Disaster Declaration, he or she formally requests the governor to declare a disaster. The mayor's request is accompanied by supporting data from the damage assessment records. The governor may have a state agency check the damage estimates before deciding whether it is appropriate for the state to declare a disaster and/or to request a Presidential Disaster Declaration. If the governor does request a disaster declaration, the U.S. Department of Homeland Security (DHS)/FEMA will dispatch a damage assessment team to survey damages, state and local capabilities, and required assistance. (If the governor does not think such a declaration is warranted, the state has an obligation to provide resources, but the process stops. A mayor may not directly approach the federal government for a disaster declaration.) The director of FEMA reviews the report of the assessment team and makes a recommendation to the secretary of DHS, who advises the president about the appropriateness of a declaration. The president makes the final decision.

As noted above, depending on the type of federal disaster declaration, a variety of different types of federal assistance may be made available. The U.S. Conference of Mayors has prepared a detailed listing of the assistance available for municipalities under different levels of Stafford Act disaster declarations. Presidential declarations also make assistance available directly to individual citizens and/or families. Such assistance includes such services as low-interest recovery loans, temporary housing, counseling, tax refunds, and food stamps. It is important for the local emergency manager to monitor this type of assistance, whether private or federal, to help ensure that victims are aware of the range of services available and are able to pursue the appropriate assistance.

Communications interoperability

The fifth response-related challenge is that of communications interoperability. Communications interoperability "refers to the ability of public safety personnel to communicate by radio with staff from other agencies, on demand and in real time."[8] Effective communication is the foundation of effective disaster operations. However, many small local government emergency response agencies have inadequate internal radio communications, and even large agencies may have substantial interagency problems of communication. In a 1998 study of 1,045 fire and emergency medical service (EMS) agencies, 30 percent of participants agreed that the

Hurricane preparedness program, Sanibel, Florida

Sanibel, Florida (pop. 6,000), sits on a barrier island in the Gulf of Mexico, vulnerable to natural disasters. Cars can access or leave the island only via a causeway bridge. Although hurricanes are rare there, the city revises its hurricane preparedness plan annually to maximize its use of new technology. Thus, when a Category 4 hurricane struck on August 13, 2004, the city was able to implement a model evacuation, reentry, and recovery plan that saved lives and minimized economic losses.

Each year, Sanibel provides hurricane education to residents and businesses through workshops, the city Web site, and direct mailings to every household. Police officers speak to civic clubs, and the city invites community partners to a hurricane planning meeting. In addition, the city prebids contracts for debris cleanup; conducts training with police, fire, and structural safety inspectors; identifies an off-island facility for use as a temporary city hall in case of evacuation; and trains a volunteer amateur radio team to provide a secondary means of communication. The skills of the city's nonessential employees are identified for reassignment should a hurricane occur.

Per the city's plan, the preevacuation phase was activated five days before Hurricane Charley was due to arrive. Each department consulted its hurricane checklist. Employees installed hurricane shutters and conducted team conference calls with the county emergency operations center (EOC) to verify that preparations were on target. Two days before hurricane landfall, evacuation notices were issued and voluntary evacuations began. The next morning, evacuation became mandatory, and by the evening of August 12, city hall had been relocated to an off-island facility.

Charley struck the following evening, uprooting thousands of trees, blocking every roadway, destroying the electrical distribution system, damaging 90 percent of structures, and cutting off the city's water supply. At dawn, designated city teams returned by boat, establishing a temporary EOC to coordinate activities. Simultaneously, a flyover assessment was completed. Police and firefighters conducted a search-and-rescue mission to check for injured residents. A team of city staff and volunteers reviewed every structure to assess the level of damage incurred and its safety for entry; another team worked to clear trees from main roadways.

The EOC housed police, fire, utility, and forestry representatives; U.S. Fish and Wildlife workers; and emergency relief agencies. With public and private agencies under one umbrella, the process moved efficiently. Sanibel police coordinated with the National Guard, Florida Department of Law Enforcement, Florida Wildlife Commission, Salvation Army, and Red Cross to provide security as well as food and water to emergency crews. One staff member stayed at the EOC, maintaining constant contact with city officials to obtain and dispatch supplies, and a pet rescue and medication program operated from the temporary city hall.

During the evacuation, the city used a hotel lobby as an information station where citizens could view photographs of the damage while the city manager and council conducted daily briefings for evacuees and the media. Computer technicians used emergency generators to post real-time Web site updates, and portable generators kept sewer lift stations functioning. With no cell phone service, the volunteer amateur radio network was invaluable.

The well-coordinated effort made it possible to deem the island safe for public return five days after the hurricane struck, and at 7:00 AM, on August 18, the island officially reopened. Every returning citizen was greeted at the bridge tollgates with a ten-page "welcome home" packet, including a new refuse collection schedule, a list of licensed contractors, and the location of Red Cross shelters.

City hall reopened with extended hours, seven days a week. The city Web site's real-time updates became a vital tool to convey information about the status of utility restorations, the latest refuse pickup schedule, where to obtain ice, how to get rid of spoiled food, how to find medical help, and more. With worldwide interest, the number of Web site hits increased from a prehurricane average of 4,000 daily to 172,578 daily.

Damage estimates to private property ran more than $720 million, but thanks to a comprehensive, well-executed preparedness plan, the city suffered no loss of life during the storm or recovery process, and its use of new technology under adverse conditions kept not just the community but the entire world informed.

The Sanibel Hurricane Preparedness Program and city manager Judith Ann Zimomra received the Public Safety Program Excellence Award in the less-than-10,000 population category from ICMA in 2005.

Source: Excerpted from "2005 ICMA Annual Awards," *Public Management (PM)* (September 2005): 13, 29.

Lessons learned from the response to the attack on the Pentagon

When the 757 jetliner slammed into the Pentagon on the morning of September 11, 2001, everyone on the ground knew it was a terrorist attack: Two planes had already struck the twin towers of the World Trade Center in New York City.

Even so, the first people at the scene were not from the U.S. military, the Federal Bureau of Investigation, or the Central Intelligence Agency. They were from the fire department in Arlington County, Virginia. While the Pentagon sits within view of the nation's capitol, it lies squarely within Arlington's borders.

And that's the way it will be in any terrorist attack or disaster. The locals will get to the scene first. Whether they will know what to do and have the resources to do it is more of an open question.

By most accounts, Arlington County did a pretty good job at the Pentagon, and that certainly was no accident. The county was well prepared. "If there was any luck on September 11, it was that the plane crashed on the Virginia side of the Potomac River," says Donald Kettl, who teaches political science at the University of Pennsylvania and studies homeland security issues. Arlington's response wasn't perfect, but what it did right—and how it continues to upgrade its capabilities—stand as a model for other communities struggling with the confusing issue of homeland security.

Working together

Long before 9/11, the county had considered what kinds of disasters it might face and what agencies within the county and in adjoining jurisdictions could provide help. The county then coordinated, trained, and exercised with the appropriate agencies.

For starters, the county police and fire departments were used to working together and were part of the county's emergency management team, which also includes public works and public health officials. Team members operate under preestablished policies and procedures and have run exercises to polish up those procedures.

In addition, fire and emergency medical service (EMS) agencies in Arlington and its neighboring Virginia counties have operated under an automatic aid agreement for nearly thirty years. They share resources without regard to jurisdictional boundaries, as well as use the same equipment. That means they can literally hook up to each other's equipment and tune to each other's radio frequencies. This helped immeasurably at the Pentagon. In New York City, some first responders from other jurisdictions had trouble communicating with their NYC counterparts—or even connecting their water hoses.

Arlington also had participated in numerous field exercises with the Pentagon that simulated chemical attacks and mass casualty situations. The 1995 sarin gas attack in a Tokyo subway also spurred Arlington officials to beef up their emergency response program.

Crucial to the coordinated response at the Pentagon, however, was the system of command. Arlington County and its Virginia neighbors use the same incident management system, so on 9/11, responders knew who was in charge and what their role was.

Under a unified management system, the nature of the incident determines which agency takes charge. Since the Pentagon disaster was first of all a building fire and collapse with hazardous materials and mass casualties, the police department recognized right away that the fire department would be the lead agency.

"It was readily apparent that this was a crime with a law enforcement component," said James Schwartz, Arlington County's fire chief who commanded the fire-rescue operation at the Pentagon. "But that had to take a back seat to the life-saving effort." At the same time, fire and EMS responders understood that evidence needed to be collected even as they were driving trucks, putting up ladders, and stretching hoses.

Arlington uses an "all hazards" approach in its disaster planning. That means it doesn't matter, for example, what causes a building to collapse—gas main explosion, earthquake, or terrorist bomb. If people are trapped, they're trapped. The Federal Emergency Management Agency has employed "all hazards" for years, and it has been useful to Arlington County—at the Pentagon and in subsequent emergencies, such as an anthrax threat and Hurricane Isabel. The anthrax incident made public health the lead agency.

Such extensive planning and training clearly is what helped Arlington perform as well as it did on September 11. "We've been at this so long that we probably know more than the average community," Schwartz said. "And we're probably more aware of what still needs to be done."

Too many cooks

Chief among the things that didn't work well was controlling access to the disaster scene. Ideally, first responders come when called by the incident commander. At the Pentagon, the response from neighboring Virginia counties was "very disciplined," Schwartz said. Thanks to the automatic aid agreement, the jurisdictions didn't send anybody who wasn't requested, whereas some units from Washington, D.C., and nearby Maryland raced to the site without being summoned.

The unexpected arrival of units became a tremendous problem: It prevented Schwartz from deploying resources efficiently and could have endangered those responders since he didn't know they were there and therefore couldn't communicate with them. Moreover, not knowing what additional units were at the Pentagon and where they were made it more difficult to evacuate the site, which had to be done three times during the operation as unidentified aircraft headed in the direction of the Pentagon.

It's hard for responders to stand by during an emergency, Schwartz acknowledged, "because they're not people used to sitting on their hands." But discipline is crucial. "If we overdeploy for one incident with unnecessary resources, we may leave ourselves vulnerable elsewhere in the region," Schwartz said. In addition, some responders could be lost unnecessarily if there's a secondary attack at the initial incident site.

Arlington County didn't turn away any emergency responders at the Pentagon "but we should have," Schwartz said. The need for better scene-access control was identified in the county's *After-Action Report,* which was released in July 2002.[1] Arlington is currently working on an access-control plan.

A related problem was the absence of a system to check credentials or otherwise identify legitimate responders. "The challenge is how to protect a large incident site like the Pentagon," Schwartz said. "How do you ensure that the next ambulance or fire truck arriving at the scene isn't driven by terrorists and laden with explosives?"

The United States can learn lessons from other countries, especially Israel, which uses markings on first-responder vehicles that are changed frequently. Schwartz thinks the problem may be solved by technology—biometrics and other ways of identifying people. In any case, he said the federal government needs to take the lead by developing protocols to help localities secure incident scenes.

Talking at each other

In many communities around the country, fire and police personnel can't communicate with each other, much less with their counterparts in other jurisdictions. Communication at the Pentagon site was not a huge problem, but there were glitches. For example, Virginia responders could all use the same radio frequency, but, initially, D.C. and Maryland responders could not. Within hours of the crash, units from D.C. and Maryland were given 800-MHz radios, which solved the problem. Now all metro-area jurisdictions use that system. But having everyone on the same system is not the end of the problem at a disaster scene. "The last thing we want is for everybody to be able to talk on a common operating frequency," Schwartz said, because that could clog the airwaves and make it difficult for the principals to connect. And that's another instance in which an incident management system is key to a disciplined response. Under such a system, the various principals know their roles, and they do the talking.

While equipment and technology are vital pieces in an emergency, Schwartz emphasized that using a common command system is even more important. "If you're not operating by the same incident management system, then the rest of this stuff doesn't work at all."

The long haul

Normally, if there's a big fire in Arlington County, many units are dispatched to the scene, and it's usually over in a few hours. The Pentagon operation, however, lasted ten days. That meant the larger community was thinly protected for a prolonged period. It also monopolized equipment and consumed materials. Arlington has used federal grant money to buy more equipment in order to sustain longer operations.

(continued)

Lessons learned from the response to the attack on the Pentagon *(continued)*

Similarly, each responder in every jurisdiction that could be called to a disaster needs certain basic equipment that is instantly available: a chemical suit, respiratory protection, and a chemical antidote. Additional resources can be available at the regional level. "We need to remember that redundancy in this business is good," Schwartz said, noting that if a second airliner had hit Washington, a lot more resources would have been needed. "In the new world of terrorism, we can't resource for just one big response."

Since 9/11, the federal government has predeployed equipment packages around the country which can be at an incident scene in a couple of hours. Packages include the firefighting, rescue, and law enforcement equipment necessary for 150 responders–all of which, Schwartz said, are necessary. "I believe that in the future we'll face multiple incidents that will really challenge the resources of a region."

Source: Adapted from Carol Anderson, "Pentagon in Peril," *Securing the Homeland,* a special report from *Governing Magazine* and *Congressional Quarterly* (October 2004): 22-24, available at homelandsec.org/publications.asp?pubid=495 (accessed September 9, 2007).

[1]Arlington County Fire Department, *After-Action Report on the Response to the September 11 Terrorist Attack on the Pentagon* (Arlington County, Va.: Titan Systems Corporation, July 2002), available at arlingtonva.us/departments/Fire/edu/about/docs/after_report.pdf (accessed September 9, 2007).

lack of wireless communications interoperability hampered their ability to perform response duties.[9] Because the problem has persisted for so long, most agencies have devised ways of "working around" difficulties. For example, in the response to the 1995 Oklahoma City bombing, "first responders had to use runners to carry messages from one command center to another because the responding agencies used different emergency radio channels, different frequencies, and different radio systems."[10]

Working around problems is not a solution. Simply to execute day-to-day operations in the same or adjoining jurisdictions, police and fire departments need to have radio command communication. When the events being dealt with are not day-to-day operations but large emergencies, disasters, and particularly incidents involving weapons of mass destruction (WMDs), the number and variety of responder agencies multiply and so do the issues of interoperability. On September 11, 2001, police helicopters noticed that the World Trade Center towers were showing signs of impending collapse. Police commanders issued a radio warning to police responders to begin evacuations, but because firefighters used a different radio frequency, the message did not reach them. Thus, "totally unaware of the impending collapse, at least 121 firefighters, most within striking distance of safety, according to the *New York Times,* died."[11]

Although mutual aid systems and other types of communications governance systems have substantially reduced problems of interoperability, the issue persists. Not even programs promoting interoperability in the public safety arena have erased the problems, which the National Task Force on Interoperability has identified as

- Outdated and incompatible communications equipment
- Limited and fragmented funding
- Poor planning coordination and poor cooperation among agencies
- Insufficient radio spectrum assigned to public safety
- Limited equipment standards.[12]

In 2004, DHS consolidated its existing programs on interoperability and supplemented them by issuing a comprehensive statement of requirements (standards) for interoperability for first responders; to implement the requirements, DHS created the SAFECOM program. The Federal Communications Commission has eliminated one obstacle by assigning more frequencies to public safety to support voice, data, and video communications, and the SAFECOM program is attempting to eliminate the remaining four problems. It has devised a comprehensive plan for increasing awareness of the interoperability issue, and it promotes coordinated planning and partnerships at local, regional, and state levels; it also outlines funding strategies to pay for equipment to achieve interoperability, and it urges state and local agencies to adopt comprehensive standards for interoperability. SAFECOM, like other DHS initiatives, is too new for evaluation, but it appears to supply a workable plan. Thus, the elements of a nationally successful interoperability program are present although the problem of funding is not unequivocally solved.

The federal context of disaster response after 9/11

Attending to the resources and guidance available from the federal government is another way that local emergency managers try to prevent their EOP from being inadequate to the demands of a disaster. At the federal level, the terrorist attacks on September 11 made a major difference in emergency management. Following those attacks, the federal government moved to systematically include terrorist incidents with natural and technological events as appropriate targets of comprehensive emergency management. (Local governments, with and without support from the federal government, had been preparing for terrorist events for some years; those efforts were intensified after 9/11.)

The inclusion of terrorism in the threat environment creates a need for changes in thinking as well as in practice. For natural and technological events, the strategies, tactics, and equipment used in emergencies are often appropriate for, or readily adaptable to, disasters and catastrophes, but when one introduces the terrorist use of WMDs—which may include CBRNE agents—the nature of the environment changes and the response capability of all responders is compromised. With any disaster response operation, fundamental success rests on planning, training, and exercising, and the need for planning, training, and exercising does not change when terrorism is added to the threat environment. However, the strategies for response, the equipment with which to address the threat, and the personal protective equipment needed to preserve the responders do change.

It has always been the case that the trained responders who are the first to arrive at the scene of an emergency, a disaster, or a catastrophe are most likely to be fire, EMS, and police personnel, although public works, transit, and engineering staff, as well as private sector organizations (e.g., personnel from private ambulance systems or hazmat response teams from chemical companies) may also perform specialized roles. (As the demands of the incident escalate, these first responders are supported by more specialized personnel, equipment, and resources from outside the community.) These "first on-scene" responders rely on plans, training, equipment, and skills to formulate a response to any demand, whether the demand is generated by the hazard or by the response itself. Effective response combines flexibility and improvisation, particularly in the face of low-probability events that produce consequences far outside the norm.

When terrorism is added to the threat environment, the strategies for response, the equipment with which to address the threat, and the personal protective equipment needed to preserve the responders all change.

In the late 1990s, with the growth of the Metropolitan Medical Response System (MMRS, created by the Nunn-Lugar-Domenici Act of 1996),[13] the concept of operations for local government terrorism response changed slightly. Historically, local response plans had addressed terrorism through small strike forces—specially trained and equipped personnel who would be dispatched at impact. These teams are separately organized and quartered, and there is an inherent delay in their arrival at any given scene. This approach works as long as the strike team and necessary resources can be deployed immediately and can arrive after only a very brief delay. The approach has several disadvantages, however: first, a delay in response may increase potential losses; second, the first on-scene responders may be exposed to the same agents as victims, possibly without appropriate protection; third, terrorists sometimes use secondary devises, such as explosives, timed to disrupt response or to injure or kill the first on-scene personnel.

By the late 1990s, the approach to antiterrorism had evolved into one that embeds incident response in the local incident management system (IMS). This more recent approach requires that all first responders be trained to identify the signs and symptoms of CBRNE agents without necessarily being able to identify the precise agent; that they be equipped with generic protective gear for the full range of agents; and that they have sufficient training and equipment to know when to call for specialized help (instead of trying to abate the event themselves) and to be able to initiate a protective response for victims and bystanders. Thus it is not necessary for every police officer and firefighter to be trained in the operation of specialized detection equipment and to carry specialized protective gear and chemical or biological antidotes.

The federal bureaucratic framework

The 9/11 attacks were a watershed for U.S. emergency management policy. Many of the policy weaknesses that had been identified during previous decades, as well as the response challenges identified on 9/11 itself, became salient to policy makers. A window of opportunity opened when the political will to change, the funding for change, and an apparently continuing threat all coincided. The principal federal response was to create DHS in 2003. DHS absorbed FEMA and more than a dozen other agencies and programs. The department's budget is large enough for it to be able to impose federal expectations on state and local governments by mandating participation in federal programs as a condition for granting funds.

The original DHS mission encompassed three goals: prevent terrorist attacks within the United States, reduce vulnerability to terrorism, and minimize the damage and speed the recovery from terrorist attacks.[14] But DHS soon adopted all-hazards responsibilities, although the emergency management community has expressed reservations about the department's allocation of resources to nonterrorism issues. In 2005, the U.S. Government Accountability Office (GAO) found that DHS had directed that "the majority of first responder grant funding be used to enhance . . . capabilities [related to] terrorist attacks."[15]

Concerns about the Department of Homeland Security

GAO has expressed concerns about the ability of DHS to become fully operational, about whether the department is doing less than it might to affect municipal preparedness, and whether the resources that should support preparedness are being diluted.[16] Some concern is linked to internal organizational and interdepartmental issues in policy implementation. That is, the department brings together very diverse functions—for example, terrorism intelligence gathering, the Secret Service, the Coast Guard, and infrastructure protection—that all compete for parts of the same DHS budget. Also, some DHS functions overlap with the functions of other agencies and programs and must compete in the larger U.S. budget and policy environment; DHS has programs for computer security, for example, but so do the National Security Agency and other organizations in the intelligence community.

There is also persistent policy uncertainty about the way to involve law enforcement in the system without undercutting other DHS sections, such as FEMA, that must respond to natural and technological threats as well as to terrorist threats.[17] For local governments, law enforcement is the principal arm for the intelligence-gathering function, which constitutes the only real approach to terrorism mitigation. Emergency managers and fire services do not execute this function—but since they do engage in preparedness, response, and recovery, there must be a way to formally connect the three disciplines in the comprehensive emergency management process.

GAO defines the prime internal organizational challenge for DHS as focusing management efforts and connecting current policy aspirations with accumulated knowledge and existing and historical programs.[18] Although this advice is constructive, at some level there must also be a dialogue about the role of DHS functions in national policy on homeland security. DHS can manage its way out of internal chaos, but the larger national policy problem—agencies with overlapping responsibility—remains.

Problems at the Federal Emergency Management Agency

Problems affect FEMA in particular. When Hurricane Katrina hit in 2005, FEMA largely failed to implement the National Response Plan (NRP).[19] Discussion continues on plans by Congress and the DHS secretary to reorganize FEMA. In 2006, seeking a nominee for the position of FEMA director, the Bush administration contacted seven former state emergency management directors and current FEMA officials,[20] and all seven declined, citing similar reasons: no realistic path to improving FEMA within its DHS context and low likelihood of adequate funding. Although the DHS budget is generally seen as substantial, much of it has historically been allocated to other functions (e.g., specialized terrorism-directed programs) and other formerly stand-alone agencies (such as the U.S. Coast Guard). Eventually Acting Director David Paulison accepted, bringing many years of professional fire services management experience to the directorship.

It appears that turbulence at DHS and FEMA will continue, however, increasing the likelihood of strategic uncertainty. This uncertainty stems from overall challenges to policy—both to its continuity and to its implementation—combined with new organizational structures, fluctuating priorities, and personnel newly assigned to the programs they administer.

Federal and local emergency management relations

Although the future is hard to predict under any circumstances, the trend is toward higher levels of federal intervention in local emergency management. Federal interventions to develop and impose consistent planning and response models on state and local government have created a rapidly changing environment characterized by a largely "top-down" flow of communications and requirements.[21] DHS has developed guidance for state and local agencies that extensively defines appropriate capabilities and includes a "universal" list of tasks for agencies to accomplish.[22] In addition to these operational and emergency management prescriptions, the NRP addresses state and local access to federal resources, and the National Incident Management System (NIMS) specifies practices in local planning, incident management, and resources. Federal mandates are being enforced through legislation and administrative rule making, incentives, and arrangements for financial support. No systematic attempts have been made, however, to measure the success of these attempts to influence local planning and response practices.[23]

There are numerous federal programs that fund local emergency management, and although many of them are now located in DHS,[24] major programs also exist outside DHS—notably in the U.S. Department of Health and Human Services (DHHS) (especially the biological threat initiatives supported by the Centers for Disease Control and Prevention and by MMRS), the U.S. Environmental Protection Agency, and the Nuclear Regulatory Commission. Among the national programs, three have significant and direct impacts on local government disaster operations. Two of them—MMRS, created in 1996, and the Urban Areas Security Initiative (UASI)—are specifically related to federal attempts to combat terrorism. The third, NIMS, mandates a specialized (and centralized) approach to all emergency planning and response. The creation of NIMS and UASI and the continuation of MMRS have taken place in the turbulent post-9/11 environment.

UASI, MMRS, and NIMS were selected for review here because of their size and scope, their direct impact on disaster operations, the size of their budgets over time, and their demands for a high level of federal involvement in local government emergency management.

The Urban Areas Security Initiative

In 2003, the DHS Office for Domestic Preparedness inaugurated UASI as part of the National Strategy for Homeland Security. Part of the objective of UASI was to develop a capacity to target specific resources to urban areas whose probability of suffering terrorist attacks was judged to be highest. The Homeland Security Appropriations Act for fiscal year (FY) 2004 continued and expanded UASI at a funding level exceeding $4 billion. Seven urban areas were approved for funding in 2003, with the number growing to fifty in 2004. For 2005, DHS added seven new UASI jurisdictions, while—without public explanation—not funding seven urban areas from 2004. In FY 2005, the financial awards were substantial, ranging from a high of more than $207 million to New York City to a low of $5 million for Louisville, Kentucky. The 2006 UASI guidance reduced the number of continuing cities to thirty-five—the thirty-five with the highest calculated risk levels—and reduced future program funding. In 2007, urban areas were divided into two tiers for basic funding. One tier consists of six very high-risk urban areas: the San Francisco Bay Area, the Chicago area, the Houston area, the Los Angeles–Long Beach area, the National Capital Region, and New York City–Northern New Jersey; the other tier consists of thirty-nine urban areas with lower risk.

UASI is discussed here in terms of its aims and funding mechanisms, its concept of operations, and its policy and prospects.

Aims and funding mechanisms

In 2004, then DHS secretary Tom Ridge charged UASI with "creat[ing] a sustainable national model program to enhance security and overall preparedness to prevent, respond to, and recover from acts of terrorism."[25] UASI does not impose a generic emergency response model on participating urban areas but, instead, requires that local governments—organized around a designated core city—cooperate in developing a strategic plan that either creates new disaster plans or supplements existing ones. The program has also developed an all-hazards emphasis. UASI authorizes program expenditures across five areas: planning, equipment acquisition, training, exercises, and management and administration (capped at 3 percent of the total allocation).

The funding mechanism is intergovernmental, with federal money being allocated to states (which can retain up to 20 percent), and the states then distributing funds to local governments. Local government allocations are based on the local governments' strategic plans and the mutual agreements entered into by a core city's urban area administrator, the participating municipal governments, and the county and state emergency authorities. All expenditures are subject to federal review.

Concept of operations

Strategic and operational plans developed by urban areas are "authorized use only" and not subject to public scrutiny. Despite the claims of all-hazards intent, much of the publicly available UASI application material focuses on terrorism, although many "disaster functions" promoted by UASI are applicable to disaster agents other than terrorist attacks. Anecdotal evidence indicates that many of the UASI urban areas have adopted response plans that follow the organizational model used by the National Urban Search and Rescue program (US&R).[26] This is an established and tested model for responding to natural, technological, and terrorist events that require the movement of special personnel and resources to a potentially distant incident site. The US&R approach relies on storing specialized equipment and resources to be used by a specially trained team; both the team and the resources are capable of being moved to an incident site on very short notice. For example, teams from the US&R program were dispatched from as far away as Washington State to work on rescue and recovery in New York City on September 11. UASI technically defines the response area as the urban area, but many states require their UASI operations to extend statewide.

The UASI concept of operations follows the US&R model but usually creates multiple mobile teams (rapid response teams [RRTs]) of mixed law enforcement and fire services personnel (firefighters cross-trained as paramedics; technical rescue technicians and hazmat technicians; and police officers specializing in special weapons and tactics, or SWAT, with bomb technicians). Many UASI systems also borrow the incident support team (IST) concept from US&R; ISTs are mobile, assess the needs at an incident site, and subsequently serve as a command and resource structure for the deployed teams. The RRTs and ISTs are usually dispersed throughout the urban area to reduce travel time and achieve distribution of resources. UASI capability can be requested by any jurisdiction facing an incident that will overwhelm the local response system. The teams can quickly deliver expertise, response vehicles, and equipment while simultaneously maintaining a reserve to ensure the security of the larger urban area.

Once UASI is activated, an interjurisdictional command-and-control system is required. In a large incident, every jurisdiction centers its command strategy in a local emergency operations center (EOC), and all federal resources reside in, or are available through, the Federal Bureau of Investigation's Joint Operations Center, an NRP joint field office, and/or the state

EOC. Local EOC decision makers report to and act in accord with the elected leaders of their jurisdictions. The deployed IST supports the local incident commander and the local jurisdiction's EOC. The local incident commander—supported by the jurisdictional EOC—controls incident response and initiates and directs the deployment of nonlocal UASI resources as part of the command function. The IST (which, as noted, supports the local command structure) serves as the interface among the local jurisdiction, UASI resources, state resources, and federal resources.

Policy and prospects

Local emergency managers view UASI, at least in part, warily. UASI brings substantial funding to local needs and allows a degree of local choice in planning, administration, and funding allocations; however, there are complaints that the federal authorities tightly define authorized expenditures within each predetermined budget category and that local governments bear a substantial financial accounting load. Since the 2003 initiation of UASI, DHS's reporting demands have substantially increased, but these demands have not been matched by funding to support the generation of the required reports. There is also concern that the pass-through mechanism from federal to state to local agencies is complex and administratively demanding and that it siphons funds from emergency response to other uses. Finally, for UASI to succeed in creating a functioning local emergency management capability, there must be high levels of continuing cooperation among federal, state, county, and municipal governments. Given the inherently problematic nature of intergovernmental relations, however, this is a serious challenge. In any case, if budgets continue to shrink the number of UASI cities, the program's ability to provide protection on a national scale will be seriously compromised.

At present, there is little basis on which to judge the success of UASI. The program is new, and plans are kept secure to avoid having their contents divulged to potential terrorists. All urban areas funded in the FY 2005 cycle have obtained federal approval of strategic plans. Evaluations of functional or full-scale UASI exercises are not available in open literature. Although many of the UASI urban areas have MMRS programs that provide emergency management system models, it is not clear if UASI strategic plans build on these capabilities, revise them, or change them entirely.

The Metropolitan Medical Response System

The first program of national scope to address WMD/CBRNE threats was MMRS, established under the Nunn-Lugar-Domenici Act (the amendment to the National Defense Authorization Act for FY 1997). Overseen by DHHS, this program created coalitions of public, private, and non-profit organizations to address mass casualty consequences of terrorist attacks. The program was originally based completely on the strike team concept and was called Metropolitan Medical Strike Teams. But as described above, strike teams for local response have disadvantages, and the national system changed within a year to one based more in the IMS and contingent on broad-based training and equipping of police officers and firefighters. Of the twenty-five MMRS cities established before 1999, only two (Atlanta and Washington, D.C.) retained a strike team framework through 2007, and these two did so largely because their response areas are geographically widespread, have special needs (District of Columbia), and have an elevated likelihood of multiple simultaneous attacks.

Although only modestly funded, MMRS has produced high levels of response activity, particularly following 9/11, when the anthrax threats were prevalent. Its programs are concentrated in areas of high population density and areas with high-probability targets. By March 2004, 124 city and regional MMRS programs had been established. With forty-three states having at least one MMRS program, geographic coverage is very broad. The programs have been exercised for years and have independent positive evaluations.

MMRS enhances local efforts to manage mass casualty incidents arising from terrorist use of WMDs. Its mission is driven by the realization that, for local governments, specialized federal assets for terrorist attacks are forty-eight to seventy-two hours away. MMRS ensures that cities can operate independently until external support arrives. It also fosters a strong local IMS that incorporates specialized extra-community resources. Each MMRS city maintains a pharmaceutical cache (to federal standards for type and quantity) of antidotes and prophylaxis for chemical, radiological, and biological threats. In addition, since the program's inception, the MMRS focus has expanded beyond CBRNE agents and now includes explosive threats and any other agents (natural or technological) that could produce large numbers of casualties.

The most significant organizational feature of MMRS is that it links multiple response systems. Horizontal linkages connect first responders, public health, emergency management, law enforcement, mass fatality, and medical and behavioral health services. There also are vertical linkages inasmuch as public health participation involves city, county, and state agencies. In addition, links are formed between local government and private and nonprofit organizations such as hospitals, environmental cleanup companies, ambulance systems, and funeral director associations. MMRS cities plan for the receipt and integration of important federal assets by building a relationship with the National Disaster Medical System and developing a capacity to receive pharmaceuticals from the strategic national stockpile, along with other specialized assets from federal programs like CHEMPACK and BIOWATCH. (CHEMPACK places one or more caches of pharmaceuticals for chemical, biological and nuclear agents in high-risk cities for immediate access by emergency responders, and BIOWATCH involves both stockpiling biological agent antidotes and creating detection systems for biological agents.)

MMRS requirements address some issues of mitigation and recovery but emphasize preparedness and response. As a condition of declaring an MMRS program "fully operational," each city conducts a full-scale exercise with federal evaluation and maintains an exercise calendar.

The topics discussed here in connection with MMRS are the program's concept of operations, its system responsibilities, and policy and funding.

Concept of operations

There are two models for MMRS operations. The 1997 strike force model builds specially trained and equipped mobile forces that respond to confirmed WMD/CBRNE incidents. The majority of participating municipalities, however, use the MMRS IMS model, which requires all first responders to be able to detect WMD/CBRNE incidents and initiate a response. This model requires training in the recognition of CBRNE agents and wide issuance of personal protective equipment. The training and equipment permit safe initial response as the IMS expands to meet incident demands.

The MMRS IMS model assumes that if a geographically defined incident scene exists, threat management should be directed from that location. If there is no distinct incident scene, MMRS focuses incident management in the jurisdictional EOC with links to other governmental EOCs. In any case, the individuals commanding the response operations are always those geographically closest to the incident. If no geographically defined incident scene exists, or if there are multiple incident scenes, operations are overseen by the relevant (jurisdictionally closest) EOC.

The MMRS IMS model has two components that take response actions. The first consists of trained and equipped responders guided by an incident commander. The second is the EOC, which MMRS augments with special administrative staff for terrorist incidents and with technical experts from private, nonprofit, and public organizations who have special skills related to CBRNE agents. The MMRS concept of operations emphasizes the integration of planning and response efforts, with many agencies working together to achieve common response objectives.

System responsibilities

At the level of the IMS for incidents with a defined scene, a fire service hazmat response model is often used. When there are agents other than hazardous chemicals—particularly biological agents—or other system activation paths, the organizations involved in the MMRS response may be different (i.e., may not be dominated by hazardous materials). For example, a potentially terrorist-based outbreak of botulism poisoning might first be noticed by emergency department physicians, who would notify the public health system rather than a fire department. The organizations involved in MMRS response operations vary by the needs of the incident, but the model was created specifically to deal with the special demands of terrorist incidents. These special demands are identified in Figure 9–1, which shows the response organizations and their primary functions for terrorist incidents that generate a scene for operations.

Policy and funding

Before 2003 and the creation of DHS—in other words, for most of the MMRS program's history—funding came directly to the cities from DHHS, so concerns about the loss of funding to intermediate government levels did not arise. In March 2003, responsibility for MMRS passed to FEMA/DHS—but

Figure 9-1 MMRS response organizations and functions for terrorist incidents

Organization/agency	Primary functions
First on-scene fire, police, emergency medical service	Secure the incident area Assess victim needs Conduct situation assessment Collect casualties and initiate victim management Conduct emergency gross decontamination Preserve evidence
Technician responders: Hazardous materials, technical rescue	Adjust scene layout Inspect impact area Perform agent identification (scene measures) Conduct victim extrication Conduct technical decontamination
Medical responders: Emergency medical technicians, paramedics, behavioral health specialists	Triage victims Administer medical treatment at scene Provide mental health support Organize patient transport to definitive care
Law enforcement	Gather intelligence and coordinate with the Federal Bureau of Investigation Control evidence Secure scene, treatment areas, emergency operations center, mass shelters, jurisdiction critical facilities Manage bomb operations Manage special weapons and tactics operations
Hospital mass patient care: Hospital physicians, nurses, personnel	Conduct technical decontamination Provide full patient diagnosis and treatment Triage victims for the National Disaster Medical System Manage mass surge Extend treatment capacity
Mass fatality management: Medical examiner's personnel, mortuary association personnel	Receive human remains Safeguard personal property Identify the deceased Complete and maintain case files Preserve chain of evidence Release remains for final disposition
Public health: State and local health departments	Provide specialized medical expertise Manage epidemiological investigation Support agent identification Recommend preventive health measures Coordinate with the Centers for Disease Control and Prevention Advise agent control (mass prophylaxis, quarantine) Coordinate with national pharmaceutical programs

then in 2006 it passed back to DHHS. The continuing organizational challenge for MMRS cities has rested in the requirement that they integrate participation by municipal departments, county and state government agencies, and the private sector.

The availability of federal funds has been erratic, and the money has often failed to cover the costs to cities sustaining MMRS programs, as the following list indicates:

- 1997: Twenty-five MMRS cities given $350,000 each plus equipment loans from the U.S. Department of Defense to establish programs
- 1998: No new programs and no continuation funding for existing programs
- 1999: Twenty new MMRS cities given $600,000 each, with twenty-five existing programs given $200,000 each to enhance biological preparedness
- 2000: Twenty-five new programs established with $600,000 each, and no continuation funding for existing programs
- 2001: Twenty-five new programs at $600,000, with no continuation funding for existing programs
- 2002: Twenty-five new programs at $600,000, with $50,000 continuation funding for existing programs
- 2003: Four new regional MMRS programs at $600,000; existing programs were assigned $280,000 in FY 2003 and $400,000 in FY 2004
- 2004–2007: No new cities; overlapping performance periods make funding difficult to track; funds to sustain existing MMRS cities in 2004, 2005, and 2006 averaged $235,000 each year, with $258,145 allocated in 2007.

One detail may give some perspective to the magnitude of the funding: for a moderate-sized city, the cost to repurchase the MMRS-required pharmaceutical cache (repurchase is at expiration, or about every three years) can exceed $100,000. But despite modest or sometimes no continuation funding, none of the cities has dropped out of the program.

Ultimately, aside from the National US&R program,[27] MMRS represents the only federally devised model for disaster operations that has been tested through repeated exercises and deployments. The challenge for MMRS is sustainability, or the need to maintain "adequate funding and effective management of preparedness and efforts to keep domestic preparedness as a policy priority."[28] DHS has provided low-level funding to MMRS, but future federal support is not guaranteed. No new MMRS programs have been established since 2003. The FY 2006 allocation for the national MMRS program decreased to slightly more than $29 million, down from $50 million in FY 2005. In both years the program was assigned no funding in the Bush administration's proposed budgets, but congressional efforts restored some of it. For FY 2007, the allocation to the national program was just over $32 million.

The National Incident Management System

DHS Presidential Directive Number 5 (HSPD 5), a direct response to the September 11 attacks, established NIMS; it requires all federal agencies to adopt NIMS immediately, and all state and local organizations (including Indian Nations) to adopt it as a condition for accepting federal homeland security funding. Despite the name, NIMS is much more than a specification of command structure.[29] It prescribes a nationwide, FEMA-based planning process and defines preparedness practices for state and local jurisdictions.

The elements of NIMS

NIMS is intended as a nationwide framework for all-hazards planning and response. This goal duplicates (but does not cross-reference) the aim of long-standing FEMA programs: to engage in *comprehensive emergency management* through *integrated emergency management systems.* To achieve this goal, NIMS has six components, but much of the current NIMS documentation addresses only one of them—command and management—and specifically only one part of it: the Incident Command System (ICS). "Other aspects...will require additional development and refinement to enable compliance at a future date."[30] Detailed descriptions of NIMS are available elsewhere,[31] so this section merely summarizes the six components: command and management, preparedness, resource management, communications and information management, supporting technologies, and ongoing management and maintenance.

The command and management component addresses incident response organization and public information, defining the principal issues as those centering on the incident command system, "multiagency coordination systems," and "public information systems." The NIMS ICS is very similar to the traditional IMS widely used for decades by fire services. In the conversion from IMS to NIMS ICS, federal planners have changed commonly used names for some IMS components (IMS "sectors" are NIMS "divisions" or "groups"), have changed standard protocols for naming IMS assignments (the IMS "branch chief" is the NIMS "director"), and have relegated some functions to different levels. For example, the IMS safety section is a subfunction under command in NIMS, and NIMS command lacks the IMS roles of senior advisor and support officer. The NIMS ICS, like IMS and other incident command systems, otherwise provides a structure for incident response that emphasizes organization, accountability, and command, using both unified command, which is established across disciplines in the same jurisdiction, and area command, which is established across multiple jurisdictions.

Preparedness, the second NIMS component, "involves an integrated combination of planning, training, exercises, personnel qualification and certification standards, equipment acquisition and certification standards, and publication management processes and activities."[32] The program defines the planning process and structures to be used. There is an explicit admonition to address mitigation, whereas recovery is mentioned but not elaborated on. The guidance admits that the specifics of preparedness are jurisdictional functions (and not federal functions), but the guidance nonetheless issues demands. Much of NIMS preparedness reflects conventional procedures for planning, training, exercising, and mutual aid pacts. The different and constraining parts of NIMS include DHS standards for the testing and certification of local personnel and for the certification of jurisdictional equipment.

NIMS is intended as a nationwide framework for all-hazards planning and response. This goal duplicates the aim of long-standing FEMA programs: to engage in *comprehensive emergency management* through *integrated emergency management systems.*

The NIMS resource management component also is complex and intrusive. Of the principal responsibilities, three are routine parts of any emergency response system: resource activation before and during an incident, resource dispatch capability, and recall and deactivation protocols. The other responsibility under resource management requires a resource inventory that uses a DHS-devised "resource typing system." The system defines each resource individually and specifies how it is to be categorized, acquired, and tracked by local jurisdictions. DHS also demands that local personnel who handle resources be certified and credentialed through a centralized process. And DHS/FEMA has rules for determining what resources are needed for an incident as well as how they are to be ordered, mobilized, tracked, reported, and recovered. DHS/FEMA has developed partial definitions (e.g., not all types of self-contained breathing apparatus are defined) for 120 resources.[33] (To put this in perspective, it is usually estimated that as many as ten thousand resources were used in the response to Katrina.)

The communications and information management component develops standards for incident communications (including for interoperability) and defines processes for managing incident information. Relative to intra- and interagency communications, NIMS specifies that "effective communications processes and systems exist" that will follow unnamed standards "designated by the NIMS Integration Center in partnership with recognized standards development organizations."[34] The center, staffed by federal NIMS specialists, maintains a Web site where NIMS documents and requirements are posted and where locals are allowed to post questions about NIMS and NIMS compliance. Interoperable communications are required, but no milestones are suggested, and none of the existing federal programs on interoperability[35] is mentioned—not even the existing programs established by DHS.

The supporting technologies component requires that local governments continually review the availability of new technology. This requirement directs local jurisdictions to adopt strategic planning practices that make it possible for scientific advances to be identified, assessed, and incorporated into the preparedness process. What is not clear is how much of the cost burden for this is to be assumed by local governments. DHS does commit itself to a science and technology research program whose results will be shared with other governments and the private sector, and the NIMS Integration Center will "issue appropriate guidelines as part of its standards-development and facilitation responsibilities."[36] There is no explanation of how these standards will relate to the testing and certification of equipment described under other NIMS components.

The ongoing management and maintenance component establishes strategic direction for, and oversight of, NIMS. This responsibility is assumed by DHS and delegated to the NIMS Integration Center. DHS commits itself to creating a feedback function to receive comments about NIMS from other governments and the private sector.

NIMS as policy

As emergency management policy, NIMS is very hard to assess meaningfully. Multiple features have been cited as problematic, beginning with its creation. Many practitioners and disaster researchers agree that "NRP and NIMS have been developed in a top down manner, centrally coordinated by DHS. Views differ on the scope and intent of stakeholder involvement in developing NRP and NIMS."[37] It appears that the academic disaster-research community was minimally involved in the process of generating NIMS. The record is not clear about how or from whom DHS solicited other guidance or how such other guidance was incorporated. Of even greater concern to municipal emergency agencies is the detail in which processes and protocols are specified, superseding local practice. The significant question is whether such detailed specification promotes or retards the effective and efficient management of disasters.

On the matter of certification and standard setting, the demands placed on local governments have their counterpart in the demands that the requirements impose on DHS, FEMA, and the NIMS Integration Center themselves. The clearest certification standards to date require that virtually all administrative and response personnel in emergency-relevant organizations throughout the United States be trained and tested in NIMS and NIMS ICS—requirements that initially meant that personnel would individually take Internet classes through FEMA and be examined, graded, and certified. But the technology was not always adequate or able to handle the demand. After much frustration at state and local levels, DHS announced a new program for 2006 to partner with local entities and to simplify and speed the certification process.[38] The elaborate partnership program requires "Master Trainer" training and certification, instructor training and certification, and facility requirements. It is not clear whether the new program relieves demands on the federal sector, but it certainly increases demands on DHS training partners. Since the partnering is not required, it also is not clear how willing localities will be to participate. What may seem confusing is that on the one hand, compliance (certification) is required, but on the other hand, partnership (the road to compliance) is optional.

In summary, much of NIMS is "interim," much is simply not yet elaborated (five of the six components), and many demands for standardization and certification are without a specified process. Those certification processes that have been implemented have proved hard to operate with the DHS control (computer) systems. The NIMS formulation tends to ignore existing federal and local programs that have the same or similar goals, and NIMS documentation is not well integrated with that of other federal plans and programs. Faced with programs that do not operate, DHS has devised solutions that themselves are unworkable and has continued to issue demands for compliance.

The demands on local jurisdictions are overwhelming, yet the federal Homeland Security Grant Program (HSGP) either explicitly forbids the acquisition of personnel or minimally funds administrative support. Only the largest municipalities can fully staff an emergency management agency, and therefore many of the extensive NIMS requirements have to be fulfilled by small committees or by fire services or police departments whose primary functions are public safety rather than emergency planning. The likelihood that NIMS can be successfully implemented is therefore hard to estimate. Although DHS can impose NIMS adoption as a condition of accepting federal disaster preparedness funding, official adoption is quite different from effective implementation. In principle, a national IMS (NIMS ICS) is highly desirable, but the implementation process is slowed when so many elements that centrally define local emergency planning are put into the federal framework.

Summary

Response operations have been made increasingly complex and challenging by changes over time: for example, greater numbers of people moving into areas prone to natural hazards, the invention and use of more and different chemicals, and the rapid rise of terrorist threats. Looked at specifically in terms of disaster agents, different CBRNE agents demand different detection equipment, different types and levels of personal protective gear for responders, and different medical approaches to victims. In terms of the scale of the event and the size of the region affected, different resources and approaches to management will be required. This combination of variable features—agent, scale, and size of region—means that effective response operations are most likely to flow from an effective emergency planning process. The message to administrators is that the route to good disaster operations is a firm (and financial) investment in emergency planning.

Disaster operations are also improved when knowledge of patterned behavior and demands is incorporated into the design of response, as well as when measures are taken to meet administrative mandates. Disaster research demonstrates that, contrary to conventional wisdom, people faced with disaster tend to take constructive actions and are prone to (temporarily) increase their trust in government. This knowledge tells us that citizens can reasonably be expected to take part in response operations and that when emergency managers share reasoned expertise and protective action recommendations with citizens, trust will enhance compliance. Particularly in the case of terrorist incidents, knowledge that disaster experiences can have longer-term consequences for citizens reminds emergency managers to anticipate demands for behavioral health care. It is also known that volunteers come to emergency scenes and others send donations (cash, food, clothing, and other tangibles). This pattern cues emergency managers to devise a system and a place for sorting and assigning both personnel and other donations.

The route to good disaster operations is a firm (and financial) investment in emergency planning.

Research also demonstrates that concern over role abandonment by first responders is largely mythic; only under rare circumstances do first responders leave their assigned posts. Indeed, measures are available—"welfare checks and reports" on family members, for example—that reduce the anxiety of first responders during operations. Finally, the key to response and recovery funding from local, state, and federal government is the disaster declaration. For mayors, governors, and the president of the United States, the release of jurisdictional funds for emergency and recovery purposes is contingent on an official jurisdictional declaration. Knowing this, local government officials should streamline their information-gathering and damage assessment processes during operations in order to produce the information needed for successive (local, state, and federal) disaster declarations.

To support disaster operations, DHS administers the HSGP, two of whose five separate programs are examined above. (These five, in the order in which they are mentioned in the HSGP application packet for 2007, are the state homeland security program, UASI, the law enforcement terrorism prevention program, MMRS, and the Citizen Corps.) In 2007, UASI and MMRS together were allocated nearly $800 million. UASI, initiated in 2003, currently covers forty-five urban areas that have been determined to have high or special risk characteristics. As policy, UASI has the advantage of targeting areas on the basis of risk calculations, and it allows local jurisdictions some latitude in devising an operational approach. At the same time, it does not provide for new personnel, and it imposes immense accounting demands on both state and local governments. There has been no formal or independent (of DHS) comprehensive evaluation of UASI. The jurisdictions have completed multiple exercise cycles since the program started, but security demands limit access to information about outcomes.

MMRS began in 1997 and currently includes 124 jurisdictions. A large component of MMRS is managing mass casualties, so in that sense it is less global in operational scope than UASI. Funding for starting and continuing local MMRS programs has been sporadic, but the jurisdictions involved have largely remained loyal to the program and the concept. MMRS programs have been successfully activated for many major disasters, including the rush of anthrax "emergencies" that arose across the United States in the four months following 9/11. Together, UASI and MMRS offer emergency planning and operations coverage for a large proportion of the population of the country and for virtually all large urban areas.

Finally, the national government has engaged in a significant effort to improve response operations through NIMS. NIMS is administered by DHS/FEMA and is required for all federal executive departments and agencies and for all local and state governments that accept homeland security funding. The basic goals of NIMS—to have a common incident command system, promote effective communications and emergency planning, embrace a common system for operational resources and information, and support continuing attention to technology—are very positive in principle. Indeed, almost all emergency managers and disaster researchers would concur that both locally and nationally, the emergency response system would be stronger if all NIMS dimensions were implemented. However, there appear to be three major difficulties that interfere with achieving the positive effect. First, the full NIMS program has not been laid out by DHS/FEMA; there are many gaps in definitions of what is expected and how it should be achieved. Second, there is generic concern that NIMS as a single plan, devised largely by federal officials, may not fit all jurisdictions equally well and therefore may not be capable of delivering the full benefits promised. Third, NIMS imposes many administrative demands and training demands on local jurisdictions that translate into serious budget expenditures. To date, DHS has allowed some NIMS compliance costs to be covered under the constituent programs of the Homeland Security Grant Program. If such costs are not defrayed through this or some other mechanism in the future, it is not clear how local jurisdictions can continue to progress in implementing NIMS.

Notes

1 E. L. Quarantelli, *Catastrophes Are Different from Disasters* (Newark: Disaster Research Center, University of Delaware, 2005).

2 Michael K. Lindell, Carla Prater, and Ronald W. Perry, *Introduction to Emergency Management* (New York: John Wiley & Sons, 2007).

3 Jim Dwyer and Christopher Drew, "Fear Exceeded Crime's Reality in New Orleans," *New York Times*, September 29, 2005, A1, available at nytimes.com/2005/09/29/national/nationalspecial/29crime.html (accessed September 7, 2007).

4 Michael K. Lindell and Ronald W. Perry, *Communicating Environmental Risk in Multiethnic Communities* (Thousand Oaks, Calif.: Sage, 2004).

5 Emmanuel Skoufias, "Economic Crises and Natural Disasters: Coping Strategies and Policy Implications," *World Development* 31 (July 2003): 1087–1102.

6 Charles Fritz, "Disaster," in *Contemporary Social Problems*, ed. Robert K. Merton and Robert A. Nisbet (New York: Harcourt, Brace and World, 1961), 651–694.

7 Russell R. Dynes and E. L. Quarantelli, "Family and Community Context of Individual Reactions to Disaster," in *Emergency and Disaster Management*, ed. H. J. Pared, H. L. P. Resnik, and L. G. Parad (Bowie, Md.: Charles Press, 1976), 244.

8 Public Safety Wireless Network Program (PSWNP), "Critical Issues Facing Public Safety Communications" (Washington, D.C.: PSWNP, 2003), 2.

9 PSWNP, *Fire and EMS Communications Interoperability* (Washington, D.C.: PSWNP, 1998).

10 National Task Force on Interoperability, *Why Can't We Talk?* (Washington, D.C.: National Task Force on Interoperability, 2003), 4, available at homelandsecurity.alabama.gov/PDFs/why_cannot.pdf (accessed September 8, 2007).

11 Ibid.

12 U.S. Department of Homeland Security (DHS), *Statement of Requirements for Public Safety Wireless Communications and Interoperability: The SAFECOM Program* (Washington, D.C.: DHS, March 2004), iv.

13 Public Law 104-201, formally known as the Defense against Weapons of Mass Destruction Act of 1996.

14 G. W. Bush, *The Department of Homeland Security* (Washington, D.C.: The White House, June 2002), available at dhs.gov/xlibrary/assets/book.pdf (accessed September 8, 2007).

15 U.S. General Accountability Office (GAO), *DHS' Efforts to Enhance First Responders' All-Hazards Capabilities Continue to Evolve* (Washington, D.C.: GAO, 2005), 21, available at gao.gov/new.items/d05652.pdf (accessed September 8, 2007).

16 GAO, *Major Management Challenges and Program Risks: Department of Homeland Security* (Washington, D.C.: Government Printing Office, January 7, 2003), available at gao.gov/pas/2003/d03012.pdf (accessed September 8, 2007).

17 Office of Inspector General, *Review of the Status of Department of Homeland Security Efforts to Address Its Major Management Challenges* (Washington, D.C.: DHS, 2004), available at dhs.gov/xoig/assets/mgmtrpts/OIG_DHSManagementChallenges0304.pdf (accessed September 10, 2007).
18 GAO, *Overview of Department of Homeland Security Management Challenges* (Washington, D.C.: GAO, 2005), available at gao.gov/new.items/d05573t.pdf (accessed September 8, 2007).
19 GAO, "Statement by Comptroller General David M. Walker on GAO's Preliminary Observations Regarding Preparedness and Response to Hurricanes Katrina and Rita," February 1, 2006, available at gao.gov/new.items/d06365r.pdf (accesses September 8, 2007); see also DHS, *National Response Plan* (Washington, D.C.: DHS, 2004), available at dhs.gov/xlibrary/assets/NRP_FullText.pdf (accessed August 29, 2007).
20 Eric Lipton, "FEMA Calls, but Top Job Is Tough Sell," *New York Times,* April 2, 2006, A1, available at nytimes.com/2006/04/02/washington/02fema.html (accessed September 8, 2007).
21 Kathleen J. Tierney, *The Truth about Homeland Security* (Boulder: Natural Hazards Center, University of Colorado, 2005).
22 DHS, *Universal Task List: Version 2.1* (Washington, D.C.: DHS, 2005), available at ojp.usdoj.gov/odp/docs/UTL2_1.pdf (accessed September 10, 2007); and DHS, *Target Capabilities List: Version 1.1* (Washington, D.C.: DHS, 2005), available at scd.hawaii.gov/grant_docs/Target_Capabilities_FINAL_05_23_05.pdf (both accessed September 10, 2007).
23 GAO, *Combating Terrorism: Evaluation of Selected Characteristics in National Strategies Related to Terrorism* (Washington, D.C.: GAO, 2004), available at gao.gov/new.items/d04408t.pdf (accessed September 8, 2007).
24 National League of Cities, *Homeland Security: Federal Resources for Local Governments* (Washington, D.C.: National League of Cities, 2002).
25 DHS, *Fiscal Year 2004 Urban Areas Security Initiative Grant Program* (Washington, D.C.: Office of Domestic Preparedness, DHS, 2004), ii, available at ojp.usdoj.gov/odp/docs/fy04uasi.pdf (accessed September 10, 2007).
26 FEMA, *National Urban Search and Rescue Response System: Field Operations Guide* (Washington, D.C.: FEMA, September 2003).
27 The National US&R program is a successful, long-established program for special mobile all-hazards emergency response teams (with capabilities in CBRNE, hazardous materials, and technical rescue). The teams are hosted by cities that accept FEMA funding for partial support of the teams. The teams have been deployed to most major disasters and catastrophes in the United States, including New York for September 11, 2001, and the Gulf Coast for Hurricanes Katrina and Rita in 2005.
28 David Grannis, "Homeland Security: Federal Re-sources for Local Governments," in *First to Arrive: State and Local Responses to Terrorism,* ed. Juliette Kayyem and Robyn Pangi (Cambridge, Mass.: MIT Press, 2003), 209.
29 DHS, *National Incident Management System* (Washington, D.C.: DHS, March 2004), available at dhs.gov/xlibrary/assets/NIMS-90-web.pdf (accessed September 6, 2007).
30 Ibid., x.
31 Lindell, Prater, and Perry, *Introduction to Emergency Management.*
32 DHS, *National Incident Management System,* 4.
33 FEMA, *Resource Definitions: 120 Resources* (Washington, D.C.: FEMA, 2004).
34 DHS, *National Incident Management System,* 50.
35 DHS, *Resource Definitions: 120 Resources.*
36 DHS, *National Incident Management System,* 57.
37 Charlie Hess and Jack Harrald, "The National Response Plan: Process, Prospects and Participation," *Natural Hazards Observer* 28 (July 2004): 2, available at colorado.edu/hazards/o/archives/2004/july04/july04.pdf (accessed September 10, 2007).
38 DHS, "Cooperative Training Outreach Program (CO-OP)," *ODP Information Bulletin* 193 (October 20, 2005), available at ojp.usdoj.gov/odp/docs/info193.pdf (accessed March 19, 2007).

CHAPTER 10

The role of the health sector in planning and response

Erik Auf der Heide and Joseph Scanlon

This chapter provides an understanding of

- The importance of coordination between the health sector and other organizations and agencies in disaster planning and response
- The role of the health sector in planning and response
- The role of the public health system in disasters
- Formal planning for disaster health care.

Unlike routine emergencies, disasters tend to cross political, jurisdictional, functional, and geographic boundaries. As a result, disasters often generate responses from multiple levels of government (city, county, special district, state, tribal, and federal), and from multiple organizations and entities in both the public and private sectors. To mount an effective response, public and private organizations that normally operate autonomously and independently must work together.[1]

Unfortunately, the policies and procedures that should be in place to ensure multiagency coordination and communications are often lacking. Many of those involved in the response may never have worked together before. They may have different training, organizational structures, equipment, and procedures, and they may use different terminology and communications frequencies. Moreover, local authorities may not be aware of all the entities and resources that are available, much less be able to integrate them into the response effort.[2] Under these circumstances, the usual means of coordinating routine responses will be ineffective.

The purpose of this chapter is to familiarize emergency managers with the role of the health sector in coordinated disaster planning and response. For the purposes of this discussion, the health sector includes medical institutions (e.g., hospitals, clinics, pharmacies), medical practitioners (e.g., doctors, nurses, emergency medical technicians), and public health agencies (e.g., state, tribal, and local health departments). After a brief discussion of the context in which coordination occurs, the chapter describes the role of the health sector in disaster-related activities. It then outlines the role of the public health system in disasters and provides an overview of formal planning for medical response.

Coordination in disaster planning and response

The health sector is not alone in providing assistance to the ill and injured in disasters. Public safety agencies, including fire and police departments, take part in search and rescue, transport of casualties, decontamination of casualties, and provision of emergency medical care and first aid. In addition, there is often substantial involvement on the part of the military and, most importantly, spontaneous volunteers (including the survivors themselves). Thus, it would seem obvious that local government—including emergency managers—and the health sector should communicate and coordinate with one another in disaster planning and response. But the challenges posed by disasters often make this difficult. In fact, health sector organizations often fail to coordinate even among themselves.[3] For example, a 2004 survey of hospitals in Los Angeles County revealed not only that 67 percent had never participated in joint training with local police and fire departments, but also that only 16 percent of them had written mutual aid agreements with other hospitals.[4]

Coordination between local governments and the health sector is complicated by the fact that most health care in the United States is provided by the private sector (e.g., private hospitals, private physicians, and private ambulance services) and is therefore largely outside of the direct operational and fiscal control of government. As a result, coordination with these private entities requires skills in negotiation, mediation, facilitation, and compromise—skills that, while not generally associated with the top-down, command-and-control approach characteristic of the civil defense era, have become increasingly important in modern emergency management.[5]

The emergency manager may be the only person in the community whose constant focus is on preparation for disaster.[6] Thus, he or she is in a unique position to promote interorganizational planning and preparedness, and to ensure that the local government and health care organizations and agencies work together. The emergency manager must also ensure that all organizations, including those in the health sector, understand that they *need* to cooperate with one another. Readiness for medical emergencies depends on the ability of local government, public safety agencies, *and* the health sector to plan, train, drill, respond, and evaluate jointly.

In some communities, established groups (such as local emergency planning committees or the local Metropolitan Medical Response System) may serve as the focal point for

---

Disclaimer: The findings and conclusions in this report are those of the authors and do not necessarily represent the views of the Agency for Toxic Substances and Disease Registry.

Failure of coordination: The Station Nightclub Fire

The 2003 response to the Station Nightclub Fire in West Warwick, Rhode Island, illustrated the inadequate coordination between the local government and health care organizations. Shortly after 11 PM, during a rock band performance, a pyrotechnic display set off a fire in the nightclub, injuring more than 200 patrons and killing 100 others. While at the individual tactical level the response was remarkable and heroic, at the strategic level there were numerous system problems:

- Private ambulances in Rhode Island do not normally respond to 911 calls, nor was there a plan for using them in disasters. Thus, when they were needed to supplement the number of available public ambulances, the fire department had to look in the Yellow Pages to get contact information. Moreover, because of the lack of coordination, the police and fire departments independently requested private ambulances at the same time, and local officials made no attempt to periodically assess the location and safety of those ambulances while they were being used to transport patients from the nightclub to hospitals. Moreover, none of the ambulances in use—whether fire department, municipal, or private—knew how to contact the local hospitals to which they were transporting patients.
- No plan existed to guide the use of medical helicopters in disasters.
- Although both police officers and firefighters from all over the state provided care for the injured, they set up separate command posts and did not coordinate their efforts.
- The state emergency management agency did not declare a disaster and did not activate the statewide mass casualty plan, nor were any emergency operations centers activated in this incident. In the absence of a declaration, and concerned about patient privacy protection, hospitals were reticent to share desperately needed victim information with the emergency management agency so it could compose accurate lists of the missing.

No one was designated to coordinate the overall responses of the various hospitals. For example, hospitals did not coordinate with one another or with the emergency management agency in requesting medical help through the U.S. Department of Homeland Security. Such coordination was also absent when it came to the transfer of patients to out-of-state hospitals.

Source: Titan Corporation, *The Station Club Fire: After-Action Report*, Contract no. GS10F0084K2001F_341 (Washington, D.C.: Office for Domestic Preparedness, U.S. Department of Homeland Security, 2004); available at llis.dhs.gov/member/secure/getfile.cfm?id=10806 (accessed August 22, 2007).

community-wide emergency planning.[7] If a community-wide emergency planning committee does not exist, the emergency manager should establish one and ensure that it includes government and health sector representatives. One approach is to establish a health sector planning subcommittee as part of a larger group; examples of who should be represented on this subcommittee are shown in Figure 10–1. Provisions for a community-wide joint emergency operations center, the use of joint incident command following disasters, and the establishment of an intercommunity and interagency radio system and mutual aid plan also facilitate collaborative, strategic responses.

Hospitals
Clinics
Mental health agencies, clinics, and hospitals
Urgent care centers
Emergency medical service agencies, providers, and dispatchers
Medical societies (private physicians)
Poison centers
Dialysis (artificial kidney) centers
Environmental protection agencies
Public health agencies
Nursing associations
Nursing registries[1]
Red Cross
Medical Reserve Corps
Psychiatrists, psychologists
Disaster medical assistance teams
Jail infirmaries
Physician paging services
Social workers
Rehabilitation centers
Rescue teams
Hospices
Assisted living facilities
Nursing homes

[1]A nursing registry is an organization that registers local nurses who are available for work. This work could be providing temporary coverage at hospitals, making nurses available for private work in the home, or other situations where nurses are needed on a temporary basis.

Figure 10-1 Recommended representation on a community- or areawide disaster health planning subcommittee

The role of the health sector in disaster planning and response

As discussed in Chapter 1 of this volume, disasters differ from emergencies qualitatively as well as quantitatively: disasters often require health sector responders to carry out tasks that are different from those called for in routine emergencies, and to do so under conditions of great urgency and uncertainty. This is why specific planning and training for disasters is necessary. Among the most important tasks are

- Communication
- Search and rescue
- Triage
- Hospital and nursing home evacuations
- Victim tracking and handling public inquiries
- Managing medical donations
- Managing contamination
- Ensuring access to nonhospital sources of medical care
- Dealing with loss of response infrastructure.

Communication

As noted earlier, disasters require different organizations—often from different jurisdictions and levels of government as well as from the private sector—to work collaboratively. One of the keys to successful collaboration is clear, accurate, and timely communication. Hospitals need prompt notification of disasters so that they can, for example, transfer or discharge stable patients to make room for incoming victims; expedite the transfer of patients from the emergency department to floor beds; open up and staff auxiliary rooms to treat disaster victims; call in off-duty staff; and, if needed, set up decontamination areas and distribute personal protective equipment. Often, however, hospitals are taken completely by surprise: the earliest notification they receive is from the first arriving casualties or the news media, rather than from officials in the field.[8]

In the event of disaster, the following information needs to be shared:

- The nature and scope of the disaster
- Whether hazardous substances are involved
- Whether physician teams are needed at the site
- What arrangements are being made to get health care staff and suppliers through police security lines
- Estimates of the number, types, and severities of casualties
- The number and types of casualties that each hospital can accommodate
- The number and types of casualties being sent by ambulance to each hospital, along with estimated times of arrival
- Whether outside health sector resources are needed
- Where incoming resources should report (e.g., at a designated staging or check-in area, as shown in Figure 10–2)
- What hospitals and health facilities need traffic control or security assistance from police
- What hospitals or health facilities need to be evacuated
- Whether hospitals need fire department assistance because of hazardous material (hazmat) spills, or loss of fire alarm systems or sprinklers
- What hospitals need water, food, supplies, supplementary staff, or electrical generators (or fuel for them)
- When the last casualty will be being transported
- What precautions fire and police need to take to protect themselves (e.g., against exposure to biological, chemical, or radioactive substances contaminating victims or the disaster site) during response and recovery activities.

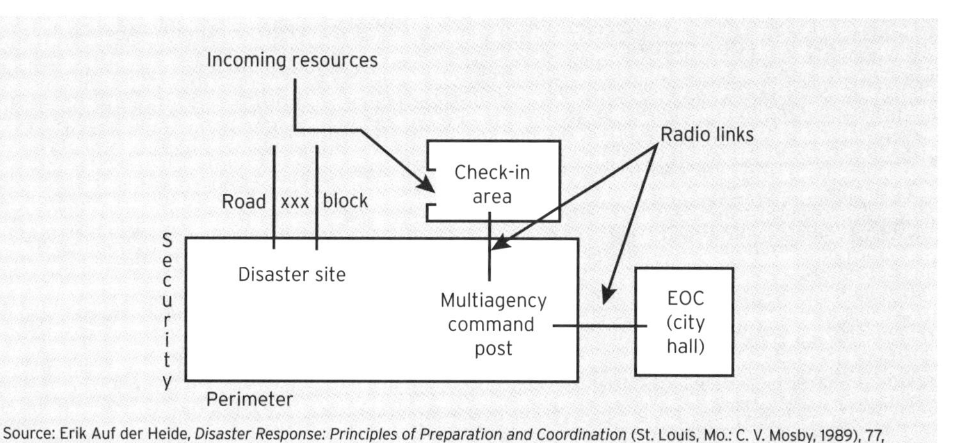

Source: Erik Auf der Heide, *Disaster Response: Principles of Preparation and Coordination* (St. Louis, Mo.: C. V. Mosby, 1989), 77, available at orgmail2.coe-dmha.org/dr/index.htm (accessed August 22, 2007).

Figure 10-2 Check-in areas for managing unrequested volunteers and outside responders

Telephone and radio technologies More than half a century of experience has shown that even if the usual means of communication—land lines and cellular phones—are not damaged, circuits or networks rapidly become overloaded and unusable. Pagers, two-way radios, and Internet connections that rely on telephone connections are also prone to failure.[9] One of the most common observations in disaster after-action reports is the need for interoperable radio communication networks to tie the various emergency response organizations together.[10] This need was recognized as far back as 1983 in a report titled *California's Emergency Communications Crisis,* published by the California legislature's Joint Committee on Fire, Police, Emergency and Disaster Services.[11] But twenty-five years later, the need has still not been met,[12] largely because of a lack of funding.[13]

Dealing with overresponse and self-dispatch

A disaster often sets off a larger response than may actually be desirable. For example, those who first arrive on the scene may request assistance ("It's a big one. Send everything you've got") before fully assessing the need. Both trained and untrained volunteers may hear of the disaster on police scanners or via the news media, assume that help is needed, and "self-dispatch." Although it is unlikely that overresponse, or convergence, and self-dispatch can be eliminated in disasters, some measures can be taken to reduce it:

- Avoid appeals to send "all available ambulances" or have "anyone with first aid training" go to the scene. The response is often far greater than intended, and the request is difficult to cancel once issued.
- Be proactive in announcing over the media and public safety radio frequencies when additional help is *not needed.*
- Initiate a needs assessment (or "situation assessment") and mutual aid plans as rapidly as possible. With mutual aid plans to guide dispatch, ad hoc mobilization and self-dispatch of unneeded resources may be reduced. Mutual aid plans can also specify staging or check-in areas near but not in the impact zone, thereby allowing the command post to identify what resources have responded and to keep them out of the way until they're actually needed. Mutual aid plans should include provisions for proactively notifying mutual aid resources when they are not needed. In the absence of mutual aid plans (and associated mutual aid communications systems), outside responders have difficulty finding out if their help is needed and, assuming that too much help is better than too little, will self-dispatch.
- Have emergency personnel and vehicles respond to a staging or check-in area where they can be held until needed, released if unneeded, or assigned a response task as required. These areas can be situated adjacent to, but outside of, the impacted zones to reduce unnecessary emergency vehicle congestion (see Figure 10-2 above).

Ultimately, however, planners will need to train local responders to expect unsolicited assistance and should designate persons to help coordinate their efforts.

Emergency managers can take a number of steps to improve disaster communications. For instance, they can

- Promote the creation of a community-wide emergency communications committee with the responsibility to establish (1) common communication frequencies for mutual aid requests and (2) procedures to link health sector organizations with one another and with public safety organizations and dispatchers
- Work with surrounding communities or the state to establish mutual aid systems (with associated communications networks) with those entities
- Work with cellular and telephone companies to give electronic priority ("essential service" designation) to health sector organizations when circuits are overloaded
- Make arrangements to distribute mobile cellular towers ("cells on wheels") and antenna towers, portable and mobile two-way radios, and portable radio repeaters
- Find other creative ways to communicate, such as using motorcycle club members as couriers.

The role of the media Emergency management has traditionally focused on what local government and other disaster responders should do to protect the public. However, increasing attention is being given to the role of the media in assisting and informing the public.

In the event of disaster, many factors can compromise the media's ability to reach the public; some of these factors are related to the inherent limitations of warning systems, others to the effects of the disaster itself, and still others to inadequate preparedness. Sirens are a notoriously poor means of warning the public: they may not be heard, and even if they are, they cannot provide specific information on the nature of the threat or on steps the public should take to protect itself. Although tone-alert weather radios can now be used for all-hazards disaster notifications, few people own these devices. If a disaster occurs at night, when most people are asleep, warnings broadcast over commercial television and radio stations may not be heard; moreover, a disaster may lead to the loss of electrical power, which means that most people will lose television, satellite, and cable services. Commercial broadcast stations may themselves lose power as the result of a disaster and may not have emergency generators; these stations may also have no one on duty to receive warnings from local officials, especially in smaller communities where they merely rebroadcast satellite feeds during nighttime hours.

Emergency managers need to work with media outlets to ensure that those outlets can survive and function after disasters—for example, that they have emergency backup power, backup antennas, food and provisions for long staff stays at the station, a means for emergency officials to contact station staff after normal business hours, and agreements with surrounding stations to broadcast information into the local area if the local station cannot broadcast. Stations should be educated about local disaster threats, such as high winds, flooding, and seismic risks, so that they can take protective action for their facilities and staff. In the event that communication through mass media fails, emergency managers need contingency plans for providing door-to-door warnings and information on protective actions.[14]

The role of citizens Increasing attention is also being given to how members of the public can help themselves and others. They can, for example, evacuate the area, shelter in place, help locate victims, decontaminate themselves, obtain vaccinations, maintain sanitary conditions, take antibiotics, and exercise proper precautions when using emergency generators to protect against carbon monoxide poisoning. During the first outbreak of West Nile virus in the United States, for example, the public was enlisted to identify and eliminate the breeding areas of mosquitoes that spread the disease.[15] During the 2003 pandemic of severe acute respiratory syndrome (SARS), studies showed that transmission of the disease was significantly reduced by simple measures taken by the public, such as frequent hand washing and use of surgical masks.[16]

In order to take self-protective actions, the public needs accurate and consistent information on

- The nature of the health threat
- The location and geographical extent of the threat (e.g., in the case of a plume of hazardous material, where it is, where it is going, and how fast it is moving)

Public response to a bioterrorism attack

Federal, state, and local planners are developing strategies to rapidly vaccinate the public in the event of a smallpox attack. Such strategies include procedures for quickly obtaining smallpox vaccine from national stockpile locations around the country, setting up clinics in schools and other locations, and triaging and vaccinating large numbers of people within a few days. Strategies should also include provisions for notifying the public about the availability of vaccines and about priorities for who should receive the vaccine, especially if there are shortages.

However, a 2004 report by the Center for the Advancement of Collaborative Strategies in Health points out the problems that can occur with planning for catastrophes with which we have no previous experience,[1] such as a smallpox bioterrorist attack. Planners have assumed that in the event of an attack, the public would be very worried about catching smallpox and would follow instructions to go to clinics and be vaccinated. In fact, plans detail crowd control measures to deal with the anticipated flood of people who would show up at the vaccination sites. But interviews with citizens revealed that only 24 percent would rush to vaccination sites, 19 percent would go later when it was most convenient, 2 percent definitely would not go, and 55 percent would need more information before they decided to go.

Interviewees gave a number of reasons for not getting a vaccination, including the mistaken beliefs that they were already immune because of a previous vaccination or that they were not at risk because their lifestyles did not bring them into contact with many people. Other reasons included such fears as that they might contract smallpox at the triage-and-vaccination clinics, that they might be carrying the disease and would expose others at clinics, or that the government would not tell them about dangerous side effects or risks of the vaccine.[2]

[1]Roz D. Lasker, *Redefining Readiness: Terrorism Planning through the Eyes of the Public* (New York: Center for the Advancement of Collaborative Strategies in Health, New York Academy of Medicine, September 2004), available at redefiningreadiness.net/pdf/RedefiningReadinessStudy.pdf (accessed July 20, 2007).

[2]Ibid.

- How people can determine whether they have been exposed to a communicable disease or hazardous substance
- What symptoms the disease or exposure to it causes
- How the illness or contamination spreads
- What to do in case of illness or exposure—for example, where to get treatment or preventive care (e.g., antibiotics or immunizations)
- The risks of illness or exposure at hospitals or other treatment locations
- The long-term effects of exposure
- The effectiveness of recommended self-protective measures
- If evacuation is ordered, what specific areas need to be evacuated, what the evacuation routes are, where shelter is available, and how to obtain assistance in complying with evacuation orders.

When it comes to providing information about a disaster, *how* communication occurs is often as important as *what* is communicated. The following guidelines will help emergency managers improve the effectiveness of risk communication:

- Before disaster strikes, work with health sector organizations and other agencies to develop procedures for ensuring that the public receives information that is consistent and up-to-date. Assign subject matter experts to listen to and watch all media broadcasts to identify inaccuracies and inconsistencies and bring such problems to the attention of those who can resolve them.
- Communicate openly and honestly. Officials are sometimes reluctant to share information with the public, either because they don't want to cause panic or because they feel that the public will not comprehend the complexities of the situation. But in fact, panic is extremely rare in disasters, and the public usually responds in an adaptive and productive manner. Officials should be candid about how much they do or do not know about health risks, and should give the public some indication of when additional information will be available.

- When disaster strikes, provide accurate, consistent information through multiple sources trusted by the public, recognizing that different segments of the public differ in what languages they understand, where they routinely seek information, and whom they trust. The effectiveness of a warning is enhanced if it is issued repeatedly.
- Tailor disaster-related communication to the specific needs of the community. For example, in Hurricane Katrina, those in low-income neighborhoods needed to know how to access public transportation for evacuation, while those in more affluent neighborhoods who owned private vehicles needed to know what roads were open.
- Ensure that communication is two-way. Establish call centers to expeditiously address inquiries from concerned citizens (which, of course, presumes an intact telephone system—by no means a given), and make sure those call centers receive continual updates. Plan and talk with formal and informal community groups.
- Use technology effectively to get meaningful information to the public. Warning sirens, for example, carry the least specific information but may be successful in getting the public to turn to sources of more specific information, such as broadcast radio or television.[17]

Search and rescue

Initial post-disaster search and rescue is not usually well-coordinated—in part because most of it is usually carried out by the survivors themselves.[18] However, because the need for widespread search and rescue is rare in routine emergencies, and because the roles and responsibilities related to this task often have not been previously worked out or practiced, this lack of coordination often persists even when official emergency response agencies become involved.[19]

It is unrealistic to believe that massive spontaneous citizen involvement in post-disaster search and rescue can be controlled or prevented. (There is even some question as to whether preventing it—if it were possible—might result in the loss of life because of delays in locating trapped victims.) However, the effectiveness of disaster search and rescue can be improved by training law enforcement officers and firefighters to coordinate the work of spontaneous volunteers and by agreeing on who should be responsible for coordinating the volunteers.[20]

Because survivors of a disaster are the ones most likely to know where the missing are located, it is beneficial to involve them in the search and rescue effort. Those coordinating search and rescue should make it a point to interview survivors—both those in the field and those who have been transported to hospitals—to find out where victims were last seen before disaster struck. Registration clerks in hospital emergency departments should have protocols for obtaining this information (or, at least, contact information for persons who have this information) and relaying it to rescue officials in the field. It is important for hospitals to note not just who the victims are but who brought them to the hospital, because these persons may also have information on other missing persons or may be reported as missing themselves. In the same way, Neighborhood Watch, Citizen Corps, and other community organizations can also assist with search and rescue.

Triage

The word *triage* is derived from the French verb *trier,* which means "to sort." The fundamental principle of triage is to do the greatest good for the greatest number. Generally, care is provided first to the most serious casualties and to those most likely to survive as a result of that care (see Figure 10–3).

During the triage process, casualties are assigned to priority categories—often designated by a number or color—for treatment and transport. Because categories are not currently governed by a universally accepted standard, they may vary depending on the locality. However, many localities use a four-category system, such as red (critical), yellow (urgent), green (minor), and black (dead, or "unsalvageable" given the resources available).

Despite its importance, field triage is often not carried out in disasters. Among the reasons for this may be the lack of training for emergency medical service (EMS) providers on how to carry out and coordinate triage in multiagency, multijurisdictional responses; a lack of medical authority and direction in the field; and the tendency of untrained survivors who are involved

Figure 10-3 Triage after sarin attack, Tokyo subway, 1995

Source: T. Okumura et al., "The Tokyo Subway Sarin Attack: Disaster Management, Part 1: Community Emergency Response," *Academic Emergency Medicine* 5, no. 6 (June 1998): 813–617.

in search and rescue to transport casualties to the closest hospital, bypassing any field triage station either because its existence and location are unknown or because such care is considered inferior to that available in hospitals.[21]

Doing the greatest good for the greatest number in a disaster involves more than assigning priorities for treatment and transport. It also means using available medical resources as efficiently as possible. For example, patient destinations should be coordinated on a community-wide basis so that no hospital receives a disproportionate share of the casualties. Further, casualties should be sent to those hospitals that are best suited to provide the care needed. For example, it is preferable to send patients with minor injuries to facilities other than the trauma center, and to send severely burned victims to facilities that are specifically equipped to handle burn injuries.

Unfortunately, this is not what usually happens in disasters. Instead, the majority of casualties—including the most serious cases and those who are dead on arrival—often end up at the hospitals that are closest to the high-impact areas, that are more familiar to those performing the transport, or that are renowned for providing emergency care. A study conducted by the Disaster Research Center, at the University of Delaware, found that in 75 percent of the disaster events studied, more than half of the casualties were transported to the closest hospital.[22] Numerous other disaster case studies have reported similar findings.[23]

Uneven or inappropriate distribution of casualties occurs for a number of reasons, including lack of planning and training, failure to assign responsibility for coordinating casualty distribution, and lack of interagency radio communications. Another reason is the ad hoc transport of casualties by nonambulance vehicles such as private cars, police cars, buses, and taxis.[24] Many individuals who are searching for and rescuing victims believe that the best emergency care is transport to the closest hospital by any means as quickly as possible. If enough ambulances are not readily available, volunteer rescuers don't tend to sit by idly awaiting their arrival but instead use the most expedient transportation means at hand to get casualties to medical care. After the 2001 World Trade Center attack, for example, only 7 percent of the casualties transported to area hospitals arrived by ambulance.[25] Similarly, after a sarin gas attack on the Tokyo subway system in 1995, Tokyo's St. Luke's hospital, one of the medical facilities closest to the attack, received the most casualties, only 7 percent of which arrived by ambulance; of the rest, 35 percent arrived on foot, 24 percent by taxi, 14 percent by private car, 13 percent by nonambulance fire department vehicle, just over 1 percent by police car, and 6 percent by other means.[26] Nonambulance transport also helps to explain why, during disasters, hospitals rarely receive timely notification from authorities in the field or estimates of the numbers, types, and severities of casualties they will receive.[27]

Doing the greatest good for the greatest number in a disaster involves more than assigning priorities for treatment and transport. It also means using available medical resources as efficiently as possible.

Another pattern commonly seen in disasters is a sort of "reverse triage": that is, the first casualties to arrive at hospitals are the least serious ones. This happens because the least injured are those who are the most able to rapidly transport themselves, while the most serious casualties tend to be those who are entrapped in the rubble and require the use of complicated rescue techniques and heavy equipment—which untrained survivors cannot provide—before they can be transported.[28]

Even though most disaster casualties bypass field triage efforts and the attempts to distribute the patient load among area medical facilities, there are still some who will be transported by ambulance. Therefore, procedures should be established to poll area medical facilities when disasters strike to identify which facilities are overloaded, which can receive more patients, and which are damaged or evacuating. (Such procedures, of course, require the existence of an interoperable radio communications network.) The information that is gathered should be used to guide ambulance destinations. Even without such information, however, ambulance dispatchers and drivers can be fairly certain that hospitals closest to the impacted areas are most likely to become overloaded and so should consider bypassing those facilities. Medical mutual aid plans and joint training and exercises should address triage and casualty distribution,[29] and police and firefighters, who carry out much of the nonbystander search and rescue in disasters, should be familiarized with area triage plans and procedures.[30]

Hospital and nursing home evacuations

Sometimes disasters or hazard threats necessitate the evacuation of hospitals, nursing homes, and other congregate care facilities. Even medical care facilities that are not damaged may have to evacuate stable patients to make room for expected disaster cases. This, of course, requires that these facilities receive timely information about impending threats or incoming casualties. At that point the following tasks need to be undertaken:

- Communication with other hospitals to assess how many transferred patients they can receive
- Coordination with ambulance services to transport patients (and to make sure that ambulances do not transport casualties to hospitals that are themselves being evacuated)
- Coordination with public safety agencies to assess the most appropriate evacuation routes
- Implementation of effective patient tracking systems
- Notification of evacuated patients' family members.

These tasks should be coordinated at the community (or intercommunity) level, for example, by a designated "disaster coordination hospital," the 911 dispatch center, the local EMS agency, or the hospital association. And often these tasks must be undertaken in the face of serious damage to the hospital; spillage of hazardous materials; leaking water; blocked roads; loss of power, communications, and elevator function; and the continued arrival of casualties. Evacuations of facilities may be especially challenging because ambulances may be tied up at the disaster. In addition, coordination of evacuation may be difficult because cellular and telephone lines are often damaged or jammed.

Unfortunately, community-wide plans for hospital and nursing home evacuations are not always in place (and even when they exist, they are not always implemented), and facilities sometimes carry out evacuations independently and in an uncoordinated manner. Thus, emergency managers should work with community health sector organizations to develop areawide plans, training, and exercises for evacuating hospitals, nursing homes, jail infirmaries, and other health facilities. This planning should

- Include the availability of backup portable generators, sump pumps, water, and supplies, so that hospitals may be able to avoid evacuating and shutting down
- Identify the individuals or entities that will coordinate evacuation (both within the community and with neighboring communities)
- Specify the criteria for evacuation and identify who has on-site authority to make the decision
- Provide for local authorities to receive rapid notification of any facilities that are threatened by hazards and thus might require evacuation
- Provide for engineers to rapidly evaluate a facility to determine whether it is safe for continued occupation
- Establish protocols for the transfer of patient records and medications
- Provide for the transfer of staff (including temporary medical privileges) to the host facility
- Arrange for incoming ambulance patients to be detoured to other facilities
- Provide for the notification of patient families
- Provide for the return of patients and staff to evacuated facilities after the disaster
- Assign responsibility and provide guidelines for the management of those who spontaneously volunteer to assist with evacuation or patient care
- Include arrangements for alternative means of transportation (e.g., boat, helicopter) when roads are impassible
- Include alternative arrangements for patients who receive regular outpatient treatment at the hospital (e.g., dialysis patients).

In a widespread destructive event, however, it may be necessary for facilities to keep open even if they have suffered damage.

Handling inquiries and victim tracking

Because it is common to have family members living in other parts of the country or world, or for family members living in the same household to spend part of the day apart from each other, family members may not be together when a disaster occurs. In a 2005 randomized survey of 680 adult evacuees from Hurricane Katrina who were temporarily residing in shelters in Houston, 40 percent said that they were separated from immediate family members—those with whom they normally lived—but knew where they were, while 13 percent indicated that the whereabouts of immediate family members were unknown; moreover, 32 percent did not know the whereabouts of other close relatives or friends.[31] After the 2004 Indian Ocean tsunami, call centers at foreign ministries all over the world were deluged with inquiries from persons who thought their loved ones or other people they knew may have been victims. The Canadian Office of Foreign Affairs received 100,000 calls although only twenty of the victims were Canadian.[32] And efforts to locate and communicate with family members are intensified by the fact that with modern mass media communications, even a relatively small disaster can quickly become international news. Unfortunately, news reports often fail to provide specific details about the disaster that would allow viewers to tell whether their loved ones were involved.[33]

Typically, the first response of families after a disaster is to phone their loved ones. If they cannot make contact this way, they will call hospitals, police departments, fire departments, relief agencies, 911 operators, or government offices in their search for information. Even for disasters involving a few hundred homeless, injured, or killed, inquiries may number in the thousands. To prevent hospitals and other health sector organizations from becoming overwhelmed by phone calls, it can be helpful to anticipate key information that the public will need and, where possible, provide it through recordings or menus. For example, callers seeking information on the missing could be given a phone number to call and/or a Web site to check for a list of disaster casualties, their locations, and their general conditions. Emergency managers should ensure that press releases about disasters contain specific information (perhaps using maps) about areas affected. This will reduce the number of people who think their loved ones might have been involved. It is also essential that hospitals share information with call centers so that both can provide people with the information they seek.

Even for disasters involving a few hundred homeless, injured, or killed, inquiries may number in the thousands.

If no answers are available by phone and the disaster site is within driving distance, people will inquire in person. This can be very disruptive to hospitals, health departments, and government and public safety agencies, many of which have not developed procedures for victim tracking in disasters—procedures that can facilitate responses to inquiries and help determine who needs to be searched for and where they might be found. Victim tracking requires collecting information on the missing from hospitals, jails, shelters, morgues, and other locations; once collected, the information should be centralized for transmission to and dissemination from a toll-free number located in a different part of the country (so as to reduce phone traffic in the affected community), and thereby made available as appropriate to the public and public safety agencies. However, because telephone and cellular circuits are likely to malfunction, area- or statewide encrypted two-way radio or satellite networks must be established to successfully carry out victim-tracking efforts. Overall, it should be the responsibility of the emergency manager and/or the community-wide emergency planning committee to ensure that formal arrangements are made to have such tasks carried out, and

Massive inquiries about the missing on 9/11

"It happened throughout the United States on September 11, 2001, but nowhere like lower Manhattan. A busy signal, 'Please try your call again later,' or complete lack of dial tone met the ears of landline callers and cell phone users. And nowhere was it more crucial that a dial tone exist than in lower Manhattan, as people tried to locate family members, hospitals tried to contact reserve staff, and emergency management and public health agencies tried to coordinate a response to the day's tragedy.... Several landlines serving New York University Downtown Hospital still functioned after the collapse, but became immediately congested as people frantically called to inquire about family and friends. The effects were wide-reaching, preventing hospital staff from communicating within the hospital and with the community at large."

Not only was a major cellular antenna damaged, but cell phone demand doubled. Many New Yorkers found their cell phones useless. "At the end of the first day..., there was a flood of relatives in the emergency departments, walking from hospital to hospital, trying to find their family members."[2] At the same time, patient accountability was lacking. To some extent, the problem was attributed to the fact that triage tags were not used and ambulances did not report to staging areas upon arrival in the impact area. In addition, civilians flagged them down and crews initiated treatment and transport without reporting their actions to incident command. Another contributing factor was the failure to designate a central hospital whose responsibility would be to coordinate the dissemination of information on patient dispositions. However, at least in New York City, one of the main factors was that only 7 percent of the victims were transported by ambulance.[3] The rest completely bypassed the emergency medical services system, getting themselves to hospitals on their own.

[1]Bo Emerson, "Cellphones Come through When Emergencies Strike," *Atlanta Journal-Constitution*, September 23, 2001, G7.

[2]Susan Waltman, of the Greater New York Hospital Association, as quoted in Lara Misegades, "Phone Lines and Life Lines: How New York Reestablished Contact on September 11, 2001" (Washington, D.C.: Association of State and Territorial Health Officials, 2002), available at astho.org/?template=1bioterrorism.html (accessed August 24, 2007).

[3]M. G. Guttenberg et al., "Utilization of Ambulance Resources at the World Trade Center: Implications for Disaster Planning [Abstract]," *Annals of Emergency Medicine* 40, no. 4 (October 2002): S92.

ideally, the system should be run by an organization like the American Red Cross, so that hospitals and other health facilities are relieved of the burden of dealing with the problem.

At field triage sites, numbered triage tags with multiple "carbon" copies can be used to help keep track of patient destinations. As noted earlier, however, most casualties will bypass field triage, and overwhelmed rescuers often abandon record keeping in favor of patient care. Therefore, it is often necessary to collect information after the fact from places where those thought to be missing might be located (e.g., hospitals, morgues, shelters, and jails). Also as noted earlier, another useful strategy is to have hospital clerical staff collect information, not just on the casualties but also on those who transported the casualties to the hospital.

Hospitals are governed by the confidentiality guidelines of the Health Insurance Portability and Accountability Act (HIPAA) of 1996[34] and are therefore reluctant to release information. However, the act does allow health care providers to disclose protected health information to a person or entity that will assist in notifying a patient's family member of the patient's location, general condition, or death.[35]

Managing medical donations

Disaster-impacted communities will often receive a massive amount of donations, many of which are unsolicited. The situation can be aggravated when local elected officials or newscasters assume that medical volunteers or donations are needed and issue public appeals for them without first confirming the need with the intended recipients. Host communities, already burdened by the disaster impact, may then have to spend considerable amounts of money and time properly disposing of the donated items. When the Murrah Federal Building in Oklahoma City was bombed in 1995, medical donations came by the truckload from all across the nation. However, except for dressings and bandages, little of it was used; the Oklahoma Hospital Association conservatively estimated that more than $1.5 million worth of medical supplies and equipment was wasted.[36] Another problem may arise when massive numbers of blood donors congregate at hospitals, causing congestion and diverting the attention of facility personnel.[37]

Emergency managers should work with local health authorities, government officials, and the news media to ensure that public appeals for medical donations, including blood, are not made without the approval of the potential recipients. In fact, if donations are not needed, it may be wise to proactively notify the mass media of this fact. Once issued, appeals for donations are almost impossible to rescind. Pharmaceutical donations, in particular, cause problems when they are perishable, are sent in amounts greater than needed, include inappropriate or out-of-date medications, or have not been sorted and categorized before being sent.

If donations are requested, it is important to have them sent to a location outside of the impact area and well away from medical centers or traffic routes leading to those centers (e.g., to a school or warehouse), where they can be sorted and distributed in a manner that does not interfere with emergency operations. Serious consideration should be given to appeals for cash rather than material goods. Cash can be used more flexibly, contributes more to the local economy, is easier to transport, and does not require sorting or disposal. An essential element in disaster planning is deciding who will coordinate the overall management and distribution any donations.

Managing contamination by hazardous substances

During mass casualty incidents that do not involve hazardous materials, rescuers may suffer cuts and bruises, or worse, if they are operating under treacherous conditions, but the victims themselves usually do not pose a threat to those trying to help them. This changes when victims are chemically contaminated. For example, after the 1995 sarin attack in the Tokyo subway, 135 (nearly 10 percent) of the emergency medical technicians involved in the response developed acute symptoms and had to be treated at hospitals, and some of the hospital medical staff developed nausea, pupillary constriction, and burning in the eyes and throat from exposure to sarin vapors emanating from patients' clothing. Thus, one of the greatest concerns in dealing with hazardous substances is protecting emergency responders, hospital staff, and other patients from secondary exposure to contaminants. The arrival of even one contaminated patient can close down an entire hospital emergency department.

There have been very few large-scale chemical incidents involving mass casualties. As suggested by the limited data available, the first agency to identify the problem is often the first hospital to receive contaminated patients, which implies that the hospital does not have time to establish decontamination procedures and that hospital staff, the receiving area, and the emergency area are likely to become contaminated. (While the ideal is to have first responders decontaminate casualties in the field, this frequently does not happen.)[38] The best thing that a hospital can do is to try and warn other hospitals so that they can set up decontamination units outside their doors. Provision for such warnings should be institutionalized in the community disaster plan, written into individual hospital plans, and addressed in disaster training. To the extent practicable, contaminated patients should be diverted to the hospital already contaminated while other patients are sent to hospitals not receiving contaminated patients. Although in other disaster situations it may be best to spread the patient load among all available hospitals, this is not necessarily the case with contaminated casualties. If the presence of contaminants leads to hospital closure, it may be better to have this happen at only one facility.

It is also necessary to warn ambulance and other emergency personnel who may be responding to the incident. Such warnings should go out over all components of the public warning system, which again underscores the necessity of having an interorganizational communications system that will function even when cellular and telephone systems do not. Making such a system work requires coordination between specially trained hazmat units, poison control centers, an array of public and private entities, the mass media, and the health sector. In short, the contamination issue is a perfect illustration of why local entities cannot afford to undertake emergency planning in isolation from one another.

Effective management of contaminated casualties depends on the following factors:

- Recognition that contamination has occurred
- Identification of the contaminant so that appropriate decontamination and treatment can be carried out
- Provision of advanced notice from the field so that hospitals can prepare to receive contaminated patients

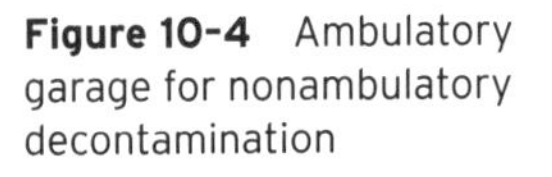

Figure 10-4 Ambulatory garage for nonambulatory decontamination

Photo courtesy of the Noble Training Center, Anniston, Alabama

- Provision of rapid notice from receiving hospitals to other hospitals, which might otherwise unknowingly admit contaminated patients
- Use of personal protective equipment by hospital staff to prevent secondary exposure
- Establishment of decontamination facilities (see Figures 10–4 and 10–5)
- Isolation of patients pending decontamination
- Safe removal of contaminants (e.g., by showering)
- Prevention of hypothermia if outdoor showering is required during inclement weather.

Emergency managers should not assume that EMS providers and hospitals are adequately prepared for hazmat disasters, particularly when such events involve "exotic" contaminants. If it has not already been done, emergency managers should work with local health care organizations to assess health sector preparedness. The following questions may be helpful to ask:

- Does the hospital have a plan for dealing with contaminated casualties?
- Have the staff members who will come into contact with contaminated casualties received training in the plan?
- Is there a community-wide plan for hazmat response that has been jointly developed by police departments, fire departments, EMS providers, hospitals, health departments, poison centers, hazmat teams, environmental agencies, private industry, and 911 dispatchers?
- Do these agencies and personnel train and drill together?
- Is the hospital able, on short notice, to prevent contaminated casualties from entering the hospital except through the decontamination area?
- Are staff members provided with appropriate personal protective equipment (i.e., chemical suites with booties, gloves, and respirators)?
- Are there facilities for isolating and decontaminating casualties while protecting them from the elements?

Figure 10-5 Dining hall patio for ambulatory decontamination

Photo courtesy of the Noble Training Center, Anniston, Alabama

Protecting the public and health care workers

Health care risks associated with various types of bioterrorism attacks, as well as with naturally occurring epidemics such as avian flu, have led policy makers and planners to focus increasingly on methods that might be needed to protect exposed or potentially exposed populations. Key issues in risk management for such events include quarantine and isolation. *Quarantine* is the separation of asymptomatic persons who might have been exposed to a contagious disease until the time has passed during which they would be expected to come down with the disease (the incubation period). *Isolation,* which is often confused with quarantine, is the separation of persons who have actually come down with the disease. Planning for and implementing such measures require a focus on legal, logistical, and other issues.

Regarding the legalities, quarantine and isolation became important issues during the severe acute respiratory syndrome (SARS) epidemic, which occurred between November 2002 and July 2003. But while public health laws were on the books in Canada, China, Hong Kong, Singapore, Taiwan, and Vietnam, the legal authority to require quarantine and isolation existed only for *specific,* previously known diseases. Thus, laws had to be amended to cover this new illness.

Law enforcement personnel were used to serve quarantine orders and to enforce quarantine.[1] Although the SARS epidemic showed that most people comply voluntarily with quarantine provisions, a few individuals–such as those with mental illnesses and illicit drug users–needed to have activity restrictions enforced.

Logistically, large-scale quarantine requires processes for delivering food, medicine, supplies, and financial support to those who are confined but still have car payments, mortgages, electrical bills, and other ongoing financial obligations. It also requires provisions for those who need health care for preexisting chronic medical conditions or those who need to arrange for the care of pets or livestock.[2]

Epidemics pose unique problems for health care workers: because they may work or visit more than one location, these workers could conceivably become carriers during a pandemic. (Like physicians, clergy may also visit the ill at more than one hospital.) A 2004 report by the Institute of Medicine suggested that separate hospitals (isolated from one another) be used to treat epidemic patients and other patients.[3] This would help ensure the continued availability of treatment facilities, not only to epidemic patients but also to those without contagious disease.

[1]David M. Bell, "Public Health Interventions and SARS Spread, 2003," *Emerging Infectious Diseases* 10, no. 11 (2004): 1900-1906; Mark A. Rothstein et al., *Quarantine and Isolation: Lessons Learned from SARS* (Louisville, Ky.: Institute for Bioethics, Health Policy and Law, University of Louisville School of Medicine, 2003), available at archive.naccho.org/documents/Quarantine-Isolation-Lessons-Learned-from-SARS.pdf (accessed August 24, 2007).

[2]Rothstein et al., *Quarantine and Isolation.*

[3]Stacey Knobler et al., eds., "Summary and Assessment," in *Learning from SARS: Preparing for the Next Disease Outbreak–Workshop Summary* (Washington, D.C.: National Academies Press, 2004), available at nap.edu/catalog.php?record_id=10915 (accessed August 24, 2007).

- When contaminated casualties arrive unannounced, is there an interim process for rapid decontamination outside (e.g., a fire hose with spray nozzle and warm water) until decontamination showers can be set up?
- If decontamination occurs indoors, is the area ventilated to the outside and not into other areas of the hospital where staff, visitors, or other patients might be exposed?
- Do ambulances and hospitals have an adequate stock of antidotes (e.g., 2-PAM, atropine)?
- Does the hospital emergency department have immediate access to medical information on specific chemicals?
- Have law enforcement agencies worked with hospitals to provide adequate traffic control and security?

Ensuring access to nonhospital sources of medical care

Because disaster medical planning has traditionally focused on hospitals, the threat of mass casualties from terrorism or naturally occurring epidemics has increased concern about the need for hospital surge capacity. However, case studies suggest that disasters also interrupt the public's access to sources of routine health care,[39] which, in turn, can create risk for people who have chronic health problems such as asthma, high blood pressure, diabetes, heart disease, mental illness, and kidney ailments.[40] Moreover, many disaster victims have minor injuries, such as lacerations, cuts, bruises, puncture wounds, sunburn, heat exposure, and animal bites and stings, most of which can be addressed by nonhospital sources of medical care. To the extent that these sources of care can survive and function after a disaster, the burden on hospitals may be reduced.[41] Unfortunately, however, the disaster may render these routine sources of medical care inaccessible.

Emergency planning for nonhospital medical care should focus on two goals: preventing damage and loss of function, and enhancing capacity for patient care. Among the actions that communities can take to prevent damage and loss of function are (1) securing zoning ordinances that discourage the construction of health care facilities in or near hazard zones such as floodplains, chemical plants, or earthquake faults; and (2) establishing building codes to protect the health care facilities from local hazards, such as earthquakes, tornadoes, and hurricanes. In addition, health care facilities can take a number of precautions on their own. For example, in seismically active areas, they can take simple measures (e.g., Velcro straps, cabinet latches, lips on shelves) to prevent key equipment and supplies from falling and breaking. In flood-prone areas, they can avoid storing important equipment and supplies in basements.

Emergency managers should encourage, assist, and support the development of disaster plans by nonhospital medical service providers. Plans should include provisions for

- Calling in off-duty staff and making sure they can pass through police security lines
- Extending office hours
- Providing on-site sleeping arrangements for staff
- Providing on-site feeding and sanitation for staff as well as for their family members and pets
- Obtaining assistance as needed to reopen and recommence patient care activities
- Obtaining supplementary office and medical supplies
- Notifying patients when the office must be relocated
- Providing alternative transportation for patients needing medical care
- Requesting an "essential service" designation from telephone and cellular providers so that medical facilities have electronic priority when circuits are overloaded
- Making plans to relocate when the facility can no longer function at its current location
- Making provisions for backup power, alternative sources of water, data backup, and surge protection
- Having waterless hand cleaner.

Emergency managers should also work with the health sector to ensure that nonhospital medical facilities are integrated into community disaster plans and that there is a means of communicating with these facilities when telephone and cellular circuits are not functioning. To help prevent inaccurate or distorted messages from being broadcast to the public, emergency managers should prepare announcements for the mass media before a disaster—ideally in both audio and video formats—which the media can then be instructed to release upon notification. The news media can also help during a disaster by notifying patients when and where their physicians or other routine sources of medical care may have relocated and by reminding them, in the event of an evacuation, to bring with them enough of their medications for an extended stay.[42] And emergency managers can encourage health insurers to allow people to get prescription refills wherever possible—and, if necessary, to allow pharmacies to refill prescriptions temporarily without a physician's authorization.

Emergency managers should see that systems are in place to rapidly assess which hospital and nonhospital health care facilities are functioning, which are destroyed, and which are in need of assistance to restore services, as well as to ensure that such assistance is provided (e.g., by loaning them portable electrical generators and helping them obtain priority for the repair of phone lines, power, and other utilities). Regulations and insurance incentives or disincentives should be used to discourage the construction of medical facilities in areas subject to disasters (e.g., floodplains, seismic zones, hurricane surge areas). Local government should also ensure that building codes require disaster-resistant design for medical facilities. Government plans may include provisions to deal with facilities that have been identified as "vital services" or that carry some similar designation.

Dealing with the loss of response infrastructure

"Response infrastructure" refers to the resources used to carry out the disaster response. This includes trained emergency personnel, buildings (e.g., fire departments, hospitals), communications systems, supplies, equipment, emergency vehicles, transportation routes (e.g., roads, rail lines, airports), and transportation systems (e.g., cars, buses, aircraft, boats, trains).

While infrastructure damage has occurred in U.S. disasters, the extent of loss has usually not been enough to paralyze response. Generally speaking, hospital damage has not been so extensive as to shut down all hospitals in affected areas or seriously compromise the delivery of medical services. When massive loss of infrastructure does occur, the event falls into the definition of a *catastrophe* (see accompanying sidebar). Hurricane Katrina, which was categorized as a catastrophe, paralyzed the medical infrastructure in the impact region.

While medical supply shortages can happen, this does not appear to happen in many U.S. disasters. In one study of twenty-nine U.S. disasters, only 6 percent of hospitals experienced supply shortages.[43] (With the increasing reliance on just-in-time inventory, however, shortages could be more frequent in future disasters.) When shortages do occur, communities can request support, via their state emergency management agency, from the Strategic National Stockpile (SNS), which is maintained by the Centers for Disease Control and Prevention (CDC) headquartered in Atlanta. The SNS contains medications, medical supplies, and equipment to supplement and resupply state and local health responders in the event of a national emergency. Beginning within twelve hours of a federal decision to deploy them, stockpile assets can be delivered anywhere in the United States and its territories.[44]

In the event of disaster, hospitals and ambulance services can generally double or triple their staffs by sending out a call for off-duty personnel, and large numbers of medical volunteers will offer their services. Nevertheless, health care operations may be seriously compromised by the loss of lifeline services such as water and electrical power. Working with health care facility administrators, emergency managers should request inspections by qualified engineers to ensure that lifeline services will not be lost during disasters, because of either problems with outside

Disasters versus catastrophes

Just as there are important differences between disasters and routine emergencies, there are important differences between disasters and catastrophes. In disasters, there may be temporary shortages of personnel, supplies, and materials, or shortages of very specialized resources (e.g., neurosurgeons), but usually enough infrastructure survives to carry on. Moreover, vast amounts of help rapidly pour in from neighboring areas unaffected by the disaster. In fact, it is not uncommon for disaster-stricken communities to get much more help than is actually needed.

In catastrophes, on the other hand, there is massive infrastructure damage. Local health care personnel and emergency responders are themselves often rendered homeless, injured, or too ill to join in the response. Health care facilities suffer so much damage that they are unable to provide care and must evacuate their patients to other cities or states. Ambulances are either damaged or unable to operate because roads and bridges are inaccessible. Large numbers of citizens with ongoing health problems have to seek shelter and care in distant cities and states. Furthermore, damage is so widespread that neighboring areas cannot send help.

Because catastrophes are so rare, they are extremely difficult to plan for—particularly in times of budget constraints. Moreover, the effectiveness of preparedness plans for domestic catastrophes has been hard to assess because the plans have been so rarely tested. And the impacts associated with catastrophic events are so difficult to predict that there is a risk of creating elegant "paper plans" based on invalid or unproven assumptions[1]—which, while appearing adequate, may create a false sense of security. The terrible toll taken by Hurricane Katrina is a reminder of the dangers involved in failing to plan realistically for low-probability, high-consequence events. Because experience with routine emergencies provides an inadequate basis from which to plan for both disasters and catastrophes, emergency managers need to avail themselves of the important lessons and guidance derived from the study of such events.

[1]Lee Clarke, *Mission Improbable: Using Fantasy Documents to Tame Disaster* (Chicago: University of Chicago Press, 1999).

supply or damage to internal systems. Specifically, emergency managers should work with hospitals ensure that

- Essential equipment and hospital rooms that will be used for patient care and ancillary services in the event of disaster are on backup power circuits
- Generators are tested for sustained operations
- Cooling systems for generators can function even when municipal water systems lose pressure; generators (as well as their batteries, switching equipment, and fuel lines) are protected from disaster damage.

Emergency managers should also arrange to obtain high-capacity portable or mobile generators in case generators at hospitals or other health care facilities fail, and they should encourage hospitals to establish mutual aid networks to handle shortages of supplies, personnel, and equipment in disasters.

The role of the public health system in disasters

In contrast to medical practitioners and hospitals, public health agencies are concerned with the health of the *community as a whole.* In public health, a major focus is on prevention rather than cure. Preventive measures are particularly important in diseases for which there is no cure and in diseases for which vaccines are available.

Increasingly, federal funding is being channeled through public health departments to support preparedness for terrorist and other threats, and the involvement of public health agencies (and the EMS providers under their authority) in this effort is growing. Some examples of what public health agencies do in emergencies include

- Investigating disease outbreaks caused by improperly prepared or refrigerated food and taking the food item off the market
- Recommending to state or local mosquito abatement agencies (should an increase in mosquitos carrying infectious diseases [e.g., dengue fever, West Nile virus] be discovered) that spraying for disease-carrying mosquitoes be conducted
- Conducting "public health surveillance"—that is, collecting reports from medical laboratories and health care practitioners and facilities on new cases of diseases to provide early warning of outbreaks
- Undertaking "epidemiological investigations"—that is, using case investigations and statistical techniques to identify sources of disease outbreaks and how they spread
- Tracking down those who have been exposed to dangerous infectious diseases, so that they can be provided with early treatment and preventive vaccines or antibiotics
- Ensuring that those with serious contagious diseases, such as bacterial meningitis or tuberculosis, are isolated until they are no longer contagious
- Quarantining those who may have been exposed to a dangerous contagious disease until the incubation period (the length of time it takes from exposure until symptoms appear) has passed
- Making certain that persons living in temporary shelters, such as school gymnasiums or churches, are not exposed to health risks (e.g., from inadequate food-handling procedures)
- Ensuring that community water supplies are safe and uncontaminated
- Providing information to the public on health hazards and on ways to prevent illness and injury in disasters[45]
- Providing laboratory resources in support of public health (e.g., in case of infectious diseases or environmental exposures).

In disasters caused by epidemics, toxic releases, and other public health threats—either natural or man-made—public health and law enforcement authorities will be carrying out investigations simultaneously. Each may collect information and evidence that will be of importance to the other. Therefore, it is crucial that they plan, train, and respond together. Representatives

Challenges to the public health system

Recent public health emergencies, such as severe acute respiratory syndrome (SARS) and anthrax, have posed some problems not faced during previous infectious disease emergencies. During the anthrax attacks of 2001, for example,

- The Centers for Disease Control and Prevention (CDC) was called upon for the first time to respond to disease outbreaks occurring nearly simultaneously in five separate geographic areas.
- Few public health or medical practitioners had ever seen or treated a single case of this highly uncommon disease, much less dealt with a terrorist attack. Moreover, it was eventually discovered that some of the anthrax was of a special weaponized grade, which had different characteristics from the naturally occurring variety.
- Investigations were being carried out simultaneously by public health and law enforcement agencies, so that public health officials who lacked security clearances sometimes had difficulty getting information they needed.[1]
- Federal and state public health agencies were bombarded with information from disparate sources that included local and state health departments, postal distribution sites, unions, physicians, hospitals, clinics, and laboratories. At the same time, health agencies were being bombarded with calls for information about anthrax from physicians and the public.
- Public health laboratories had great difficulty keeping up with the demands of the outbreak. One state laboratory, which normally processed one anthrax analysis a year, analyzed more than 2,000 samples in two months' time.[2] Overall, more than 120,000 laboratory tests for anthrax had to be carried out during the response, and the need was generated to develop procedures for prioritizing which samples needed to be tested.[3]

The anthrax attacks of 2001 revealed that few health departments had planned how they would procure critical resources in a disaster. The normal procurement procedures were too lengthy and cumbersome for the demands of the crisis. One public health lab director indicated that, while he had enough money in the budget to purchase the additional equipment he needed for analyzing anthrax samples, it would take the state office of general services two months to process the request.[4] The ability of public health agencies to respond to the attacks was also hindered by a lack of mutual aid planning. For example, many states had no reciprocity provisions enabling public health workers to be credentialed to work in adjoining states.[5]

[1]U.S. General Accounting Office (GAO), *Bioterrorism: Public Health Response to Anthrax Incidents of 2001* (Washington, D.C.: GAO, 2003), available at gao.gov/new.items/d04152.pdf (accessed August 24, 2007).

[2]Elin Gursky, Thomas V. Inglesby, and Tara O'Toole, "Anthrax 2001: Observations on the Medical and Public Health Response," *Biosecurity and Bioterrorism: Biodefense, Strategy, Practice, and Science* 1, no. 2 (2003): 97-110, available at upmc-biosecurity .org/website/resources/publications/2003_orig-articles/2003-06-15-anthrax2001observations.html (accessed August 24, 2007).

[3]GAO, *Bioterrorism: Public Health Response*.

[4]Gursky, Inglesby, O'Toole, "Anthrax 2001."

[5]Ibid.

from public health agencies should also be represented on community disaster planning bodies and at community emergency operations centers.

Formal planning for disaster health care

In the United States, formal planning for disaster health care varies widely, especially at the local and state levels. However, there are several programs—most notably those funded or managed at the federal level—with which emergency managers should be familiar.

The Joint Commission

Traditionally, health care planning has focused largely on hospitals. The Joint Commission (until 2007, the Joint Commission on Accreditation of Healthcare Organizations, or JCAHO) is a key player in hospital preparedness. The commission was first established in 1951 as the Joint Commission on Accreditation of Hospitals; two years later it established hospital fire protection standards. In 1965, standards were added requiring hospitals to have written mass casualty plans and drills. Currently, the commission also promulgates preparedness standards for long-term care facilities, home health care providers, behavioral health institutions, ambulatory clinics, and outpatient surgery centers.

According to Joe Cappiello, vice president of Accreditation Field Operations at the Joint Commission, the decision to accredit is based on compliance with a number of standards, not just preparedness. Currently, each facility must have an inspection score that meets a certain threshold (which differs according to such factors as institutional size). Although failing to meet the emergency management standards would not, in itself, cause a failing score, it could certainly contribute to a facility's inability to get over the threshold.[46]

The National Response Plan

In recent years, terrorist attacks and concerns about pandemic diseases such as SARS and avian flu have caused public health agencies at all levels to become more involved in disaster planning. Under the National Response Plan (NRP), Emergency Support Function (ESF) 8—Public Health and Medical Services—is concerned with providing supplementary assistance to local, tribal, and state governments in identifying and meeting public health and medical needs during incidents of national significance. The key elements of this assistance include assessment of public health, medical, and behavioral health needs; public health surveillance;[47] and provision of medical personnel, equipment, and supplies.

Under the NRP, the lead agency for federal health care response to disasters is the U.S. Department of Health and Human Services (HHS); HHS is assisted by a number of component agencies, including the CDC, the U.S. Public Health Service, the Occupational Safety and Health Administration, and the Food and Drug Administration. ESF 8 coordinates with other emergency support functions, including transportation (ESF 1); mass care, housing, and human resources (ESF 6); urban search and rescue (ESF 9); and agriculture and natural resources (ESF 11).[48]

The National Disaster Medical System

The National Disaster Medical System (NDMS) was originally established in 1984 within HHS; its mission was to supplement state and local responses to disaster and to provide backup support to the U.S. Department of Defense and Veterans Administration medical systems during times of overseas conflict. In recent years, its mission has been expanded to provide medical response during terrorist attacks and to conduct advance staging for special events, such as political conventions, that are associated with potential risks to national security. Responsibility for NDMS had been reassigned to the U.S. Department of Homeland Security but now has been returned to HHS.

NDMS has three core components: (1) disaster medical teams, supplies, and equipment; (2) movement of patients from the disaster to unaffected areas of the nation; and (3) provision of definitive medical care at NDMS-participating hospitals in unaffected areas. Several types of medical teams are available through NDMS, including disaster medical assistance teams (DMATs); disaster mortuary operational response teams; veterinary medical assistance teams; national medical response teams, which deal with chemical, biological, and nuclear incidents; and international medical surgical response teams. DMATs have been used in numerous disasters, but prior to Hurricane Katrina there was little experience with the intrastate, interhospital movement of the ill and injured.[49]

The Medical Reserve Corps

The Medical Reserve Corps (MRC), which is administered by HHS, was formed in 2002 as a partner of the Citizen Corps.[50] Its mission is to identify, organize, credential, and train local medical and public health volunteers for disasters. Members include practicing and retired physicians, nurses, and other health professionals, as well as other citizens interested in health issues, who are willing to provide disaster health services. The exact structure, mission, and sponsorship of each MRC unit vary according to local needs.[51]

The Community Emergency Response Team program

The Community Emergency Response Team (CERT) program, another partner of the Citizen Corps, trains citizens to support first responders, to provide immediate assistance to victims, and to organize spontaneous volunteers in disasters. CERT training includes disaster preparedness, disaster fire suppression, basic disaster medical care, and light search and rescue.[52]

The Metropolitan Medical Response System

The Metropolitan Medical Response System (MMRS) was formed in 1996 in response to the Tokyo sarin attack and the bombing of the Alfred P. Murrah Federal Building in Oklahoma City, both of which occurred in 1995. The program was originally designed to strengthen and coordinate local and regional response capabilities in highly populated areas that might be targets for attacks with weapons of mass destruction. However, it was thought that the program concepts, organizing principles, and human and material resources could also be applied to other types of large-scale disasters. As of April 2007, the Federal Emergency Management Agency (FEMA) operates the program.

MMRS supports local jurisdictions in planning, developing, equipping, and training regionalized networks of first responders drawn from law enforcement, the medical community, the public health system, the fire service, and hazmat response teams. Five areas of planning are emphasized: early recognition, mass immunization and prophylaxis, mass patient care, mass fatality management, and environmental issues.

The Emergency Management Assistance Compact

The Emergency Management Assistance Compact (EMAC) is an interstate mutual aid agreement that facilitates the sharing of resources, personnel, and equipment across state lines in times of emergency or disaster. While not specifically focused on health care, EMAC has been used successfully for sharing medical, evacuation, and search and rescue assets, most notably during Hurricane Katrina.[53]

State emergency response commissions and local emergency planning committees

The Emergency Planning and Community Right-to-Know Act, passed in 1986, requires that each state establish a state emergency response commission (SERC) to oversee planning for hazmat emergencies. The SERC, in turn, establishes planning districts within the state, each with a local emergency planning committee (LEPC). Each LEPC must have, at a minimum, representation from elected state and local officials; the local law enforcement, emergency management, emergency medical services, fire services, first aid, public health, environmental, hospital, and transportation sectors; broadcast and print media; community groups; and owners and operators of hazmat facilities. Although these committees are formed for the purposes of hazmat planning, their members are often involved with state and local all-hazards planning as well.

Summary

Some of the principal responsibilities of the health sector in disaster are communications, triage, hospital and nursing home evacuations, handling inquiries and victim tracking, managing medical donations, managing contamination by hazardous substances, ensuring access to nonhospital sources of medical care, and dealing with the loss of response infrastructure. However, the health sector cannot undertake any of these activities on its own. Disaster planning and response require coordination across political, geographical, jurisdictional, and functional boundaries.

Notes

ICMA's copyright does not extend to Chapter 10, which was co-written by a federal government employee as part of his official duties and therefore considered a work of the United States government; Chapter 10 may be reproduced without permission. Nonetheless, the book in its entirety is copyrighted by ICMA.

1 Arlington County Fire Department, *After-Action Report on the Response to the September 11 Terrorist Attack on the Pentagon* (Arlington County, Va.: Titan Systems Corporation, 2002); Kathleen J. Tierney, Michael K. Lindell, and Ronald W. Perry, *Facing the Unexpected: Disaster Preparedness and Response in the United States* (Washington, D.C.: Joseph Henry Press, 2001).

2 M. Nordberg, "United Flight 232: The Story behind the Rescue," *Emergency Medical Services* 18, no. 10 (1989): 14–15, 22, 24.
3 Arlington County Fire Department, *After-Action Report;* California Association of Hospitals and Health Systems, *Hospital Earthquake Preparedness: Issues for Action: A Report on the Loma Prieta Earthquake Issued October 17, 1990* (Sacramento: California Association of Hospitals and Health Systems, 1990); C. Freeman, C. Van Ness, and J. E. Morales, *Hurricane Andrew: Lessons for California* (Sacramento: California Emergency Medical Services Authority, 1993); J. S. Klein and J. A. Weigelt, "Disaster Management: Lessons Learned," *Surgical Clinics of North America* 71, no. 2 (1991): 257–266; Gus Koehler et al., *Medical Care for the Injured: The Emergency Medical Response to the April, 1992, Los Angeles Civil Disturbance,* EMSA #393-01 (Sacramento: California Emergency Medical Services Authority, 1993); K. J. McGlown and M. D. Fottler, "The Impact of Flooding on the Delivery of Hospital Services in the Southeastern United States," *Health Care Management Review* 21, no. 3 (1996): 55–71; McKinsey and Company, *The McKinsey Report—Increasing FDNY's Preparedness* (New York: Fire Department of the City of New York, 2002), available at nyc.gov/html/fdny/html/mck_report/toc.shtml (accessed August 22, 1007); Robert A. Olson et al., *Critical Decisions: Evacuating Hospitals after the 1994 Northridge Earthquake* (Sacramento, Calif.: Robert Olson Associates, Inc., 1998); Robyn Pangi, *Consequence Management in the 1995 Sarin Attacks on the Japanese Subway System,* ESDP-2002-01, BCSIA-2002-04 (Boston: John F. Kennedy School of Government, Harvard University, 2002), available at bcsia.ksg.harvard.edu/BCSIA_content/documents/Consequence_Management_in_the_1995_Sarin_Attacks_on_the_Japanese_Subway_System.pdf; Kathleen J. Tierney, *Project Summary: Disaster Analysis: Delivery of Emergency Medical Services in Disasters* (Newark: Disaster Research Center, University of Delaware, 1993); Tierney, Lindell, and Perry, *Facing the Unexpected.*
4 Amy H. Kaji and Roger J. Lewis, "Hospital Disaster Preparedness in Los Angeles County, California," *Annals of Emergency Medicine* 44, no. 4 (2004 Supplement).
5 Gary A. Kreps, "Organizing for Emergency Management," in *Emergency Management: Principles and Practice for Local Government,* ed. Thomas E. Drabek and Gerard J. Hoetmer (Washington, D.C.: ICMA, 1991), 44, 49.
6 Because disasters are rare events, however, those who gain experience managing them may have moved on by the time the next one occurs. The emergency manager must see to it that the lessons learned are not forgotten and that newcomers are briefed and trained.
7 For more information on the Metropolitan Medical Response System, go to fema.gov/mmrs/; see also dhs.gov/xnews/releases/press_release_0535.shtm (accessed August 22, 2007). For more information on local emergency planning committees, see rtknet.org/resources.php (accessed August 22, 2007).
8 Erik Auf der Heide, "Principles of Hospital Disaster Planning," in *Disaster Medicine,* ed. David Hogan and Jonathan L. Burstein (Philadelphia: Lippincott Williams and Wilkins, 2002).
9 Erik Auf der Heide, *Disaster Response: Principles of Preparation and Coordination* (St. Louis, Mo.: C. V. Mosby, 1989), available at orgmail2.coe-dmha.org/dr/index.htm (accessed August 22, 2007); Auf der Heide, "Principles of Hospital Disaster Planning"; Oklahoma City Document Management Team, *Murrah Rescue and Recovery Operation: Final Report to the Mayor and City Council* (Oklahoma City, 1996).
10 Auf der Heide, "Principles of Hospital Disaster Planning."
11 Joint Committee on Fire, Police, Emergency and Disaster Services, *California's Emergency Communications Crisis* (Sacramento: California State Senate and Assembly, 1983).
12 Donald A. Lund, "Learning to Talk: The Lessons of Non-Interoperability in Public Safety Communications Systems" (Durham: Justiceworks, University of New Hampshire, 2002); Viktor Mayer-Schönberger, "Emergency Communications: The Quest for Interoperability in the United States and Europe," ESDP-2002-03, BCSIA 2002-7 (Boston: John F. Kennedy School of Government, Harvard University, 2002); Association of State and Territorial Health Officials, *Phone Lines and Life Lines: How New York Reestablished Contact on September 11, 2001* (Washington, D.C.: Association of State and Territorial Health Officials, 2002); National Task Force on Interoperability, *Why Can't We Talk? Working Together to Bridge the Communications Gap to Save Lives: A Guide for Public Officials* (Washington, D.C.: National Task Force on Interoperability, 2003), available at homelandsecurity.alabama.gov/PDFs/why_cannot.pdf (accessed August 22, 2007); U.S. General Accounting Office, *Project Safecom: Key Cross-Agency Emergency Communications Effort Requires Stronger Collaboration* (Washington, D.C.: U.S. General Accounting Office, 2004).
13 U.S. Conference of Mayors, *United States Conference of Mayors Interoperability Survey: A 192-City Survey* (Washington, D.C.: U.S. Conference of Mayors, 2004), available at usmayors.org/72ndannualmeeting/interoperabilityreport_062804.pdf (accessed August 22, 2007).
14 Joseph Scanlon, Gillian Osborne, and Scott McClellan, *The 1992 Peace River Ice Jam and Evacuation: An Alberta Town Adapts to a Sudden Emergency* (Ottawa: Emergency Communications Research Unit, Carleton University, 1996).
15 Vincent T. Covello et al., "Risk Communication, the West Nile Virus Epidemic, and Bioterrorism: Responding to the Communication Challenges Posed by the Intentional or Unintentional Release of a Pathogen in an Urban Setting," *Journal of Urban Health: Bulletin of the New York Academy of Medicine* 78, no. 2 (2001): 382–391, available at centerforriskcommunication.com/pubs/crc-p1.pdf (accessed August 22, 2007).
16 Joseph T. F. Lau et al., "SARS Transmission, Risk Factors, and Prevention in Hong Kong," *Emerging Infectious Diseases* 10, no. 4 (2004): 587–592, available at cdc.gov/ncidod/eid/vol10no4/pdfs/03-0628.pdf (accessed August 22, 2007); Jiang Wu et al., "Risk Factors for SARS among Persons without Known Contact with SARS Patients, Beijing, China," *Emerging Infectious Diseases* 10, no. 2 (2004): 210–216, available at cdc.gov/ncidod/EID/vol10no2/03-0730.htm (accessed August 22, 2007).
17 Auf der Heide, *Disaster Response.*
18 Auf der Heide, "Principles of Hospital Disaster Planning."
19 One reason for the lack of coordination among providers of emergency services is the traditional independence and autonomy of these agencies. See Auf der Heide, *Disaster Response;* Freeman, Van Ness, and Morales, *Hurricane Andrew: Lessons for California;* Oklahoma City Document Management Team, *Murrah Rescue and Recovery Operation;* Tierney, *Project Summary: Disaster Analysis;* Tierney, Lindell, and Perry, *Facing the Unexpected;* and Alberta Public Safety Services, *Tornado, a Report: Edmonton and Strathcona County, July 31st, 1987* (Edmonton, Ontario: Alberta Public Safety Services, 1991).
20 H. E. Moore, *Tornados over Texas: A Study of Waco and San Angelo in Disaster* (Austin: University of Texas Press, 1958).
21 Auf der Heide, *Disaster Response.*
22 E. L. Quarantelli, *Delivery of Emergency Medical Care in Disasters: Assumptions and Realities* (New York: Irvington Publishers, 1983).
23 Erik Auf der Heide, "The Importance of Evidence-Based Disaster Planning," *Annals of Emergency*

Medicine 47, no. 1 (January 2006): 34–49, available at atsdr.cdc.gov/2p-emergency-response.html (accessed August 21, 2007).

24 Auf der Heide, "Principles of Hospital Disaster Planning."

25 M. G. Guttenberg et al., "Utilization of Ambulance Resources at the World Trade Center: Implications for Disaster Planning," *Annals of Emergency Medicine* 40, no. 4 (2002): S92.

26 T. Okumura et al., "The Tokyo Subway Sarin Attack: Disaster Management, Part 1: Community Emergency Response," *Academic Emergency Medicine* 5, no. 6 (1998): 613–617.

27 Auf der Heide, "Principles of Hospital Disaster Planning"; Quarantelli, *Delivery of Emergency Medical Care in Disasters*.

28 Eric K. Noji, *The Public Health Consequences of Disasters* (New York: Oxford University Press, 1997); Quarantelli, *Delivery of Emergency Medical Care in Disasters*.

29 Planners may wish to refer to the Incident Command System Multi-Casualty Branch description for an outline of tasks associated with prehospital triage. This is currently the most up-to-date published Incident Command System Guidance related to field management of medical casualties. See FIRESCOPE, "Multi-Casualty Branch," chapter 14 of *Incident Command System, Fire Service Field Operations Guide*, ICS 420-1 (Riverside, Calif.: OES FIRESCOPE OCC, June 2004), available at firescope.org (accessed August 22, 2007).

30 For further information on triage, see Auf der Heide, *Disaster Response;* and J. Lairet, "Triage," in *Disaster Medicine*, ed. David E. Hogan and Jonathan L. Burstein (Philadelphia: Lippincott Williams & Wilkins, 2002), 10–15.

31 *The Washington Post*/Henry J. Kaiser Family Foundation/Harvard School of Public Health, "Survey of Hurricane Katrina Evacuees" (2005), available at washingtonpost.com/wp-srv/politics/polls/katrina_poll091605.pdf (accessed August 22, 2007).

32 Joseph Scanlon, "Convergence Unlimited: Call Centres and the Indian Ocean Tsunami" (unpublished report, 2006).

33 In an air crash, for example, if information about the flight number, destination, and exact location of the crash site is provided by the media, it will allow many viewers to determine that their loved ones were not on the plane and were not present where the crash occurred.

34 For a brief description of the act, go to the U.S. Department of Health and Human Services, "Privacy and Your Health Information," at hhs.gov/ocr/hipaa/consumer_summary.pdf (accessed August 21, 2007).

35 See 45 CFR164.510(b)(1)(ii) at hhs.gov/hipaafaq/notice/491.html (accessed August 22, 2007).

36 David Hogan, "The Oklahoma City Terrorist Blast: A Case Study of Disaster," in *Emergency Preparedness in Health Care Organizations*, ed. Linda Young Landesman (Oakbrook Terrace, Ill.: Joint Commission, 1996).

37 Auf der Heide, "Principles of Hospital Disaster Planning."

38 R. J. Geller et al., "Nosocomial Poisoning Associated with Emergency Department Treatment of Organophosphate Toxicity—Georgia, 2000," *Morbidity and Mortality Weekly Report* 49, no. 51 (2001): 1156–1158; D. Kevin Horton, Zahava Berkowitz, and Wendy E. Kaye, "Secondary Contamination of ED Personnel from Hazardous Materials Events, 1995–2001," *American Journal of Emergency Medicine* 21, no. 3 (2003): 199–204; Okumura et al., "The Tokyo Subway Sarin Attack"; B. M. Vogt and J. H. Sorensen, *How Clean Is Safe? Improving the Effectiveness of Decontamination of Structures and People following Chemical and Biological Incidents* (Oak Ridge, Tenn.: Oak Ridge National Laboratory, 2002).

39 Tesfaye Bayleyegn et al., "Rapid Assessment of the Needs and Health Status in Santa Rosa and Escambia Counties, Florida, after Hurricane Ivan, September 2004," *Disaster Management and Response* 4, no. 1 (2006): 12–18; W. G. Hlady et al., "Rapid Health Needs Assessment following Hurricane Andrew—Florida and Louisiana, 1992," *Morbidity and Mortality Weekly Report* 41, no. 37 (1992): 685–688; N. George-McDowell et al., "Surveillance for Injuries and Illnesses and Rapid Health-Needs Assessment following Hurricanes Marilyn and Opal, September–October 1995," *Morbidity and Mortality Weekly Report* 45, no. 4 (1996): 81–85; Frank Sabatino, "Hurricane Andrew: South Florida Hospitals Shared Resources and Energy to Cope with the Storm's Devastation," *Hospitals* 66, no. 24 (1992): 26–28, 30.

40 Sources of routine medical care that can become nonfunctional following disasters include private physician offices, pharmacies, urgent care centers, clinics, dialysis centers, hospices, nursing homes, mental health clinics, home health care agencies, assisted living facilities, and dental offices.

41 Auf der Heide, "Principles of Hospital Disaster Planning."

42 In 2004, following Hurricane Charley, a survey of Florida residents over age sixty with preexisting medical conditions revealed that 28 percent in Charlotte County, 21 percent in DeSoto County, and 18 percent in Hardee County reported having lost access to routine care for those conditions.

43 Quarantelli, *Delivery of Emergency Medical Care in Disasters*, 96.

44 More information on the Strategic National Stockpile is available via the Internet at bt.cdc.gov/stockpile/ (accessed August 22, 2007) or from the local or state public health agency.

45 Fact sheets and press releases covering the various types of public health threats can be obtained through the U.S. Agency for Toxic Substances and Disease Registry at atsdr.cdc.gov and through the CDC at cdc.gov (accessed August 2007).

46 Joe Cappiello, interview with author, August 2007. For more information on the Joint Commission, see jointcommission.org/.

47 Public health surveillance involves monitoring communities to detect significant changes in health conditions.

48 For more information on the National Response Plan, go to dhs.gov/dhspublic/interapp/editorial/editorial_0566.xml (accessed August 22, 2007).

49 For more information on the National Disaster Medical System (NDMS), go to hhs.gov/aspr/opeo/ndms/index.html.

50 For more information on the Citizen Corps, go to citizencorps.gov/programs/.

51 For more information about the Medical Reserve Corps (MRS), go to medicalreservecorps.gov/About.

52 For more information about the Community Emergency Response Team (CERT) program, go to citizencorps.gov/cert/index.shtm.

53 For more information on the Emergency Management Assistance Compact (EMAC), go to emacweb.org/.

CHAPTER 11

Recovery

Brenda D. Phillips and David M. Neal

This chapter provides an understanding of

- The process of recovery planning for a disaster
- The dimensions of recovery from a disaster
- The resources for recovery from a disaster.

Recovery is viewed here as a social process in which the local government manager creates crucial partnerships to guide the affected community toward a multifaceted recovery from disaster. The partnerships link government, local citizens, and an array of organizations and agencies, with the aim of proceeding through a series of decisions that will result not only in the return of residents and the restoration of economic stability, but also in a recognition that certain kinds of risks will remain. As the community attempts to meet those (and other) broad societal needs in the wake of a disaster, it is critically important for the government manager to know who is involved in decision making and how decision making occurs. Accordingly, this chapter reviews the role of government in the recovery process and offers guidance on how to navigate that process. The perspective is that of the government manager. The broad topics covered are the recovery planning process, the dimensions of recovery, and the resources for recovery.

The problems of recovery begin even with the very term *recovery.* Historically the term has been loosely taken to mean restoration, rehabilitation, reconstruction, or even restitution. However, each of these terms connotes something different for those who are affected by recovery efforts. *Restoration,* for example, suggests a return to pre-disaster levels, although such a return may not mitigate future risks. *Rehabilitation,* in contrast, hints at improvements. *Reconstruction* emphasizes the built environment, whereas *restitution* connotes legal action.[1]

In short, the words used to describe post-disaster activities are often used inconsistently and without awareness of the precise roles and responsibilities of government, the interests and needs of stakeholders (e.g., residents and businesses), or the practice of emergency management. Such an inexact cluster of terms may lead to disillusionment at best, or to a faltering or even disrupted recovery at worst. Imagine, for example, the likely conflict when city officials are planning restoration but residents expect the outcome to be rehabilitation. The differences in expectation—on the one hand, expecting to return to pre-disaster levels, and on the other hand, expecting to improve conditions in the community—can be very divisive.[2] An important part of the local government manager's responsibility, therefore, must be to bring all parties into agreement on just what their aims are.

Although recovery is a distinct phase in the life cycle of emergency management, it is affected by activities that take place during the mitigation, preparedness, and response phases.

Recovery usually has two phases, short term and long term. Short-term recovery overlaps with response and may include "search and rescue, damage assessments, public information, temporary housing, utility restoration, and debris clearance."[3] Short-term recovery also includes the handling of donations. Long-term recovery, which is what the bulk of this chapter focuses on, addresses the basic dimensions of a community's existence: permanent housing, economic conditions, the environment, the infrastructure (e.g., roads and bridges), and lifelines (e.g., water, power, telephone service). Each dimension may be affected by social and psychological conditions that affect individual and collective abilities to move through the long-term recovery period.

And it must not be forgotten that although recovery is a distinct phase in the life cycle of emergency management, it is affected by activities that take place during the mitigation, preparedness, and response phases. For example, the more a household or community prepares for a disaster or attempts to mitigate the disaster's consequences, the easier recovery will be.[4] Similarly, the decisions made during the response phase can influence the recovery effort. For example, setting aside bricks from destroyed buildings can make it easier to restore a historic downtown and reduce landfill use. Completing the circle, the recovery phase often presents unique opportunities to initiate new mitigation programs because the public's attention is now centered on risk.[5]

It must also be noted that although a chapter such as this seems to tie up loose ends and present a neat package, in reality recovery is not a streamlined, bureaucratic process, and there is no single model of how to do it. The process varies considerably across different contexts: jurisdictions vary by size, and so in some communities the emergency management people will be paid staff, whereas in others everyone will be volunteer—including the emergency manager.

Five perspectives on recovery

1. "Recovery cannot succeed if the aims, priorities and processes do not have community support; this can require considerable community participation both pre and post event (in the short, medium and long-term), which includes building relationships with groups within local authority areas. It is important therefore to recognize the diverse needs of different communities such as those within a rural or urban context, high or low socio-economic groups and the implications of differing cultural diversity and belief systems."

   Source: Sarah Norman, "Focus on Recovery: Holistic Framework for Recovery," in *New Zealand Recovery Symposium Proceedings*, ed. Sarah Norman (Wellington, New Zealand: Ministry of Civil Defence and Emergency Management, 2004), 40, available at civildefence.govt.nz/memwebsite.nsf/Files/Recovery_symposium2004/$file/Recovery_symposium2004.pdf (accessed August 10, 2007).

2. "Since exercise of power is usually a reflection of important value differences in a society, it would be surprising if such a factor did not surface during the disaster recovery of individuals, organizations, communities and societies. The operations of political considerations may be more subtle in social systems with strong democratic ideologies. But it would be naive to think that even in such societies, like the United States, no political factors enter into the relevant decision making and the providing of recovery aid."

   Source: E. L. Quarantelli, "Disaster Recovery: Research Based Observations on What It Means, Success and Failure, Those Assisted and Those Assisting," Preliminary Paper 263 (Newark: Disaster Research Center, University of Delaware, 1998), 11, available at dspace.udel.edu:8080/dspace/bitstream/19716/282/1/PP%20263.pdf (accessed August 10, 2007).

3. "The emphasis of recovery planning is shifting away from recreating damaged places to creating communities that are *better* (in the sense of being more sustainable) than the ones that preceded them. This goal involves researchers, managers and public policy makers in the difficult art of anticipating the future and the circumstances that will govern public policies for years and decades to come."

   Source: James K. Mitchell, "Reconceiving Recovery," in *New Zealand Recovery Symposium Proceedings*, ed. Sarah Norman (Wellington, New Zealand: Ministry of Civil Defence and Emergency Management, 2004), 66; available at civildefence.govt.nz/memwebsite.nsf/Files/Recovery_symposium2004/$file/Recovery_symposium2004.pdf (accessed August 10, 2007).

4. "Poor families and large families have the most trouble acquiring adequate aid and recovering from disaster, and are consequently more vulnerable to a disaster. Members of ethnic minorities, particularly blacks and Hispanics, are typically more likely to belong to such families. These families have greater numbers of non-productive dependents, poorer insurance coverage, less money in savings accounts, and fewer personal resources.... [S]uch families are under stress even prior to a disaster.... [R]ecovery policies should recognize this fact so that social inequities will not be magnified by a disaster.... [C]learly, programs for outreach to such groups must be expanded and used in disasters involving significant numbers of blacks or Hispanics."

   Source: Robert C. Bolin and Patricia Bolton, *Race, Religion and Ethnicity in Disaster Recovery* (Boulder: Institute of Behavioral Science, University of Colorado, 1986), 222.

5. "Economic recovery from disaster is about the resilience of local economies, but ... it is also about scale in space and time, the magnitude and distribution of the losses and benefits flowing from the disaster, the impacts on assets and on flows of goods and services, adoption of new technologies, marketing of recovery and shifts in local power.... Is the aim of economic recovery simply to restore the pre-disaster state? Or should disasters be embraced as opportunities to make local economies more resilient?"

   Source: John Handmer and Marnie Hillman, "Economic and Financial Recovery from Disaster," in *New Zealand Recovery Symposium Proceedings*, ed. Sarah Norman (Wellington, New Zealand: Ministry of Civil Defence and Emergency Management, 2004), 235, available at civildefence.govt.nz/memwebsite.nsf/Files/Recovery_symposium2004/$file/Recovery_symposium2004.pdf (accessed August 10, 2007).

Disasters vary as well, both by type and by frequency, and so some places will be facing their first and worst disaster whereas others will be quite experienced. Moreover, only a few places will have the luxury of preplanning recovery; most will plan for it in the aftermath of a disaster and in a rather ad hoc manner. All these qualifications form the backdrop for this chapter.

Recovery planning: Timing—before and after the event

Recovery planning may take place either before or after the event but is best handled before, in a relatively calm, stress-free environment, with time available for thinking through all possible avenues.

Recovery planning before the event

Ideally, the pre-disaster recovery plan provides a generic roadmap of mutual aid agreements, prewritten emergency ordinances (these vary in nature, ranging from ordinances that address road closures for snow removal to those that are disaster specific), policies for landfill use, contingency procedures for databases and computer records, a protocol for staffing during a long-term recovery period, strategies for involving the public, and a means of expediting the (re)construction permitting process. The pre-disaster recovery process also includes identifying recovery partners and laying the foundations for recovery actions. The challenge, though, lies in motivating stakeholders (all potential partners) to participate, especially in communities that lack prior experience with disasters or that do not perceive the risks to be high.

To initiate or update their recovery plans before the event, local emergency management officials should collaborate with partners elsewhere in local government. Working together, the parties can leverage their resources, skills, and ideas and reduce risks to the community.

Specifically, emergency managers should work in concert with community planners to address both hazards and recovery. Comprehensive planning, which is a routine function of an urban planning office, often addresses hazards such as floodplains; it can also enhance protection of fire zones, foster seismic codes, and prioritize infrastructure needs. Thus, linking it to recovery planning can result in a safer, better-managed community. Moreover, by working with community planners before a disaster, emergency managers can reduce the impact on a community when a disaster does hit. Essentially, integrating recovery planning into the existing comprehensive planning effort enables two goals to be met with one effort. In fact, some states already mandate that hazard management be part of a community's comprehensive planning.

Integrating recovery planning into the existing comprehensive planning effort enables two goals to be met with one effort.

In addition, linking comprehensive and recovery planning—although not a prerequisite for receiving federal hazard mitigation grants—allows municipalities to compete for the grants more effectively.

Recovery planners may also find it useful to work with those who are involved in routine disaster planning, especially the exercises and regular updates that are part of the planning and preparedness phase (see Chapter 7). These exercises and updates can be integrated into—or can improve—the disaster recovery planning process. As part of general disaster planning, the requisite risk and hazard analysis can identify areas and neighborhoods at greater risk, and—knowing the populations involved—local officials can then gauge, for example, the types of shelter and housing likely to be needed under particular scenarios. Specifically, neighborhoods at higher risk are those in or next to hazardous areas (e.g., floodplains, chemical or other potentially dangerous facilities, and major transportation routes that carry hazardous cargo), and research has shown that people living close to higher-risk areas are often the poor, ethnic minorities, and/or the elderly. Other aspects of recovery can also be anticipated—the need to mobilize people with specific language skills or to provide further assistance to those who need help maneuvering through the bureaucratic maze of recovery. In short, during the planning and preparedness phase, which should include routine updates of the basic emergency plan, disaster planners can anticipate patterns of victimization and plan accordingly, looking ahead not only to response but also to recovery.[6]

Finally, local officials can focus explicitly on recovery planning within their own disaster plan. In keeping with the National Response Plan (NRP), they may wish to make their disaster recovery plan part of their own local emergency support functions. A disaster recovery plan may be a stand-alone plan, or it may be an annex to an emergency operations plan. In the current NRP, which is organized into functional areas for emergency support, recovery is located under Emergency Support Function 14 (ESF 14). The recovery planning section of the NRP addresses the long-term impacts on housing, business and employment, community infrastructure, and social services,[7] and since ESF 14 focuses on coordination among relevant (federal) organizations for recovery planning, local recovery plans can do the same.[8]

One city that has done recovery planning before the event is Los Angeles, which lives under the constant threat of a major earthquake. In 1987, members of the city's Emergency Operations Organization brought together a wide range of experts to initiate recovery planning. In a series of workshops, participants identified actions that the city should take before, during, and after a disaster to assist with recovery. Examples include promotion of structural and nonstructural mitigation programs, identification of disposal areas, and consideration of economic recovery strategies. The plan that emerged from the workshops stresses mitigation activities both before and after a disaster as a cornerstone of effective recovery planning. Los Angeles focuses on issues related to residential, commercial, and industrial rehabilitation; public sector services; economic recovery; land use and reuse; organization and authority; psychological rehabilitation; vital records; interjurisdictional relations; and traffic mitigation.[9] The Los Angeles plan demonstrates how recovery plans can vary by context and jurisdiction. The dimensions that may be addressed will likely emerge through a visioning or discussion process.

Recovery planning after the event

Experts agree that despite the advantages of developing a recovery plan before the event, most communities in fact handle recovery planning after the event because few jurisdictions have the capability, resources, or staff to do it before. Accordingly, recovery plans are typically developed when stress is high and decisions must be made quickly, with stakeholders exerting pressure to get things back to normal.

During the disaster response, a disaster recovery group must be established or (if it was established beforehand) convened to begin dealing with recovery policy issues. Many communities design task forces or committees to guide the recovery process. Such entities typically include local government representatives, social service agencies, citizens, local businesses, and other relevant stakeholders. Their work may vary from designing new codes and ordinances to conducting the visioning processes that lay the groundwork for rebuilding. The local government representatives should come from a wide array of departments (e.g., planning, public works, city manager/county administrator) to coordinate initial and long-term recovery activities. Key officials for both the response activities (mainly fire and police) and the recovery activities (e.g., planning, building, redevelopment) will be stationed in the emergency operations center, and in fact the response group may discuss some issues related to recovery; nevertheless, the activities and issues of the two groups are distinct, and the recovery group should be a separate operation.[10]

Although each community faces recovery challenges that are specific to it and specific to the particular disaster, it is advisable to envision a holistic recovery that addresses all dimensions valued by a community.

Experts also agree that leaders should first facilitate a stakeholder-based consensus vision,[11] while avoiding hasty commitments for rebuilding either "like we were before" (because the way things were before might have exacerbated the situation) or "better than we were" (because better might be cost prohibitive).[12] After the 1989 Loma Prieta earthquake that devastated downtown Santa Cruz, California, the newly formed group Vision Santa Cruz convened nearly three hundred meetings, opened a downtown office, and built small-scale models of an entirely rebuilt downtown, which were open for public viewing. Sometimes the Federal Emergency Management Agency (FEMA) provides funds for a consultant to help with the process; cities also have sponsored their own forums. The process happens in widely divergent ways. How things are done in New York City is different from how they are done in rural Ohio, for example. What is important is to convene the stakeholders, facilitate discussion, and emerge with a vision of how to proceed.

Because pressure to launch the rebuilding effort will emerge fairly soon after the disaster, the visioning process must be set in motion rather quickly—ideally, within thirty days of impact.[13] Notifying the public of the process and inviting participation, particularly through attendance at meetings, will require dedicated staff and volunteer time as well as creative ways of reaching a displaced populace. Social service agencies, local colleges, utility companies, the post office, faith-based organizations, and volunteers can all help disseminate information before the meetings; in addition, they can provide language support, assist persons with disabilities, offer child care, and provide refreshments at the meetings to ensure a good rate of participation and interest. (Involvement of the public is discussed in greater detail below.)

Although each community faces recovery challenges that are specific to it and specific to the particular disaster (see Figure 11–1), it is advisable to envision a holistic recovery that addresses all dimensions valued by a community. (As previously noted, the general dimensions addressed by post-disaster recovery plans are housing needs; economic development; environmental concerns; reconstruction of major infrastructure, including roads and bridges; and protection of lifelines from future disasters.) For all the dimensions, but particularly for infrastructure and lifelines, municipalities must weigh the benefits and costs of rapid versus slower recovery, especially if mitigative actions are to be incorporated.

And mitigative actions may well be incorporated because issues of disaster recovery, land use planning, and mitigation are all interrelated. Following a disaster, the recovery period provides an excellent opportunity to promote hazard mitigation, especially through the use of land use planning, codes, and ordinances. With the disaster event fresh in the minds of everyone—local officials, members of the business community, and residents—opportunities develop to show the benefits of effective land use planning, codes, and ordinances that will lessen the impact of the next disaster, helping to preserve businesses and neighborhoods.

In fact, addressing disaster issues within the context of land use planning is a strategy popular in cities today. (As has been explained, land use planning can be done either separately or within a comprehensive planning context.) For example, municipalities have zoned areas

Figure 11–1 Relationship between magnitude of event and type of recovery challenges

Recovery challenges	Small-scale event (earthquake magnitude 5.0 or less; F2 tornado of limited length)	Normal disaster (F3 tornado of several miles; flash flooding of less than 5 feet)	Catastrophic event (Hurricane Katrina, massive damage to broad region)
Social and psychological needs	Usually short-term effect; opportunity to involve schools and agencies in proactive response.	Usually short-term effect, but stress debriefing and counseling should be made available.	Potential for significant impact on employees and staff (of government, organizations, businesses, schools) and on families.
Housing	Dozens of homes with minor damage; weaker structures badly damaged; up to 2 years for recovery.	Thousands affected; most residents able to return home in 2–3 years. Federal, state, and local governments able to handle most housing with help from voluntary sector.	Massive and widespread losses; hundreds of thousands of homes destroyed or damaged; 5–10 years anticipated for recovery. All governmental levels overrun and unable to assist all households.
Economic sector	Limited impact to economic sector unless a direct hit.	Larger businesses return most quickly; smaller businesses challenged to return.	Massive disruption to employees and businesses; small businesses unlikely to return; all businesses hit heavily.
Environment	Minimal impact; opportunity to improve tree density, address stormwater, increase public awareness.	Opportunity to significantly improve local environmental conditions: increase open space, preserve floodplains, consider density transfers.	Massive damage, undermining local ecosystems; disaster often exacerbated by neglect to environment before disaster.
Infrastructure and lifelines	Usually a rapid return.	Up to years depending on the event; most resources quickly recover.	Months to years to repair roads, bridges, hospitals and other key institutions; some never recover.

to preserve sand dunes, safeguard coastal marshes, and restrict land use on barrier islands; and simple ordinances disallowing building in floodplains can avert significant damage, just as restoring wetlands can reduce flood costs. Wildfires have devastated thousands of homes in recent years, and city government could reduce recovery costs and needs by zoning land in ways that enhance fire protection and respect urban interface with forest ecosystems. Hurricanes can be mitigated by both structural and nonstructural measures: structural measures would be to erect sea walls, require that buildings be elevated, and build or reinforce dikes; examples of nonstructural measures would be to limit development or require setbacks.[14]

In most communities, existing plans, codes, and ordinances may suffice. However, post-emergency ordinances must be considered; such ordinances may result from the recovery planning process itself, or they may be separate from it. Overlay districts, which add an additional layer of regulation, may be used to protect urban and wildland interface, riverfronts, and seismically vulnerable soils, or to specify design elements for business zones.[15] Locally specific sections may upgrade stormwater and drainage, beach erosion, and slope stabilization; restore barrier islands; mandate congregate shelters; and manage debris.

After a disaster, cities often write ordinances to increase safety or update codes, but cities pass ordinances at varying speeds, and the speed of passage influences the speed at which residents can rebuild. The speed at which residents can rebuild will also be affected by any moratoriums that may be put into place for hazardous areas; after the Loma Prieta quake, for example, some homeowners could not rebuild for years because of unstable hillsides made even more unstable by drought conditions. In addition to new ordinances and moratoriums, new zoning may be considered in order to restrict rebuilding in floodplains, protect environmentally sensitive areas, or preserve historic locations. Then, too, in locations where a lack of code compliance compromises dwellings, cities may press for renewed enforcement. But enforcing codes may reduce affordable housing and increase homelessness, and communities should be prepared to address this issue. Elected officials and staff should also be prepared to consult with legal counsel about their obligations to protect the public from future events.

Recovery planning: Public participation and plan topics

Disasters provide a window of opportunity for municipalities to energize a public now acutely aware of hazards and highly motivated to recover and to reduce future risks. Experts concur that the best recoveries involve the broader community: just as disaster recovery offers local government the chance to change conditions for the better, so a process that is open to the public offers stakeholders the opportunity to invest in those changes. Furthermore, by encouraging public involvement and heightening the buy-in from stakeholders, local government can make its own job easier in the long run. Public participation can provide key input on matters such as new codes and zoning (both of which affect rebuilding), relocation of businesses near revitalized transportation sectors, and preservation of green space.

As noted above, local government should convene key stakeholders—residents, business owners, community organizations—in a process that generates a shared vision for recovery.[16] In other words, community leaders and residents should interact directly with decision makers and city staff. Research has found that this type of locally based effort is more effective than recovery efforts directed by the top or from outside the affected area.[17]

Experts recommend that cities go beyond merely asking for verbal input from the community and, instead, proactively manage decisions so as to ensure equity and monitor long-term community needs. Proactive management means taking charge to organize and fund the effort. It requires financial, public, and political commitment to participatory processes—to an effort in which citizens, community leaders, and officials collaborate to produce a vision for recovery on which all of them agree. The effort will take concrete form in the resultant recovery plan.

Strategies for involving the community

To achieve community involvement, public officials can use several strategies that range from traditional representation to grassroots efforts. That is, some communities may desire a formal approach, in which people go through their elected leaders; other communities may base their community participation on grassroots efforts.

The city of Oakland has used a combination of strategies. The city created the Oakland Hills Fire Prevention and Suppression Benefit Assessment District two years after the community's devastating 1991 wildfires. To accomplish this, Oakland started with *standardized representative policy making*, using public officials to represent their constituents' interests. The city then used referenda to fund capital improvements and equipment; and mayors from Oakland and Berkeley used *nonbinding direct involvement* by citizens when they convened the Task Force on Emergency Preparedness and Community Restoration to make a series of proposals on volunteer development, residential density codes, and citizen-based fire crews. However, this type of nonbinding direct involvement may not result in the acceptance of all recommendations because they are only advisory; for recommendations to be binding, they have to be voted on. South Dakota used a voting approach after the 1993 Vermillion River Basin flood, and the approved mitigation plans were put in place.

Techniques to involve the public can include public meetings with presentations by experts, panel discussions, workshops, field trips, live call-in radio, small meetings between citizens and elected officials, scenario planning with best and worst views, and structured charettes designed to foster community brainstorming and problem solving.[18]

Initial steps

Public involvement requires several steps. To get started, the local government may need to gather information through interviews, focus groups, surveys, or Web sites. Once public concerns have been identified, government leaders should then convene a diverse group of stakeholders, decide on ground rules, and select a strategy for managing the process.

Identifying stakeholders is crucial in order to form an effective working group, generate innovative ideas, and represent the full community. A matrix, map, or checklist of possible participants may help organizers initially identify the types of persons, neighborhoods, or organizations that should be included. Stakeholder participation should be considered fluid and open to further inclusions; avoiding exclusiveness is a way to lessen charges of favoritism.[19] And several different types of group members or consultants may prove valuable: for example, content experts provide information, whereas process experts focus on how the group moves toward consensus.[20] Content experts and process experts may be members of the community or hired consultants.

A key part of the public involvement process involves notification. When the decision is being made on the languages in which to disseminate information about meetings, it will be

Do's and don'ts for encouraging public involvement

- Anticipate issues rather than having them be imposed.
- Define issues in terms amenable to resolution.
- Avoid "either/or" terms.
- Avoid seeing public involvement as good or bad.
- Know what you are trying to get from involving the public.
- Recognize that public involvement requires sharing decision-making authority.
- Define ahead of time what can and cannot be negotiated.
- Define ahead of time which "publics" to involve.
- Consider citizen attitudes toward institutional goals.
- Select an appropriate decision-making form.
- Use more than one approach.
- Work to build relationships.
- Keep an eye on the public interest.
- Accept and learn from failure.

Source: Natural Hazards Research and Applications Information Center, *Holistic Disaster Recovery* (Boulder, Colo.: Institute for Behavioral Science, University of Colorado, 2002), 3-8, based on J. C. Thomas, *Public Participation in Public Decisions: New Skills and Strategies for Public Managers* (San Francisco: Jossey-Bass, 1995). Copyright © 1995 by John Wiley & Sons, Inc. Reprinted with permission of John Wiley & Sons, Inc.

helpful to know the demographic makeup of the community. Demographic information can also provide insights into where to post messages—for example, at Spanish-speaking and African American churches, senior centers, laundromats, chambers of commerce, and community centers. Officials are advised to develop good working relationships with local media and to consider developing their own Web pages, e-mail lists, and hard-copy newsletters. Libraries and organizations may be able to assist in distributing CD-ROMs of public meetings, and a local cable station may be willing to broadcast videotaped events.

The location of any meetings involving the public is an important consideration as well. Because people may be displaced, multiple meetings in multiple locations may be necessary. After Hurricane Katrina, virtual chat rooms, televised town halls, and other unusual forms of communication were used to reach constituents.

Those who are planning meetings should also consider the shifts that people work, the need for child care, transportation issues, and language needs (including sign language). Going to neighborhoods where people live and work may prove feasible; understanding that not everyone can attend all the time but that many may still want to be informed and involved is crucial to success.

Meetings may have different purposes, such as problem solving, visioning, or strategic planning. Regardless of the particular purpose, though, the aim is to build credibility and trust, so it is important to foster open dialogue without criticism and to manage conflict. For more effective communication, throughout the process stakeholders should get to know each other and work in small groups.

Also throughout the process, groups should document their efforts and identify diverse ways to disseminate the documentation—ways that range from putting notices on poster paper to sending messages electronically.[21] Local leaders should be called upon for help in reaching historically marginalized or disenfranchised neighborhoods or ethnic groups, or people who are either isolated or in congregate care facilities.

Topics for a recovery plan

A consultant in the field of planning has suggested a generic set of topics for local officials to use as a basis for their recovery planning. Among the topics are authority, purposes, definitions, recovery organization, recovery plan, general provisions, temporary regulations, demolition of damaged historic buildings, temporary and permanent housing, a hazard mitigation program, and recovery and reconstruction strategies.[22] The model can be very helpful to local officials but should be molded to fit local needs.

Recovery plans may take a variety of forms. Most communities envision a rebuilt community, and through a community consultation process they identify strategies for a rebuilt or even reinvigorated community. In 1999, Hurricane Floyd left Princeville, North Carolina, flooded to

Using geographic information systems

Whether before or after an event, local officials may find their disaster recovery planning enhanced by geographical information system (GIS) technology. GIS provides a unique way to present spatial and geographic data combined with social, economic, and political data.[1] This technology has already found an important niche in disaster management. It could also be effective for disaster recovery planning. Before an event, GIS can estimate or predict areas of vulnerability and the type of physical and social damage that could occur after a specific event.[2] Models can then provide estimates of physical, social, and economic losses. After an event, GIS can be used to map areas for search and rescue, locate debris and track its removal, show which residences and businesses are considered clean and safe for reentry, and locate temporary mobile homes and other temporary housing settlements.[3]

[1] Nicole Dash, "The Use of Geographic Information Systems in Disaster Research," in *Methods of Disaster Research,* ed. Robert A. Stallings (Philadelphia, Pa.: Xlibris Corporation, 2002), 320-321.

[2] Arleen A. Hill and Susan L. Cutter, "Methods for Determining Disaster Proneness," in *American Hazardscapes: The Regionalization of Hazards and Disasters* (Washington, D.C.: John Henry Press, 2001), 16-24.

[3] Dash, "Use of Geographic Information Systems," 324-325.

its rooftops, with its built environment almost completely destroyed. Because of Princeville's historic legacy as the first city in the United States incorporated by African Americans, extensive efforts went into envisioning a recovered community. The Princeville recovery plan (for excerpts, see the accompanying sidebar) illustrates how one community addressed recovery, and other communities might want to consider some of its dimensions. The plan spells out an effort linked to Princeville's historic character and specific hazards, ties recovery to mitigation, and identifies a staffing framework to put the plan into action. Other community recovery plans may see other items as crucial to their recovery efforts. For example, recovery plans might include a newly configured transportation flow, designs for a revitalized downtown, efforts to build energy-efficient housing, or repairs to infrastructure and lifelines.

The dimensions of recovery

What are the most challenging dimensions of need during recovery, and how can local government address them? Clearly, these dimensions are housing reconstruction, economic recovery, environmental protection, restoration of infrastructure and lifelines, and, in the short term, removal of debris and disposition of donations. In addition, underlying these several dimensions of need are social and psychological conditions, which strongly affect the recovery process as a whole and should therefore not be ignored, even though local government may or may not become directly involved in them.

Social and psychological conditions

Fortunately, in addressing the social and psychological aspects of recovery, city government can be proactive rather than reactive.[23]

Social conditions are reflected at the level of households and neighborhoods, and the focus here is on preventing the loss of the human resources and riches that are essential to a vibrant community life. Thus, local government managers must think through how their communities, neighborhoods, and cultures can be sustained: how government can enable people to return to homes that are culturally, environmentally, and structurally familiar, and how it can enable people to remain vitally connected to opportunities for health care, employment, recreation, and transportation.

An important consideration in sustaining households and neighborhoods is ensuring that residents know how—and are able—to obtain the aid available to them. Residents will vary in their understanding of how to access aid, deal with bureaucracies, or challenge denials of aid, and for some survivors, accessing aid may present special difficulty. This may be the case particularly for elderly single women; people who have been, or are newly, homeless; or people with disabilities.[24] In working with federal aid, for example, local government can serve as an advocate for such groups and ensure that appropriate outreach is conducted to all segments of the public.

Support services are another aspect of the effort to sustain neighborhoods, and support services for families must be part of recovery. Typically, however, they are nongovernmental; the local government's role is primarily to identify need and to facilitate and support such services. For example, to ensure widespread and diverse public involvement in recovery planning, government might see that child care is provided at meetings that involve the public (although it would not itself provide the care), and might direct that meetings be held where residents are currently living or working. Local government may also have to establish a liaison with voluntary organizations to address unmet needs, such as ensuring that victims of domestic violence can retain their temporary housing while the perpetrators are evicted, extending park and recreation services into trailer parks, and providing programs for teenagers during the summer and after school. Homeless shelters supported by government resources (fire, police, emergency medical, and so forth) may need increased support. It may be beneficial to give priority to residential treatment facilities and other types of facilities that provide housing for larger numbers of people.

The psychological conditions of recovery are equally fundamental, although generally after a disaster, psychological trauma is found to be not as widespread as anticipated. Nevertheless,

Post-disaster recovery plan: Princeville, North Carolina

The town of Princeville (pop. 1,668) lags far behind many other cities and towns devastated by the floods spawned by Hurricane Floyd during the middle of September 1999. This recovery plan is part of a very simple but powerful mission: to bring all of the citizens of Princeville home as quickly as possible while rebuilding toward a better and more disaster-resistant community.

[Author's note: To restore the historical and cultural core of the community, the plan called for key changes, as indicated by the following items.]

- Bold steps
  - Return town hall to historical center of town.
  - Restore and adapt original town hall as an African American cultural museum.
  - New and replacement facilities: senior center, post officer, fire station.
  - Develop parks and recreation: trails, parks and recreation to "capture the spirit of place and family that is such an important part of African-American history and culture."
  - Upgraded and expanded infrastructure.

[Author's note: The housing recovery strategy began by contrasting pre-flood housing characteristics with post-flood conditions, determining the extent of housing needs, and specifying issues of concern, including the historic character of the housing and the fact that the majority of residents are elderly people with low incomes.]

- Housing recovery strategy: "the most important element of the mission is the return of Princeville's residents to their town."
  - Overview of pre-flood housing characteristics and needs
  - Post-flood characteristics
  - Housing need
  - Issues

[Author's note: The recovery plan stated, "Flood recovery also presents an opportunity to take bold steps to recognize Princeville's heritage and the cultural legacy it can preserve for future generations." To protect Princeville from further flooding, mitigation efforts included cleaning out ditches and strengthening the dike along the Tar River. As part of the overall mitigation assessment, the plan specified the five areas of concern shown below.]

- Hazard mitigation improvements: restored dike along the Tar River and address storm water flooding.
  - Existing conditions
  - Flood hazard
  - Soils and wetlands
  - Wind
  - Mosquito abatement

[Author's note: To launch the recovery effort, the plan suggested hiring a recovery project manager and housing specialist. To ensure that the recovery proceeded in a way that preserved local resources, an environmental review was warranted.]

- Implementation: organizational strategy and environmental review
  - Organizational strategy
- Recovery project manager.
- Housing recovery: intake counselors, case managers, construction coordinator, rehabilitation specialist and administrative assistant.
  - Environmental review
- Environmental review of housing projects
- Environmental review of community recovery projects

[Author's note: Appendices provided information on several other aspects of the plan, as shown below.]

- Appendices
  - Timetable
  - Grant programs
  - Mitigation tools and techniques
  - Conservation easements
  - Mitigation strategies for historic buildings
  - Municipal building cost estimates
  - Transition housing

Source: Excerpted from Emergency Response Program Management Consultants, *Princeville Recovery Plan* (Arlington, Va., May 2000). This plan was prepared for the Federal Emergency Management Agency following multiple rain events, including Hurricane Floyd. Additional information can be found in the statement by President Bill Clinton, available at clinton4.nara.gov/WH/New/html/20000229_1.html (accessed September 7, 2007).

depending on the extent and scope of the event, or its type, it may generate a range of mental health needs. Residents may experience heightened stress and anxiety; perhaps personal trauma due to loss of family, neighbors, or property; or psychological problems due to disrupted medications, routines, or resources. Typically, psychological needs for residents are addressed by private insurance or through local social services and mental health outreach services, such as those provided through the American Red Cross or other mental health providers. City staff may need to support the mental health community as the latter struggles to restore social and psychological services and access to medications while simultaneously assisting its own members who may have lost their homes. In addition, it is important for local government managers to understand that although most citizens respond fairly well to disaster situations and launch rebuilding, the public's patience and coping skills will be weakened by the disaster. For local government staff, therefore, working with the public may demand new skills, and government managers may want to offer stress debriefing sessions for their employees.

At a time when local government can least afford to lose personnel, stress of one kind or another may lead to higher-than-normal turnover. The city of Santa Cruz, which normally experiences an annual workforce turnover of 11 percent, faced a turnover of 26 percent after the Loma Prieta earthquake. Turnover may result from damage to an employee's home, as when the homes of scores of workers in Gulf Coast areas were destroyed after Katrina. Or staff may be lured by a location with fewer risks, or may be under too much stress from a heavy or unfamiliar workload if they have to take on new tasks and lack the training for working with a stressed-out public.[25]

Housing reconstruction and household recovery

Housing is the largest part of any community's building stock, and in addition it represents a considerable link to local economies.[26] (The terms *reconstruction* and *recovery* are used interchangeably in this context, but the former means rebuilding and the latter means people are back home and don't have to move again.) Housing recovery will require that government become involved in a variety of efforts and activities, beginning with helping the community envision a rebuilt residential sector. Other activities may include siting temporary housing, writing ordinances that encourage builders to rebuild affordable housing, designing a permit process, inspecting new construction, watching for unscrupulous contractors, securing funding, and working with voluntary organizations to help the under- and uninsured to return home. The effort as a whole demands that the local government prioritize need, involve the public, build partnerships, secure financing, and support families through a long, difficult, and cumbersome process.

Government may have to address up to four phases of housing recovery: emergency shelter, temporary shelter, temporary housing, and permanent housing.[27]

Emergency shelter The first phase of housing recovery, emergency shelter, is citizen based and occurs when families use tents, cars, overpasses, or other marginally adequate locations for shelter. Depending on the type of disaster, the availability of sheltering options, and the particular population, tent cities or other emergent shelters may develop spontaneously.[28]

Temporary shelter Historically the American Red Cross has been mandated by Congress (although at the time of this writing the mandate is under review) to provide the second phase, temporary shelter. The time spent in temporary shelter can vary. In general, low-income and marginalized populations usually populate such shelters, and the lower the income, the longer it usually takes to move through a phase. In most disasters, the majority of evacuees will shelter with family and friends or in hotels, but after a catastrophic event like Hurricane Katrina when facilities are overrun, government (local, state, and federal) is likely to play a role in providing temporary shelter.

Temporary housing Before the third phase, temporary housing, can be provided, emergency management agencies and other response organizations will need to make damage assessments. In fact, to secure any FEMA funding at all, it is necessary to submit a request through the governor to FEMA, and the request must be supported by proper documentation of dam-

Preparing for recovery: Lessons learned after Katrina

Through a program sponsored jointly by ICMA and Fannie Mae, a number of local government professionals from Florida traveled to Mississippi to help their peers after Hurricane Katrina. They returned with personal insights into how to prepare for disaster and recovery.

For building and code enforcement officials, the time spent in Mississippi confirmed the importance of moving fast on damage assessment and inspections. With any disaster, people will quickly try to restore their homes, despite any risks with electrical wiring, mold, or other potential problems. "Folks will start to do repairs and some of the necessary work, whether the city is there to permit it or not," observed Nestor Abreu, the building and code administration director in Palm Coast, Florida, who spent time in Pass Christian, Mississippi, after that town was leveled by Katrina.

Information technology professionals found they had to think about their work quite differently. "You really can't rely on computers," noted James Majcen, who went to Pass Christian from Palm Coast immediately after Katrina and helped develop a geographic information system (GIS) tracking system. "When you get to that level of devastation, you need the traditional pen and paper, and a strategy that does not rely on electricity." Majcen and his colleagues built an application that allowed field inspectors to record damage in specific locations on paper and then map the damage by linking the inspection sheets with a GIS. They came to see the value of having preprinted forms that can be used until computers are back up.

A finance department member from Palm Coast, Joan Dunn, saw a worst-case scenario in Pass Christian. She remembers municipal staff members trying to fill out Federal Emergency Management Agency paperwork at a time when most of them had lost everything and city hall had been destroyed. "I don't know how they were functioning," she said. Dunn recalls one staff member going through a paper bag full of mostly handwritten forms, trying to pull together payroll information. "It took my breath away to see her sitting there struggling and trying to get her co-workers paid," she said. When Dunn returned to Palm Coast, she met with her boss to talk about the records for which they needed to have backup copies, and how and where to store everything.

Dunn's observation about the importance of good record keeping was a recurring theme. Matt Efird, the risk manager in Deland, Florida, traveled to Pascagoula, Mississippi, to perform insurance audits. He combed through pictures, estimates, purchase orders, and other records to determine if the city was eligible for additional reimbursement. Fortunately, he found everything he needed. "Paperwork is everything. If it's not written down, it didn't happen," he said. "They were very meticulous about their paperwork. That made everything a lot easier."

Another theme that emerged was noted by Keith DeWise, of Pensacola, who went twice to Mississippi, first as a police officer immediately after Katrina hit and then in the fall of 2006—having changed jobs to work in code enforcement—to help with later-stage recovery needs. Although the two trips were different, he learned the same lesson in both cases: "The value of teamwork," he said, "that's the biggest thing. No matter how big or small a project, if everybody pitches in, it goes so much easier."

That message is echoed by William Whitson, the former assistant city manager in Port Orange, Florida, and now the city manager in Cairo, Georgia. He wasn't surprised to hear that local government professionals in Mississippi say they'll gladly volunteer to provide disaster recovery assistance somewhere else when it's needed. "Come crunch time, we're there for each other," he said. "If there's anything I've learned, that's what it is—we have to be there for each other."

Source: Adapted from Christine Shenot, "Lessons Learned in Mississippi," *Public Management (PM)* (April 2007): 9.

ages. FEMA typically sends inspectors to the damaged area to verify assessments and then submits the request to the White House for approval. If the president issues a Presidential Disaster Declaration, funding for individual and public assistance can be secured.

Typically, for the third phase, the recovery task forces that communities organize have subgroups dedicated to housing. After the Santa Cruz earthquake, for example, the Housing Recovery Task Force generated two subcommittees, one on short-term and one on long-term housing, and specific efforts emerged to help displaced low-income elderly people. These specific efforts were coordinated with local organizations—both community and nonprofit—on aging.

Formal aid mechanisms are available for temporary housing: both FEMA and the Red Cross will provide rental assistance, and FEMA grants will help with the finding of temporary locations.

In addition, FEMA has historically, although reluctantly, developed mobile home parks or placed travel trailers on lots undergoing reconstruction (see Figure 11–2). "Reluctantly"

Figure 11-2 Rows of FEMA-provided travel trailers installed in Plaquemines Parish, Louisiana, for temporary housing after Hurricane Katrina

Photo courtesy of FEMA/Robert Kaufmann

because mobile home parks are problematic in that they require pads, infrastructure, and basic amenities if they are to serve families that may be displaced for up to eighteen months or longer, as in a Katrina-like event. In the case of catastrophic events, "temporary" housing may require FEMA and/or local government involvement for as long as five to ten years. Typically, however, FEMA funding lasts only up to eighteen months.

The relief funds can be used to design creative solutions. Indeed, after large-scale events especially, local needs to house families temporarily will require that government be creative in exploring partnerships (a number of organizations and agencies can serve as partners) and also in exploring ways of meeting the needs. After the Loma Prieta earthquake, Santa Cruz reopened a closed nursing home to shelter low-income elderly residents and residents with mental disabilities. Santa Cruz also created temporary housing through a motel voucher program—a program allowing families that met certain criteria to regain normal household functions, such as cooking and doing laundry. In Arkadelphia, Arkansas, universities have provided dormitory rooms to families after tornadoes, although moving the families out of the dormitory rooms proved difficult when there was no place for them to go. And after the May 3, 1999, tornadoes, the Oklahoma City office of the U.S. Department of Housing and Urban Development (HUD) generated a list of rental facilities and collaborated with local, state, and other federal agencies to match households to vacancies. (HUD may also be able to assist with Section 8 certificates and vouchers.)[29] HUD in Los Angeles was drawn "to the center of one important sector of the disaster recovery process, partly doing what it normally does—make sure that affordable housing is available in the city. Partly, however, the Housing Department improvised, defining a unique need within the recovery process and creating programs to address it."[30]

Permanent housing The final phase of housing reconstruction and household recovery is the provision of permanent housing, when families do not face any further moves. Rebuilding can take place in a variety of ways. The Arkadelphia 2025 Recovery Task Force worked with the University of Arkansas to design affordable energy-efficient housing in what had been a predominantly mobile home neighborhood. Community-based organizations rebuilt farmworker housing lost after Hurricane Andrew in South Florida cities. Faith-based organizations worked individually and collaboratively to rebuild most of the small town of Princeville, North Carolina. Santa Cruz expedited a public housing project already under way before the city's earthquake. Dozens of faith-based efforts engaged in cleaning and rebuilding after Hurricane Katrina.

Through design ordinances, the permitting process, and code inspections, local government plays a role in facilitating rebuilding. Although citizens—especially homeowners with insurance or the means to rebuild immediately—often express frustration with the lengthy delays entailed by government requirements, government must establish a process that meets three objectives simultaneously: facilitate rebuilding, make the community far safer, and ensure compliance with local codes and ordinances. Communicating this set of objectives to the public is key. Government will find that informing homeowners and inviting them to participate in writing and reviewing ordinances may reduce local conflict and misinterpretation. Some local governments have offered housing fairs and workshops to inform both homeowners and builders about the rebuilding process. In addition, staff involved in the permitting and inspection processes may benefit from training in working with beleaguered homeowners and builders.

Four housing recovery models for restoring or creating permanent housing characterize government options: redevelopment, capital infusion, market, and limited intervention.[31]

1. *Redevelopment:* Redevelopment occurs when the national government completely redevelops an area.
2. *Capital infusion:* Capital infusion brings in outside capital and expertise. In the United States, redevelopment and capital infusion are unusual, but these models may be adopted when the affected communities are economically marginalized or historically significant, as was the case with Princeville, North Carolina, the first U.S. town chartered by African Americans.
3. *Market model:* The market model relies on the real estate sector, assuming that individuals will use their own resources (such as insurance, savings, or other assets) to recover.
4. *Limited intervention:* Far more common in the United States, the limited intervention model involves government programs providing individual assistance in the form of income-based grants or loans. FEMA provides the bulk of these funds, although HUD has offered assistance to low-income households, recovery organizations, and housing projects. FEMA will provide grants to low-income households up to a maximum amount; additional assistance is nearly always required from faith-based and other rebuilding organizations. FEMA also provides low-interest loans to higher-income households through the Small Business Administration (SBA). (A disaster victim will fill out an SBA application and, if the application is rejected, the applicant can get a FEMA grant; but if the application is accepted, the applicant can get an SBA loan.) Although higher-income families usually have insurance, a recent trend after some events has been the denial of insurance benefits. The bankruptcy of insurance companies after a disaster has also been a problem, one that may occur more often in low-income communities. In such communities, people may have insurance but they are usually underinsured or their policies are with companies that are most likely to go bankrupt.

Finding or creating permanent housing may be especially problematic in municipalities where there is a lack of affordable housing—where a shortage of vacancies is aggravated by high rents for the vacancies that do exist. In an effort to meet local need, the city of Watsonville (near Santa Cruz) enacted an ordinance after the Northridge earthquake, specifying that 25 percent of all new construction must be affordable. Ideally, though, municipalities will bring attention to bear on affordable housing before a disaster, knowing that there will be a payoff when disaster strikes. Creative partnerships and financing from agencies (such as HUD appropriations) may be necessary.[32]

After the 1989 Loma Prieta earthquake, the Multidisciplinary Center for Earthquake Engineering Research in Buffalo, New York, and the Bay Area Regional Earthquake Preparedness Project in San Francisco convened a group of experts to identify post-disaster housing recommendations. The first recommendation was that addressing affordable housing issues, beyond the disaster context, is critical. Other recommendations were as follows:

- Avoid costly and time-consuming temporary housing
- Tie housing to employment
- Identify, reduce, or retrofit vulnerable housing before disaster
- Prepare municipalities for an onslaught of permitting, loan applications, and financing

- Be aware that financing low-income housing is the most difficult local government responsibility
- Involve local citizens in all dimensions of the recovery process.[33]

A wide variety of organizations can assist with rebuilding affordable housing, and local government should make it a priority to dedicate staff and volunteers to coordinating this effort. Staff will be whoever is available and is designated—perhaps new hires, perhaps some who are reassigned. Tasks will include identifying qualified households; securing funding; moving projects through the permit process; and coordinating, feeding, and housing volunteers. The National Voluntary Organizations Active in Disaster (NVOAD), which has strong links with many state and local organizations, can link local government to organizations that build low-income housing. Some organizations (such as Habitat for Humanity) have rebuilt significant portions of damaged communities. Others concentrate on the housing needs of specific populations (the Mennonite Disaster Service, for example, concentrates on the housing needs of elderly women and other marginalized populations).

Economic recovery

Economic impacts and recovery following disaster have not been fully researched, but what we do know is that there is probably both good and bad news. The good news is that although disasters may create short-term problems, over longer periods they usually do not cause major economic changes except potentially in a catastrophic situation.[34] The bad news is that business disruption can have ripple effects by undermining tax revenues: businesses that are disrupted pay less in taxes, forcing the local government to cut personnel, services, and programs. To minimize the bad news, local government will have to understand the need for financial capital to be invested in rebuilding (and should encourage such investment) and will also have to understand the importance of governmental attention to the commercial sector. In particular, recovery efforts should be linked to capital from local and external sources (such as foundations, state and federal grants, private and corporate donations, capital improvement funds, and special taxes, as discussed in the section on financial resources later in this chapter).[35] Local government managers must also understand the availability and limitations of public assistance programs, and the crucial need to maintain careful records in order to receive federal financial reimbursement. Furthermore, these managers will have to make choices during a period when the community is pushing for the return to normalcy. In that situation, local government will simply have to try to strike a good balance between competing needs.

Business disruption may of course occur when the business itself is damaged. However, sometimes even if the business itself is not damaged, damage to the surrounding infrastructure or lifelines can prevent a business from opening because employees cannot get to work or may have lost their housing. After Hurricane Katrina, even though many businesses could have reopened in the central business district and historic sections of New Orleans, floodwaters had destroyed much of the housing used by employees.

A number of factors can make it possible to predict whether a specific business can survive a specific disaster. Disruptive physical damage alone is not necessarily the determining factor. The length of the post-disaster closure may matter (and may be partly a matter of size, for larger chains usually get up and running fairly fast, whereas smaller businesses usually struggle). Following both the Loma Prieta earthquake and Hurricane Andrew, businesses that failed had actually been in better financial condition than others just before the disaster—but they were forced to close for longer periods following the disaster. Besides the three factors of amount of damage, financial condition, and length of closure, other factors that may determine business survival are reliance on local markets, previous disaster-related experience, and the emergence of disaster-caused operational problems.[36] And the 2001 Seattle Nisqually earthquake revealed that small businesses that are in rental space and rely on foot traffic are particularly vulnerable, regardless of physical damage.[37]

Local government can spur business recovery by helping to relocate businesses and thereby reduce the long closures that threaten business viability. Santa Cruz opened the "Phoenix

Pavilions" to house displaced businesses, and in the case of one such business—a bookstore—citizens showed up to move thousands of volumes to the relocated store. Nearby Watsonville assisted downtown merchants with mobile homes to use as temporary facilities.

Local officials can assist businesses and business owners in both preparing for and recovering from disaster. A key element in preparing for disaster is encouraging public-private partnerships between the emergency management agency, chamber of commerce, and specific businesses to do business continuity planning (see Chapter 3). In a post-disaster context, the immediate period after a disaster is the best time to focus businesses on recovery. Task forces on economic recovery should begin by generating "lessons learned" and listing the changes necessary within businesses as well as in policies and/or legislation.[38] Local government can assist these efforts by using economic revitalization tools such as redevelopment authorities, adding aesthetic improvements to a commercial sector, promoting local businesses through public relations campaigns, and supporting business continuity efforts. Portions of property taxes can be redirected toward infrastructure improvements. And capital improvement programs can target economic revitalization.[39] The key actions in jump-starting post-disaster economic recovery include

- Ensuring a liaison between general recovery task forces and the business sector
- Establishing strong lines of communication with each company's executives
- Designating a recovery team manager
- Assigning specific recovery responsibilities to individuals
- Collecting critical information for planning purposes
- Identifying loans, grants, insurance options, and government programs
- Ensuring strong public relations
- Providing administrative support for recovery tasks.[40]

Federal funding through post-disaster programs offered by the U.S. Department of Labor can provide temporary assistance to individuals experiencing job loss due to the event. Creative financing may make it possible for those who are unemployed because of the disaster to work on various dimensions of the recovery—in housing, environmental efforts, and infrastructure repair, for example.

Environmental recovery

Ironically, appreciation for America's rich natural resources underlies some of the nation's disaster risks. Coastlines and waterways are considered desirable locations to live. Accordingly, cities have developed along major rivers, and the rich soils from floodplains and volcanic areas have encouraged agricultural production.[41] But as the U.S. population has grown and density in popular locations has increased, so too have our personal and collective risks. A way to reduce risks is to live in concert with natural resources; doing so encourages environmental protection and makes it possible for people to enjoy the recreational and aesthetic aspects of the natural environment.

Recovery offers a window of opportunity to protect the environment in ways that enhance the community's environmental quality while promoting a long-term reduction of risks. For example, during recovery local government can address such preexisting environmental problems as stormwater runoff, pollution, and loss of wetlands and watersheds, as well as problems associated with development in general. Recovery efforts can reduce erosion by targeting rivers and beaches, and can require the repair of sewer systems and storm drains or ruptured tanks containing chemical wastes.[42] Other opportunities to improve the environment as part of recovery would include purchasing open space, using federal funds for buyouts (see sidebar on page 224), and allowing density transfers for developers. Because environmental conservation programs prove popular with citizens, local government can use recovery as a way to build support for such programs.

Recovery also offers a window of opportunity to cut costs while enhancing the community's environmental quality. In Garland, Texas, planting trees reduced stormwater runoff by 19 million cubic feet per major storm and saved millions of dollars in infrastructure, air

Federal acquisitions: The Missouri Community Buyout Program

The largest federal buyout occurred in Missouri after the Midwest floods of 1993, described as a "sea change" by those examining benefits versus costs. Missouri governor Mel Carnahan established the Missouri Community Buyout Program using funds from the Federal Emergency Management Agency (FEMA) and the Community Development Block Grant program. Fifty-seven communities in Missouri participated in the program, an effort labeled a great success primarily because subsequent flooding resulted in significantly fewer Presidential Disaster Declarations and lower costs to the government.

The Missouri Buyout Program cost FEMA $100 million overall, with $54.9 million dedicated to acquiring 4,800 properties. Success was measured in Arnold, Missouri, when repeated floods occurred albeit with dramatically reduced costs:[1]

Measures of success	1993 flood	1995 flood	2002 flood (as of 6/25/02)
Sandbagging sites	60	3	0
Public assistance	$1,436,277	$71,414	$0
Individual assistance applications	52	26	1

[1]Federal Emergency Management Agency (FEMA), "Success Stories from the Missouri Buyout" (Washington, D.C.: FEMA, August 2002), available at fema.gov/library/file?type=publishedFile&file=mo_buyoutreport.pdf&fileid=a5736520-2d48-11db-ad64-000bdba87d5b (accessed September 7, 2007).

quality, and residential energy costs. A cost-benefit analysis from New York City revealed that environmentally based savings would result if the city purchased Hudson River watershed land rather than build an $8 billion filtration plant.[43]

In the event of a Presidential Disaster Declaration, funds for projects that enhance environmental quality can be secured from the Hazard Mitigation Grant Program of the Stafford Act (Section 404). Projects that can be funded under the program include construction of detention ponds, riverbank stabilization, land purchase, stormwater and wastewater improvements, beach nourishment, sand dune restoration, controlled burns, and brush clearing. Section 406 of the Stafford Act authorizes Public Assistance funds for such projects as replacement of a water treatment plant.[44] After a 500-year event, Princeville, North Carolina, secured funds from state and national park services for a historic trail along its riverbank and through the town. The National Oceanic and Atmospheric Administration offers Section 309 grants for coastal flood and hurricane hazard mitigation. The Natural Resources Conservation Service program can give communities help (sometimes funding) to protect wetlands.

During recovery, local government managers are advised to work closely with the U.S. Environmental Protection Agency (EPA) to ensure that hazardous materials do not contaminate—or further contaminate—local environmentally sensitive areas.

Infrastructure and lifelines

Infrastructure usually includes roads, bridges, rail systems, mass transit systems, waterways, and air systems. Lifelines usually include power, pipelines, fuel, telecommunications, and water treatment facilities.[45] When infrastructure is damaged, there may be delays in the movement of supplies for building and reconstruction, of products for commerce, and of people seeking to go to and from work. As a result, recovery slows.[46] But without lifelines, some forms of transportation are not able to function and, more importantly, neither is the community as a whole.[47] Just as with the other dimensions of recovery, however, infrastructure and lifelines can involve an opening of doors for change—transportation lanes can be rerouted and facilities relocated, for example, or power lines can be put underground—which may mitigate future risk and even create new partnerships to meet key societal needs.

Certainly, the effort to restore physical infrastructure and lifelines will require a wide variety of partners, both private and public. Although partners are always important, in the case of infrastructure and lifelines the local government is dependent on the companies that construct or produce them. This effort could take considerable time, depending on the extent and type of disaster. In Santa Cruz, California, for example, repairs to the earthquake-damaged downtown water and sewage lines went on for years. After Katrina, utility companies began a struggle—one that may continue for years—to provide service to customers and simply to survive economically. Also after Katrina, the Gulf Coast area faced telecommunications losses for up to nine months.

The lifeline that usually comes to mind first when one thinks of the effect of outages is electrical power. The Midwest outage in August 2003 nearly brought the region to a standstill. In the parts of South Florida that were without electrical power for nearly two months after Hurricane Andrew, again, life came to a near standstill. After Katrina, even though utilities restored power to neighborhoods within months, the companies were without customers and faced bankruptcy—another way in which a disaster may lead to lack of function. Local officials will have to decide to restore power systems quickly while also hoping to mitigate future risks.

Gas and liquid fuel provide the energy to manufacture items, move the vehicles that transport people and goods, and generate electricity. Local government may need to work with energy suppliers to ensure rapid and steady streams of energy resources. In some cases, damage or destruction to gas and liquid fuel lines in one part of the nation can lead to regional or even national crises for fuel, as was the case after Katrina.[48] Since the rebuilding or repairing of fuel systems may take some time, local officials must be prepared to find other means of acquiring and distributing fuel—a pressing need after Katrina.

The disruption of telecommunications, whether telephone, computer, or other mechanism, can create a wide array of problems. Information needed for transportation, commerce, and even national security can be compromised if the telecommunications system is damaged, overloaded, or destroyed. Thus, backup communication systems both inside and outside local government must be in place before the disaster. In the past, rural areas have turned to creative means of communication, including the use of amateur radio groups. And for months, the area around Homestead, Florida, even used tethered balloons to convey locations of food, water, first aid, and shelter.

Without water and sewage, most businesses, schools, and hospitals will not be able to open, health concerns may arise, and fire suppression will be impossible. Not until both water and sewer systems are brought back online can more than rudimentary recovery activities begin.[49]

By working with state and federal officials, local communities can obtain Public Assistance funds from FEMA for remediation projects. Typically, the federal government pays about 75 percent of the cost for these projects (sometimes more), with local and state entities paying the remaining 25 percent.

Before disaster, a risk and hazard analysis can help local officials find frail spots in their infrastructure and lifelines. Post-disaster, local government must collaborate and communicate closely with companies and the municipal offices that manage critical infrastructure and lifelines. Although most communities will experience fairly rapid return of lifeline functions, infrastructure needs may linger for some time. Routes between work and home and access to commercial sectors may be compromised, and detours may last for years. As noted in the section on the recovery planning process (and in the discussion just above about restoring power), municipalities will have to decide whether it is better to move faster or slower in reestablishing infrastructure and lifelines; if mitigative actions are to be incorporated, moving with more deliberation might be wiser.

Debris management

Disasters generate thousands of tons of debris. It can come from animal corpses and food remnants, vegetation, appliances and furniture, personal items, and hazardous waste. After Katrina, hundreds of thousands of refrigerators became hazardous, requiring householders to tape them shut, move them to the curb, and await vendor removal to appropriate facilities.

Because the removal and safe disposal of debris is costly, local government needs to include debris management in its planning process. Fortunately, in areas where disasters can be anticipated, debris planning can take place pre-disaster.[50] Identifying appropriate contractors or establishing a relationship with them in the pre-disaster planning process will make debris removal and disposal more efficient during the recovery phase. In most cases, debris removal may last from a couple of weeks to a few months.[51]

In some areas, debris removal and disposal could put great strains on existing landfills.[52] The U.S. Army Corps of Engineers can assist local municipalities with debris removal; a number of experienced vendors can be contracted with; and FEMA may provide guidance. After Hurricane Andrew, local government permitted "Mount Trashmore" to be built, creating an artificial hillside in southern Florida. Other communities have accessed landfills in nearby jurisdictions. FEMA and EPA have urged that great care be taken to dispose of debris properly, including separating woody materials from hazardous materials because woody debris that has not been polluted with toxic chemicals can be mulched.

Debris removal also has a human side. When cities are quick to cart away debris so that the recovery process can occur more rapidly, victims and their families can be adversely affected if their cherished possessions are part of what is carted away. After the Loma Prieta earthquake in 1989, workers helped victims salvage items during debris removal. The situation is even more complex when the disaster results from a terrorist attack: debris may contain evidence as well as human remains, and the integrity of both must be preserved.

Donations management

Generally in the wake of a disaster, well-meaning individuals, groups, and organizations will cause large numbers of volunteers and physical items (used clothing, shoes, toys, furniture, water, and food) to converge on the affected area.[53] Although the volunteers can be put to work (see below), the donations may not always be appropriate—and may simply create problems. After Hurricane Andrew, for example, well-meaning donors dropped off clothing all over the area, which required troops to bulldoze the piles of fabric rotting in the humid climate.

Immediately after a disaster, local officials should use the media to promulgate two related messages about donations: (1) physical donations should not be sent because of the problems they cause and (2) cash has a much greater practical value than physical items. Cash is much easier to "store" than physical items, and officials can spend it to buy urgently needed items from businesses close to the disaster area, thus stimulating the local economy, regenerating the tax base, and contributing to the area's economic recovery.[54]

Volunteers can help sort and distribute the donations. Donated clothes that cannot be used can be sold or donated to companies that produce rags, with the proceeds going to the recovery effort. Extra food can be donated to the local food bank. In large-scale disasters, groups such as the Seventh Day Adventists can be relied on to operate massive warehouses. However, finding temporary warehouses may be difficult. Arkadelphia, Arkansas, used the county fairgrounds building as a distribution center. A high school in Oklahoma opened "Bridge Creek SuperCenter" in the school gym. Lancaster, Texas, used a closed grocery store building and brought victims in to "shop" for necessities. After the 1999 tornadoes, Oklahoma City accepted a corporate offer of a massive (empty) food warehouse, and the warehouse filled up with donations within forty-eight hours.

Most volunteers converge within the first weeks after the disaster; however, most of the help is needed during rebuilding, so care must be taken to ensure that volunteers are managed to their most effective levels. One strategy for doing this may be to develop a plan to bring those volunteers back during the recovery period. Working with voluntary organizations (discussed further on under "Community Resources") is one way to ensure that work crews will return at the appropriate time with needed skills. Local government may facilitate that effort by identifying locations for volunteer tent cities or other types of accommodation, such as a church camp, a local dormitory, or even a formerly closed nursing home.

Financial and community resources for recovery

Although the scope and magnitude of a disaster may make recovery seem overwhelming, local government should be aware that help for affected communities is available from a number of sources. Such help ranges from local through federal assistance and comes from government, voluntary organizations, and the private sector. To access the range of help available and put it to productive use, local officials have to think imaginatively. Drawing on the federal government should be only the start—not the beginning and end—of mobilizing resources for recovery.

Local officials should also be aware that the kinds of resources they should secure to help support recovery efforts range widely. Depending on preexisting resources and preevent planning, the local government may need to secure and manage financing (grants, loans, donations); personnel (both individuals and organizations), including contractual vendors and consultants; and state and federal assistance. The grants and loans may be offered by independent foundations, agencies, organizations, and banks. The state programs may be specific to housing, business recovery, or public infrastructure. (Government managers should work with state officials to determine which state laws are applicable in the particular context, including the laws newly enacted in the aftermath of the disaster.) The primary piece of federal legislation influencing recovery efforts is the Stafford Act, which tends to be amended after major disasters and which provides parameters for what can and cannot be funded. (FEMA bases individual and public assistance on this legislation when it designs programs to assist local government. For information on FEMA's three types of assistance, see "Financial Resources" below.)

Knowing where to find the needed resources—both financial and community—and how to secure and manage them can expedite recovery, just as not knowing can generate heavy burdens, especially for future generations. Local government managers must stay on top of possible funding opportunities by keeping connected to local, state, and federal representatives and by continually searching for funding.

Financial resources

In the event of a Presidential Disaster Declaration, financing may come from the federal government. As previously noted, FEMA permits federal funding for recovery efforts for up to eighteen months, although individual household funding may last no longer than a few months; in extreme cases, funding can be extended beyond the eighteen months. Or local, regional, and state foundations may be tapped. Disaster donations made to cities or organizations could be set aside as well. And the local government may use tax policy to help. In short, a wide range of funds can be secured, sometimes through creative partnerships and ideas.[55] A few of the key resources include

- *FEMA Public Assistance:* This program provides monies to repair, rehabilitate, or replace damaged public facilities, such as a wastewater treatment plant, and limited facilities of private nonprofit organizations that must be open to the public. (FEMA can also provide funding and assistance for debris removal or protective measures to safeguard the public.) FEMA can provide up to 75 percent of the costs, with state and local governments covering the remainder. This program has forms and accounting requirements, and care must be taken to follow FEMA procedures exactly. The best advice is to keep careful records—including receipts—of every single expenditure, using a local government accountant or similar person, so that the 75 percent/25 percent match can be verified. Local government managers are also advised to work closely with their assigned state liaison officer to the FEMA public assistance coordinator and to attend official kick-off meetings. The FEMA project officer can help the applicant design a scope of work and draw up an appropriate budget. A FEMA specialist may be able to provide specific insights, such as those related to infrastructure issues. FEMA's public assistance officer coordinates communication between the state and the federal staff to ensure compliance. FEMA Independent Study Course 630 is recommended.[56]
- *FEMA Individual Assistance:* This program for households may include temporary housing (rental assistance, trailers), funds for repair and replacement of damaged items, and grants or loans for permanent housing. FEMA can provide only funds that return the home to functional status and to a condition equivalent to its former level, rather than upgrading the home. Individuals must first apply for an SBA loan and, if rejected, will be certified for a FEMA grant. Maximum dollar amounts are set annually and cannot be exceeded. Also

included are funds for medical, dental, mental health crisis counseling, funeral costs, and transportation assistance (for health care, grocery shopping, and so forth—e.g., from a trailer park).

- *FEMA Mitigation Funding:* When a Presidential Disaster Declaration is issued, FEMA's Hazard Grant Mitigation Program can supply funds for projects that reduce future risks. Projects may include funding retrofits, buying disaster-prone properties to create open space, funding elevations, supporting flood control projects, and facilitating efforts to review and revise building codes (see Chapter 6 on mitigation).
- *U.S. Department of Housing and Urban Development:* HUD may allow some funds to be used toward special rebuilding programs, or it may reassign Section 8 certificates.
- *Small Business Administration:* The SBA can make property loans available for homeowners and businesses.
- *U.S. Department of Labor:* The Labor Department can ensure that those who have lost their jobs because of the disaster receive unemployment insurance benefits.
- *Community Development Block Grants (CDBGs):* Disaster recovery managers should work with local government agencies responsible for CDBGs to secure funding.
- *National Park Service:* The National Park Service may provide funds for open space and trails in floodplains.
- *U.S. Department of Agriculture:* The Department of Agriculture offers assistance with harvesting, farmland rehabilitation, purchase of floodplains, mitigation of drought conditions, and farm loans.[57]
- *National Flood Insurance Program (NFIP):* The NFIP identifies, maps, and insures flood-prone areas for communities that adopt floodplain management ordinances.

Other loan programs that involve public or private assistance may develop if state or other agencies and foundations choose to fund something special as a result of the disaster. For example, following the four hurricanes that struck Florida in 2004, the state of Florida and banks in the state instituted a short-term loan program for small businesses. Loans of up to $25,000 became available to provide financial assistance for businesses until federal loans were approved and available.

Local government may help with disaster recovery by using varying types of taxing methods (e.g., special taxing or assessment districts, tax increment financing, impact fees, differential taxation). Contra Costa County, California, and Lee County, Florida, have used special taxing and assessment districts to assist with their hazard mitigation and disaster recovery efforts.[58] California approved a quarter-cent sales tax to be dedicated to recovery and reconstruction efforts after Loma Prieta. Monies from the sales tax were used for matching federal grants, providing state agencies with grants toward earthquake recovery, and retrofitting some state buildings.[59]

A municipality's best friend in a disaster is the accounting staff.

Media partners can be asked to donate time or space for public service announcements. And communities themselves are rich in resources if government takes the time to build and use local capacities. For example, many businesses, malls, colleges, places of worship, and union halls may be able to provide meeting space, kitchens, and even refreshments (given that feeding people at meetings is critical).[60] Additional guidance on the financial aspects of recovery is provided throughout this chapter.

Without a doubt, a municipality's best friend in a disaster is the accounting staff. By documenting expenses, the municipality will facilitate reimbursement from the federal government. In addition, while many agencies, organizations, and foundations offer financial and technical assistance with disaster recovery, securing those funds requires the skills of proposal writers; accounting staff is probably best at helping with the budget part of the proposals. Once projects are funded, the accounting staff will need to manage and document the funds appropriately.

Community resources

A number of community resources are either already in place before the disaster or formed after it. These include other communities with similar experiences, voluntary organizations, pre-disaster community groups, new post-disaster groups and organizations, and consultants and vendors.

Representatives and officials from experienced communities can provide advice on such issues as obtaining federal, state, and other types of aid and on issues related to housing, economic development, the environment, and infrastructure reconstruction.

Volunteer organizations that can be a major resource for communities in the recovery period range from national organizations with local chapters (see sidebar on page 230) to local

Peer help after Katrina

In August 2006, Pascagoula city manager Kay Kell was working with Fannie Mae and other groups to try to jump-start the redevelopment of housing a full year after Katrina, and she kept coming back to one inescapable truth. "The more we talked, it came out that what we needed were people," she said.

Fannie Mae had been working with ICMA on various efforts to support local governments in neighborhood revitalization, and responding to Pascagoula's recovery needs emerged as an excellent opportunity for a partnership. The two organizations launched a pilot peer professional/loaned-executive program in Pascagoula that was designed to recruit and direct local government professionals to assist their peers in cities recovering from a disaster. The idea was to supplement staff resources with professionals from unaffected areas, who could help address many of the needs that follow the initial response to a disaster during the long-term recovery process.

Nine cities and Bay County ultimately sent local government professionals from Florida to Pascagoula. The key to their effectiveness was their spirit of teamwork, said Carol Westmoreland, of the Florida League of Cities. "It was a real people-to-people, very practical effort. They didn't build a lot into the system that wasn't essential."

In addition to helping the city with building inspections, Kell said that the Florida staff identified $400,000 to $600,000 in additional federal and state reimbursements due to the city. They also helped develop and publicize a citizens' needs survey and helped Pascagoula put the finishing touches on a presentation about redevelopment opportunities for an Urban Land Institute conference in the fall of 2006 that attracted developers from across the country.

The most significant contribution came in the realm of inspections, permitting, and code enforcement. Steve Mitchell, who is Pascagoula's building official and in 2006 was named director of planning and zoning, said his crew showed up within two days of the storm and immediately fanned out to start assessing damage. Once they had finished the preliminary inspections, they had to do more in-depth assessments in heavily flooded areas to determine which structures would have to be raised, moved, or demolished. The more detailed damage assessments were required by the National Flood Insurance Program.

In the end, about 1,200 structures needed more careful damage assessments before any permits could be issued for repairs. At the peak, the city was issuing some 130 to 140 permits a day, Mitchell said, compared with a pre-Katrina norm of about 8 to 10 a day. In the code enforcement area, his team also had to inspect about 460 swimming pools that, because they had been left untended for months, had become breeding grounds for mosquitoes.

"We knew that if we could make it over the hump, we would be in relatively good shape and could even things out. But we just could not get over the hump," he said.

Everything changed when code enforcement staff and building inspectors started arriving from Florida. "They knew what they were looking for. All I had to do was point them in the right direction," Mitchell said. "When you can bring somebody in here who knows what they're doing and you don't have to hold their hand, it makes all the difference in the world."

Although Pascagoula had the extra help in the building department for only a couple of weeks, that help enabled Mitchell and his team to get back to a manageable level of activity. By the end of 2006, they were down to about 35 to 50 permits a day, mostly for new construction. "We're busy, no doubt about it, but it's a routine that we can live with," he said.

Source: Adapted from Christine Shenot, "Lessons Learned in Mississippi," *Public Management (PM)* (April 2007): 9.

National organizations with local chapters

Among the national organizations that have local chapters are the following:

- **American Red Cross:** Mandated by Congress; provides mass care (food, shelter) and assists with clothing, household furniture, rental assistance, mental health outreach, first aid, and more at no cost; see redcross.org
- **Salvation Army:** Provides food, clothing, direct aid, home repairs, funds, volunteers, mobile feeding vehicles; see salvationarmy.org
- **The National Voluntary Organizations Active in Disaster (NVOAD):** Coordinates planning efforts by many voluntary organizations responding to disaster; see nvoad.org.

Key experienced organizations and their specialties:

- **Seventh Day Adventists:** Manages donations
- **Mennonite Disaster Service:** Rebuilds low-income homes
- **Church of the Brethren:** Provides child care
- **Southern Baptists:** Provides mobile kitchens, showers, and labor
- **Habitat for Humanity:** Rebuilds low-income homes.

faith-based organizations. For volunteer organizations to provide the most effective and efficient recovery resources for victims, the groups' leaders must coordinate closely with each other and with local officials. If the local government has a plan, coordination is much easier to do.

Local organizations represent important resources with local knowledge and links to affected populations. Ideally, a community's pre-impact planning effort will include networking to link these organizations to each other and to local government. Strategies for working with local organizations might include serving on each other's boards, conducting disaster drills, cohosting information fairs during tornado awareness month, and cross-training. Local churches, synagogues, mosques, and temples often manage donations, provide shelter and food, link city officials to at-risk populations, and facilitate communications. It is not unusual for various interfaith coalitions to emerge and address unmet needs. "Unmet needs" committees are typically established to address households that fall through the cracks of available aid programs (and FEMA may provide a liaison to assist in forming an unmet needs committee). These ad hoc entities have become increasingly common and often feature, as key players, many members of NVOAD (for information on local VOADs, see just below). Other local resources may include

- Chambers of commerce and civic groups
- Social service agencies that link government to the elderly, the poor, and persons with disabilities—for example, the Area Agency on Aging, Children's Services, schools for persons who are blind or deaf, food kitchens, community centers, and senior centers
- Universities, colleges, and schools, which can provide a cadre of expert personnel.

One type of pre-disaster community group that volunteer organizations in some communities have formed is a local counterpart of NVOAD. VOADs help coordinate any assistance that volunteer organizations may offer during all phases of a disaster, including short- and long-term recovery efforts. VOADs bring in voluntary organizations to rebuild, and they focus on people who fall through the cracks of assistance programs—people such as the underinsured, who would not be able to rebuild even with the maximum FEMA individual grant. VOADS may also help to make sure that activities by volunteer organizations are not duplicated (either among themselves or by government) and are not neglected. If a community does not have a VOAD, community leaders can promote the formation of one. (See the sidebar on national associations with local chapters for contact information on the national organization.)

As noted above, most communities still do not have formal recovery organizations or plans in place before a disaster occurs. As a result, new ad hoc groups and organizations may form to assist with the recovery process. Some of these will be neighborhood groups seeking to promote their issues by working with government and volunteer organizations. Govern-

Women will rebuild

When Hurricane Andrew devastated South Florida in 1992, a group of powerful leaders formed "We Will Rebuild" in order to manage relief funds, designate projects for recovery, and support the area over the long term. But We Will Rebuild quickly came under fire for appointing "insiders" and excluding the diverse community that characterizes Miami–in particular, African Americans and non-Cuban Hispanics. And although the hurricane had damaged most of the public and subsidized housing and therefore dramatically affected low-income women and children, funds intended for child care were diverted to the chamber of commerce. Criticism of We Will Rebuild centered on its priorities, including setting a director's salary at $150,000 and focusing on economic recovery over and above housing issues.

As research predicts will happen in such situations, the contentiousness generated a counter-organization. This one was named "Women Will Rebuild." Its members expressed concern over being shut out of the "good ole boy network...running things when they had no real idea of what the problems were, especially the problems of women" (p. 190). Fifty women's groups joined Women Will Rebuild, which made a special effort to diversify its base culturally. The organization's accomplishments included advocating for women and children, increasing diversity on the board of We Will Rebuild, creating committees on families and children as well as on domestic violence, and funding social services and affordable housing.

Source: Summarized from Elaine Enarson and Betty Hearn Morrow, eds., *The Gendered Terrain of Disaster: Through Women's Eyes* (Miami, Fla.: International Hurricane Research Center/Praeger Publishers, 1998), 185–200. Copyright © 1998 by Praeger Publishers. Reproduced with permission of Greenwood Publishing Group, Inc., Westport, Conn.

ment representatives would be wise to accept and work with these groups, for they may facilitate financial assistance, speed up housing reconstruction, raise concerns about and provide solutions to equity issues, and serve as important links to stakeholders and constituents.[61]

Consultants and contractual vendors can be another source of expertise to help communities, local officials, and neighborhood victims through the recovery process. The consultant, however, should be contracted with before the disaster so that he or she can assist with the recovery planning process and can become familiar with the community and its hazards. Officials should carefully and closely investigate the potential consultant's expertise, credentials, experience, capabilities, and reputation in similar municipalities. One source of consultants may be local or regional universities, whose faculty members may have expertise in a wide range of issues related to disaster recovery.

Government may decide to contract through competitive or, when appropriate, no-bid contractors. Care should be taken that such vendors possess appropriate credentials and expertise in disaster-specific work environments and can deliver as promised. State attorney general offices may wish to assist local governments in certifying vendors and monitoring them for unscrupulous practices.

Summary

Recovery is a process that involves local, state, and federal government in partnership with one another and with citizen stakeholders, businesses, voluntary organizations, utilities, transportation departments, and more. To rebuild and recover from disaster, each jurisdiction will need to access varying kinds of resources from this group of partners.

Ideally, before a disaster, local government managers will develop both short- and long-term recovery plans. In reality, though, many jurisdictions lack the personnel to do so. Consequently, most communities attempt to recover without a concrete plan or without key partnerships in place. In the recovery process, local government plays a key role by launching a visioning process for a rebuilt, safer community and by involving the public in that process. It also serves as a channel through which the rebuilding process flows, as local officials pass new ordinances that affect rebuilding, review and revise building codes to make the community safer, and facilitate residents' passage through the permitting process.

The type of disaster that has affected a specific geographic location will influence the dimensions of disaster that must be addressed. Communities may face a massive tornado that destroyed the downtown business sector or a flood that inundated residential sectors. Regardless, the challenges faced by many communities during the recovery period are similar. Local government managers can enhance the future of their communities by working with appropriate state and federal agencies to remove debris appropriately and to secure and use monetary and in-kind donations effectively. They can facilitate the rebuilding of damaged residences by working with homeowners, voluntary organizations, builders, planners, and inspectors. To restore economic vibrancy, managers may need to work closely with the local business community, from assisting with relocation to temporary facilities to engaging it in a new vision for revitalized commercial and industrial sectors. Infrastructure and lifelines may need repair or complete replacement. The local government's role may range from funding projects, such as new roads and bridges, to working with utility companies to restore services appropriately and expeditiously.

Recovery represents an opportunity to rebuild in a way that reduces the impact of future disasters. Incorporating mitigation into recovery—for example, by relocating homes and businesses outside of the floodplain—could reduce requests for future federal aid as well as decrease the number of damaged properties and lives lost. Linking recovery planning to existing comprehensive and land use planning provides a chance to advance the community's efforts to expand green space (often formerly floodplains), implement new ordinances (e.g., for seismic retrofit), or strengthen the relationships between government and citizens participating in the planning. Under a Presidential Disaster Declaration, funds can be secured for mitigation projects as well as Individual Assistance and Public Assistance projects—and an important function of the local government manager is knowing how these programs work and what accounting procedures are necessary for funding and/or reimbursement.

Recovering from disaster requires various commitments from local government, ranging from financing recovery efforts to remaining patient with frustrated residents. Managers may need to reconfigure staffing positions to participate in planning and recovery activities, to expedite the permitting process, to secure grants and monitor funded projects, and to generate innovative projects. Recovery is a social process that, ideally, engages a full range of partners to rebuild a safer community for all.

Notes

1 E. L. Quarantelli, "Disaster Assistance and Socioeconomic Recovery at the Individual and Household Level: Some Observations," Preliminary Paper 171 (Newark: Disaster Research Center, University of Delaware, 1991), available at dspace.udel.edu:8080/dspace/handle/19716/543 (accessed August 21, 2007).

2 E. L. Quarantelli, "Disaster Recovery: Research Based Observations on What It Means, Success and Failure, Those Assisted and Those Assisting," Preliminary Paper 263 (Newark: Disaster Research Center, University of Delaware, 1998), available at dspace.udel.edu:8080/dspace/bitstream/19716/282/1/PP%20263.pdf (accessed August 21, 2007).

3 Natural Hazards Research and Applications Information Center (hereafter, Natural Hazards Center), *Holistic Disaster Recovery: Ideas for Building Local Sustainability after a Natural Disaster* (Boulder, Colo.: Institute of Behavioral Science, University of Colorado, 2001), 2-4, available at colorado.edu/hazards/publications/holistic/holistic2001.html (accessed August 10, 2007).

4 David M. Neal, "Reconsidering the Phases of Disaster," *International Journal of Mass Emergencies and Disasters* 15, no. 2 (1997): 239–264, available at colorado.edu/hazards/research/qr/qr79.html (accessed August 21, 2007).

5 David R. Godschalk et al., *Natural Hazard Mitigation: Recasting Disaster Policy and Planning* (Covelo, Calif.: Island Press, 1999), 13.

6 James Schwab et al., *Planning for Post-Disaster Recovery and Reconstruction* (Chicago: APA Planning Advisory Service, 1998), 75–111.

7 U.S. Department of Homeland Security (DHS), *National Response Plan* (Washington, D.C.: DHS, 2004), 14-1, available at dhs.gov/xlibrary/assets/NRP_FullText.pdf (accessed August 21, 2007).

8 Ibid., 14-2.

9 *Recovery and Reconstruction Plan* (Los Angeles: Emergency Operations Organization, 1994), 1–20. The full plan can be viewed at lacity.org/epd/pdf/mpa/r&r%20annex%20plan.pdf (accessed August 21, 2007).

10 James Schwab, "A Planner's Tool Kit," in James Schwab et al., *Planning for Post-Disaster Recovery and Reconstruction* (Chicago: APA Planning Advisory Service, 1998), 113–148.

11 FEMA may provide funding for visioning, typically by hiring a consulting firm for the damaged community.

12 Schwab et al., *Planning for Post-Disaster Recovery.*

13 Ibid., 85.

14 Ray Burby, "Natural Hazards and Land Use: An Introduction," in *Cooperating with Nature: Confronting Natural Hazards with Land-Use Planning for Sustainable Communities,* ed. Ray Burby (Washington, D.C.: Joseph Henry Press, 1998), 1–26.

15 Schwab et al., *Planning for Post-Disaster Recovery,* 127.

16 Natural Hazards Center, *Holistic Disaster Recovery.*

17 Jeanine Peterson, "A Review of the Literature and Programs on Local Recovery from Disaster," Natural Hazards Research Working Paper 102 (Boulder: Natural Hazards Center, University of Colorado, 1999), available at colorado.edu/hazards/publications/wp/wppubs.html#2000 (accessed August 21, 2007).

18 Natural Hazards Center, *Holistic Disaster Recovery,* 3-2 to 3-3.

19 David Chrislip, *The Collaborative Leadership Fieldbook: A Guide for Citizens and Civic Leaders* (San Francisco: Jossey Bass, 2002). See lists of potential planning

participants in Schwab et al., *Planning for Post-Disaster Recovery.*

20 Chrislip, *The Collaborative Leadership Fieldbook.*

21 Ibid.; see also Michael K. Lindell and Ronald W. Perry, *Communicating Environmental Risk in Multiethnic Communities* (Thousand Oaks, Calif.: Sage, 2004).

22 Kenneth C. Topping, "The Text of a Model Recovery and Reconstruction Ordinance," in Schwab et al., *Planning for Post-Disaster Recovery,* 149–167.

23 Natural Hazards Center, *Holistic Disaster Recovery.*

24 Quarantelli, "Disaster Recovery: Research Based Observations," and Richard C. Wilson, *The Loma Prieta Quake: What One City Learned* (Washington, D.C.: ICMA, 1991). See also Brenda D. Phillips, "Homelessness and the Social Construction of Places: The Loma Prieta Earthquake," *Humanity and Society* 19, no. 4 (1996): 94–101.

25 Quarantelli, "Disaster Recovery: Research Based Observations."

26 Mary Comerio, *Disaster Hits Home: New Policy for Housing Recovery* (Berkeley: University of California Press, 1998), 15.

27 E. L. Quarantelli, "Sheltering and Housing after Major Community Disasters—Case Studies and General Observations," Final Project Report 29 (Newark: Disaster Research Center, University of Delaware, 1982); see also E. L. Quarantelli, "General and Particular Observations on Sheltering and Housing after American Disasters," *Disasters* 6 (1982): 277–281.

28 Brenda D. Phillips, "Cultural Diversity in Disaster," *International Journal of Mass Emergencies* 11, no. 1 (1993): 99–110.

29 Robert A. Stallings, *The Northridge Earthquake "Ghost Towns,"* Final Report to the National Science Foundation, NSF Grant No. CMS-9416196 (Los Angeles: Program in Public Policy, School of Public Administration and Department of Sociology, University of Southern California, 1996).

30 Stallings, *The Northridge Earthquake "Ghost Towns,"* 40.

31 Comerio, *Disaster Hits Home,* 121.

32 Stallings, *The Northridge Earthquake "Ghost Towns."*

33 National (now Multidisciplinary) Center for Earthquake Engineering Research, *Findings and Recommendations: Symposium on Policy Issues in the Provision of Post-Earthquake Shelter and Housing* (Buffalo, N.Y.: National Center for Earthquake Engineering Research and the Bay Area Regional Earthquake Preparedness Project, 1992).

34 Gary R. Webb, Kathleen J. Tierney, and James Dahlhamer, "Predicting Long-Term Business Recovery from Disaster: A Comparison of the Loma Prieta Earthquake and Hurricane Andrew," *Environmental Hazards* 4, nos. 2–3 (2002): 45–58.

35 Natural Hazards Center, *Holistic Disaster Recovery,* 5-2 to 5-3; see also Stephanie Chang, "Reconstruction and Recovery in Urban Earthquake Disasters," in *Proceedings of the Fifth United States/Japan Workshop on Urban Earthquake Hazard Reduction* (Oakland, Calif.: Earthquake Engineering Research Institute, 1997).

36 Webb, Tierney, and Dahlhamer, "Predicting Long-Term Business Recovery from Disaster."

37 Stephanie Chang and Anthony Falit-Baiamonte, "Disaster Vulnerability of Businesses in the 2001 Nisqually Earthquake," *Environmental Hazards* 4, nos. 2–3 (2002): 59–71.

38 John Laye, *Avoiding Disaster: How to Keep Your Business Going When Catastrophe Strikes* (Hoboken, N.J.: John Wiley & Sons, 2002).

39 Natural Hazards Center, *Holistic Disaster Recovery,* chap. 5.

40 Laye, *Avoiding Disaster,* passim.

41 Ray Burby, ed., *Cooperating with Nature: Confronting Natural Hazards with Land-Use Planning for Sustainable Communities* (Washington, D.C.: Joseph Henry Press, 1998).

42 Natural Hazards Center, *Holistic Disaster Recovery.*

43 Ibid.

44 Ibid., 7-17.

45 Kathleen J. Tierney and Joanne K. Nigg, "Business Vulnerability to Disaster-Related Lifelines Disruption," in *Lifeline Earthquake Engineering: Proceedings of the Fourth U.S. Conference,* ed. Michael J. O'Rourke (New York: American Society of Civil Engineers, 1995), 72–79.

46 Federal Emergency Management Agency (FEMA), *Plan for Adopting and Developing Seismic Design Guidelines and Standards for Lifelines* (Washington, D.C.: FEMA, 1995), 16–17, available at fema.gov/plan/prevent/earthquake/pdf/fema-271-part-1.pdf (accessed August 21, 2007).

47 Ibid., 12–13.

48 Ibid., 13–14.

49 Ibid., 19–21.

50 Robert C. Swan, "Debris Management Planning for the 21st Century," *Natural Hazards Review* 1, no. 4 (November 2000): 222–225.

51 Gabriela Y. Solis et al., "Disaster Debris Management" (Vancouver: Disaster Preparedness Resources Centre, University of British Columbia, 1996).

52 Ibid.

53 David M. Neal, "The Consequences of Excessive Unrequested Donations: The Case of Hurricane Andrew," *Disaster Management* 6, no. 1 (1994): 23–28.

54 James M. Kendra and Tricia Wachtendorf, "Rebel Food . . . Renegade Supplies: Convergence after the World Trade Center Attack," Preliminary Paper 316 (Newark: Disaster Research Center, University of Delaware, 2002), available at udel.edu/DRC/pp316.pdf (accessed August 21, 2007); see also Neal, "The Consequences of Excessive Unrequested Donations."

55 A full list can be found at FEMA, "Response and Recovery," at au.af.mil/au/awc/awcgate/frp/frprecvm.htm (accessed August 21, 2007).

56 Information on FEMA's Independent Study Program can be found at training.fema.gov/IS/crslist.asp.

57 Schwab et al., *Planning for Post-Disaster Recovery,* 331; for a list of programs and resources, see Appendix C.

58 Schwab et al., *Planning for Post-Disaster Recovery.*

59 Wilson, *The Loma Prieta Quake,* 52–53.

60 Natural Hazards Center, *Holistic Disaster Recovery,* 3–8.

61 E. L. Quarantelli, "Emergency Citizen Groups in Disaster Preparedness and Response Activities" (Newark: Disaster Research Center, University of Delaware, 1985).

PART THREE

Major Issues in Emergency Management

CHAPTER 12

Legal issues

William C. Nicholson, Esq.

This chapter provides an understanding of

- Laws that are specific to the emergency management function
- The role of tort law in emergency management
- Governmental immunity from liability
- Evolving standards that have legal implications for emergency management
- Legal issues in preparedness, response, recovery, and mitigation.

The emergency manager is charged with the responsibility of protecting the community from disaster. He or she does so within the context of emergency management law—a set of legal responsibilities that derive from federal and state statutes, local ordinances and policies, and case law.[1] Some of these responsibilities are unique to emergency management; others apply to all local government functions; and still others pertain to all emergency response personnel. The local emergency management agency must therefore comply with a sometimes bewildering variety of legal enactments.

Protecting the local government from liability is no less important than protecting the community from hazards.[2] To avoid liability, emergency managers and their legal advisors should work together closely throughout all phases of emergency management. Although not every bad choice can be prevented, the close involvement of legal counsel—an approach known as "litigation mitigation"—can result in a much better set of options.[3] In the mitigation phase, legal advisors may assist with revisions of fire and building codes. During preparedness, attorneys help draft plans and evaluate plan revisions; assess training standards; and analyze exercises for potential legal issues. During response, attorneys advise the emergency manager and local elected officials on the legal implications of various alternatives. During recovery, attorneys can help to ensure that expenses are properly documented and that the shift to mitigation is properly performed.

This chapter covers the legal context within which emergency management operates, including federal, state, and local laws that apply specifically to emergency management and those that have broader application. The chapter also addresses tort law, the principal source of legal liability in the emergency management context, and immunities, which are the various shields that may protect units of government and their employees. Evolving standards that have legal implications for emergency management are described, as are the legal issues associated with each stage of emergency management: preparedness, response, recovery, and mitigation.

Emergency management–specific law

The daily activities associated with emergency management are affected by federal, state, and local laws specific to the emergency management function; these include, for example, legal obligations to plan, train, and exercise.

Federal law

The most important federal laws affecting emergency management are the Robert T. Stafford Disaster Relief and Emergency Assistance Act (Stafford Act)[4] and the Homeland Security Act of 2002 (HSA).[5] Taken together, these two acts provide the broad federal context within which emergency management operates. The Stafford Act established the programs and processes through which the federal government provides disaster assistance to states, local governments, tribal nations, individuals, and nonprofit organizations. In addition to providing disaster relief, it provides funding for warning and preparedness programs and is the source of various grants, including Emergency Management Performance Grants (EMPGs).

Under the Stafford Act, if a disaster exceeds the combined response capabilities of both state and local governments, and if a federal-state-local preliminary damage assessment (PDA) finds the damage to be sufficiently severe and widespread, the president may declare a major disaster or emergency. (Generally, an emergency is considered to be of lesser scope than a disaster.) If an event is widespread or local authorities are not available, the state may act on its own without waiting for a request from the local government; however, local officials may not appeal directly to the federal government for disaster assistance.[6]

Disclaimer: This chapter does not constitute legal advice. The issues discussed herein are largely matters of state and local law, which vary considerably among jurisdictions. Emergency managers and other local officials who need legal assistance must consult competent local legal counsel.

The author wishes to acknowledge the generous assistance of Michael S. Herman, Esq., senior counsel for the Subcommittee on Economic Development, Public Buildings and Emergency Management of the Committee on Transportation and Infrastructure in the U.S. House of Representatives, for his insightful reading and comments on the text. Any errors that may remain are the author's responsibility.

The decision to issue a Presidential Disaster Declaration is based on two factors: the exhaustion of state and local resources and the extent of damage. There are no fixed requirements for damage per capita, although there have been efforts to institute them. If a disaster develops exceptionally quickly or is unmistakably overwhelming, the PDA may be postponed until after the declaration. In addition, under the Catastrophic Incident Annex of the National Response Plan (NRP), the secretary of the U.S. Department of Homeland Security (DHS) may unilaterally declare a disaster and immediately begin positioning federal response assets: this is the most rapid method available to the federal government for responding to disaster.

The HSA, which President George W. Bush signed into law on November 25, 2002, established DHS, whose principal mission is to combat terrorism. Thus, one of the key challenges posed by the law was to integrate the work of DHS with the efforts of local, state, and emergency management agencies—that is, to create a national system for all-hazards emergency preparedness and response. To address this issue, among others, President Bush issued Homeland Security Presidential Directive 5 (HSPD 5),[7] which instructed all federal agencies to collaborate with DHS to establish the NRP and the National Incident Management System (NIMS), which is the operational arm of the NRP.

The purpose of HSPD 5 is to ensure that all levels of government use a single, comprehensive approach to domestic incident management. HSPD 5 requires DHS to create and enforce standards that apply to federal, state, and local emergency response agencies. Moreover, beginning in fiscal year 2005, state and local agencies that fail to comply with these standards will lose eligibility for federal grants and other funding. These requirements affect fire protection, law enforcement, and emergency medical service agencies, as well as emergency management agencies, all of which receive significant funding from the federal government. NIMS directs the establishment of "guidelines, protocols, and standards for planning, training, and exercises, personnel qualification and certification, equipment certification, and publication management."[8] National preparedness benchmarks will be "maintained and managed" through a NIMS Integration Center (NIC). The secretary of DHS has authority to revise NIMS standards.

NIMS, the NRP, and a series of HSPDs further delineate how the federal government prepares for and responds to terrorism and other events. If an event is terrorism related or occurs on federal property, the federal government has primary jurisdiction and control. For other types of events, control remains with the authorities who would normally be in charge—generally, the governing body of the local government in question—although local officials may choose to delegate part of their authority for day-to-day operations to others, such as the local emergency manager.

State law

The state constitution provides the bedrock on which the legal system is grounded: all state laws must comply with the state, as well as the federal, constitution. State statutes, and sometimes rules promulgated under statutory authority, are the most important sources of emergency management law. Such enactments, often referred to as the state's emergency management code, provide the immediate context within which local emergency management operates.

The state emergency management code typically describes the powers of the governor, the duties of the state emergency management agency,[9] and the responsibilities of local emergency managers. In Indiana, for example, the "definitions" section of the state code sets forth the broad scope of emergency management responsibilities: "'Emergency management' means the preparation for and the coordination of all emergency functions, other than functions for which military forces or other federal agencies are primarily responsible, to prevent, minimize, and repair injury and damage resulting from disasters."[10] The code also contains a list of itemized duties, including (1) preparing and keeping current the emergency plan, (2) preparing and distributing an inventory of emergency duties for all officials, and (3) detailing the chain of command for continuity of government purposes.[11] Thus, failure to keep a "current plan," for example, may expose the emergency manager and the jurisdiction to liability.[12]

Local ordinances

Every local jurisdiction should have in place a local ordinance that fills in the gaps in state emergency management law.[13] Of course, the emergency manager must understand and comply with all local ordinances, particularly those directed specifically at the emergency management function.

Tort law

While violation of federal, state, or local law may result in liability, the major basis for municipal liability is state law—specifically, the category of tort known as negligence. Fortunately, local elected officials and local government staff who act according to established standards are generally protected from liability.

A tort is a civil (as opposed to criminal) wrong. Examples of torts include medical malpractice, auto accidents, and defamation. Torts may also damage property interests, as when a hazardous materials spill renders a neighboring property uninhabitable. The purpose of tort law is to allow those who are wronged by others to seek compensation for their injuries. Damages for a civil wrong are measured in financial terms: a successful plaintiff receives monetary compensation. Corporations, which the law views as "artificial people," may sue and be sued for tort damages. Similarly, units of government may be sued in tort.

For a court, the intention of the actor—that is, the mental state of the person who has been accused—is a key factor in determining liability in tort. There are three types of intent in tort: negligence, gross negligence, and intentional wrongdoing. *Negligence* is a common-law principle that has developed over many years.[14] The principle assumes that everyone has a general obligation to behave in a reasonable manner toward others, given the particular circumstances. If someone acts unreasonably (which may include failing to act at all), and that act or failure to act is the legal cause of injury to a person or property, liability may result.

The purpose of tort law is to allow those who are wronged by others to seek compensation for their injuries.

In the emergency management context, negligence usually results from the failure to adequately perform particular governmental duties. For example, a local government that fails to properly train or supervise emergency management workers may be held liable for injuries that result from that failure. Other sources of liability for negligence include failure to create or adhere to a plan,[15] and failure to comply with legal duties, such as those that are embedded in workplace safety laws.[16] It is important to note that a violation of law may be used as proof in a civil suit: if the elements of the violation are the same as the elements required for civil liability, and the burden of proof is the same for both, the only issue in a civil trial may be the amount of damages.[17]

Gross negligence is one step beyond negligence: it reflects a higher level of intent and involves actions that are extremely unreasonable under the circumstances. In some cases, gross negligence may be required for a finding of liability. For example, under the Volunteer Protection Act,[18] discussed later in this chapter, negligent acts are shielded from liability, but grossly negligent acts may not be. In some states, gross negligence is referred to as "recklessness."

Elements of negligence

When all of the following elements are present, the result is liability.

- **Duty:** The requirement to act in a reasonable manner, given the circumstances
- **Breach of duty:** Failing to act, or taking action that is unreasonable under the circumstances
- **Legal causation:** Refers to the fact that the harm occurred as a reasonably close-at-hand consequence of the act or failure to act (often called *proximate cause*)
- **Personal injury or property damage:** "No harm, no foul."

Intentional wrongdoing involves deliberate harm. An example is battery: the wrongful touching of another person, which may vary in its intensity. Intentional torts like battery are also crimes, which may be prosecuted by the government and could subject the wrongdoer to fines, imprisonment, or both. (Not all intentional torts are crimes, however; the deliberate infliction of mental distress, for example, is a tort but is generally not criminal.)

Immunities

Emergency management often entails difficult decisions that may result in property damage or personal injury. Disaster response statutes and common law provide customary defenses and immunities that apply to actions taken in response to emergencies.[19] Thus, under certain circumstances, governmental units, governmental employees, and those who are acting in the capacity of governmental employees (including contractors doing government work and properly rostered volunteers) may be immune from liability for actions that might otherwise be the basis for a lawsuit.[20]

Immunities do not, however, act as a complete bar to liability. They are almost never available if injury or damage results from conduct other than negligence—that is, from willful conduct, gross negligence, wanton disregard, or bad faith.[21]

Immunity under state law

To protect governmental units from litigation, state law provides immunity from tort for official acts—that is, acts that are within the employee's scope of duties. In addition, state emergency codes often include specific provisions to shield government officials—including, for example, elected officials, the emergency manager, and the heads of other governmental units—who would be engaged in vital decision making during emergencies. In some states, the law goes further: instead of being limited to officials who are engaged in decision making, it provides broad immunities for a variety of individuals and entities, including the state, political subdivisions, local governments, and governmental employees.[22]

In some states, particular immunity provisions are directed at emergency workers, whether volunteers or employees. In addition, state law may immunize owners of real estate or motorized vehicles who voluntarily authorize the use of their property during an emergency. Finally, some states have enacted "Good Samaritan" statutes, which may provide immunity to certain classes of emergency medical responders;[23] however, such laws usually do not apply if the responder is operating in an official capacity.

Immunity under federal law

On the federal level, two immunity doctrines—governmental function and discretionary action—may insulate governmental activity from liability. The governmental function doctrine protects activities that (1) are undertaken pursuant to constitution or statute, (2) are performed exclusively by a governmental entity for the benefit of the public, and (3) have no equivalent in the private sector. Law enforcement and tax collection are examples.[24] (The government derives income from most governmental functions.) It is through its status as a traditional or inherent governmental function that emergency management obtains its general immunity from tort liability.

The discretionary action exclusion is found in the Federal Tort Claims Act (FTCA). For federal government employees performing discretionary emergency management duties, the Stafford Act, at Section 305, provides an additional level of immunity. Discretionary immunity is not based on the type of action but on the specific act or action. Under this test, an act is discretionary, and therefore immune, if it involves a choice of the type that should be protected from lawsuits. In the emergency management context, discretionary immunity protects policy-level government employees from agonizing about lawsuits during disaster planning and response. In *Berkovitz v. United States,* the U.S. Supreme Court fashioned a two-part test for determining whether discretionary immunity is available.[25] The first part evaluates the nature of the conduct: if a "specific and mandatory" course of action was dictated by statute, regulation, procedure, or policy, then the actor had no choice and there is no discretion to protect: thus, the discretionary immunity exception does not apply (although governmental function may apply). The second part applies only if the act entailed judgment; it must then be determined whether it is the kind of judgment that the discretionary immunity exemption was designed to shield.

The exemption protects only governmental actions and decisions that are based on public policy; if the decision or action was not founded on public policy, the suit may proceed. For example, federal policy favors the prevention of terrorism, and Congress appropriates funds for that purpose. When local governments apply for terrorism security grants, their applications are evaluated according to guidelines promulgated by the federal government; however, when two applicants both satisfy the general application requirements, the granting agency is not compelled to select one applicant over another. The selection of one applicant over another is the sort of act that would be protected by discretionary immunity.

State statutes typically incorporate a variety of the Supreme Court test. However, where state law does not include such tests, the approach of the particular state—that is, case law—must be examined to determine how a particular act will be judged under state law. Where state courts have mentioned repeatedly that the discretionary immunity exemption provided under state law is fundamentally the same as the exemption included in the FTCA, the discretionary immunity test applies.[26]

In *Commerce and Industry Insurance Company v. Grinnell Corporation,*[27] the Fifth Circuit Court of Appeals reversed a lower court's summary judgment, finding instead that the discretionary immunity test had been incorrectly employed. The lower court had held that the policies and procedures controlling the fire department's approach to a warehouse fire were suggestions rather than mandates, which allowed room for discretion. However, the circuit court found that the regulations and fire department policies were mandatory, and that the firefighters failed to comply with them. Because these regulations and policies left no room for variation, there was no discretion to protect.

If the courts find that established regulatory standards apply to a given action, they may hold the organization and its employees responsible for failure to comply with such standards.

The *Commerce and Industry Insurance Company* decision is important because it may represent the future direction of cases brought against emergency management organizations: if the courts find that established regulatory standards (such as the requirement to have a current plan or to be in compliance with NIMS) apply to a given action, they may hold the organization and its employees responsible for failure to comply with such standards. In such cases, emergency management agencies will find it difficult to rely on discretionary immunity for protection from liability. One somewhat reassuring aspect of this issue is that courts construe statutory waivers of government immunity quite narrowly. (A statutory waiver is a legislative enactment that states, for example, that willful conduct, gross negligence, wanton disregard, and bad faith are not protected from being the basis for liability—which, by omission, assigns immunity only to simple negligence.) Thus, courts will look carefully into the facts underlying the waiver: do the facts clearly demonstrate gross negligence, for example, or was the action merely negligent? If the facts do not clearly fall within the region where immunity does not apply—that is, do not clearly support application of the waiver—the court will protect the unit of government.

Under the FTCA, the defenses provided for the acts of federal agencies or their employees mirror those of the state in which the acts occur. Thus, within a given state, defenses for torts committed by the federal government are treated the same way as if they had been committed by any member of the public. For example, under the FTCA, a federal employee who has an auto accident while on official business is viewed by the law in the same way as anyone else who is involved in an auto accident.

Apart from the protections afforded under the FTCA, separate and additional immunities are provided under federal regulatory laws. A prime example of such a federal regulatory structure is nuclear power generation, over which states have no control. If a state does not provide immunity where a federal scheme allows it, the federal law prevails.

Other applicable laws

In addition to complying with laws pertaining specifically to emergency management, local officials and emergency managers must obey other laws of general application, including the following:

- Occupational Safety and Health Administration laws (applicability varies by state)
- Contract law, including state and local government contracting requirements
- Personnel law, including statutes; the rules, regulations, and rulings of administrative agencies; common-law tort and contract rules; and state and federal constitutional specifications that delineate the rights of employees
- Government ethics law, including state and local laws, as well as the federal Hatch Act (which applies to entities that receive federal funds)
- Procedural laws (such as state administrative procedures acts and any local procedural requirements).

Evolving standards

Emergency management is subject to a variety of standards that are constantly evolving. Failure to comply with these standards may expose a local government to liability.

The National Response Plan and the National Incident Management System

As noted earlier, the NRP and NIMS assign specific and detailed duties to local governments during all phases of emergency management. Although adoption of NIMS is not mandatory, failure to do so will result in loss of eligibility for federal funds.[28] Apart from the impact on funding, however, failure to implement NIMS has other potential consequences. First, in a jurisdiction that is not part of NIMS, lack of interoperability could slow response. Second, a local government that has failed to adopt NIMS may have difficulty obtaining resources because NIMS-compliant jurisdictions may receive priority. Third, NIMS appears to be well on its way to becoming a standard within the field; thus, if failure to comply with the standard is the proximate cause of death or injury, liability may result. In fact, in the case of improved safety standards, the adoption of standards need not be complete throughout an industry to create a potential for liability.[29]

NFPA 1600

On April 29, 2004, the American National Standards Institute (ANSI) urged the National Commission on Terrorist Attacks Upon the United States (also known as the 9/11 Commission) to put in place NFPA 1600, the National Fire Protection Association's Standard on Disaster/Emergency Management and Business Continuity Programs, as the national preparedness standard.[30] On July 22, 2004, the 9/11 Commission formally endorsed NFPA 1600 and recommended that DHS promote its implementation. The commission also recommended that compliance with NFPA 1600 be considered in assessing a company's insurance rating and creditworthiness. Finally, the 9/11 Commission stated that "compliance with [NFPA 1600] should define the standard of care owed by a company to its employees and the public for legal purposes."[31]

NFPA 1600 creates a shared set of standards for emergency management and business continuity. It also incorporates, by reference, techniques for exercising plans, and it includes a list of resource organizations within the fields of emergency management and business continuity. From a legal perspective, one critical aspect of NFPA 1600 is the requirement that all emergency management and business continuity programs comply with all relevant laws, policies, and industry practice.[32]

NFPA 1600 has been incorporated into the Intelligence Reform and Terrorism Prevention Act of 2004 as a standard to be promoted by DHS.[33] This step does not directly give NFPA 1600 the force of law; however, as is the case with NIMS, NFPA 1600 is likely to be enforced by the courts as a general safety standard. Additional laws that require the use of the incident management system—including state fire codes and the Occupational Safety and Health Administration's "Hazardous Waste Operations and Emergency Response" standard[34]—provide further support for NFPA 1600 as an enforceable national norm.

The National Emergency Management Association also supports NFPA 1600 as the standard for emergency management. Along with other standards, such as the Federal Emergency Management Agency's (FEMA's) State Capability Assessment for Readiness (CAR), NFPA 1600 provides the basis for the Emergency Management Accreditation Program (EMAP).[35] Both NFPA 1600 and EMAP are well on their way to becoming the industry standard—which creates pressure on all local emergency management agencies to be compliant with NFPA 1600 and to seek accreditation through EMAP.

The Performance Partnership Agreement and the Cooperative Agreement

The Performance Partnership Agreement (PPA) and the Cooperative Agreement (CA) detail the requirements that must be met before the federal government, through the state emergency management agency, will issue funds to support local emergency management programs. These agreements are a variety of private law, under which localities that receive federal funds must obey federal mandates.[36]

The PPA is a strategic plan that is revised every five years; the CA is modified yearly. The PPA incorporates training objectives and may require the state and local governments to participate in training exercises. The requirements incorporated into the CA are designed to help state and local governments move toward the objectives outlined in the PPA. Governmental units that fail to meet these objectives will lose EMPG funding. In the event of a civil suit, loss of EMPG funding would be persuasive evidence that the emergency management function is inadequate; the only questions would be whether the inadequacy of the function caused the loss and what was the extent of the damages.

Out of a desire for federal funding, the vast majority of local governments adhere to the PPA and CA. Although jurisdictions that do not receive funds are not bound by these documents, the requirements outlined in both constitute industry benchmarks related to NFPA 1600. As noted earlier, a court may hold NFPA 1600 as a standard of care, creating potential legal liability for local governments that do not sign a PPA and a CA. In other words, any local unit of government that does not perform to these standards is exposing itself to liability.

Legal issues in preparedness

From the perspective of protecting against liability, preparedness—which entails planning, training, and exercising—is perhaps the most important phase of emergency management.

The duty to plan

The accepted national standard is to create and keep current an emergency operations plan (EOP). Every state has a statutory requirement mandating the creation of such plans at both the state and local levels. In addition, federal environmental law requires the creation of response plans for extremely hazardous substance (EHS) releases, and states, which are required by federal law to oversee EHS plans, have incorporated this requirement into their own legal structures.

If creating and maintaining a plan improves response, does it follow that failure to plan exposes a local government to potential liability? Under *Berkovitz v. United States,* failure to perform a "specific and mandatory" statutory or other duty would not be subject to discretionary immunity. Thus, failure to prepare a required plan will result in liability if that failure causes harm. In a case involving the Nuclear Regulatory Commission, the court found that an agency employee had made a choice in failing to follow agency rules, which required the creation of a plan: this choice resulted in a lack of immunity. Under the principle of *respondeat superior,* the master answers for the acts of the servant; thus, the agency was responsible for the actions of its employee. Where a challenged governmental activity does not involve public policy considerations and is instead a matter of established policy, the agency will be held responsible for the negligence of its employees.[37]

Planning requirements for working with the private sector

Under Homeland Security Presidential Directive 8, private sector entities that control critical infrastructure must prepare protective plans. As the general coordinator for local emergency management, the local emergency manager should review these plans and ensure that they are consistent with the EOP. Private industry generally lacks the resources to fully protect critical infrastructure and relies on local law enforcement as the first line of defense. Thus, the law enforcement annex of EOP must reflect this potential responsibility. In addition, important resources, such as heavy equipment, are often obtained from the private sector. Such arrangements must also be incorporated into the plan.

A negligently drafted plan can also result in liability. State statutes typically require a "current" plan. As noted earlier, the planning portion of NFPA 1600 not only requires emergency management agencies to undertake specific tasks, but also imposes a duty to adhere to current laws, policies, and industry practices.[38] For emergency management agencies that wish to receive federal funds, NIMS requires EOPs, corrective action and mitigation plans, and recovery plans; NIMS also has rather detailed requirements for EOPs. The standards embodied in NFPA 1600 and NIMS are complementary and set the benchmark for planning. Plans that fail to meet these standards—including the requirement to take account of evolving laws, policies, and practices—would likely be deemed negligent. Similarly, failure to revise a plan to reflect lessons learned from an exercise or from actual events may result in liability.

Experienced emergency managers often refer to the EOP as a "living document": a plan is never final.[39] From a liability perspective, this means that the determination of whether a plan has been negligently drafted is very much a matter of time and circumstance.

Failure to follow the plan

Even if a proper EOP is in place, failure to comply with it may result in liability, both for the emergency manager and for the unit of government. Under *Berkovitz,* liability would depend on whether the plan is interpreted as setting down a compulsory course of action. Despite the fact that an EOP is an internal document with no direct legal consequences for the public, it would most likely would be interpreted as a "specific and mandatory" responsibility under *Berkovitz,* which refers to "statute, regulation or policy"[40] as the basis for determining whether a responsibility is specific and mandatory. Violating an internal policy may serve as the basis for a suit if the policy "leaves no room for implementing officials to exercise independent policy judgment."[41]

Deciding whether a document like a plan limits an official's options or only informs his or her discretionary judgment requires close scrutiny. EOPs usually address a range of matters and include both general guidelines and extremely specific requirements. A statement tasking the county commissioners with general responsibility for protecting the public, for example, clearly contemplates the use of discretion. On the other hand, if a plan contains detailed lists of actions, there may be no occasion to perform any acts other than those listed, unless the plan permits choices.

Exercising and revising the plan

The purpose of exercising is to test the plan and help improve it. One of FEMA's most unswerving requirements is a means of identifying and correcting any shortcomings exposed by exercises. With that goal in mind, a "hot wash" debriefing typically takes place immediately after an exercise: the hot wash permits lessons learned to be recorded while they are still fresh. Later examinations permit a more detailed analysis of where the plan worked correctly and where it needs improvement.

If a local government fails to revise the plan and to improve training in order to reflect lessons learned from exercises and that failure then results in property damage or personal injury in an actual event, the local government may be liable. For example, in 2004, FEMA, the state of Louisiana, the city of New Orleans, and other local government units conducted a joint

Planning for special risks

Hazardous materials releases

Responses to hazardous materials (hazmat) emergencies are strictly controlled by both the Occupational Safety and Health Administration (OSHA)[1] and the U.S. Environmental Protection Agency (EPA).[2] The emergency manager must ensure that (1) the emergency operations plan (EOP) reflects the detailed planning requirements applicable to hazmat releases and that (2) responders comply with these safety mandates.[3]

Generally, in-plant response teams will be the first to arrive at the site of an industrial incident. These teams are required by OSHA's rule for "process safety management of highly hazardous chemicals,"[4] which is designed to prevent or minimize the consequences of a release of toxic, reactive, flammable, or explosive chemicals. In the case of spills or airborne releases that occur on public property such as highways, or that go beyond the boundaries of an industrial facility, the first response organization is typically the fire service.

The extreme danger associated with hazardous materials calls for responders with sophisticated technical expertise. The Emergency Planning and Community Right-to-Know Act (EPCRA) mandates a separate planning structure for the release of extremely hazardous substances.[5] States commonly incorporate EPCRA's requirements into their own laws to provide a supporting state structure for the program.[6]

Terrorism

Terrorism is a regrettable fact of modern life. Terrorism planning requirements, however, did not suddenly arise after the attacks of September 11, 2001. In 1995, President Bill Clinton issued Presidential Decision Directive 39,[7] which addressed terrorism mitigation, planning, deterrence, and response. In 1999, to ensure that the Federal Response Plan (FRP) would provide an adequate structure to respond to terrorism, the Federal Emergency Management Agency promulgated the Terrorism Incident Annex to the FRP.

Under Homeland Security Presidential Directive 5, all EOPs must now include a terrorism annex that complies with the standards in NIMS.[8] As is the case with other widely accepted standards, a state or local government's failure to include this annex in its plan could result in liability if that failure leads to, or exacerbates, personal injury or property damage.

Under the Homeland Security Act of 2002, all terrorist attacks are federal events: that is, after a terrorism incident, the federal government is actually in charge. While this relieves local officials of the burden of command, the emergency management agency must still function in a manner that is consistent with the terrorism annex of the EOP if it is to avoid potential liability.

exercise called Hurricane Pam—a mock Category 3 storm whose aftermath closely resembled that of Hurricane Katrina, including massive flooding, overtopped levees, stranded residents, and civil unrest in the Louisiana Superdome. Failure to incorporate the lessons of Pam into planning and training will likely be used to support claims of federal, state, and local liability.

Mutual aid agreements

Taking proactive steps to ensure access to as many resources as possible is an important part of planning. Mutual aid agreements (MAAs), which ensure support from neighboring jurisdictions, play a crucial role in emergency management and are mandated or strongly suggested by a number of standards, including NFPA 1600 and NIMS.[42] The value of MAAs is clear: they multiply available resources.[43] Many states have either already made provisions for an intrastate agreement or are in the process of doing so. In Indiana, for example, an intrastate mutual aid program applies to every political subdivision of the state that does not opt out by ordinance or resolution.[44] At the state level, the Emergency Management Assistance Compact (EMAC) is the model for interstate agreements.[45]

Jurisdictions that request assistance must understand that potential legal claims may arise from doing so. Although the model interstate and intrastate agreements recommended by EMAC cover such issues as who is responsible for injuries to members of the assisting unit, in the absence of a preexisting written agreement or applicable statute, case law indicates that the requesting entity may be responsible for workers' compensation claims if members of an assisting unit are injured during the response.[46]

School violence

The federal Safe and Drug-Free Schools and Communities Act of 1994 (Safe Schools Act) is a federal grant program focused on the prevention of violence and drug abuse in schools.[9] Under this act, schools are required to create safety plans to address in-school violence. The program funnels federal aid to states, which make grants to local educational agencies and community organizations. The funds may be used also to support the private sector, nonprofit organizations, and colleges and universities that develop antiviolence and antidrug initiatives.

School safety plans must take into account the school's particular characteristics. State statutes typically require state and local emergency management agencies to review school safety plans to ensure that they are consistent with the local EOP. But even if this were not the case, local officials, who have ultimate responsibility for safety in schools, would be well advised to require the emergency manager to review the sufficiency of such plans in order to protect the unit of government from liability.

[1]U.S. Department of Labor, Occupational Safety and Health Administration (OSHA), "Hazardous Waste Operations and Emergency Response," *Code of Federal Regulations,* title 29, section 1910.120 (2007), available at osha.gov/pls/oshaweb/owadisp.show_document?p_table=standards&p_id=9765 (accessed July 24, 2007).

[2]"Protection of Environment: Environmental Protection Agency: Toxic Chemical Release Reporting: Community Right-to-Know," *Code of Federal Regulations,* title 40, section 372.18 (2006), available at law.justia.com/us/cfr/title40/40-27.0.1.1.13.html (accessed July 24, 2007).

[3]See, generally, William C. Nicholson, "Legal Issues in Emergency Response to Terrorism Incidents Involving Hazardous Materials: The Hazardous Waste Operations and Emergency Response ('HAZWOPER') Standard, Standard Operating Procedures, Mutual Aid and the Incident Command System," *Widener Law Symposium Journal* 9, no. 2 (2003): 295.

[4]U.S. Department of Labor, OSHA, "Process Safety Management of Highly Hazardous Chemicals, *Code of Federal Regulations,* title 29, section 1910.119, available at osha.gov/pls/oshaweb/owadisp.show_document?p_table=STANDARDS&p_id=9760 (accessed July 24, 2007).

[5]"Establishment of State Commissions, Planning Districts, and Local Committees," *U.S. Code* 42 (2007), §§ 11001-11050, available at law.justia.com/us/codes/title42/chapter116_.html. EPCRA is contained in Title III of the Superfund Amendment and Reauthorization Act of 1986 (SARA Title III).

[6]See, for example, "Emergency Planning and Notification," Indiana Code § 13-25-2 (2007), available at in.gov/legislative/ic/code/title13/ar25/ch2.html (accessed July 24, 2007).

[7]"U.S. Policy on Counterterrorism," Presidential Decision Directive 39, June 21, 1995, available at ojp.usdoj.gov/odp/docs/pdd39.htm (accessed July 24, 2007).

[8]"Management of Domestic Incidents," Homeland Security Presidential Directive/HSPD-5, February 28, 2003, available at whitehouse.gov/news/releases/2003/02/20030228-9.html (accessed July 24, 2007).

[9]*Safe and Drug-Free Schools and Communities Act of 1994 (Safe Schools Act),* U.S. Code 20 (2007) §§ 7101 *et seq,* available at access.gpo.gov/uscode/title20/chapter70_subchapteriv_parta_.html (accessed July 23, 2007).

Legal issues in response

A number of issues arise during the response phase that have legal implications: declaring an emergency, requesting a disaster declaration, evacuation and quarantine, standard operating procedures, and the use of volunteers.

Declaring a state of emergency

As a general matter, a state or local jurisdiction should declare a state of emergency when doing so will be to its advantage. In the majority of states, the chief executive will issue the declaration, with counsel from the emergency manager.

A local emergency declaration triggers the disaster response and recovery portion of the emergency plan. Furthermore, the declaration usually activates specific powers for the chief executive officer, allowing a variety of actions for the duration of the declaration. At the state level, authorized actions on the part of the governor may include any of the following:

- Deploying and using any forces to which the plan or plans apply
- Using or distributing any supplies, equipment, materials, and facilities assembled, stockpiled, or arranged to be made available under any law relating to disaster emergencies
- Serving as commander-in-chief of the organized and unorganized militia and of all other forces available for emergency duty
- Suspending the provisions of any regulatory statute that prescribes the procedures for conduct of government business, or the orders, rules, or regulations of any agency, if

strict compliance with any of these provisions would in any way prevent, hinder, or delay necessary action in coping with the emergency

- Using all available governmental resources that are reasonably necessary to cope with the emergency
- Transferring the direction, personnel, or functions of departments, agencies, or units to perform or facilitate emergency services
- Subject to any applicable requirements for compensation, commandeering or using any private property deemed necessary to cope with the emergency
- Assisting in the evacuation of all or part of the population from any stricken or threatened area in the jurisdiction
- Prescribing routes, modes of transportation, and destinations in connection with evacuation
- Controlling ingress to and egress from a disaster area, the movement of persons within the area, and the occupancy of premises in the area
- Suspending or limiting the sale, dispensation, or transportation of alcoholic beverages, firearms, explosives, and combustibles
- Making provision for the availability and use of temporary housing
- Allowing people who hold a license to practice medicine, dentistry, pharmacy, nursing, engineering, and other similar professions to practice their professions in the state during the period of the state of emergency if the state in which a person's license was issued has a mutual aid compact for emergency management with the state
- Giving specific authority to allocate drugs, foodstuffs, and other essential materials and services.[47]

Local ordinances typically contain similar provisions, although local powers do not include control of the militia or the authority to allow out-of-state medical practitioners to practice.

Although an unnecessary declaration of emergency has the potential to create liability, a successful lawsuit would have to demonstrate gross negligence or malicious intent. Liability might result, for example, if the declaration were clearly unnecessary, and if the special powers granted to the executive—such as ignoring normal contracting requirements or appropriating private property—were used in an objectionable way. It is conceivable, however, that postponing a declaration would result in liability—if, for example, a delay in obtaining timely assistance from higher jurisdictions led to injury to persons or property.

As noted earlier, states have various approaches to emergency management immunities. In some states, immunities will not be in effect until an emergency is declared; in others, every action related to emergency management, regardless of whether a declaration has been made, is protected by immunity statutes.

Requesting a disaster declaration

When a disaster occurs, the emergency manager is typically aware of the limits of readily obtainable resources. Moreover, the type and severity of an event often make it immediately clear that local resources will not be sufficient. Although, as a rule, local resources must be exhausted—or in imminent danger of exhaustion—before a successful request for aid can be made, in some circumstances an appeal for assistance may be made even before an event occurs: this allows response resources to "lean forward" so that assistance will not be delayed. Moreover, as noted earlier, under the Catastrophic Incident Annex of the NRP, the secretary of DHS may unilaterally begin marshaling federal resources when a disaster is imminent. This authority is independent of the normal channels for requesting assistance provided by the Stafford Act.

The request for assistance must be made through the appropriate channels: local governments may apply for help from the state, and the state may then apply for federal assistance. (However, as previously noted, the state may make a direct request for federal aid even if the local government has not appealed to the state for assistance.) The procedures for making requests to the state are detailed in state law, and the procedures for applying for federal assistance are detailed in the Stafford Act; under this act, the governor cannot apply for federal assistance until certifying that the state EOP has been implemented.

Normally, before a request is made, FEMA and the requesting local jurisdiction, and often the state, jointly conduct a preliminary damage assessment (PDA). All requests involve a written declaration that contains the following elements:

- A statement that a disaster is imminent or occurring
- A description of the disaster
- A statement that resources are or soon will be exhausted
- An assessment of the actual or anticipated extent of damage
- The types of assistance requested
- A request for assistance.

At the state level, a request for federal assistance begins when the governor tenders the application to FEMA's regional office;[48] the regional office forwards the governor's request to FEMA headquarters, where it is evaluated. Headquarters then passes the request on to the White House, along with a recommended course of action, and at the same time notifies the secretary of DHS.

When a Presidential Disaster Declaration is made, the director of FEMA appoints a federal coordinating officer, and the agency specifies the types of support to be made available and the counties qualified to receive assistance. In the case of large-scale or catastrophic events, the declaration process can be accelerated.

Evacuation and quarantine

Evacuation is one of the best illustrations of the essential challenge of emergency management: how to adequately protect people and property while avoiding unnecessary intrusions on daily life. Although evacuation may be the best protective step available, it is expensive and inconvenient for people and businesses. Furthermore, the evacuation itself may create risk. At the same time, however, failure to evacuate in a timely manner may result in deaths and injuries that might not otherwise have occurred. Controversy in the aftermath of the Asian tsunami in 2004 and of Hurricane Katrina in 2005 has raised the profile of legal issues associated with evacuations: in the case of both events, observers have criticized the evacuation plans, the evacuation steps taken before and after the events, and the timing of evacuation decisions, and lawsuits have been filed.

State and local emergency management legislation provides the legal authority for declaring an evacuation. Under these laws, the chief executive generally has the power to forcibly evacuate people from their homes and businesses. There is a tremendous amount of uncertainty in making such a decision, so the choice to evacuate must be the product of informed discretion. One might therefore expect that a decision about whether and how broadly to evacuate would be protected by discretionary immunity. As noted earlier, however, immunities are not absolute. As the lawsuits arising from Katrina and the Asian tsunami go forward, new legal precedents may be set in the area of evacuation law.

Sometimes people refuse to evacuate, even in response to a mandatory evacuation order.[49] At that point, difficult choices must be made. How much time should be devoted to attempting to persuade unwilling residents to evacuate when there are many other people to be saved? Should limited resources be devoted to arresting and confining unwilling citizens (at the risk of a possible suit for false arrest)? One successful tactic has been to compel those who wish to stay behind to execute a "next of kin" form so that the government can notify their family members after their demise (the implication being that refusal to evacuate will make death inescapable). The document also should include a waiver of liability so that the government will be able to argue that it is not responsible for the person's loss of life.

Occasionally, residents or responders may need to reenter an evacuated area during an emergency. For example, residents may want to check on pets or to ensure that their homes are secure; responders may need to deal with problems such as gas leaks. The nature and importance of the task must be evaluated to determine whether the resulting risks—in terms of both lives and legal liability—are worthwhile; typically, the incident commander makes this decision. For maximum legal protection, reentry should be permitted only when the danger has receded.

The 2001 anthrax attacks and several copycat events, as well as the SARs (severe acute respiratory syndrome) outbreak of 2003, highlighted the issue of whether existing powers for quarantine and evacuation are sufficient for the aftermath of a bioterrorism attack. The possible scope of isolation and quarantine—and the logistical difficulty of large-scale quarantine—are daunting.

Generally, public health is a matter of state law.[50] To promote uniformity in state law, the Centers for Disease Control and Prevention supported the creation of a broadly accepted standard. The result was the Model State Emergency Health Powers Act (MSEHPA),[51] whose intent is to promote a "renewed focus on the prevention, detection, management, and containment of public health emergencies."[52] After the release of MSEHPA, the National Association of Attorneys General passed a resolution that encouraged public health authorities, and their legal advisors, to learn more about the statutory and constitutional provisions regarding public health response; the association also endorsed reassessments of state public health statutes in light of the act. After a review of MSEHPA, a substantial majority of states modified their public health statutes.

The Model Act provides helpful guidance on quarantine law. Under MSEHPA, governors have broad powers to declare and enforce public health emergencies. The act also allows adopting states to confiscate medical supplies and drugs, without consideration of ownership, during a declared public health emergency, and to appropriate, make use of, and, if need be, destroy property. With respect to infectious disease, MSEHPA's two goals are effective detection and efficient response. For purposes of detection, the act places pharmacies, hospitals, and outpatient service providers on the front line of the fight against bioterrorism. For example, it requires these entities to report unusual clusters of symptoms or prescriptions, and to assist in tracking outbreaks. With respect to response, the Model Act preserves the traditional authority of most states and many local governments to isolate or quarantine anyone who has been in contact with, or has been diagnosed with, certain contagious diseases. It also contains provisions, based on existing state laws, that authorize officials to examine, vaccinate, and treat people during a public health emergency, and to isolate or quarantine individuals who decline. A written order from public health officials is necessary for isolation or quarantine. The act also authorizes strong steps if required, including forcible testing for illness and administration of vaccinations.

Evacuation is one of the best illustrations of the essential challenge of emergency management: how to adequately protect people and property while avoiding unnecessary intrusions on daily life.

Because any mandatory examination, isolation, quarantine, or treatment is a constraint on individual liberty that triggers constitutional due process protections, the Model Act (1) enumerates the rights to a hearing and (2) requires officials to show that the proposed restraint is the "least restrictive means" to prevent the spread of infectious disease. The appeals process, which is designed to be rapid (no more than a period of days) and to be conducted through the courts, includes access to free legal counsel. The process begins with a "show cause" hearing, in which the state must demonstrate that its actions were proper. To prevail, a plaintiff would have to establish that the state breached the conditions of the isolation or quarantine order (e.g., that the state's actions resulted in wrongful isolation or wrongful quarantine, or that the state failed to follow proper procedures).

Critics have portrayed the broad powers embodied in the Model Act as infringing on constitutionally guaranteed freedoms. The Privacy Rule of the Health Information Portability and Accountability Act of 1996,[53] for example, acknowledges the right to privacy by creating classes of protected health information. MSEHPA, in contrast, employs a "disclose now, obtain consent later" approach to personal health information. Indeed, the Model Act allows authorities to obtain, use, and even publicly release such information before complying with privacy regulations. In an emergency, the government must strike a balance between individual liberties and the common good; public health decisions are likely to be important entries in a long line of cases involving preventive jurisprudence.

Standard operating procedures

During response, local officials and emergency managers must comply with any standard operating procedures (SOPs) that apply to them. (These procedures are usually created by local emergency management officials and emergency response agencies.) In general, rules such as SOPs are characterized as "internal private law": they are internal because they are established by the institution in question; they are private because they apply only to those who work for the entity that established them; and they are law because violation can result in penalties, including reprimands, loss of pay, suspension from work, and firing. As previously noted in the discussion of the *Commerce and Industry Insurance Company* decision, failure to comply with mandatory SOPs may result in civil liability. Criminal prosecution for a failure to comply would be highly unusual, but where the violated SOP establishes a specific duty to a specific person, failure to perform that duty might be a criminal act—which could result not only in criminal charges but also in a private suit for money damages.

One way to deal with liability concerns is to create "operating guidelines" rather than "operating procedures." This approach reflects the fact that emergency managers and responders are often more highly trained than they were in the past, and thus more capable of exercising discretion. The substitution of "guidelines" for "procedures" may enlarge the application of discretionary immunity. It is important to ensure, however, that the instructions are fundamentally different from those they replace: merely changing the title from "Standard Operating Procedures" to "Standard Operating Guidelines" is unlikely to pass muster in court.

It is also possible to implement alternative procedures—measures that can be used during an emergency when existing procedures do not apply—but such procedures must comply with all applicable laws. From the perspective of liability, the safest approach is to ensure that all procedures, including any alternative procedures applicable in an emergency, have been properly scrutinized by legal counsel and put in place prior to an event.

Using volunteers

Although volunteers can be a critical asset for emergency management, numerous legal issues may arise from taking advantage of volunteer resources. The first question to address is the matter of definition: who is a volunteer? Simply showing up at the scene of an event does not render one a volunteer. On the contrary, one of the incident commander's greatest challenges is the throng of well-meaning individuals who assemble at a disaster scene: experience shows that "emergent responders" will arrive at the scene individually or as a group, even if assistance has not been requested, and even after active attempts have been made to prevent would-be volunteers from converging on the scene. These individuals are often not affiliated with any organizations that are formally integrated into the EOP, their skills and training are unknown, and they make the job of the incident commander—who is responsible for securing the scene—all the more difficult.

The incident commander must use both determination and diplomacy to preserve control of the site. One of the incident commander's first responsibilities is organization of a perimeter, which should be controlled by law enforcement. Anyone who tries to enter the perimeter without proper authorization must be prevented from doing so and removed to a distant staging area. There, training and capabilities can be assessed, and the individual's proper role, if any, can be assigned.

To protect against liability, volunteers must be properly integrated into the emergency management team. In the event that emergent volunteers are found to be trained responders with needed skills, a record of the assessment must be made. Only then can they be officially registered as volunteers. Once they have been registered, the volunteers are available for the incident commander, or his or her delegated representatives, to employ. Formally assessing and registering volunteers will help to shield the emergency management agency and other responding entities from liability, as it demonstrates that reasonable care has been taken to determine the competencies of the volunteers. Of course, the incident commander has an accompanying duty to assign the volunteers to duties for which they are qualified.[54]

The best approach is to establish long-term relationships with the most important volunteer groups well before a crisis. The role of volunteer organizations should be described in the plan, and these organizations, like all other emergency response entities, should participate

The Volunteer Protection Act of 1997

Congress enacted the Volunteer Protection Act of 1997 (VPA) to address the concern that fear of litigation would make citizens less likely to volunteer for public service.[1] The VPA creates statutory immunity for volunteers for negligent acts. This means that volunteers' liability protection is limited to situations in which they are found to have engaged in simple negligence; they are still potentially liable for gross negligence or intentional wrongdoing.[2] Although states may avoid coverage by passing a law to opt out of the VPA, none has done so. In addition to being protected from negligence lawsuits, a volunteer who is affiliated with a nonprofit organization and is acting within the scope of his or responsibilities may not be liable for punitive damages, even in the case of negligence or gross negligence. Punitive damages are available only when the volunteer acts maliciously. (However, this immunity does not extend to the volunteer's organization.)

In an important exclusion, the VPA does not exempt volunteers from liability for any harm caused while driving a motor vehicle. This exclusion is important, given that research indicates that half the claims involving emergency response organizations arise from vehicle accidents. Also, although the VPA alters the basis for a lawsuit, it probably does not affect administrative actions that are based on negligent activity by the regulated person. Therefore, state laws that provide for administrative penalties in cases of negligent conduct that endangers others continue to be valid. So, for example, if state laws covering emergency medical services provide for administrative penalties, such as loss of certification, for a volunteer emergency medical technician who treats a patient negligently (e.g., by failing to take vital signs as frequently as training dictates), the volunteer (defined in the VPA as someone who is reimbursed only for actual expenses of up to $500 a year) could still lose certification, even though he or she would be protected under the VPA from being sued by the patient.

[1]*Volunteer Protection Act of 1997*, Public Law 105-19, available at frwebgate.access.gpo.gov/cgi-bin/getdoc.cgi?dbname=105_cong_public_laws&docid=f:publ19.105.pdf (accessed October 8, 2007).

[2]As noted earlier in the chapter, gross negligence means that the person acted in a highly unreasonable manner considering the circumstances.

in planning, training, and exercising. Typically, the sponsoring organization ensures that volunteers are properly trained and provides them with credentials that identify them as people who may be on the scene during an event. Credentials make the safety officer's job less onerous and remove a significant administrative burden from the incident commander.

Legal issues in recovery and mitigation

Emergency management laws and ordinances incorporate standards for terminating a state of emergency. Normally, action of law terminates a disaster; in other words, the event is terminated automatically by state or local statute after a certain period of time unless the governor, the legislature, or analogous local authorities opt to extend it. Depending on the requirements of the applicable law, renewal or extension of a state of emergency requires either a simple declaration by the local government's chief executive officer or ratification by the governing body.

The federal government provides three categories of grant for disaster recovery: mitigation grants; Individuals and Households Program (IHP) grants; and Public Assistance grants, which provide reimbursement for certain emergency costs for nonprofit groups and support the reconstruction of public and some nonprofit infrastructure. If a Small Business Administration (SBA) emergency has been declared, low-interest loans may be available to help businesses and homeowners undertake reconstruction.[55] (An SBA emergency is often declared when the extent of damage is not sufficient to trigger a Presidential Disaster Declaration.)

NIMS has enlarged the traditional requirement for an EOP by requiring recovery plans, corrective action plans, and mitigation plans. Planning for short-term recovery addresses measures such as life support for injured people, damage assessment, search and rescue, restoration of services, provision of temporary housing, and clearance of debris. Long-term recovery plans identify strategic priorities for restoration, improvement, and growth, and incorporate measures to help the community better withstand future calamities. Corrective action plans put into practice lessons learned from training and exercises or from actual incidents. Mitigation plans address actions that can be taken prior to, during, or after an event to reduce or eliminate risk, or to lessen the actual or potential effects of an incident. In effect, emergency

Proactive preparation for federal assistance

Proper planning on the legal side—including the proper documentation of expenses and the wording of local ordinances—can bring about significant financial benefits. However, the federal government will reimburse only for properly documented expenses, so good financial controls will make a difference when the time comes to request federal reimbursement.

Both building and fire codes should conform to current best practices, and all new or rebuilt structures should be required to achieve compliance. New construction of—or repairs to—infrastructure, such as bridges and highways, must conform to current best practices. Inasmuch as federal assistance will pay for reconstruction or repair only to the degree that the unit of government imposes the same obligations on itself or on private parties, local ordinances addressing the construction or replacement of infrastructure can have a significant effect on the amount of funding received from the federal government.

managers now have a legal obligation to ensure that, in the wake of a disaster, reasonable steps are taken to make the jurisdiction safer.

From a legal perspective, important mitigation efforts include (1) the adoption of zoning and subdivision regulations that can help reduce hazard impact and (2) the adoption of fire and building codes to reflect current best practices. Although fire and building codes typically apply only to new construction, the local government should consider requiring retrofits to comply with specific aspects of codes that have been shown to significantly improve safety. In particular, these include earthquake resistance and sprinkler systems for public buildings.

Summary

Emergency management takes place within a complex and constantly evolving legal environment. To avoid liability, emergency managers and their legal advisors must engage in proactive collaboration to address all four phases of emergency management. Only when local officials, emergency managers, and their legal advisors fully understand one another's responsibilities and expertise can they effectively mitigate legal hazards.

Notes

1 State and local legislation affecting emergency management also includes ethics rules and procurement policies. In some states, ethics rules are advisory only; in others, they are legal obligations. Even in states where ethics rules are advisory, however, failure to comply may result in discipline by the employer.

2 An important principle that must be mentioned is that of *respondeat superior*, which refers to the obligation of the master to be responsible for the mistakes of the servant. Under this principle, government employees who are negligent in the course of their work may expose the unit of government to liability.

3 See William C. Nicholson, "Litigation Mitigation: Proactive Risk Management in the Wake of the West Warwick Club Fire," *Journal of Emergency Management* 1, no. 2 (2003): 14–18.

4 *Disaster Relief Act of 1974*, Public Law No. 93-288, 88 Stat. 143 (1974).

5 *Homeland Security Act of 2002*, Public Law 107-296, 116 Stat. 2135 (2002), available at dhs.gov/xlibrary/assets/hr_5005_enr.pdf (accessed October 8, 2007).

6 Local governments may, however, appeal directly to local federal facilities, such as military bases. Such facilities are permitted to assist only if they are unable to obtain the approval of a higher authority.

7 "Management of Domestic Incidents," Homeland Security Presidential Directive/HSPD 5, February 28, 2003, available at whitehouse.gov/news/releases/2003/02/20030228-9.html (accessed July 24, 2007).

8 U.S. Department of Homeland Security (DHS), *National Incident Management System* (Washington, D.C.: DHS, March 2004), 33, available at dhs.gov/xlibrary/assets/NIMS-90-web.pdf (accessed July 25, 2007).

9 In the wake of 9/11, some states changed the names of their emergency management agencies to include the words *homeland security*, or they created additional agencies that have specific powers and authorities with respect to terrorism.

10 "Emergency Management and Disaster Law," Indiana Code § 10-14-3-2(a) (2007), available at ai.org/legislative/ic/code/title10/ar14/ch3.html (accessed July 23, 2007). The definition does not itemize all duties.

11 Ibid., § 10-14-3-17 (h-i) (2007).

12 For general information about the content of state emergency management laws, see Keith Bea, L. Cheryl Runyon, and Kae M. Warnock, *Emergency Management and Homeland Security Statutory Authorities in the States, District of Columbia, and Insular Areas: A Summary* (Washington, D.C.: Congressional Research Service, Library of Congress, 2004), especially Table 1, which includes references to reports on each state's laws. Local emergency managers and the attorneys who advise them should have copies of the relevant Congressional Research Office publications and state laws; the publications can be obtained from members of Congress.

13 Localities must tailor their law to complement their state structure.

14 "Common law" refers to the unwritten body of law that is based on custom and case law precedent.
15 For a detailed discussion of liability for failure to plan properly, see Ken Lerner, "Governmental Negligence Liability Exposure in Disaster Management," *Urban Lawyer* 23, no. 3 (1991): 333.
16 Depending on the intent and circumstances, violation of OSHA laws may be criminal. See usdoj.gov/usao/eousa/foia_reading_room/usam/title9/crm02011.htm (accessed July 25, 2007).
17 See, for example, *Meridian Ins. Co. v. Zepeda*, 734 N.E. 2d 1126, 1130–31 (Ind. App. 2000). The elements of the violation are the specific things that must be proven in order to find guilt in a criminal context or liability in a civil context.
18 *Volunteer Protection Act of 1997*, Public Law 105-19, available at frwebgate.access.gpo.gov/cgi-bin/getdoc.cgi?dbname=105_cong_public_laws&docid=f:publ19.105.pdf (accessed October 8, 2007).
19 Customary defenses are those that apply to anyone. Immunities are additional defenses available to government employees.
20 Lerner, "Governmental Negligence Liability Exposure," 335.
21 Wanton disregard and bad faith are varieties of intentional acts. In state emergency management statutes and in laws that concern suing the state, gross negligence, wanton disregard, and bad faith are often listed as being excluded from immunity.
22 As stated in "Emergency Management and Disaster Law," Indiana Code § 10-14-3-15(a) (2007), "any function under this chapter and any other activity relating to emergency management is a governmental function."
23 "Certain classes" refers to people who come upon an injured person and give help without being under any obligation to do so; this could include an off-duty doctor or emergency medical technician, unless that person's employment agreement specifies that he or she is always on duty or on call.
24 "Municipal, County, School, and State Tort Liability," in *American Jurisprudence*, 2nd ed. (Eagan, Minn.: Thomson West, 2001), chapter 57.
25 *Berkovitz v. United States*, 486 U.S. 531 (1988).
26 Jim Fraiser, "A Review of the Substantive Provisions of the Mississippi Governmental Immunity Act," *Mississippi Law Journal* 68 (1999): 703.
27 *Commerce and Indus. Ins. Co. v. Grinnell Corp.*, 280 F. 3d 566 (5th Cir. 2002).
28 Small, local emergency response entities that are not compliant with NIMS—such as a volunteer emergency medical services group in a town that lacks the necessary resources to comply—may still receive funding.
29 See *The T. J. Hooper*, 60 F. 2d 737 (2nd Cir. 1932), cert. denied, *Eastern Transportation Co. v. Northern Barge Corp.*, 287 U.S. 662 (1932), where Judge Learned Hand found a vessel unseaworthy for lack of a radio receiving set, despite the absence of statutes, regulations, or even custom about their use on sea vessels. Two barges had been lost in a storm, and the tugs and their tows might have sought shelter in time had they received weather reports by radio.
30 NFPA 1600: Standard on Disaster/Emergency Management and Business Continuity Programs (Quincy, Mass.: National Fire Protection Association, 2007), available at nfpa.org/assets/files/PDF/NFPA1600.pdf (accessed July 24, 2007).
31 National Commission on Terrorist Attacks upon the United States, *The 9/11 Commission Report* (New York: Norton and Co., 2004), 398, available at 9-11commission.gov/report/911Report.pdf (accessed July 24, 2007).
32 NFPA 1600, Chapter 5.2.
33 *Intelligence Reform and Terrorism Prevention Act of 2004*, Public Law 108-458, available at nctc.gov/docs/pl108_458.pdf (accessed October 8, 2007).
34 U.S. Department of Labor, Occupational Safety and Health Administration, "Hazardous Waste Operations and Emergency Response," *Code of Federal Regulations*, title 29, section 1910.120 (2007), available at osha.gov/pls/oshaweb/owadsp.show_document?p_table=standards&p_id=9765 (accessed July 24, 2007).
35 See "Emergency Management Accreditation: An Overview" available at emaponline.org/?16. FEMA is currently developing an online self-assessment support tool to augment the EMAP Standard; this tool will replace the CAR as FEMA's state emergency management capability assessment instrument; see fema.gov/news/newsrelease.fema?id=3712 (July 23, 2003) (accessed July 25, 2007).
36 Private law is an enforceable agreement between individual entities; a contract is an example.
37 *Roberts v. United States*, 724 F. Supp. 778, 791 (D. D.C. 1989).
38 NFPA 1600, Chapter 5.2.
39 Since 9/11, specific legislation addressing terrorism has been added to state emergency management codes. In particular, state laws that list requirements for planning commonly include planning for terrorism. See, for example, Florida Statutes, Chapter 252.34 (2006), which defines emergency management responsibilities, and Florida Statutes, Chapter 395.1056 (2006), which requires hospitals to plan for terrorism events.
40 *Berkovitz v. United States*, 486 U.S. 531 (1988) at 536.
41 Ibid., at 547.
42 For a model mutual aid agreement, see the Emergency Management Assistance Compact (EMAC) Web page at emacweb.org.
43 The appropriate contents for intrastate and interstate mutual aid agreements may be found on the National Emergency Management Association (NEMA) Web site at nemaweb.org. The EMAC Web site (see note 42) provides model legislation for intrastate legislation and county-state mutual aid legislation.
44 "Emergency Management and Disaster Law," Indiana Code § 10-14-3-10.6 (2007) (see note 10).
45 NEMA, *Model Intrastate Mutual Aid Legislation* (Lexington, Ky.: NEMA, 2004), available at emacweb.org/docs/Wide%20Release%20Intrastate%20Mutual%20Aid.pdf (accessed July 23, 2007).
46 See, for example, *Thomas v. Lisbon*, 550 A.2d 894 (Conn. 1988). One issue that EMAC does not address is the credentialing of visiting emergency responders such as doctors and emergency medical technicians. This was reportedly a matter of such debate when EMAC was written that the decision was made to allow each state to deal with the issue separately, which many have done. EMAC itself grants automatic credentialing across state lines only to state employees who have credentials from their home jurisdiction. For those individuals and emergency response organizations that become federally certified under NIMS, nationwide credentialing will be included in the package; see *National Incident Management System*, 46 (full citation at note 8). This approach will do a great deal to lessen uncertainty in the aftermath of a disaster.
47 This list is adapted from "Emergency Management and Disaster Law," Indiana Code § 10-14-3-12 (2007) (see note 10).
48 In some cases, the Stafford Act authorizes declarations without a governor's request.
49 One reason that people may resist evacuation is that their companion animals are not allowed to go with them to Red Cross shelters. Advance agreements with

veterinarians for assistance in caring for such animals help many jurisdictions to deal with this matter.

50 W. Rudman and R. Clarke, *Emergency Responders: Drastically Underfunded, Dangerously Unprepared* (New York: Council on Foreign Relations, 2003).

51 MSEHPA was created by a team of experts from the Center for Law and the Public's Health at Georgetown and Johns Hopkins Universities. See the first draft of this act, available at publichealthlaw.net/MSEHPA/MSEHPA2.pdf.

52 MSEHPA Preamble, 6.

53 "Standards for Privacy of Individually Identifiable Health Information: Final Rule," *Federal Register* 67, no. 157 (August 14, 2002): 53182–53239, available at hhs.gov/ocr/hipaa/privruletxt.txt.

54 See William C. Nicholson, "Legal Issues in Emergency Response to Terrorism Incidents Involving Hazardous Materials: The Hazardous Waste Operations and Emergency Response ('HAZWOPER') Standard, Standard Operating Procedures, Mutual Aid and the Incident Management System," *Widener Symposium Law Journal* 9, no. 2 (2003): 295, 321–322.

55 The answers to basic questions and referral to more extensive materials will be found on the FEMA Web site at fema.gov.

CHAPTER 13

Identifying and addressing social vulnerabilities

Elaine Enarson

This chapter provides an understanding of

- The concept and complexity of social vulnerability
- Approaches to reducing social vulnerability: strategies and tools.

An abiding commitment to public safety keeps emergency managers awake at night and on the go all day. This core value motivates them to reduce avoidable harm, giving special attention to those people most exposed to hazards and least able to cope with disasters. As the practice of emergency management as a whole becomes more holistic, participatory, and community based, local managers are reaching out more to vulnerable populations—groups that, for a variety of reasons, have fewest defenses against a disaster and are least resilient in its aftermath.

Addressing the needs of these populations is not a new concern in emergency management. Planning ahead to assist residents with disabling physical or mental conditions, the very young, the frail elderly, and those who are not fluent in English has long been part of community education, outreach, and preparedness in emergency management offices across the nation. However, the overall national shift in focus from response to mitigation gives more prominence to the need for approaches that also, in the long run, help reduce social vulnerabilities—social and economic conditions that make it hard to cope. Mitigation aims to reduce the risk of disasters, and its foundation is planning ahead to assist and develop the capacities of those whose resources are not as great as their exposure.

The everyday patterns of life put some people more than others at risk. Emergency managers who understand those patterns will be better able to direct and manage scarce resources efficiently and equitably. By the same token, emergency managers who do not pay careful attention to the living conditions, needs, and resources of population groups at high risk may not be able to ensure that vital preparedness, response, and recovery resources reach those most in need. An approach that strives to reduce social vulnerabilities as well as respond to them challenges emergency managers to move beyond obvious needs—for sign language interpretation, for example, or life-preserving medical equipment or services—to consider the less obvious potential of vulnerable populations as partners throughout the disaster cycle. For instance, when emergency managers strive to reach non-English-speaking and recent immigrants, their best allies may be other immigrants living in poverty in hazardous places who are seen as trustworthy, can translate local languages, are informal opinion leaders in their neighborhoods, and have gained know-how navigating government bureaucracies.

A vulnerability-reducing approach also means that throughout the disaster cycle, emergency managers will be able to act on their knowledge of the resources as well as on the needs of highly vulnerable groups. For example, they will be able to offer decision makers expert testimony about the implications that choices made on issues of social policy, land use, housing, and transportation have for social vulnerability. They can advocate for community-led post-disaster relief and recovery approaches that reflect the capacities and vulnerabilities of those affected. (For a good discussion about involving the community in mitigation, see Chapter 6.) What happened in New Orleans after Hurricane Katrina underlines the need for this kind of informed advocacy.

After Hurricane Katrina, the iconic photo of an elderly African American woman wrapped in the American flag (see Figure 13-1) bore witness to the moral imperative in emergency management of equal protection for all—before, during, between, and after catastrophic events. Local emergency management can advance this goal by capitalizing on the nation's heightened awareness of social vulnerability in disasters. Now more than ever, a clear analysis of the forces that increase vulnerability is imperative. What are the characteristics of everyday life that put people in harm's way, and what can be done to make people safer? What are the fault lines in the community that will enable some more than others to meet the challenges of a disastrous earthquake, oil spill, biological attack—or devastating Gulf Coast hurricane? These questions about social vulnerability are more present than ever before in communities large and small. Focusing on high-need populations is sometimes seen as an "add on" to an already complex job or is mistaken for political advocacy, but it is neither. For only if emergency management professionals and other emergency responders understand the social as well as physical aspects of vulnerability—and work in tandem with high-need populations

This chapter benefits enormously from the work of Cheryl Childers, Betty Hearn Morrow, Deborah Thomas, and Ben Wisner, with whom I prepared the Instructor's Guide to *A Social Vulnerability Approach to Disasters* for the FEMA Higher Education Project. My thanks to all and to our external reviewers, members of the advisory committee, and Wayne Blanchard.

Photo courtesy of Alan Chin

Figure 13-1 Waiting in the Superdome for help to arrive after Katrina

and their advocates—will the nation be better prepared for the next catastrophe, whether it be a pandemic, a political attack, or an extreme environmental event.

This chapter begins with an extended discussion of social vulnerability as a concept and of the difference that an understanding of the concept makes in practice. It then reviews a number of planning strategies and tools available for use by emergency managers, who serve a complex and often divided people in an era of increased risk. The chapter concludes with a look at some implications for the profession as a whole in the future.

Social vulnerability: The theory and the reality

In the 1990s, a global consensus emerged that better preparedness, response, and relief cannot fundamentally reduce people's risk of natural, technological, or human-induced disaster. Spearheading this evolution were, first, the International Decade for Natural Disaster Reduction (1990s) and, subsequently, the International Strategy for Disaster Reduction. It has come to be understood that clear analysis of the root causes of risk is needed, along with a pragmatic approach geared to anticipating and reducing to the extent possible the specific needs of highly vulnerable people in disasters. As two leading researchers have noted: "Vulnerabilities precede disasters, contribute to their severity, impede effective disaster response and continue afterwards. Needs, on the other hand, arise out of the crisis itself, and are relatively short-term. Most disaster relief efforts have concentrated on meeting immediate needs, rather than on addressing and lessening vulnerabilities."[1]

The distinction between vulnerabilities, which are underlying conditions, and needs, which are created by the particular crisis, is one that was thrust on all Americans by the Gulf Coast hurricanes of 2005. Hurricane Katrina, especially, called into question the efficiency and effectiveness of local, state, and national emergency management systems, and it represented a clarion call for change in the way the nation protects its most vulnerable people and places. Because of Katrina, the people of the nation have a heightened awareness of their own and others' social vulnerability in disasters. Whether the costs of Katrina are measured narrowly, in terms of fatalities or physical damage, or more broadly, in terms of the suffering of survivors and the still-unknown economic, environmental, and political costs, Katrina drew attention to the social dimensions of the distribution of risk in the United States. What happened in New Orleans turned a spotlight on political choices made about economic and social development without regard to the long-term vulnerability of the city and its people in a catastrophic hurricane. The result, of course, was that countless Gulf Coast residents found themselves at the mercy not just of hurricane winds and floodwaters but also of more extreme social conditions that had marginalized many well before the storm. Katrina also made it evident that tabletop drills, media exposés, technical expertise, and science-based advance knowledge about known barriers to evacuation and other measures were, in fact, insufficient to prepare the residents of the Gulf Coast for a severe hurricane.

With these important lessons in mind, the section that follows discusses the changing concepts of social vulnerability, its complexity, "hidden" vulnerabilities, and structural trends that are increasing the nation's social vulnerability.

Changing concepts of vulnerability

Traditionally, disasters were viewed as "acts of nature" that were also "social levelers," in that they had similar impacts on those who were similarly exposed to a physical hazard. Vulnerability is still seen largely as a function of exposure to a natural hazard, with the exposure resulting from location and from attributes of the built environment deriving from such things as building codes, lifeline systems (i.e., water, power, and telecommunication services), and transportation infrastructure. Although vulnerabilities resulting from these exposures are understood to have differential social impacts (the young are not likely to be trapped in a nursing home by an earthquake), attention has been mainly on spatial vulnerability—vulnerability derived from physical space (location and the built environment) rather than from social attributes. In addition, individual attributes are traditionally seen as the cause of vulnerability—for example, the obvious needs of an infant or of a wheelchair-bound person living in a high-rise building. Special-need populations, as the term suggests, are considered to be disadvantaged by a special and individual life condition such as disability or ethnicity (see Table 13–1). Vulnerability may also simply be confused with poverty or may be viewed as a one-dimensional and stable point along a vulnerability continuum. Moreover, vulnerability has traditionally been regarded primarily as a response issue. For example, planners might ask how infants and toddlers could best be evacuated from a child care center, or seniors from high-rise buildings along Miami Beach.

In contrast, the approach to social vulnerability used here is broader. It asks which social groups across the community are more and less likely to have access to, and control over, the key assets and resources that help people "anticipate, cope with, resist, and recover from the impact of a natural hazard."[2] This highlights (1) people as members of groups with a shared social status, (2) the assets and resources that are available to the group, and (3) the possibility of increasing these capacities well before a disaster.

Relative vulnerabilities and capacities are both structural and situational. They may be shaped by structural patterns grounded in politics, economics, environmental management practices, race and class relations, the gender-based division of labor, and other factors. They may also be shaped by social status and situational or context-specific living conditions that vary over time—for example, temporary disabilities, a group's sense of safety on the streets, or degrees of functional literacy. Structural and situational vulnerabilities are often compounding; for example, Native American children living in substandard housing on isolated reservations are also likely to be exposed to contaminants from toxic-waste dumps nearby.

The significance of structural patterns was emphasized in the conclusion of the nation's second assessment of scientific knowledge about hazards and disasters, which looked at the distribution of risk and referred to disasters "by design."[3] In affluent and low-income nations alike, the distribution of risk reflects the social trends, environmental pressures, and social divisions of the larger society as well as physical differences and people's own actions. As noted in a leading text, "It is necessary to move beyond looking at disasters as simply physical events and consider the social and economic factors that make people and their living conditions unsafe or secure to begin with. Fragile livelihoods are as important as fragile buildings in understanding vulnerability to environmental hazards."[4]

Table 13-1 Selected vulnerability indicators for U.S. households, 2006

Selected characteristic	Number	Percent
Senior population, 65 years and over, as percentage of total population[a]	37,191,004	12
Children under 5 years, as percentage of total population[a]	20,385,773	7
Foreign born, as percentage of total population[a]	37,547,789	13
Speaks language other than English at home, as percentage of civilian noninstitutionalized population 5 years and over[b]	54,858,424	20
Disabled population, as percentage of civilian noninstitutionalized population 5 years and over [b]	41,259,809	15
Children with all parents in family in labor force, as percentage of parents with children 6-17 years[c]	32,019,857	70
Female householder, no husband present, with own children under 18 years, as percentage of all family households[d]	8,305,456	11
Renter-occupied housing, as percentage of all occupied housing units[e]	36,530,917	33
Nonfamily householders living alone, as percentage of total households[e]	30,496,588	27
Less than ninth-grade education, as percentage of persons 25 years and over[f]	12,743,555	7
Unemployed, as percentage of civilian population in the labor force[g]	9,702,558	6
Grandparents living with own grandchildren and responsible for grandchildren under age 18[h]	2,455,102	41
Individuals below poverty level	n/a	13
Percentage of families with female householder, no husband present, with related children under age 5 only, whose income in the past twelve months was below the poverty level	n/a	45
People 65 years and over whose income in the past twelve months was below the poverty level	n/a	10
No vehicles available	9,803,809	n/a
No telephone service available	6,571,249	n/a

Source: U.S. Census Bureau, *2006 American Community Survey*, tables for social, economic, housing, and demographic data, available at factfinder.census.gov/servlet/ACSSAFFFacts?_submenuId=factsheet_1&_sse=on (accessed October 1, 2007).

[a]Based on total U.S. population in 2006: 299,398,485.

[b]Based on total U.S. civilian noninstitutionalized population aged 5 years and over: 273,835,465.

[c]Based on total U.S. population with children aged 6-17 years: 45,942,524.

[d]Based on total number of U.S. family households: 74,564,066.

[e]Based on total number of U.S. households/occupied housing units: 111,617,402.

[f]Based on total U.S. population aged 25 years and over: 195,932,824.

[g]Based on total U.S. population aged 16 years and over in the civilian labor force: 151,203,992.

[h]Based on total number of grandparents living with own grandchildren under age 18 years: 6,062,034.

The everyday living conditions of the nation's poorest, sickest, most dependent, and most isolated residents directly and indirectly increase the exposure of these residents to physical hazards and to the social, economic, political, and psychological impacts of disastrous events. Living with old age or disabilities or both in a society designed for the young and able-bodied, for example, is a challenge in "normal" or pre-disaster times, but being elderly or disabled in substandard housing and in a risky place, or being poor and homeless and also in proximity to pollutants, translates into increased susceptibility to the impacts of disaster. According to estimates from the 2000 census, 20 percent of adults and 18 percent of children in the United States live in "distressed" neighborhoods with compounded vulnerabilities; these are places with very high levels of poverty, single-headed households, high school drop-out rates, and unemployment. In 2000, these neighborhoods were also racially distinct (55 percent were black, 29 percent Hispanic).[5] Such neighborhoods can be seen as "vulnerability hot spots."

Other factors that increase vulnerability do so by undermining community solidarity—an attribute that fosters resilience in the face of disaster. The undermining occurs when any group outside the mainstream is made to feel marginal; examples are when there is bias against new immigrants or certain religious groups, disapproval of women and men in non-traditional living arrangements, or unwarranted fear of those living with HIV/AIDS (human immunodeficiency virus/acquired immune deficiency syndrome). Social divisions along these lines are as much a part of life as their opposite—the positive social connections and social strengths that make Neighborhood Watch programs and voluntary neighborhood preparedness teams successful. Both the divisions and the connections are an important part of the social context of local emergency management.

The complexity of social vulnerability

The complexity of social vulnerability, even in American communities in the same hazard zone, has been graphically demonstrated by the Gulf Coast hurricanes and other recent events. The point is illustrated also by Chicago during an extreme heat wave in 1995. Here, two communities (one primarily African American, the other primarily Hispanic) appeared to be substantially comparable on statistical measures of social vulnerability such as poverty, single-headed households, high school drop-out rates, and unemployment. Nonetheless, very different death rates occurred, and researchers demonstrated that the difference was due to differences in the "social ecology" underlying the statistical comparability of the two communities (see accompanying sidebar). Effective emergency managers must consider the complexity of social vulnerability as they strive to reduce avoidable harm.

In particular, complexity often causes social vulnerability to be underestimated. For example, researchers were surprised to find that of the 26,000 South Carolina households affected by Hurricanes Bonnie, Dennis, and Floyd, 14 percent included physically disabled persons. These residents more often lived in mobile homes, were elderly, and owned pets; those on lower incomes suffered economic losses as high as 80 percent of their per capita income. Residents in households that included persons with disabilities had not neglected to prepare their

Heat wave: A tale of two neighborhoods

A "social autopsy" to determine why some neighborhoods were more hard-hit by Chicago's 1995 extreme heat wave than others revealed that African American seniors died more often than Hispanic seniors, although the surroundings of the two groups were more alike than different and the groups appeared equally vulnerable. Sociologist Erik Klinenberg discovered why the death rates differed by getting to know the social history of each area better, noting that significant differences were masked by statistical similarities in proportion of single-headed households, age, poverty, and minority ethnicity.

Predominantly Hispanic South Lawndale still has a "village feel," with an active street culture and a strong local business community. Never as strictly segregated as predominantly African American North Lawndale, South Lawndale had an infrastructure that had not deteriorated as severely, and new waves of immigration kept the population high and small ethnic businesses strong. Fewer seniors died there than in North Lawndale because

- Local amenities and public spaces drew seniors out of their homes, so they learned of the danger and of the help available to them.
- Seniors were accustomed to frequenting thriving local businesses, and these were air-conditioned.
- Strong ties to neighbors and lower crime rates made seniors less fearful of leaving their homes.
- An active street culture based on walking rather than driving improved seniors' overall health.
- Seniors often cared for grandchildren, so they retained close ties to adult children who could help.
- Powerful churches that were actively engaged in community life reached out to seniors.

Klinenberg concludes, "We have collectively created the conditions that made it possible for so many Chicago residents to die in the summer of 1995, as well as the conditions that make these deaths so easy to overlook and forget. We can collectively unmake them, too, but only once we recognize and scrutinize the cracks in our social foundation that we customarily take for granted and put out of sight" (p. 11).

Source: Erik Klinenberg, *Heat Wave: A Social Autopsy of Disaster in Chicago* (Chicago: University of Chicago Press, 2002).

homes, stockpile food, or rehearse plans for evacuation, but they were somewhat less likely to evacuate, citing transportation problems and a perceived lack of access to shelters.[6] For these kinds of social vulnerability patterns to be well understood and integrated into local emergency management planning, the complex aspects of social vulnerability must be investigated.

First, social vulnerability is not inevitably synonymous with lack of resilience, for a group may be vulnerable without lacking or losing the capacity to cope, adapt and bounce back from adversity. Affluent tourists, for example, may be temporarily exposed to high winds and floodwaters while at an oceanfront resort, but they are also more able than other transients or members of the resort's local staff to replace damaged possessions and resume employment. To take another example, being large or headed by a single individual can increase a household's vulnerability, but it does not inevitably do so. The Dominican married mother of one who works as a domestic for a professional single mother of three may well be less able to cope with the challenges of a major earthquake than her employer.

Second, as seen in the Chicago heat wave, apparent commonalities can mask significant differences. In one area, a neighborhood in which many Asian Americans reside may be very lightly affected because of household mitigation, good insurance coverage, and secure incomes. But in another area with the same proportion of Asian Americans, residents may be less affluent or may have recently emigrated from low-income nations. They may lack the money to improve their homes or to buy insurance, and they may be divided from one another by language or fear of gang violence. Within neighborhoods, some households include two earners and others only one. And within households, women may care for children single-handedly, may share caregiving responsibilities with their partners, or may rely heavily on child care centers that, in turn, may or may not be retrofitted or have emergency plans in place.

Social vulnerability is not inevitably synonymous with lack of resilience.

Third, physical commonalities can mask important differences. One heavily pregnant woman may move slowly but her family may own a car and be ready and able to help her prepare the household, pack belongings, evacuate, clean up, and return home. But across town, another woman in late pregnancy may live in a home for runaway teens, be without access to a car, have no contact with her family, and depend entirely on the facility manager or other residents for help. Predicting the relative vulnerability of elderly people and people with disabilities is equally complex because of the diversity and range of their life experiences.[7]

Fourth, residents of the same or nearby municipalities may have very different levels of exposure both to hazards (e.g., hazardous material spills or the flooding of low-lying lands) and to social vulnerability (the social and economic conditions that make it hard to cope). Among the factors that make some communities less resilient than others are a poor tax base, reliance on a single industry or crop, absence of strong institutions (schools, churches, social organizations), poor cooperation and coordination across institutions, ineffective government and leadership, inadequate land use planning and enforcement, minority segregation and discrimination, and a transient or unstable population. In any community with high hazard exposure, important differences exist between neighborhoods and households.

Hidden social vulnerabilities

As indicated, reducing the vulnerabilities of special populations requires an appreciation of community complexity.[8] This may include looking below the radar and behind the scenes to identify high-need groups not eager to be contacted. One example is abused women in shelters. After the 1997 flood in Grand Forks, North Dakota, crisis line calls to the domestic violence center increased by 47 percent over the same period one year earlier, and protection order requests increased by 65 percent. A domestic abuse advocate at the local shelter questioned whether designated evacuation centers would be safe for women who had left their homes in fear of violence, especially in small rural communities, fearing that this "compromises the security and safety element for women.... Grand Forks isn't a large enough place where [abusers] might not think of some of the other places that we might be putting them.... It's very easy to track somebody down, and that doesn't provide the kind of security and safety we want to be able to provide for our clients."[9]

Members of some highly vulnerable groups lay low from fear of government authorities, social discrimination, or harassment. One community organizer reported after the Loma Prieta earthquake in 1989 that "many Latinos around here think the federal government can just load them up in box cars and ship them off to Mexico, no matter how long they've lived here."[10] Runaway teens living on the streets may fear being returned by authorities to an abusive home. In the highly charged post–September 11 climate, an Arab American youth may well run from uniformed first responders, as do other youth of color in some American cities. Being required to describe their household living arrangements to relief agencies can be too threatening for gays and lesbians who already feel at risk of discrimination, harassment, and hate-motivated assault.

Members of some highly vulnerable groups lay low from fear of government authorities, social discrimination, or harassment.

Other groups that will have trouble protecting themselves in the event of a biological attack, hazardous spill, or flood may also be especially hard to reach because of stigma, transience, privacy needs, or mistrust of authorities. These include street children, homeless people who are mentally ill, severely ill AIDS patients cared for at home, substance-abusing street prostitutes, and noninstitutionalized people living with cognitive disabilities. Another important vulnerability that is hidden in plain sight is functional illiteracy. When non-English-speaking residents are excluded from the population, nearly one-quarter of U.S. adults (approximately 44 million people) are considered to be functionally illiterate. In most communities as many as one in four people (the proportion will be higher or lower depending on the social group) will need special assistance and materials written or presented in ways that are accessible to them. Like the deaf community, these residents can be challenging to identify and reach owing to the severe barriers they face in acquiring the information and other resources they need to protect their homes and families from hazards and to recover from disaster.

Another problem for local emergency managers is that the vulnerability and losses of some populations are simply not "seen" as readily as those of other populations. After the 2003 wildfires in San Diego County, for example, the media focused on the damage to houses in high-end suburbs more than on the damage to the San Pasqual Indian Reservation, where one-third of all residents had lost badly needed housing. One community leader told a reporter that whether they lose a mansion or a trailer, people are "equal when they are homeless."[11] But in fact, as emergency managers know, people are far from equally affected by the loss of a residence. Moreover, those most likely to be affected in the event of a disastrous accident, attack, or extreme environmental event may not even live in the immediate area. In the attacks of September 11, the dead included an estimated 500 undocumented workers employed in low-wage service jobs, many of whom left women and children in Central American villages without income.[12]

Structural trends that increase social vulnerability

Emergency managers and other local government officials should be aware of social changes that lead to heightened vulnerability. At the local level, for example, questions that officials need to ask about vulnerable subgroups include these: Are people here gaining or losing access to resources that can help to protect them or help them to cope and recover? Are the spaces they inhabit becoming more or less hazardous? Are they more or less well organized as a group now, and are we emergency managers more or less well connected with them? Unless emergency managers understand the significant trends that affect social vulnerability and can

Selected U.S. demographic trends increasing social vulnerability

- **Population:** According to one estimate, the nation's population as of the year 2000 is expected to double to 571 million by the end of this century.[1]
- **Migration:** Between 1995 and 2005, newly arrived unauthorized migrants have added about 700,000-800,000 a year to the U.S. population, roughly the same number as legal migrants have added.[2]
- **Coastal populations:** Population in the coastal states most at risk of an Atlantic hurricane increased by 244 percent between 1950 and 2006. In 2006, 12 percent of the nation's population (nearly 35 million people) were living in coastal communities from North Carolina to Texas, compared with 7 percent (10.2 million) in 1950.[3]
- **Minority populations:** Populations self-identified as American Indian, Asian/Pacific Islander, Hispanic, and black–groups historically subject to discriminatory treatment and social marginalization–are expected to constitute nearly half (47 percent) of the U.S. population in 2050, up from 28 percent in 1999.[4]
- **Senior population:** As the baby boomers age, the number of Americans age 65 and older will grow dramatically, from 12 percent of the nation's population in 2000 to an estimated 20 percent in 2030. The fastest rate of growth within this population is currently among those 85 and older, whose numbers are projected to more than double over this period.[5]
- **Children's poverty:** The poverty rate among children under age 18 remains high at 17 percent–roughly the same rate as in 1980; as a proportion of all Americans, this percentage is nearly twice that of elderly Americans age 65 and older, which declined from 16 percent to 9 percent between 1980 and 2006.[6]
- **Demand for caregiving:** Demand for family caregivers will outpace supply. While the population of people over 65 is expected to increase by 2.3 percent a year, the number of family members available to care for them will increase by only 0.8 percent a year.[7]
- **Health and disability:** Declines in disability related to age may be undercut by rising obesity as older Americans who are obese have more chronic illnesses and lower activity rates than older Americans had in the past. According to some research, baby boomers aged 51 to 56 report being in poorer health; having more difficulty with daily tasks; and having more pain, more chronic conditions, and more psychiatric problems than were reported by people in this age group just a decade ago. Disability rates for Americans aged 50 to 69 are projected to increase from 8 percent in 2005 to 9 percent in 2015.[8]
- **Single-parent households:** Among all U.S. households, single-parent households increased from 9 percent in 1960 to 28 percent in 2003, while family households (households including two or more people related by birth, marriage, or adoption) declined from 85 percent to 68 percent.[9]
- **Solitary living:** The percentage of adults living alone, including those in the age groups most likely to marry, increased from 8 percent in 1970 to 14 percent in 2002. Older Americans are the most likely to live alone.[10]

[1]Martha Farnsworth Riche, "America's Growth and Diversity: Signposts for the 21st Century," *Population Bulletin* 55, no. 2 (June 2000): 7, available at prb.org/Source/ACFD2C.pdf (accessed September 30, 2007).

[2]Jeffrey S. Passell, "Estimates of the Size and Characteristics of the Undocumented Population," Report of the Pew Hispanic Center (March 21, 2005), 2, available at pewhispanic.org/files/reports/44.pdf (accessed September 30, 2007).

[3]U.S. Census Bureau Press Release, "Special Edition: 2007 Hurricane Season Begins," available at census.gov/Press-Release/www/releases/archives/facts_for_features_special_editions/010106.html (accessed September 30, 2007).

[4]Riche, "America's Growth and Diversity," 16.

[5]Administration on Aging, "A Profile of Older Americans: 2003," available at aoa.gov/prof/statistics/profile/2003/4.asp (accessed September 30, 2007).

[6]Mark Mather, "U.S. Racial/Ethnic and Regional Poverty Rates Converge, but Kids Are Still Left Behind," Population Reference Bureau (August 2007), available at prb.org/Articles/2007/USRacialEthnicAndRegionalPoverty.aspx (accessed September 30, 2007).

[7]Katherine Mack and Lee Thompson with Robert Friedland, "Adult Children," *Data Profiles, Family Caregivers of Older Persons* (Washington, D.C.: Center on an Aging Society, Georgetown University, May 2001), 2, available at ihcrp.georgetown.edu/agingsociety/pdfs/CAREGIVERS2.pdf (accessed October 1, 2007).

[8]D'Vera Cohn, Mark Mather, and Marlene Lee, "Disability and Aging," Population Reference Bureau (August 2007), available at prb.org/Articles/2007/DisabilityandAging.aspx?p=1 (accessed September 30, 2007).

[9]Mark Mather, Kerri L. Rivers, and Linda A. Jacobsen, "The American Community Survey," *Population Bulletin* 60, no. 3 (September 2005), available at prb.org/pdf05/60.3The_American_Community.pdf (accessed September 30, 2007).

[10]AmeriStat Staff, "Solitaire Set Continues to Grow," Population Reference Bureau (March 2003), available at prb.org/Articles/2003/SolitaireSetContinuestoGrow.aspx (accessed September 30, 2007).

anticipate the consequences of those trends and take steps to reduce them, the nation's people and places will be at increased risk, and the burdens on emergency managers will continue to increase as well.

Paradoxically, despite the nation's wealth, high levels of education, advanced technologies, and the political capital devoted to risk management, a number of trends are increasing risk. While local emergency managers are not expected to be social scientists, they need a general understanding of how the changes in American society are affecting disaster resilience, and how the decisions and activities of politicians and leaders in the private and public sectors are affecting disaster risk. For example, when political and corporate leaders make decisions about transportation systems or wetland development, the climate is altered and natural hazards are affected in an indeterminate but certain way. Decisions made about fiscal policy, trade, and immigration also affect risk because they affect personal employment and income, the financing of affordable-housing construction, the feasibility of retrofitting critical facilities, and so forth. Clearly, the nation's population growth and economic development strategies are increasing the pressures on land development, especially along the coasts, and continuing urbanization concentrates the impacts of disaster events on people and commerce.

Social and demographic changes also contribute to the lessening of social bonds and therefore play a part in increasing vulnerability and decreasing resilience. The long hours that

Vulnerabilities and disparate impacts in Hurricane Katrina

Long-term effects are still uncertain, but the profile of those most affected by Katrina is strikingly similar to what was predicted in the Federal Emergency Management Agency's 2004 tabletop exercise, Hurricane Pam. What were the barriers to actions that might have reduced avoidable harm before Katrina hit and in the storm's aftermath?

- In Orleans Parish, where the city of New Orleans is located, 47 percent of the people whose deaths were attributed to Katrina were over age 75; men were disproportionately represented among the known fatalities relative to their age distribution.[1]
- Before the hurricane, the 464 buses available for evacuation could evacuate only 10 percent of those known not to have cars.
- African Americans constituted 44 percent of all Katrina's victims.[2]
- One-fifth of the population most directly affected by the hurricane was poor; their poverty rate of 21 percent was well above the national poverty rate of 12 percent.
- Nearly one in five (19 percent) of all residents in the affected areas—and a third of the residents aged seventy-five or older—had no car.
- Just 27 percent of poor families owned a home, compared with 62 percent of nonpoor families.
- Of persons 65 or older living in flooded or damaged areas, one in two (48 percent) lived with at least one disability, and one in four lived with two or more types of disability.
- Of community-based groups working with people with disabilities, 86 percent did not know how to link with the emergency management system. Fewer than a third of all shelters in New Orleans had access to American Sign Language interpreters, 80 percent lacked TTY (text telephone) capabilities, and 60 percent lacked televisions with open-caption capability. Just over half of these groups maintained areas for posting oral announcements.[3]
- Of those people displaced to the Houston Astrodome, 74 percent reported pretax incomes under $30,000, 72 percent did not own credit cards, and two-thirds were "unbanked" (i.e., lacking savings or checking accounts).[4]
- Louisiana ranked worst in the region (and nation) for poverty among African American women when the hurricane struck; in the city of New Orleans alone, 26 percent of women were poor, compared with 20 percent of men.[5]
- Before Katrina there were nearly 900,000 single mothers living in Alabama, Louisiana, and Mississippi, and 40 percent of them lived in poverty.
- In the city of New Orleans, 41 percent of female-headed families with children lived in poverty compared with 10 percent of families headed by married couples; more than half (56 percent) of all families with related children under eighteen were headed by women.
- Nearly 25 percent of people over age sixty-five in the city of New Orleans were poor, and nearly two-thirds of them were women (61 percent vs. 14 percent of men).

women and men now work outside the home leave fewer people with the time or energy to volunteer in traditional community education programs. Additionally, population mobility continues to be high (on average, just under half of all Americans move every five years),[13] so large numbers of community residents may be newcomers who lack knowledge about local hazards and preparedness. Demographic changes such as the growth in the Latino population in the United States make the nation more culturally and linguistically diverse—and the diversity can increase social vulnerability if linguistic and cultural barriers and disparities in living conditions are not addressed.[14]

Trends in housing play a part as well. Because of the declining availability of housing that poor and marginally employed people can afford, homelessness has increased, leaving more people unprotected and less able to get back on their feet in the aftermath of a disaster.[15] Low-income renters especially are increasingly likely to reside in low-cost manufactured homes situated in high-risk places.[16] Preparedness is far less possible for the low-income households most likely to live in flimsy housing; these populations are also more likely to rely on public transportation and to lack key economic recovery assets such as regular income, savings, insurance, and health benefits.

Aging, poor health, and sex play a part as well. More retired Americans than ever before are living on fixed incomes. Older women are less likely than older men to have been fully

- More than 40 percent of children under six lived below the federal poverty level in New Orleans before the hurricane. Because young children are more likely to live in poor families than older children or adults, they were least likely to be evacuated and most likely to spend time in the Superdome and other large shelters.[6]
- Minority residents and renters made up 74 percent and 54 percent, respectively, of area residents most susceptible to flooding. The majority (38 of 49) of census tracts characterized by extreme poverty were flooded—all located within the city of New Orleans and all predominantly African American.[7]
- Compared with the nation as a whole, the residents of New Orleans disproportionately lacked health insurance (19 percent vs. 16 percent for the nation); among those without health insurance there were more than twice as many black as white women.[8]

Nearly a quarter of a million people of Latin descent lived in the affected states, with an estimated 140,000 Hondurans in New Orleans alone. Residents who were characterized as "undocumented" on the basis of racial profiling were reportedly denied assistance and, in some cases, evicted from emergency shelters. The area before the hurricanes was also home to 115,000 Asian Americans and Asian immigrants. Some 10,000 Vietnamese settled in Houston, where they faced language and cultural barriers to accessing much-needed assistance.[9]

[1]For this and the following point, see Linda Bourque et al., "Weathering the Storm: The Impact of Hurricanes on Physical and Mental Health," *Annals of the American Academy of Political and Social Science* 604, no. 1 (2006): 138–140.

[2]For this and the following four points about pre-Katrina social life, see Thomas Gabe, Gene Falk, and Maggie McCarthy, *Hurricane Katrina: Social-Demographic Characteristics of Impacted Areas*, CRS Report for Congress, RL33141 (Washington, D.C.: Congressional Research Service, Library of Congress, 2005), available at gnocdc.org/reports/crsrept.pdf (accessed September 22, 2007).

[3]National Organization on Disability, *Report on Special Needs Assessment for Katrina Evacuees (SNAKE) Project* (Washington, D.C.: National Organization on Disability, 2005), available at nod.org/Resources/PDFs/katrina_snake_report.pdf (accessed September 22, 2007).

[4]Julia S. Cheney and Sherrie L. W. Rhine, "How Effective Were the Financial Safety Nets in the Aftermath of Katrina?" discussion paper (Philadelphia, Pa.: Payment Cards Center, Federal Reserve Bank of Philadelphia, 2006), 9, available at philadelphiafed.org/pcc/papers/2006/HurricaneKatrinaJan06.pdf (accessed September 22, 2007).

[5]For this and the following three points, see Barbara Gault et al., "The Women of New Orleans and the Gulf Coast: Multiple Disadvantages and Key Assets for Recovery: Part 1. Poverty, Race, Gender and Class," briefing paper, IWPR D464 (Washington, D.C.: Institute for Women's Policy Research, October 2005), available at iwpr.org/pdf/D464.pdf (accessed September 22, 2007).

[6]Olivia Golden, "Young Children after Katrina: A Proposal to Heal the Damage and Create Opportunity in New Orleans" (Washington, D.C.: Urban Institute, February 2006): 1, 3, available at urban.org/UploadedPDF/900920_young_children.pdf (accessed September 22, 2007).

[7]*New Orleans after the Storm: Lessons from the Past, a Plan for the Future* (Washington, D.C.: Metropolitan Policy Program, Brookings Institution, October 2005), 16–17, available at media.brookings.edu/mediaarchive/pubs/metro/pubs/20051012_NewOrleans.pdf (accessed September 22, 2007).

[8]Center for American Progress, "Who Are Katrina's Victims?" (Washington, D.C., September 2005), available at americanprogress.org/kf/katrinavictims.pdf (accessed September 22, 2007).

[9]Brenda Muñiz, *In the Eye of the Storm: How the Government and Private Response to Hurricane Katrina Failed Latinos* (Washington, D.C.: National Council of La Raza, 2006), available at nclr.org/content/publications/detail/36812 (accessed September 22, 2007).

employed or to be receiving pensions or Social Security, yet on their lower incomes they live longer than men; thus, older women are growing increasingly vulnerable to disaster and are forming the majority of the senior population that is most likely to be frail and living alone.[17] In addition, the aging of the baby boomers translates into larger numbers of residents likely to have physical and cognitive limitations. The frail elderly population, a group increasing in size, is disproportionately female and hence more likely to be poor than other seniors. In addition, this group tends to rely on caregivers who themselves are disproportionately female, of low-income, and of minority ethnic status.[18]

Household and family life, too, is changing in ways that increase risk. The percentage of female-headed households is increasing, and the women who are heads of households live in poverty at twice the rate of male heads of households; these women are also disproportionately from marginalized racial and ethnic groups.[19] Owing to maternal poverty and related factors, the children from these homes often live in substandard housing with caregivers who may lack jobs with secure benefits, not to mention reliable transportation in a disaster. High and increasing rates of child poverty in the nation also mean that growing numbers of children lack health insurance and therefore are without regular health care, so they are often facing the uncertainties of hazards and disasters while in poor health.[20] The national shift away from state-supported social services especially affects families like these that depend on the social safety net in the best of times.

Finally, because so many elements of the critical infrastructure of modern life are interdependent, the complexity of modern life itself increases the vulnerability of high-need populations. The schools, hospitals, local employers, and government social service agencies that serve the people who are at increased risk in disasters are affected by the susceptibility of some of the nation's complex electrical grids to ice storms and other environmental stresses. Vulnerability to such lifeline failures is compounded by the nation's increasing reliance on computer-based information management systems that depend on functioning grids.

Emergency managers who are working in government or the private sector must take such structural changes and trends in the nation at large into account in their work at the local level.

Strategies and tools for planning a broad approach

Clearly, effective outreach to high-need groups demands a good understanding of the major social trends increasing risk in American life, as well as specific knowledge of different forms of social vulnerability across the community and of the hidden pockets of vulnerability that may exist within a neighborhood. Working relationships are also needed with those who are most knowledgeable about the capacities—as well as the self-evident needs—of the residents who are most likely to be hard-hit in a disaster. This section offers strategies and models to promote just such a comprehensive and multidimensional approach to reducing social vulnerability.

Knowledge-building strategies

As all emergency managers know, the central tenet of the nation's mitigation strategy is "all mitigation is local." A critical element of local mitigation is striving to reduce social vulnerabilities, and this effort begins with local knowledge. Knowing their communities inside out and from the bottom up helps practitioners design and implement hazard mitigation initiatives that are tailored to local groups and conditions. With this knowledge, risk managers preparing for an impending strong hurricane or monitoring the course of an out-of-control wildfire can make the most of their time, energy, and resources to avoid costly missteps. For example, they will be better able to

- Build on local community knowledge of hazards, past disaster experiences, local resources, and local capacities
- Prepare and deliver effective warning messages that reach the intended recipients with the right message delivered the right way

- Develop and test evacuation, shelter, and recovery plans that meet the needs of all at-risk residents, with special attention given to those most in need
- Anticipate the special need for translators, child care, medical equipment, faith-based counseling, and other population-specific services
- Avoid using scarce resources in a costly and inefficient way through unintended bias or lack of information.

In particular, to increase community resilience, emergency managers need knowledge related to communication. They must know the kinds of media that specific groups perceive as credible, the languages that are commonly spoken, literacy levels, who has what degree of access to which kinds of information, the role of local opinion leaders and organizations that filter and translate information, and the alternative communication networks used by vulnerable groups.

Questions about vulnerability must be asked and answered at every phase as part of routine emergency management planning: Who in this census block or on this side of town or in this household will be least and most vulnerable today—and in ten years? How will vulnerable populations—undocumented immigrants employed off the books in reconstruction work, small businesswomen displaced to nearby states, or elderly men widowed by the tornado or flood—fare? What can be done now to increase protection of responders from the long-term health and economic effects of contaminated water or the stress of their occupations?[21] The accompanying sidebar lists other important questions.

It can be hard to gain the data or information needed to assess vulnerabilities and begin to answer these kinds of questions. For emergency managers seeking a more in-depth and bottom-up profile of their community and its most disaster-vulnerable people, a combination of strategies is vital. Quantitative information is typically the starting point—for example,

Planning questions for vulnerability reduction, by phase

Prevention or mitigation phase

- What social groups are less likely to be able to invest in making their homes safer?
- What social groups are likely to engage in occupations that expose them to higher risk from natural hazards?
- Are there locations or kinds of structures where certain social groups live that are more exposed to natural hazards than other locations or structures?

Preparedness phase

- What social groups are unlikely to have time to train in first aid and other kinds of self-protection?
- What social groups are less able to purchase critical items or supplies for self-protection?

Warning phase

- What social groups are likely not to receive warning messages or not to understand them or take them seriously? Why?

Response phase

- Are there characteristics of social groups that may make it more difficult for them to be rescued, to receive adequate emergency medical care, or to access or feel safe in an emergency shelter?
- What transportation or language barriers do some people have?

Recovery phase

- Which social groups are likely to experience problems with economic recovery? What about emotional recovery?
- Which groups will take longer to recover?

Source: Adapted from Ben Wisner, "Development of Vulnerability Analysis," in *A Social Vulnerability Approach to Disasters* (Washington, D.C.: Higher Education Project, FEMA Emergency Management Institute, 2002), 15-16, available at training.fema .gov/EMIWeb/edu/sovul.asp (accessed September 21, 2007).

tracking the proportion of elderly people in the city or the number of renters in a census tract. For more culturally-specific local knowledge—concerning, for example, the family relationships of highly vulnerable populations—additional strategies are useful. This section discusses sources of both quantitative and qualitative data, including statistical measures of vulnerability, pre- and post-disaster studies, data created within the community itself, and staff involvement with the community.

Statistical measures of vulnerability Census data for state, county, tract, and block are often used to assess vulnerabilities. Typical measures include income levels and ethnic/racial composition; homeownership and rental patterns; use of public or private transportation; percentage of elderly people, single-headed households, or high-school graduates; and other demographic characteristics of persons living in the same place. Some common sources of statistical community data are local, state, and regional planning offices; research institutes in nearby universities or colleges; law enforcement agencies; and social service agencies.

The disadvantages of using census data and other population statistics, however, are that such data need frequent updating, may be limited by methodological bias (e.g., sampling populations may consist only of people with known addresses or telephones), and may not provide all the information that emergency managers need. Knowing the proportion of renters in a flood-prone area, for example, does not tell practitioners much about the specific needs that renters may have. What is the group's general income level? Are the renters organized in a tenant's association that could work with community planners to mitigate hazards? How did they respond to past disasters? An additional drawback of population statistics is that they rarely provide data disaggregated by sex. Data on sex and data on minorities are often reported separately, even though sex and ethnicity have interdependent effects on disaster-related behavior. Planners, especially those at the local level, need more complete and accurate data.

Accordingly, statistical methods alone are less likely than community-based, multidimensional approaches to yield the grounded knowledge on which to base the proactive planning needed to reduce social vulnerability. Community-based approaches (discussed further on in this chapter) involve a variety of data-gathering strategies, including collaboration with local university research institutes, advanced students seeking internships, and researchers in the private sector. It is important to note, though, that emergency managers cannot be expected to do their own research; they should ask for what they need and should be critical consumers of statistical and other information.

Pre- and post-disaster studies Disaster events themselves are great teachers. Among the "lessons learned" following every disaster are the characteristics of those who fell through the cracks. Participating in local "unmet needs committees" (likely to arise in the aftermath of a disaster) is useful for gaining insight into both familiar and emergent concerns of vulnerable populations. Emergency management authorities can initiate post-event debriefing with key social service agencies and community-based organizations.

Action-oriented statistical research projects on actual disaster events also provide useful insight into special populations at risk, and emergency managers can initiate or participate in such projects. Case studies offer vivid evidence of the disparate impacts between and within communities, neighborhoods, and households and can help practitioners plan ahead to reduce avoidable harm in future disasters.

Social science research on vulnerability conducted before disasters is invaluable. A team of hazard researchers based in Massachusetts conducted a data envelopment analysis to help provide information for practitioners seeking to reach high-need groups. Working first with thirty-four variables for which block-level census data were available, and then with the five vulnerability clusters that emerged from their analysis (poverty, transience, disabilities, immigrants, and young families), the researchers integrated social vulnerability and hazard maps. As the authors note, "Many potential initiatives can be identified through vulnerability analysis, and the more 'proactive' or 'upstream' the step taken, the greater the downstream benefit."[22]

Data created within the community In addition to statistical data, lessons learned from actual disasters, and pre-disaster research, needs assessments, and user surveys conducted by nongovernmental or community-based groups should be used. For instance, a local coalition coordinating a crisis line might publish annual reports about factors affecting public health (e.g., rates of suicide, domestic abuse, substance abuse, homelessness, or interpersonal violence) and therefore the well-being of a community. Or the United Way or its member agencies might conduct needs assessments collecting highly relevant information.

Moreover, emergency management agencies and professionals are well positioned to get useful information through outreach to senior centers, health care facilities, and schools. They will also find that Voluntary Organizations Active in Disaster (VOAD), the network of local nonprofits that collaborate for integrated emergency response, has helpful organizational knowledge of the community.

As noted above, some special-need groups are easily overlooked through biased research methods or people's desire for social invisibility. To minimize such invisibility, practitioners may work with community researchers to modify survey questions, or with social service agencies to coordinate focus group discussions with vulnerable groups. In addition, community surveys can be used to build and update community registries of specific populations, although significant challenges would still remain, such as reaching at-risk individuals moving in and out of nursing homes.[23] Practitioners can also work directly with highly vulnerable groups to assess risk. This can accomplish two things: it can produce new information, and it can build or strengthen social networks with these important constituencies. Participatory information gathering for use in comprehensive community planning is highly useful "for the information it generates and distributes, for the sense of community it can foster, for the ideas that grow out of it, and for the sense of ownership it creates."[24]

Staff involvement in the community Still another way for emergency management staff to learn about social vulnerability is by being as actively involved at the grassroots level as possible, seeking community partnerships with members of high-need social groups and their advocates and representatives. To the extent feasible, practitioners can benefit from visibly participating in community events, seeking out high-need populations in relevant faith-based organizations; at sporting or cultural events; and through routine visits to local businesses, clinics, grocery stores, laundromats, parks, schools, and community centers. Regular consultation with people who are local experts on high-need groups is helpful; such people might include the residents, staff, and directors of halfway homes or home health care agencies, or groups operating local crisis lines or counseling new immigrants. Community experts in the social and human services and in grassroots advocacy groups are often eager to share their knowledge of the living conditions, predictable needs, and resources of people likely to be at increased risk.

Knowing the community by walking the community enables practitioners to more realistically assess the match (or mismatch) between need and resources.

Emergency managers are likely to already be working with local and regional governments, major employers and worker associations, health care and education providers, social service agencies, lifeline utilities, and such established community groups as Neighborhood Watch, VOADs and the American Red Cross. The approach to social vulnerability being urged here means also partnering with child care coalitions, street clinics, after-school programs serving ESL (English as a second language) students, grassroots women's groups, housing cooperatives working with migrants, associations of public housing residents, community groups working against violence, and other advocacy and service groups able to articulate and represent the interests of populations that will be highly vulnerable in a crisis.

Another vital resource may be new groups organized in the wake of a disaster. One example among many is the Emergency Network of Los Angeles (ENLA), a coalition of more than thirty organizations that united on behalf of the area's many low-income Central American immigrants after the Northridge earthquake in 1994. ENLA is now well established: it participates in the Los Angeles VOAD, trains other nonprofits, and contracts with city and county authorities to operate an information and referral hotline and other services. ENLA is also represented on the Emergency Preparedness Commission for the municipalities and counties that make up Greater Los Angeles.[25] It may be that many coalitions and associations emerging in the wake of Hurricane Katrina will also be significant partners for local emergency managers seeking to absorb the lessons of that disaster.

Through these relationships, emergency managers can both gather and exchange information, perhaps increasing vulnerable groups' awareness about hazards and hazard mitigation. Knowing the community by walking the community enables practitioners to more realistically assess the match (or mismatch) between need and resources—and enables them to connect with the advocacy organizations of vulnerable groups well in advance of a disaster event. Knowing the community by walking it also enhances emergency managers' ability to assess trends affecting these groups and allocate resources accordingly, and to plan ahead for preparedness and relief programs that build on community strength instead of increasing dependency. Strong networks with vulnerable communities also allow the tracking of long-term recovery among the community's most vulnerable people in the aftermath of a disaster; such tracking is a vital part of equitable and effective vulnerability reduction. Importantly, partnering with grassroots organizations that are "below the radar" not only helps practitioners assist highly vulnerable people but also builds a foundation of communication and trust for the future.

Tools for assessing community vulnerability and capacity

A number of tools for assessing community vulnerability and capacity are now available for local emergency managers, and many more are likely to become available in the future as awareness of the need to reduce social vulnerabilities in disasters grows. There is no magic bullet, but the sections below describe new ways of indexing social vulnerabilities, the expanding capabilities of computer modeling, the use of community-based risk assessments, and vulnerability resources for practitioners.

Vulnerability index Although some experts think that "vulnerability science is really in its infancy,"[26] the emphasis on vulnerability assessment in the Disaster Mitigation Act of 2000 makes it a significant and growing aspect of effective emergency management. No agreed-upon set of variables or features for assessing vulnerability exists in the United States. The Social Vulnerability Index developed by Susan Cutter and her colleagues, however, guides users to a wide range of statistical data available at the county level on such key vulnerability factors as urban density, infrastructure dependence, housing stock and tenancy, population growth, medical services, and social dependence as well as socioeconomic status, ethnicity and race, age, sex, and physical abilities.[27] A number of international efforts are also under way to develop standards, indices, and best-practice vulnerability assessment models.[28] In light of the diversity of hazards and vulnerabilities in play in modern life, a "one-size-fits-all" approach seems unlikely.

Underlying all the efforts is the view that

> [W]hat is needed is knowledge about who the most socially vulnerable people are within a population and where those less resilient reside. If we have a spatial understanding of the differences in social vulnerability, policies, procedures, and disaster management protocols can be put into place before an event occurs to minimize the impact of disaster events, thus saving lives and reducing property losses, rather than afterward. It highlights the need for proactive rather than reactive approaches to vulnerability reduction.[29]

Computer-assisted risk assessment Thanks to advances in mapping technologies, vivid portraits of risk that show an abundance of detail have entered the mainstream of emergency management. Practitioners can now create useful maps of changing "American hazardscapes"[30] by using geographic information system (GIS) mapping software that integrates data to give three types of information: (1) the spatial location of physical hazards; (2) indicators

Checklist for assessing resilience and vulnerability

- Has a vulnerability study been conducted? Has a resilience assessment been undertaken? Are the results current? Are the results useful? Should new studies be conducted?
- How can you divide your area up into localities/areas that are useful for social and community analysis?
- What data are available? What additional data or information will be required? Have appropriate data sources been identified? What methods are most appropriate to achieve practical results in assessing resilience and vulnerability?
- What risks does your area face? Are there individuals, groups of people, services, or areas that are particularly susceptible to risks?
- Are there resources, services, skills, or networks within the community that can be built on to optimize resilience and reduce vulnerability?
- What action has been taken on the findings about vulnerability?
- What action has been taken on the findings about resilience?
- Has the local emergency management plan been updated? Has a schedule to review the analysis of resilience and vulnerability been set?

Source: Adapted from Philip Buckle, *Guidelines for Assessing Resilience and Vulnerability in the Context of Emergencies* (Melbourne, Victoria, Australia: Victorian Government Department of Human Services, 2000), 5, available at proventionconsortium.org/themes/default/pdfs/CRA/Victorian_government_2000_meth.pdf (accessed September 22, 2007).

of social vulnerability, such as income level and ethnic composition, of neighborhoods subject to natural hazards; and (3) technological hazards such as waste disposal sites, train lines, oil refineries, and pollution-generating facilities. Computer-aided mapping can help emergency managers "ensure congruence between the maps of risk and the maps of preparedness."[31] Mapping both hazards and social conditions that increase vulnerability best promotes mitigation. In an area of high immigration, for example, emergency managers can assume the need for multilingual risk communications and can plan accordingly. In an area of high population mobility, emergency evacuation plans can be adjusted to ensure earlier warning and evacuation and the availability of transportation for those less physically able or mobile.[32]

The original and the multihazard (MH) HAZUS software packages developed by the Federal Emergency Management Agency (FEMA) are powerful tools using quantitative data.[33] With these tools, emergency managers can access maps and databases for their physical and social environments, which they can then use to estimate and depict the likely impacts of a range of hazards at different levels of analysis. Both direct and indirect effects can be projected for the hazard impacts on, for example, neighborhoods; facilities housing the young, sick, ill, and poor; commercial and industrial sites; critical infrastructures and lifeline facilities; structures with potential for high loss, such as dams or military installations; and other socially significant sites. HAZUS analysis, with its overlaying of physical hazard maps and social vulnerability maps, can help managers pinpoint high-need areas and locate shelters and other resources accordingly. For example, in high-income areas, residents will use shelters less often.

Computer-aided mapping can help emergency managers "ensure congruence between the maps of risk and the maps of preparedness."

As useful as GIS risk maps can be, however, practitioners' comfort levels with sophisticated mapping software vary,[34] and certainly maps are only as good as the data on which they are based. As the author of a major GIS handbook observed when addressing local emergency managers, "HAZUS creators acknowledge that coaxing some of this localized data from reluctant organizations may require considerable effort. They do not, however, include a scenario for having the software do this coaxing—that will be up to you."[35]

The data on which mapping is partly based are statistics for well-established indicators of vulnerability (e.g., percentage below poverty, racial/ethnic composition, educational level). Although necessary, these statistics are not sufficient, and practitioners need more specific and substantive knowledge. In some cases, statistical profiling is an especially weak tool. For example, some researchers have studied gender-based vulnerability by using the general sex ratio in the local population, but this ratio is relatively constant and cannot reflect the gender norms that bear on people's capacity and willingness to act in disasters. In a heat wave, for example, gender relations, relationships between the old and the young, and cultural and faith-based values all come into play but cannot be mapped with the use of readily available statistical indicators. Nonetheless, both maps and statistical profiles can be integrated into the emergency manager's broader base of knowledge about community strengths and weaknesses.

A user-friendly model used for assessing social vulnerabilities is the Community Vulnerability Assessment Tool (CVAT), supported by the National Oceanic and Atmospheric Administration and the H. John Heinz III Center. It consists of seven steps for gathering, assessing, and mapping information about hazards, vulnerabilities, and risks as part of a comprehensive community planning approach.[36] The CVAT model relies on quantitative data about infrastructure, critical facilities, economic conditions, and hazards of all kinds. The Societal Vulnerability Analysis (SVA) part of the model has three dimensions: identification of "special consideration areas," such as areas with high concentrations of poverty; identification of highly vulnerable people and areas, achieved by overlaying "special consideration" and hazard maps; and construction of a community inventory to determine where vulnerable facilities are located and what other issues need special consideration.

The SVA can be a benchmark that then supports further investigation. For example, the information-gathering strategies discussed above can be used to supplement the SVA by answering such questions as these: What community organizations (governmental, nongovernmental, public, private) are active in the community? What population sectors are well represented? What groups are not at the table? Which groups are most likely to be active after a disaster? Who is underserved? What support services (including child care, family counseling, domestic violence services, home health care, and recreational programs) are there for vulnerable families? Are there any bodies coordinating the activities of organizations that work with high-risk groups? What is the level of participation in the political process? Who speaks for minority groups? Who are the community leaders? What are the formal and informal power structures? Which people are likely to be the key players in disaster recovery?

Community-driven risk assessments Informal community mapping offers another perspective on local hazards, vulnerabilities, and capacities. Many international models exist that can be adapted for use in the United States and can be accessed through the Community Risk Assessment Toolkit on the website of the ProVentium Consortium.[37] Collaborating with or enabling local risk assessment also connects emergency managers with significant groups likely to be involved in emergency relief and long-term recovery efforts at the community level.

The Capacity and Vulnerability Analysis (CVA) matrix was originally created for development agencies responding to disasters, and it distinguishes among physical/material vulnerabilities (what productive resources, skills, and hazards exist?), social and organizational considerations (what are the relations and organization among people?), and attitudinal or motivational vulnerabilities and capacities (how does the community view its ability to create change?).[38] The matrix facilitates disaggregation by sex or economic status or race and ethnicity and can be used to highlight changes along different dimensions over time.

An additional model is *Working with Women at Risk: Practical Guidelines for Communities Assessing Disaster Risk.*[39] Women in high-risk villages in four Caribbean nations participated in a workshop providing basic training and information about hazards, disasters, and the gathering of qualitative data. Through "low-tech" risk maps, focus groups, semistructured interviewing, and community analysis, local women created community vulnerability profiles to educate community members and emergency managers about risky living conditions and resources facing the community as a whole. A step-by-step set of guidelines that are not specific to either sex and can easily be adapted is available online in both Spanish and English.

Also emphasizing the participation of local community members is a set of guidelines developed by Emergency Management Australia for assessing and reducing community vulnerability. These guidelines urge coordination with community groups in order to gain (and then use) knowledge about highly vulnerable people, and they offer useful checklists and guidance for assessing a community's strengths as well as its needs.[40]

Vulnerability resources for practitioners A growing number of informational, planning, preparedness, and good practice guides are now available to emergency managers who are oriented to a social vulnerability approach. These are useful resources for bridging the gap between the concerns of especially vulnerable residents and the concerns of emergency management authorities.

By networking with advocacy and service organizations in the local area and around the nation, emergency managers may access a range of self-help checklists, guidelines, and resource materials for working with vulnerable populations to reduce risk. Among these are good practice guides for working with people with disabilities. For example, following the 9/11 attacks, the National Organization on Disability (NOD) launched an Emergency Preparedness Initiative (EPI) resulting in the *EPI Guide for Emergency Managers, Planners & Responders;* like all NOD materials, the guide emphasizes the need to respect "the innate resourcefulness, ingenuity and determination gained through the daily challenges of disability that can help the

Working with vulnerable populations to reduce risk

The following social groups are among those that are often, but not always, at increased risk because of proximity to hazards, unsafe housing, lack of information, and limited capacity for self-protection and recovery.

Renters/public housing residents	Low-income/poor
Mobile home residents	Home workers and the self-employed
Single-headed households	Marginally employed
Pre-disaster homeless	Residents of institutions and group homes
Residents living alone	Persons with disabilities or chronic illness
Multifamily households	Religious minorities
Households with many dependents	Women
Female-headed households	Infants/young children
Newcomers	The functionally illiterate
Recent immigrants	The frail elderly
Residents dependent on state resources	Sexual minorities
Residents of unincorporated areas	Members of stigmatized groups
Migrant workers	Those who do not speak English
Undocumented residents	Marginalized racial and ethnic groups
Tourists/transients	Socially or geographically isolated residents

***In all groups, consider diversity* based on gender, sexuality, age, ethnicity, race, social class, and abilities.**

***For all groups, consider specific needs* likely to arise for**

Targeted communications	Community networking
Diversity of media outlets	Translators
Specialized equipment	Special needs shelters
Trained volunteers	Economic recovery assistance
Extended recovery period	Targeted mental health
Additional and prolonged recovery aid	Targeted reproductive health care
Child care/elder care assistance	Specialized transportation
Follow-up care and services	Assistance with evacuation

***For all groups, consider capacities* and resources that may exist:**

Strong social networks	Multilingual skills
Neighborhood bonds	Connections with advocacy groups
Extended family ties	Everyday survival skills
Valuable life experience	Prior disaster experience

***Plan ahead to reduce vulnerability* through**

Vulnerability and capacity assessments	Outreach to high-need groups
Coordination with social services	Coordination with advocacy groups
Diversity on the emergency manager's own staff	Community involvement

community at large and enhance the effectiveness of emergency operations."[41] PrepareNow, an American Red Cross initiative, is an excellent resource that addresses persons with mobility restrictions, those who depend on life support systems, chemically sensitive persons, owners of service animals, and persons with psychiatric illnesses, among others.[42] The U.S. Department of Justice Civil Rights Division also has materials available about equity in service for disabled persons in disasters.[43]

In the case of seniors, many of whom may also have mobility restrictions, the online guide *Disaster Preparedness for Seniors by Seniors* is useful.[44] It was inspired by the experiences of one Red Cross chapter after a two-week power loss. Regarding children, the Institute for Business & Home Safety offers *Protecting Our Kids from Disasters,* a kit to help parents and staff in child care centers undertake various kinds of nonstructural mitigation.[45]

Materials are also available about working with vulnerable people generally. With support from FEMA's Higher Education Project, a no-cost, online college instructor's guide *(A Social Vulnerability Approach to Disasters)* can now be downloaded.[46] Course materials include analysis of social vulnerability causes and patterns, discussion of the practical implications for emergency management, and slides and handouts that can be used in community education and training as well as in the college classroom. In addition, the California Governor's Office of Emergency Services has produced a helpful planning manual, *Meeting the Needs of Vulnerable People in Times of Disaster: A Guide for Emergency Managers.*[47] This document urges emergency authorities in local government to contract in advance with community organizations for specific services (e.g., operating a hotline or providing hot meals to low-income seniors after a flood) and to seek recovery for these contractual costs. The guide also offers ideas for locating and working with community groups; tips for developing local associations that bring emergency managers and vulnerable people together; a sample protocol for pre-disaster collaboration with local organizations representing high-risk groups; and other practical materials for working with vulnerable populations to reduce risk.

VOAD member organizations often produce publications that focus on high-need groups. Two such organizations are Collaborating Agencies Responding to Disaster (CARD), based in California, and Community Emergency Response Teams (CERTs). A Red Cross initiative of interest is the Northern California Disaster Preparedness Network, which prepared a "Guide to Organizing Neighborhoods for Preparedness, Response and Recovery."[48] This guide includes tips for organizing Neighbors with Special Needs teams and registries for special-needs groups.

Emergency managers will learn of more and better guides and resources in the future through regular communication with proactive community-based advocacy groups that understand the capacities and needs of high-risk groups. When establishing such communication is not possible at a personal level, the many online and print guides now available to local emergency managers can be useful for integrated planning to address cross-cutting vulnerabilities or to help jump-start local initiatives with vulnerable groups.

Concluding comments

As this chapter has shown, social vulnerability to hazards and disasters is as much a concern in the United States and other affluent societies as it is in the developing world, where the vulnerability is often more obvious. Understanding the patterns and trends of vulnerability is now an important part of managing emergencies and disaster, and it affects the work of emergency managers throughout the disaster cycle. This chapter has presented concrete strategies and tools for addressing these concerns in practice. More generally, the message of the chapter is to "face social and demographic reality: plan through an inclusive, deliberate process. Rely on community resources within potentially affected populations. Build relationships before disaster to mitigate physical and social effects."[49]

Addressing the needs of vulnerable populations is not a new concern in emergency management, but for reasons given above, the project has taken on more urgency and breadth. The changing face of America makes vulnerability reduction an essential component of the skill set of tomorrow's emergency managers. (Indeed, as more diverse career paths into the profession evolve, the face of emergency management will change as well.) All emergency managers of the future will find it necessary to develop or strengthen their ability to analyze, understand, and reduce the vulnerabilities of many specific populations within their communities. They must become more knowledgeable about the complexity of their communities, resisting easy stereotypes about different cultures, different kinds of abilities, different ways of organizing family life, and different relationships between women and men, the young and the old. They must be skilled in communicating with high-risk groups, not just through translation or interpretation but by understanding communication barriers of all kinds and working with community representatives to tailor messages and media as needed. They will be increasingly motivated to forge close working relationships with communities at risk and will be needed as advocates on behalf of those communities when tough political decisions must be made about the distribution and use of all the nation's resources. When the next disaster unfolds, the efforts made by local emergency managers to identify "hidden" vulnerabilities, meet critical needs, build on the capacities of even the most vulnerable, and partner creatively with high-risk groups will be well-rewarded.

Reducing social vulnerability is not a short-term or simple process—but it is also not "mission impossible." In a climate of uncertainty about the national capacity to protect all residents exposed to and victimized by disastrous events, a risk management approach geared to recognizing and reducing social vulnerabilities is essential. Local emergency managers who take the lead will be honored for their efforts by future generations of Americans at risk.

Notes

1 Mary Anderson and Peter Woodrow, *Rising from the Ashes: Development Strategies in Times of Disaster* (Boulder, Colo.: Westview Press, 1989), 10.

2 Piers Blaikie et al., *At Risk: Natural Hazards, People's Vulnerability, and Disasters* (London: Routledge, 1994), 9.

3 Dennis Mileti, *Disasters by Design: A Reassessment of Natural Hazards in the United States* (Washington, D.C.: Joseph Henry Press, 1999).

4 Robert Bolin and Lois Stanford, *The Northridge Earthquake: Vulnerability and Disaster* (London: Routledge, 1998), 42.

5 William O'Hare and Mark Mather, "The Growing Number of Kids in Severely Distressed Neighborhoods: Evidence from the 2000 Census" (Baltimore, Md.: Annie E. Casey Foundation and the Population Reference Bureau, October 2003), i, available at aecf.org/upload/PublicationFiles/DA3622H1280.pdf (accessed September 22, 2007).

6 Marieke Van Willigen, "Do Disasters Affect Individuals' Psychological Well-Being? An Over-Time Analysis of the Effect of Hurricane Floyd on Men and Women in Eastern North Carolina," *International Journal of Mass Emergencies and Disasters* 19, no. 1 (2001): 59–83.

7 A good review of research findings on age and vulnerability is Neili Langer, "Natural Disasters That Reveal Cracks in Our Social Foundation," *Educational Gerontology* 30, no. 4 (2004): 275–285. For an excellent discussion with good examples from the field, see National Council on Disability, "Saving Lives: Including People with Disabilities in Emergency Planning" (Washington, D.C.: National Council on Disability, 2005), available at ncd.gov/newsroom/publications/2005/saving_lives.htm (accessed August 21, 2007).

8 Betty Hearn Morrow, "Identifying and Mapping Community Vulnerability," *Disasters* 23, no. 1 (1999): 1–18, available at sciencepolicy.colorado.edu/about_us/meet_us/roger_pielke/envs_5120/week_12/Morrow.pdf (accessed September 22, 2007).

9 Personal interview with author, Grand Forks, North Dakota; see also Elaine Enarson, "Violence against Women in Disasters: A Study of Domestic Violence Programs in the U.S. and Canada," *Violence Against Women* 5, no. 7 (1999): 742–768.

10 Bolin and Stanford, *The Northridge Earthquake*, 27.

11 Charlie LeDuff, "In Scorched Hills, Tribes Feel Bereft and Forgotten," *New York Times*, November 5, 2003, A14.

12 Ben Wisner, "Tepeyac: Case Study of Institutional and Social Learning under Stress," *International Journal of Mass Emergencies and Disasters* 21, no. 3 (2003): 59.

13 U.S. Census Bureau, "Geographical Mobility: 1995–2000," *Census 2000 Brief* (September 2003), 1, available at census.gov/prod/2003pubs/c2kbr-28.pdf (accessed August 21, 2007).

14 Alice Fothergill surveys practical differences in vulnerability and disaster experiences in "Gender, Risk, and Disaster," *International Journal of Mass Emergencies and Disasters* 14, no. 1 (1996): 33–56. See also Alice Fothergill, Enrique Maestas, and JoAnne Darlington, "Race, Ethnicity and Disasters in the United States: A Review of the Literature," *Disasters* 23, no. 2 (1999): 156–173; and Ronald W. Perry and Alvin H. Mushkatel, *Minority Citizens in Disasters* (Athens: University of Georgia Press, 1986).

15 See the 2002 survey of twenty-five American cities conducted by the U.S. Conference of Mayors, *A Status Report on Hunger and Homelessness in America's Cities, 2002* (Washington, D.C.: U.S. Conference of Mayors, December 2002), available at usmayors.org/uscm/hungersurvey/2002/onlinereport/HungerAndHomelessReport2002.pdf (accessed August 21, 2007).

16 For more on the economic and housing pressures on low-income renters, see the Web site of the National Housing Trust Fund, including the fact sheet "A Home Is the Foundation," available at nlihc.org/doc/FactSheet.pdf (accessed August 21, 2007).

17 See Reneé Spraggins, "We the People: Women and Men in the United States," *Census 2000 Special Report* (Washington, D.C.: Bureau of the Census, U.S. Department of Commerce, January 2005), available at census.gov/prod/2005pubs/censr-20.pdf (accessed August 21, 2007).

18 Among other sources of information on unpaid caregivers, see "Caregiver Health" on the Family Caregiver Alliance Web site at caregiver.org/caregiver/jsp/content_node.jsp?nodeid=1822 (accessed August 21, 2007).

19 U.S. Census data for 2000 cited in "Poverty in the US: A Snapshot" on the Web site of the National Center for Law

and Economic Justice, nclej.org/poverty-in-the-us.php (accessed August 21, 2007); also see Jennifer Cheeseman Day, "Projections of the Number of Households and Families in the United States: 1995 to 2010," *Current Population Reports* P25-1129 (Washington, D.C.: Bureau of the Census, U.S. Commerce Department, April 1996), available at census.gov/prod/1/pop/p25-1129.pdf (accessed August 21, 2007).

20 One account of children in poverty in the United States can be found in "Who Are America's Poor Children?" on the Web site of the National Center for Children in Poverty, nccp.org/publications/pub_684.html (accessed August 21, 2007).

21 Ben Wisner, "Development of Vulnerability Analysis," in *A Social Vulnerability Approach to Disasters* (Washington, D.C.: Higher Education Project, FEMA Emergency Management Institute, 2002), 15–16, available at training.fema.gov/EMIWeb/edu/sovul.asp (accessed September 21, 2007

22 George Cook et al., "Assessing the Vulnerability of Coastal Communities to Extreme Storms: The Case of Revere, MA., USA," *Mitigation and Adaptation Strategies for Global Change* 3, no. 1 (1998): 78, available at nome.colorado.edu/HARC/Readings/Clark.pdf (accessed September 22, 2007).

23 William C. Metz et al., "Identifying Special-Needs Households That Need Assistance for Emergency Planning," *International Journal of Mass Emergencies and Disasters* 20, no. 2 (2002): 255–281.

24 Mileti, *Disasters by Design*, 34.

25 Burt Wallrich, "The Evolving Role of Community-Based Organizations in Disaster Recovery," *Natural Hazards Observer* 21 (November 1996), available at colorado.edu/hazards/o/archives/1996/nov96/nov96.html#CBOs (accessed August 21, 2007).

26 Arlene Hill and Susan Cutter, "Methods for Determining Disaster Proneness," in *American Hazardscapes: The Regionalization of Hazards and Disasters*, ed. Susan Cutter (Washington, D.C.: Joseph Henry Press, 2001), 25.

27 Susan Cutter, Bryan Boruff, and W. Lynn Shirley, "The Evolving Role of Community-Based Organizations in Disaster Recovery," *Social Science Quarterly* 84, no. 2 (June 2003): 242–261.

28 For example, information about the search for meaningful indicators of risk reduction can be found on the Web site of the ProVention Consortium at proventionconsortium.org/?pageid=32&projectid=25 (accessed August 21, 2007).

29 Susan Cutter and Christopher T. Emrich, "Moral Hazard, Social Catastrophe: The Changing Face of Vulnerability along the Hurricane Coasts," *Annals of the American Academy of Political and Social Science* 604, no. 1 (2006): 110.

30 Susan L. Cutter, ed., *American Hazardscapes: The Regionalization of Hazards and Disasters* (Washington, D.C.: Joseph Henry Press, 2001).

31 Mark S. Monmonier, *Cartographies of Danger: Mapping Hazards in America* (Chicago: University of Chicago Press, 1997), 236.

32 For a good example, see George Cook et al., "Assessing the Vulnerability of Coastal Communities."

33 Information about HAZUS can be found on FEMA's Web site at fema.gov/plan/prevent/hazus/ (accessed September 22, 2007).

34 Regular workshops and support services, including online user groups, are available for HAZUS users; see the HAZUS Web site at fema.gov/plan/prevent/hazus/hz_users.shtm (accessed August 21, 2007).

35 R. W. Greene, *Confronting Catastrophe: A GIS Handbook* (Redlands, Calif.: ESRI, 2002), 13.

36 National Oceanic and Atmospheric Administration (NOAA) Coastal Services Center, *Community Vulnerability Assessment Tutorial*, online CD-ROM, available at csc.noaa.gov/products/nchaz/htm/methov.htm (accessed September 22, 2007). Through its Web site and list server, NOAA also offers regular workshops and support services for CVAT users.

37 "Community Risk Assessment Toolkit," available from the ProVention Consortium at proventionconsortium.org/?pageid=39 (accessed August 21, 2007).

38 Anderson and Woodrow, *Rising from the Ashes*.

39 Elaine Enarson et al., *Working with Women at Risk: Practical Guidelines for Assessing Local Disaster Risks* (Miami: International Hurricane Research Center, Florida International University, June 2003), available at gdnonline.org/resources/WorkingwithWomenEnglish.pdf.

40 Philip Buckle, Graham Marsh, and Sydney Smale, *Assessing Resilience & Vulnerability*, report to Emergency Management Australia (May 2001), available at radixonline.org/resources/assessment-of-personal-and-community-resilience.pdf (accessed August 21, 2007). These and other guidelines are summarized, and guidance notes are offered through the ProVention Consortium at proventionconsortium.org/themes/default/pdfs/CRA/EMA_2001.pdf (accessed August 21, 2007).

41 Emergency Preparedness Initiative, *Guide on the Special Needs of People with Disabilities for Emergency Managers, Planners & Responders* (Washington, D.C.: National Organization on Disabilities, 2004), 2, available at nod.org/resources/PDFs/epiguide2004.pdf (accessed September 22, 2007).

42 See the Web site of PrepareNow.Org at preparenow.org (accessed September 22, 2007).

43 U.S. Department of Justice, *An ADA Guide for Local Governments: Making Community Emergency Preparedness and Response Programs Accessible to People with Disabilities* (Washington, D.C.: U.S. Department of Justice, n.d.), available at usdoj.gov/crt/ada/emergencyprep.htm (accessed August 21, 2007).

44 "Disaster Preparedness for Seniors by Seniors" is available on the Red Cross Web site at redcross.org/services/disaster/beprepared/seniors.html (accessed September 22, 2007).

45 Institute for Business & Home Safety, *Protecting Our Kids from Disasters: Nonstructural Mitigation for Child Care Centers* (Tampa, Fla.: Institute for Business & Home Safety, 1999), available at ibhs.org/publications/downloads/461.pdf (accessed August 21, 2007).

46 Information about the course "A Social Vulnerability Approach to Disasters," part of the FEMA Emergency Management Higher Education Project, is available at training.fema.gov/emiweb/edu/completeCourses.asp (accessed August 21, 2007).

47 California Governor's Office of Emergency Services, *Meeting the Needs of Vulnerable People in Times of Disaster: A Guide for Emergency Managers* (Sacramento: Governor's Office of Emergency Services, 2000), available at oes.ca.gov/Operational/OESHome.nsf/PDF/Vulnerable%20Populations/$file/Vulnerable%20Populations.PDF (accessed August 21, 2007).

48 Volunteer Center of Marin, "A Guide to Organizing Neighborhoods for Preparedness, Response and Recovery" (San Raphael, California), available at preparenow.org/marin-g.html (accessed September 22, 2007).

49 Brenda Phillips, "Cultural Diversity in Disasters: Sheltering, Housing, and Long-Term Recovery," *International Journal of Mass Emergencies and Disasters* 11, no. 1 (1993): 104.

CHAPTER 14

New information technologies in emergency management

Susan L. Cutter, Christopher T. Emrich, Beverley J. Adams, Charles K. Huyck, and Ronald T. Eguchi

This chapter provides an understanding of

- The information technologies available for emergency management
- The role of information technology in emergency management since 1989
- Tools and applications appropriate to specific phases of emergency management
- Barriers to the use of information technology
- Issues of concern.

Disaster events are becoming increasingly costly and more devastating each year: 2005's Hurricane Katrina, which incurred more than $100 billion in estimated losses, was the most costly disaster in U.S. history, while the 2004 South Asian tsunami was among the deadliest of all natural disasters worldwide. This trend underscores the need for the most sophisticated tools available that can tell us what, at any given moment, is actually happening.

At the same time, a less apparent but equally significant trend has been the growing use of information technology (IT) in the emergency management field. Over the past decade, information technologies aimed at reducing the impact of hazard events or aiding in the protection of places and populations at risk have grown from theoretical constructs and "gee-whiz" gadgetry to affordable, usable, and practical methods for emergency management. These advancements provide better data collection and management, more useful visualization capabilities, intelligent sensors for monitoring potential disasters, more robust communication systems, enhanced loss estimation techniques, improved geographic information system (GIS) technology, more robust global positioning system (GPS)-based assessment tools, and next-generation remote sensing (RS) practices and hardware. The result is an increased use of IT during all phases of disaster management—preparedness, mitigation, response, and recovery.

In many cases, these technological solutions have been around for years, but their application or deployment for the various phases of disaster management were often nonexistent or slow in coming. Part of the reason was the long lead time required for response organizations to understand and embrace their efficacy in facilitating response and recovery operations: almost always, the implementation of new technologies requires the replacement of "old, but tried and true" procedures, and attempts to update long-standing traditional solutions are met with extreme resistance. In other cases, the workforce did not have the core skills or proper training necessary to make full use of the available technology, or the organizations lacked the funding to procure technological expertise and promote knowledge transfer. In general, most if not all new information technologies that are available for emergency management today require a moderate to advanced understanding of computer systems and current data compilation and storage techniques.[1]

However, disasters act as catalysts in the adoption of new and emerging technologies, and these innovations have enabled emergency management personnel to rapidly collect vital information for disaster management, to assess the impact of large disasters more effectively and rapidly, and to track and monitor progress in critical response and recovery operations. This chapter describes these information technologies and discusses how they have proven to be effective in disaster management or are considered important components of future disaster field operations.

Overview of information technologies available for emergency management

As the technologies in computing and information transfer increase, so does the ability to use those technologies for disaster management. The information technologies available for emergency management can be classified into five broad areas: (1) advanced computing, (2) geographic information systems, (3) remote sensing, (4) expert systems and the Internet, and (5) wireless technology.

Advanced computing

Advanced computing, or the use of computer-based modeling to simulate real-world phenomena, is of major benefit to emergency management. In hazard assessment, many more possible outcomes can be explored through simulation—and usually at a fraction of the cost and time—than by simply analyzing past events. Computer modeling and simulation make it possible to explore options that are beyond current experimental capabilities as well as to study phenomena that cannot be replicated in laboratories, such as the potential impact of a Category 5 hurricane on a particular coast. The results often suggest new experiments and theories, adding to the knowledge base and enabling improved predictions for future hazard events. Advanced computing also facilitates the rapid integration, analysis, display, and dissemination of information that is essential for processing the volume of high-resolution data generated by modern remote sensing instruments.

Advanced computing is not without its drawbacks, however—notably the fact that the simulations it produces are models and, as such, are fraught with many simplifying assumptions written as mathematical equations and computer codes. More importantly, such models and scenarios can't capture all the complexities of the phenomena being modeled or simulated.

Geographic information systems

Geographic information system (GIS) technologies and methods have been in use in emergency operation centers (EOCs) and disaster field offices nationwide for more than a decade and seem to increase in popularity by the day. A GIS "uses geographic location to relate otherwise disparate data and provides a systematic way to collect and manage location-based information crucial to local government."[2] Because the ability to gather, manipulate, query, and display geographic data quickly and present it in an understandable format is paramount to the mission of emergency management, GIS technology is critical to effective, collaborative decision making during disaster situations.

Although GIS capabilities have improved the ability to respond to disaster situations in real-time or near real-time, some drawbacks to the use of this technology have become apparent in recent years. For example, the needed real-time data and products are not always available; products cannot always be delivered quickly; and efforts to train decision makers in the proper use of GIS techniques are not always successful.[3]

Remote sensing

Remote sensing (RS) takes the idea of a visual inspection from the air to the next level. With RS, image sensors are flown by plane, helicopter, or satellite over an area of interest, collecting optical and radar-based imagery and transforming it into spatial information that can then be analyzed to gain an understanding of conditions at that moment in time (e.g., areas of flood extent). This makes RS most beneficial in hazard monitoring and large-scale damage assessments (see accompanying sidebar).[4] It can also be used to compile large quantities of data for the construction of mitigation tools (e.g., vegetation moisture indexes that can be used in fire control). But although this technology is able to provide data that might not otherwise be available to emergency managers, it is often cost prohibitive and very time sensitive, which limits its use by local responders.

Logistical support using remote sensing

Following the 9/11 terrorist attack on the World Trade Center, it was widely recognized that several remote sensing technologies were underused in response efforts.[1] For example, correctly calibrated temperature readings that would have been valuable to firefighters were unavailable until early October. To facilitate the collection of appropriate and timely data in the aftermath of extreme events within the United States, the Federal Emergency Management Agency and the National Aeronautics and Space Administration have established the Remote Sensing Consultation and Coordination Team (RSCOT).[2] This team, which was in a state of readiness during the 2002 Winter Olympic Games in Salt Lake City, was tasked with identifying suitable data, coordinating its acquisition, and distributing the resulting imagery to everyone involved in the response effort. To support data collection through the RSCOT system, contractual agreements between the data providers and those entities that will use the data should be in place before an event occurs, as should the trained workforce needed to implement the system.

[1]Andrew J. Bruzewicz, "Remote Sensing Imagery for Emergency Management," in *The Geographical Dimensions of Terrorism*, ed. Susan L. Cutter, Douglas B. Richardson, and Thomas J. Wilbanks (New York: Routledge, 2003), 87-97; Charles K. Huyck, Beverley J. Adams, and David I. Kehrlein, "An Evaluation of the Role Played by Remote Sensing Technology Following the World Trade Center Attack," *Earthquake Engineering and Engineering Vibration* 2, no. 1 (2003): 1-10; Deborah S. K. Thomas et al., "Use of Spatial Data and Geographic Technologies in Response to the September 11 Terrorist Attack on the World Trade Center," in Natural Hazards Research and Applications Information Center, Public Entity Risk Institute, and Institute for Civil Infrastructure Systems, *Beyond September 11th: An Account of Post-Disaster Research*, Program on Environment and Behavior Special Publication 39 (Boulder: Natural Hazards Research and Applications Information Center, University of Colorado, 2003), 147-162, available at colorado.edu/hazards/publications/sp/sp39/sept11book_ch6_thomas.pdf (accessed June 16, 2007).

[2]Ronald J. Langhelm and Bruce A. Davis, "Remote Sensing Coordination for Improved Emergency Response," paper presented at the Pecora 15/Land Satellite Information IV/ISPRS Commission I/FIEOS 2002 Conference, November 10-15, 2002, Denver, Colorado, available at isprs.org/commission1/proceedings02/paper/00083.pdf (accessed June 26, 2007).

Expert systems and the Internet

Expert systems—decision support tools and applications founded upon expert opinion and knowledge—are designed to make information quickly available to those decision makers who need it. The Internet is often used to bridge the gap between the expert system and the stakeholder. Coupled with currently available expert-based GIS and RS tools and technologies, the Internet is already used by many federal, state, and local governments in the planning and mitigation of disasters (see sidebar on expert systems). By fundamentally altering the nature and structure of communications and information flow, the Internet has transformed emergency management, and its capabilities have proven to be vital in all four areas of emergency management. Available spatial data are only clicks away from emergency managers and decision makers, who need to know the most up-to-date information on disasters (see sidebar on GoogleEarth); essential information and sources of assistance are immediately accessible to the populations at risk from potential hazard impacts (see sidebar on the Kobe earthquake).

There are, however, just as many disadvantages as advantages to the use of Internet technology. Despite its seemingly ubiquitous nature, the Internet is not a universally embraced technology nor is it accessible everywhere. These problems are exacerbated not only in rural areas of the United States, where access is limited by lack of adequate infrastructure, but also in some of the urban impoverished communities; the result is the well-known "digital divide." Additionally, the reliability and quality of information being transmitted via the Internet often come into

Finding the right satellite: Expert systems for remote sensing technology

Satellites currently orbiting the Earth can capture data, such as information on flooded areas or transportation blockages, that are important for the immediate and longer-term response functions of emergency management. But without the expert knowledge of a remote sensing (RS) liaison, emergency managers do not know which satellite to task with what data collection, whether the needed imagery has already been collected, whether a specific "bird" has already been tasked with collecting imagery that could be multifunctional, or when the next available satellite is crossing over the area of interest.

Now, an expert system is being developed at the University of South Carolina's Center for GIS & Remote Sensing that will make all the information on available satellites and their capabilities easily accessible to anyone with an Internet connection. The National Aeronautics and Space Administration and university staff are developing the Remote Sensing Hazard Guidance System (RSHGS), a Web-based system that will help emergency managers acquire and analyze remotely sensed imagery for response to a hazard event. Employing a GIS-based modeling subsystem and a satellite query subsystem, the RSHGS will be able to automatically determine optimum sensors/satellites for the hazard of interest, the geographic study area, and the time frame (e.g., forty-eight hours). Users will be able to independently determine the current and future locations and collection abilities of any unclassified satellite-borne sensor. This interdisciplinary and multiorganizational initiative will act as the RS expert to provide the stakeholder with information that he or she would otherwise be unable to access.

GoogleEarth and Hurricanes Katrina, Rita, and Wilma

The 2005 hurricane season led to revolutionary advances in the use of Internet technologies for hurricane response and recovery. Particularly noteworthy was the effort to use GoogleEarth for immediate disaster response efforts in the hardest-hit areas along coastal Mississippi and Alabama. Free professional editions of GoogleEarth software were deployed in many Federal Emergency Management Agency geospatial intelligence units immediately following Hurricane Katrina. Additionally, the Internet giant quickly added nearly 8,000 post-disaster aerial photographs of flooded areas taken by the National Oceanic and Atmospheric Administration, allowing disaster relief workers to scan images on the computer to identify specific areas of interest before formulating specific urban search and rescue plans.[1]

[1]Manfred Dworschak, "The Mapping Revolution: How Google Earth Is Changing Science," *Spiegel Online*, August 1, 2006, available at spiegel.de/international/spiegel/0,1518,429525,00.html (accessed July 2, 2007).

Internet technology and the Kobe, Japan, earthquake

The 1995 Kobe earthquake, Japan's worst natural disaster in sixty years, was one of the prominent recent events in which the Internet was used in the initial response phase of the emergency. In the aftermath of the massive earthquake, telephone lines across the city were immediately cut, but the Internet link to Kobe never failed, and it supplied many people with the latest information regarding damages, casualties, and, most importantly, critical response information.[1] The Sony Corporation set up a Web site to provide information to the public, including announcements from the government, images and maps of damages, fatalities and current relief efforts, lists of usable phone numbers and Internet connections, and information about donations and volunteer opportunities. Internet technology has been extensively used in emergency response since then, not only in aiding the initial rescue phase, but also in relating warning information to the public and assisting the public during the response.

[1]Gunilla M. Ivefors, "Emergency Information Management and Disaster Preparedness on the Internet," *Human IT* 1, no. 2 (1997), available at hb.se/bhs/ith/2-97/gi.htm (accessed July 1, 2007).

question during disaster situations, and there is also the problem of data overload, especially in terms of network capacities and the large volumes of data that are being transmitted. Finally, the Internet requires a stable power source, which is often missing in disaster-impacted areas.

Wireless technology

Wireless technology, including communication devices and global positioning system (GPS) tools for use in disaster response, is quickly changing the way we respond to emergencies (see sidebar below). Simple tasks, such as talking to team members remotely rather than from a landline, as well as more complex tasks, such as collecting digital data with wireless tools, are becoming more and more a part of disaster response. Broadband, digital cable, DSL (digital subscriber line), and two-way wireless satellites allow faster, cheaper, "always connected" devices. On the positive side, this means that there is essentially no place on earth that is too far to travel and still be in contact with team members or a home base and remain functional as part of disaster operations. On the negative side, wireless technology such as personal digital assistants (PDAs) and personal computers (PCs) connected to Wi-Fi networks are at high risk of attack from viruses, worms, and key-loggers inserted to monitor a person's activities. Although certain security measures are in place within emergency management circles to protect vital data transfers, the threat of technological piracy and privacy protection issues are still at the forefront of problems associated with the use of wireless tools. Thus, the Federal Emergency Management Agency (FEMA) does not authorize its personnel to connect to its intranet through a wireless connection in order to ensure that data within the network are kept secure.

Global positioning system technology for data collection

An essential tool for geographic information systems (GIS), global positioning system (GPS) technology uses the position of satellites to determine physical locations on the surface of the Earth with a high degree of accuracy. In the immediate aftermath of the Columbia shuttle disaster in early 2003, the value of gathering, analyzing, and displaying accurate "in situ" data quickly and efficiently for emergency management was seen. The Federal Emergency Management Agency coordinated the response and recovery operations, leaning heavily on the National Aeronautics and Space Administration and numerous other entities for technological and logistical support. Among their goals was to retrieve as many pieces of the shuttle as possible in the hopes of finding evidence to determine what caused the tragedy. The stringent guidelines placed on data collection allowed search parties to focus on specific areas over an expanse of more than a half-million acres, and enabled GIS personnel to accurately determine where shuttle pieces might be found, thus saving time and money. Ultimately, the field-deployed teams collected more than 82,000 pieces of material.

Source: Federal Emergency Management Agency, "Recap of the Search for Columbia Shuttle Material," press release 3171-71 (May 5, 2003), available at fema.gov/news/newsrelease.fema?id=2808 (accessed July 2, 2007).

The role of information technology in emergency management since 1989

One of the defining moments in terms of the need for GIS and RS technologies for disaster response and recovery was 1989's Hurricane Hugo. Prior to Hugo, the use of GIS methods and applications had not been an integral part of response and recovery operations. However, with the need to rapidly assess damages from the storm in the beachfront communities of South Carolina, GIS became a key technology in the creation of South Carolina's shorefront retreat policy.[5] While imagery was available showing the coastal development before the storm, there were no reliable baseline datasets characterizing trends in shoreline development (e.g., data on its *structural* characteristics—specifically, its construction and ability to withstand the stresses of intense weather) that would have helped in the initial damage assessment and response efforts. Using Hurricane Hugo as an "expensive" learning moment, South Carolina has since compiled a spatial database of beachfront development. This database, which includes GIS-based maps, parcel and structure attribute data, ownership information, and geo-referenced (location-identified) photographs of beachfront property in critical areas, enables information about the damaged area to be quickly connected to historical data regarding property ownership and development.

Hurricane Andrew in 1992 was another disaster event in which the lack of rapid and reliable geospatial information (maps)—real-time geographically referenced damage assessments and a systematic and consistent methodology for conducting them—hindered the state and federal response. FEMA initiated its first-ever GIS operation in support of the Hurricane Andrew Presidential Disaster declaration, producing more than three hundred maps of various descriptions, many at large scale, containing pertinent disaster response information. Map users were interviewed to establish future mapping requirements for response and recovery.[6]

In 1994, GIS technology took center stage during the initial response and recovery periods of the Northridge, California, earthquake, providing important visual and spatial information on critical operations, including real-time loss estimation data for an emergency response decision support system, water supply and sewer damage assessments, and a rapid post-assessment of building damage. GIS continued to play a major role as response and recovery moved into mitigation and preparedness phases, including the use of advanced technologies in the development of next-generation earthquake loss models for transportation systems.

Five years later, after the devastating earthquakes in Turkey, a Web-based GIS was deployed to monitor ground movement and support information sharing; this system has facilitated emergency preparedness in the region. With recent earthquakes in the Middle East and northern Africa has come an increased use of information technologies to provide real and reliable sources of up-to-date and, more importantly, usable data for response and recovery. By defining plate motion more accurately and the location of earthquake hazards and potential damage assessments more precisely, the deployment of a complex array of seismometers, GPS tracking stations, and geospatial data in the eastern Mediterranean region has improved our understanding of the seismic hazards there, thereby improving emergency preparedness in the region.[7]

The September 11, 2001, terrorist attack on the World Trade Center occasioned the greatest use of technology yet for response and recovery efforts. Prior to the attacks, mapping for emergency management purposes was such a low priority in New York that the city's EOC, which was located in the center, had only one computer devoted to mapping needs and projects.[8] But after the EOC was destroyed in the attack, the new "makeshift" EOC on Pier 92 devoted an entire section to the use of GIS and RS technologies in disaster rescue and recovery. In addition to inventory development, the GIS section contains databases of critical infrastructure that can provide a baseline for determining the extent of damage and associated losses once an event has occurred.

The successful role played by RS and GIS technologies in the response efforts at ground zero was largely attributable to information sharing between federal, state, and local GIS stakeholders, each providing important pieces of data to the response and recovery process—for example, a very detailed base map for New York City developed outside of the official city GIS function. The activities of the GIS/RS section on Pier 92 enabled search and rescue activities to continue without interruption and helped provide such data as "hot spot" analysis to firefighters and debris pile information to urban search and rescue teams searching for survivors as efficiently as possible. With the important contributions of IT in the rescue, relief, and recovery efforts following this national disaster, the role of future information technologies was assured.[9]

More recently, the importance of information technologies in disaster recovery and mitigation was confirmed in the seemingly never-ending string of hurricanes in 2004 and 2005. For example, GIS and GPS technologies were used for the site location planning of fifteen emergency group sites, providing immediate housing for disaster victims, and for the assessment of disaster damages and subsequent housing needs in Florida's counties. FEMA and Florida's state emergency response team supplied communities with more than 14,000 short-term travel trailers and mobile home units, all monitored by recertification and maintenance personnel using GIS technology.[10] Because of the high demand for personnel and particularly for GIS and RS technology for the response and mitigation phases of disaster operations, FEMA hired and trained more than 200 new mitigation personnel and created new GIS analyst positions in many of its regional offices.

Without a doubt, technological innovations have proven to be vital to emergency responders. However, they have also produced some unintended consequences.[11] For example, advanced detection systems for tropical storms and hurricanes have lulled coastal residents into a false sense of security. The better the warning predictions become, the greater our dependence on them. This poses serious logistical problems for precautionary evacuations in heavily congested coastal areas, especially when the predicted landfall is days away (with all of the attendant uncertainties in forecasting the track).

Tools and applications

The use of existing and emerging technologies in emergency management functions is expanding, yet the tools are not always appropriate or applicable to all phases of the disaster management cycle. This section discusses the tools and applications that are specifically suitable to each phase of the cycle. Some tools (called "COTS," for commercial off-the-shelf) are ready for use straight out of the box while others require the application of multiple technologies simultaneously.

Mitigation and preparedness

In undertaking mitigation and preparedness, a community needs to produce loss estimation studies, assess its hazards and risks, determine its vulnerabilities, develop an inventory of critical infrastructure, and establish an early warning system.

Loss estimation Although loss estimation studies have been in use since the 1960s, it was not until the 1990s that loss estimation methodologies became widely implemented as part of formal hazard identification and impact assessments. A major factor in this development was the emergence of GIS technology, which allowed users to easily overlay hazard-specific data layers, such as hurricane storm tracks and wind fields, rainfall estimation grids, and toxic release plumes, onto maps of various systems (e.g., lifeline routes such as water, power, and telephone), building data, and population information.

Loss estimation methodologies are now vital parts of many hazard mitigation studies. Planners typically use them to forecast the potential impacts of different hazard scenarios. They are also used to assess the benefits of a mitigation activity such as a structural retrofit (e.g., strengthening exposed structures to withstand the impact of a disaster) and (in conjunction with near real-time sensor systems, such as the ShakeMap system)[12] to project losses in the event of an actual disaster or emergency.

FEMA has recognized the value of loss estimation modeling as a key hazard mitigation tool. In 1992, the agency began a major effort (which continues today) to develop standardized loss estimation models that could be used by nontechnical hazard specialists. The resulting tool, the software program called HAZUS-MH, currently addresses earthquake, flood, and wind (see sidebar on pages 286–287). The various modules of the HAZUS-MH software have been calibrated using existing literature and damage data from past events. HAZUS-MH was

HAZUS-MH

One of the more advanced computing tools available is HAZUS-MH, a spatial data-processing software and loss estimation tool. HAZUS-MH is built on an integrated geographic information system (GIS) platform that estimates losses incurred by earthquakes, floods, and wind events. The software program comprises seven major interdependent modules, each of which is briefly described below; detailed technical descriptions can be found in the HAZUS-MH technical manuals.[i]

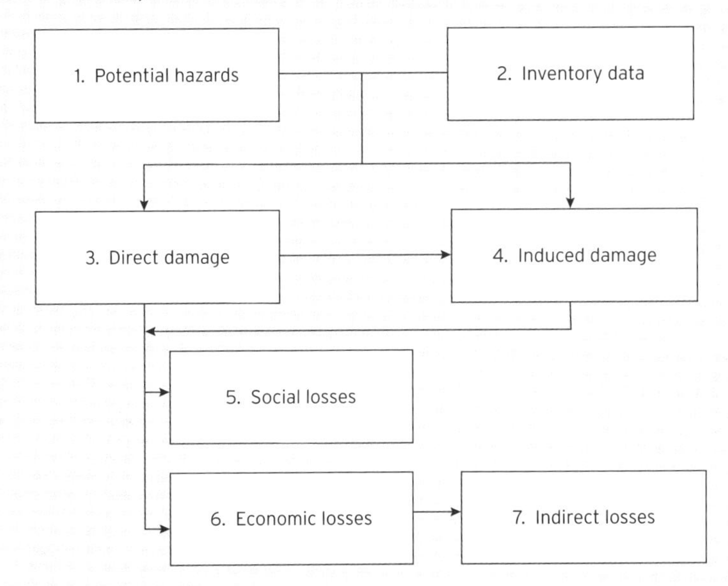

- **Potential hazards:** This module estimates the expected intensities or severities for three hazard types: earthquake, flood, and wind. For earthquake hazards, this includes the estimation of surface movement and ground fault action in the form of landslides, liquefaction, and surface fault rupture and their associated failure potentials. For flood events, HAZUS hazard potential analysis entails estimations of specific flood heights and water depths. For winds associated with tropical systems, the module includes the estimation of wind-borne debris associated with specific wind speeds. The probabalistic analysis feature includes the forecast of frequency or event probability for each event magnitude at a given location.

recently used to assess the savings from FEMA-sponsored mitigation activities; the published report concluded that a "dollar spent on mitigation saves society an average of $4."[13]

In addition to HAZUS-MH, a California-based loss estimation tool called EPEDAT (Early Post-Earthquake Damage Assessment Tool) was used following the 1994 Northridge earthquake to project losses from that event.[14] The use of loss estimation techniques in the immediate post-earthquake context was a key development, marking a significant departure from conventional "back of the envelope" calculations and preliminary damage assessment techniques. Since then, earthquake loss studies have addressed the *pre*-earthquake planning needs of utility operators, the insurance industry, and government emergency response agencies. Although the needs of these users vary, understanding and maintaining public safety have generally required the analysis of similar event scenarios—that is, those that reflect the most comparable impacts of other types of disasters on the local population and economy.

The advent of high-speed computing, satellite telemetry, and GIS has made it possible to electronically generate loss estimates for multiple scenarios and, perhaps most important, develop estimates for an actual earthquake in near real-time given the source parameters (i.e., magnitude and location) of an event. For the last ten years, Internet-based real-time earthquake data, including magnitude, location, depth, time of occurrence, and in some cases, ground motion maps or contours, have been available in California and other western states. This combination of real-time earthquake data and powerful GIS-based loss estimation tools has made the capability of near real-time loss estimates a reality in many of the seismically active regions of the world.

- **Inventory data:** This module, which contains a national-level database of elements in the built environment that are exposed to potential hazards, allows the user to run a preliminary analysis of a potential hazard area without collecting additional local information or spatial data. The default inventory database includes information on four specific inventory groups: general building stock, essential facilities, transportation systems, and utilities. The general building stock data are classified by occupancy (residential, commercial, industrial, etc.) and by building type (structural system, material of construction, roof type, and height). The provided "out-of-the-box" mapping schemes are based initially on U.S. Census and housing information, supplemented by state and substate agency data, which are often more current. The baseline data for housing types are state-specific for single-family dwellings, but for other dwellings such as commercial establishments they are more generalized, based on regions. In all cases, they are age and building-height specific.
- **Direct damage:** This element provides property damage estimates for each of the building inventory groups; estimates are based on the amount of physical exposure and degree of vulnerability of structures at different hazard-intensity levels (e.g., hurricane category, flood height, or earthquake intensity).
- **Induced damage:** This module accounts for property damages that are subsequent to the main disaster event—for example, the accumulation of debris following a flood, or a fire caused by downed power lines from a hurricane event.
- **Social losses:** Social losses include the morbidity, mortality, and impacts felt by the affected population, such as displacement and shelter needs. The casualty model provides four levels of casualty estimation (from minor injuries to deaths) for three different times of day (2:00 A.M., 2:00 P.M., and 5:00 P.M.) and for four population groups (residential, commercial industrial, and commuting). The number of displacements is calculated on the basis of the number of uninhabitable structures, which is in turn estimated by combining damage to the residential building stock with utility service outages.
- **Economic losses:** Direct economic losses are estimates of the number of damaged dwellings, the value of contents, the costs associated with relocation, the value of inventory lost by businesses, capital losses, lost wages and salary, and lost rental income.
- **Indirect economic losses:** This module evaluates the "ripple" effect in which the impact of an event on a region's economic functions are more far-reaching than the event's physical impact on the area. Estimates of indirect economic losses include decreases in sales, income, and employment by sector (i.e., commercial, industrial, and retail).

[1]National Institute of Building Sciences (NIBS) and Federal Emergency Management Agency (FEMA), *Multi-Hazard Loss Estimation Methodology, Earthquake Model: HAZUS-MH Technical Manual* (Washington, D.C.: NIBS and FEMA, 2003); NIBS and FEMA, *Multi-Hazard Loss Estimation Methodology, Hurricane Model: HAZUS-MH Technical Manual* (Washington, D.C.: NIBS and FEMA, 2003); NIBS and FEMA, *Multi-Hazard Loss Estimation Methodology, Flood Model: HAZUS-MH Technical Manual* (Washington, D.C.: NIBS and FEMA, 2003).

Hazard and vulnerability assessment Under the Disaster Mitigation Act of 2000, all localities are required to perform hazard assessments as the underlying basis for state and local hazard mitigation plans. GIS and RS technologies are used to a considerable extent for monitoring changing conditions, such as those associated with landslides or with patterns of vegetation growth as a precursor to wildfire fuel loads. Light Detection and Ranging (LIDAR) data are routinely used for fluvial and coastal flood modeling of inundation zones.

With the development of GIS-based loss estimation assessments that integrate both hazard and social vulnerability data, counties and states are better able to understand the level of hazardousness in their communities.[15] Such integrated analyses not only provide a baseline set of data on the built environment and population characteristics, but also represent a scientifically sound and methodologically robust approach to delineating those areas that are most at risk.

Inventory development Compiling a comprehensive and accurate database of existing critical infrastructure is a priority in emergency management since it provides a basis for simulating probable effects through scenario testing, while setting a baseline for determining the actual extent of damage and associated losses once an event has occurred. Demand is increasing for accurate inventories of the built environment to enable emergency and planning officials to perform vulnerability assessments, estimate losses in terms of repair costs, assess insurers' liability, and assist in real-time response and recovery.

Although the location of urban centers is generally well documented for developed and less developed nations, interest is growing in accurate, low-cost methods for characterizing the built environment and its critical infrastructure in more detail—in particular, measures of building height, square footage, and occupancy (use). To a large degree, the accuracy of loss estimates depends on the quality of input data. Default datasets are often based on regional trends rather than on local data; for example, data collected by the census are often aggregated to zip code or county level in order to ensure that privacy act information is secure. However, research done at the Multidisciplinary Center for Earthquake Engineering Research (MCEER), in Buffalo, New York, found that RS imagery offers a detailed inventory of both height and square footage which, by supplementing existing datasets such as occupancy rates, may lead to more accurate loss estimates.[16]

Remote sensing imagery (such as active interferometric SAR [IfSAR] sensors) was used to compute building heights and square footage for a case study area in Los Angeles, and the computed values corresponded closely with independently derived tax assessor data. A significant advantage of remotely derived inventories is the relative ease with which they can be updated. This is particularly important on a citywide scale, where the overview offered by satellite imagery can be used by planning departments not only to develop inventories, but also to track urban growth. Classifying imagery into vegetation, concrete, and buildings is a straightforward task that can be readily applied to coverage over a period of time: growth is detected in terms of change between the scenes.

In addition to active sensors such as IfSAR, new building inventory development techniques are emerging from the use of high-resolution optical satellite data. Research at Stanford University and ImageCat, Inc., has focused on developing an approach for rapidly obtaining spatial and structural information from a single high-resolution satellite image, called the Mono-Image Height Extraction Algorithm, or MIHEA for short. Figure 14–1 shows a three-dimensional model of Long Beach, California, that was developed using this approach.

Early warning Early warning of a disaster is possible at several different points in time. In the case of events for which ample time is provided before the actual impact on an area, as is

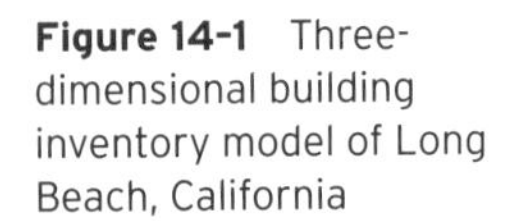
Figure 14-1 Three-dimensional building inventory model of Long Beach, California

Source: P. Sarabandi, H.-C. Chung, and B. J. Adams, "Remote Sensing for Building Inventory Updates in Disaster Management," in *Nonintrusive Inspection, Structures Monitoring, and Smart Systems for Homeland Security,* ed. Aaron A. Diaz et al., *Proceedings of SPIE* 6178 (March 2006): 95–104.

the case with hurricanes, tracking or monitoring the progress of the event is most important. Since Hurricane Camille in 1969, the National Weather Service (NWS) of the National Oceanic and Atmospheric Administration (NOAA) has been providing satellite systems to monitor the formation and movement of hurricanes and typhoons, and it has recently provided "strike probabilities" with respect to landfall as well as projected wind speeds for all major hurricane events. This information, when used with simulation models of rainfall and storm surge inundation, can provide important data for planning response and recovery efforts.

Doppler radar is another application of RS technology to warnings, especially those related to tornadoes, hail, and severe thunderstorms. The extensive use of Doppler radar systems by NWS and local news media outlets has helped to increase the warning times for tornadoes as well as improve the accuracy of predicting the tornado path. Doppler radar is also useful in estimating the amount of precipitation to help in flood warnings, especially those associated with hurricanes over land.

In the case of wildfires, fire managers can access GeoMAC (Geospatial Multi-Agency Coordination), an Internet-based mapping application that uses MODIS (Moderate Resolution Imaging Spectroradiometer) thermal imagery to help pinpoint affected areas.[17] Modeling of fire-prone areas and evacuation routes (in near real-time) is another advancement in early warning that is based on geospatial technologies.[18]

Response and recovery

Response is the most fast-paced and time-sensitive phase of the disaster management cycle, requiring data inputs and information in real-time for decision making. Once an extreme event begins, priority is given to assessing the nature, extent, and degree of damage. However, these tasks are often difficult because of the nature of the disaster, limited accessibility into the affected region, and loss of critical infrastructure such as power systems.

Structural damage detection Real-time damage detection following hazard events or disasters is a priority. In extreme events, such as natural disasters and terrorist attacks, the performance of critical transportation systems is of major concern. Given the magnitude and complexity of such systems, especially in urban areas, field-based assessment in near real-time is simply not an option. The 2004 Indian Ocean earthquake and tsunami left roads and bridges damaged and a number of villages inaccessible from the ground. Considering the critical forty-eight-hour period that urban search and rescue teams have to locate survivors, accessibility must be quickly and accurately determined in order to reroute response teams and avoid life-threatening delays. And real-time damage detection is also a priority for operational support and verification of initial damages and loss estimations for Presidential Disaster Declarations.

Structural damage is usually detected in one of two ways: (1) indirect observations based on remotely sensed data or (2) direct observations, either from field-based surveys or through aerial flyovers of the affected area. Each approach has its advantages and disadvantages. The use of remotely sensed data for assessing building damage offers significant advantages over a ground-based survey. Where the affected area is extensive and access is limited, RS imagery can present a low-risk, rapid overview. For example, Earth-orbiting RS satellites, such as IKONOS and QuickBird, can present a high-resolution, synoptic overview of the highway system in a disaster area, which can be used to rapidly assess the degree of damage. Numerous assessment techniques are documented in the literature, many of them based on the ability to "see" the building damage on the image—the bird's-eye view (see sidebar on page 290).[19] This research suggests that collapsed and extensively damaged buildings have distinct spectral signatures.

For example, the damage map created for the town of Ban Nam Khem in Thailand after the 2004 Indian Ocean earthquake and tsunami was developed through expert interpretation of high-resolution pre- and post-tsunami images (Figure 14–2). Of the 761 structures sampled, 59 percent were classified as collapsed. The most extreme damage occurred in areas bordering the open coast and inlet, where 50–100 percent of the houses were destroyed. The degree of damage as captured by the RS coverage diminishes rapidly moving inland, reaching 0-30 percent at a distance of approximately 500 meters from the shorelines.[20]

Remote sensing and the World Trade Center

Damage detection from remote sensing imagery proved particularly useful in the aftermath of the terrorist attack on the World Trade Center. IKONOS coverage acquired on September 12, 2001, and posted on the Internet provided people around the world with an early bird's-eye view of the devastation at ground zero. The first detailed pictures were captured the following day; the New York fire department recorded oblique shots from a circling helicopter, and Keystone Aerial Surveys provided vertical photographs for the New York State Emergency Management Office. From September 15-16 until mid-October, EarthData, a major provider of geospatial products and services, systematically acquired orthophotographs (digital imagery in which camera angle and topographic distortion have been removed to equalize the distances represented on the image) as well as thermal and Light Detection and Ranging (LIDAR) data. These images were routinely distributed to media outlets and appeared on television and in newspapers worldwide, aiding in the communication of true and accurate information to a public interested in knowing the latest and greatest updates.

Figure 14-2 Damage map for Ban Nam Khem, developed using high-resolution QuickBird and IKONOS imagery

The percentage of collapsed buildings is computed within zones at 100-meter intervals from the open coast and inlet shores.

% collapsed buildings

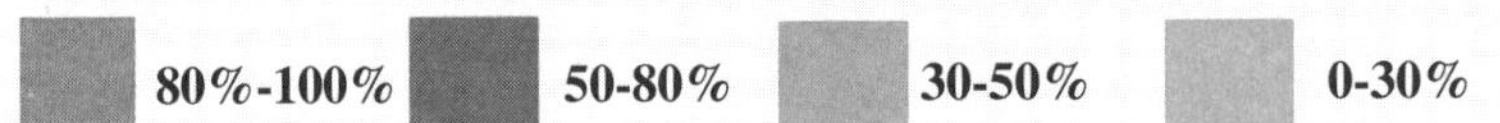

Source: Stephanie E. Chang et al., "Coastal Ecosystems and Tsunami Protection after the December 2004 Indian Ocean Tsunami," *Earthquake Spectra* 22, no. S3 (2006): S863-S887. Reproduced with permission of the Earthquake Engineering Research Institute (EERI).

However, when an event does not result in the partial or total collapse of an area's structures, RS technologies are less useful: moderate or minor damage, usually quantified in terms of the extent or density of collapsed structures, is indistinguishable from no damage. Where the roofs of the buildings remain intact, but the windows have been blown out and the interior subject to heavy rainfall and wind exposure, the bird's-eye view will not reveal damage, but an oblique view will. In this instance, ground observations or surveys are more appropriate.

Mapping impact areas RS imagery is most often used to map the geographic extent of the affected area. For example, Figure 14–3 shows the product of satellite imagery as it was used to map the extent of flooding in New Orleans after Hurricane Katrina. Other RS technologies were used to estimate the height or depth of flooding in this area. NOAA's flood depth map for August 31, 2005 (see Figure 14–4), was developed using a combination of satellite imagery from the National Geospatial Intelligence Agency and LIDAR data from Louisiana State University and the state of Louisiana; this map shows that most of New Orleans was covered by at least seven to nine feet of water, with some areas exceeding twenty feet.

What is missing from the above description is the more detailed level of analysis showing which neighborhoods were flooded and which were not, given the very subtle changes in topography in the area. For many of the residents who evacuated, specific information on their houses or their blocks was more important than the overall pattern provided by the overhead imagery.

Field reconnaissance GPS-based technologies have been one of the reasons for the significant improvement in field reconnaissance efforts after major disasters. Before this technology became available to the general public, documentation of field reconnaissance activities was cumbersome and time-consuming. Now, with GPS technology offering positional accuracies of about one to three meters anywhere in the world, it is possible to link photos and videos with actual points on the Earth. This capability becomes even more important when GPS technology is integrated with GIS systems.

One of the field-based systems that has emerged in recent disasters is the VIEWS system developed for MCEER. VIEWS is a laptop-based portable field data collection and visualization system that is used during disaster reconnaissance missions to collect geo-referenced damage observations, photographs, and video footage. The system has been deployed on foot as well as from a moving vehicle, a boat, and aircraft. Through a real-time GPS feed, the geographic location of every record is overlaid on "before" and "after" RS images and damage base maps. Inbuilt GIS functionality then enables the field team to use the high-resolution satellite scenes

Figure 14-3 Expert interpretation of flood limit (yellow lines) overlaid onto DigitalGlobe QuickBird "False Color" composite image of New Orleans, Hurricane Katrina, September 3, 2005

Source: J. Arn Womble et al., *Engineering and Organizational Issues before, during and after Hurricane Katrina*, vol. 2, *Advanced Damage Detection for Hurricane Katrina: Integrating Remote Sensing and VIEWS™ Field Reconnaissance*, MCEER Special Report Series, MCEER-06-SP02 (Buffalo, N.Y.: MCEER, March 2006), available at mceer.buffalo.edu/publications/Katrina/06SP02-web.pdf (accessed July 2, 2007).

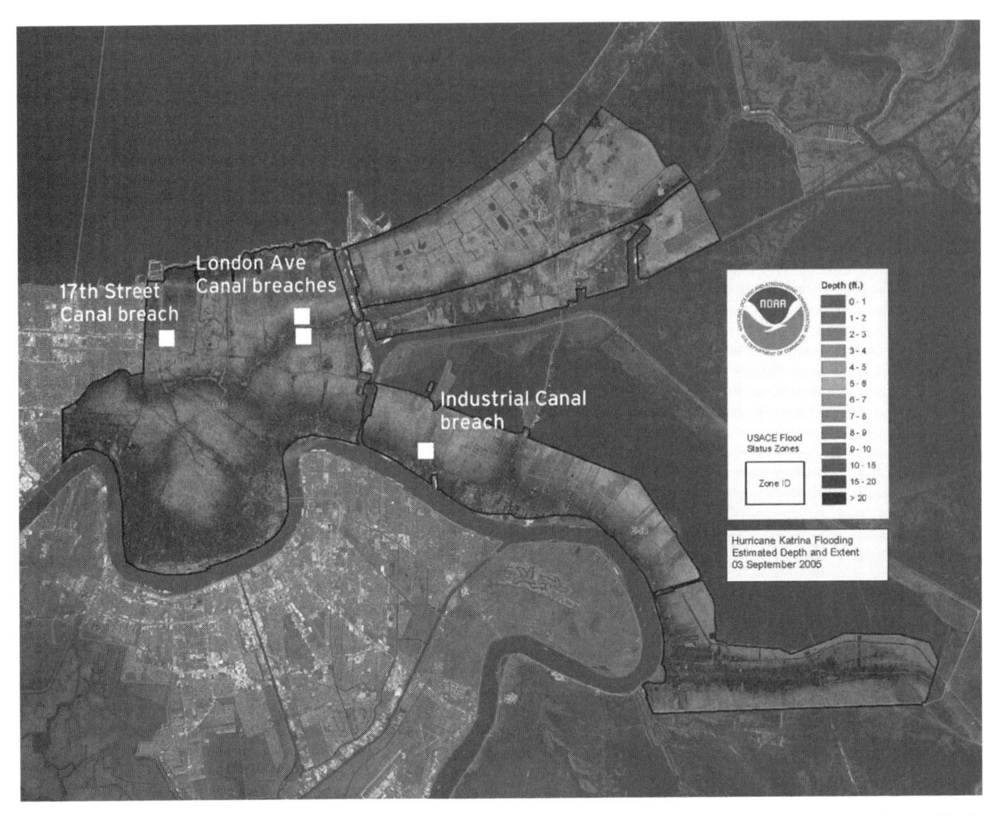

Figure 14-4 Flood depth estimation for New Orleans by NOAA after Hurricane Katrina, August 31, 2005

Source: "NOAA Aerial Mapping Assisting U.S. Coast Guard, FEMA Shows Flood Waters Receding in Regions Affected by Hurricane Katrina" (September 12, 2005), at noaanews.noaa.gov/stories2005/s2503.htm (accessed July 2, 2007).

to prioritize field survey activities, plan and track their route, and pinpoint damaged structures and features of interest. Traditional methods of post-disaster damage assessment typically involve walking surveys, whereby damage indicators together with the overall damage state are manually logged onto a spreadsheet. Past reconnaissance deployments of VIEWS—for example, following the 2003 Bam, Iran, earthquake; Hurricanes Charley and Ivan, which hit the U.S. Gulf Coast in 2004; the Niigata, Japan, earthquake in October 2004; and Hurricanes Katrina and Rita in 2005—indicate that the system significantly increases the rate at which survey data are captured and made available to end users.[21]

The Indian Ocean event in 2004 constituted the first deployment of VIEWS and high-resolution satellite imagery for post-tsunami field reconnaissance. The system was deployed to study several key sites over August 16–25, 2005, in order to establish the accuracy of preliminary RS results. VIEWS was equipped with satellite base data layers, including the Landsat land use classification, the mangrove change/loss map, and the QuickBird and IKONOS satellite imagery, which provided ground teams with a fundamental visual understanding of the area. A three-member team surveyed the affected areas from a moving vehicle, by boat, and on foot, depending on vehicular access and type of land use (e.g., mangrove). Fourteen hours of geo-referenced digital video footage were obtained along the survey route, which covered about seventy-five miles (see Figure 14–5). The team also collected a library of about 550 digital photographs.

Barriers to use

No matter how many technological advances are available for use in the disaster management community, there will always be some roadblocks that must be overcome in order to fully use existing tools.[22] Obstacles to use can be nearly anything—for example, a manager's lack of knowledge of the existence of applicable technologies, the lack of skilled labor needed to run the equipment and software, or simply a lack of monetary resources. Identifying the potential problems is the first step; knowing enough to overcome them is the second. Generally

The upper photograph on the right shows an example of the rapid reconstruction that is occurring, and the lower digital photograph shows remaining building damage.

Figure 14-5 VIEWS interface showing "before" and "after" high-resolution imagery and part of the GPS route (illustrated by the white and green dots) followed by the field team in Ban Nam Khem

Source: Shubharoop Ghosh et al., "Preliminary Field Report: Post-Tsunami Urban Damage Survey in Thailand Using the VIEWS Reconnaissance System" (Buffalo, N.Y., Multidisciplinary Center Earthquake Engineering Research, 2005), available at mceer.buffalo.edu/research/Reconnaissance/tsunami12-26-04/Tsumani-Dec2004new.pdf (accessed July 2, 2007).

speaking, the use of IT in emergency management is constrained by a lack of workforce skills; by graphical user interface design issues; and by the availability and affordability of tools, technology, and data.

Workforce skills

Lack of a skilled disaster information management team can lead to many problems, including the inability to use available technology in response, recovery, and mitigation. The key to overcoming these problems is educating and training the workforce in the use of the technology. And even then, there are challenges in keeping those workforce skills and knowledge honed and up-to-date.

A number of mechanisms are available to assist in keeping emergency personnel current in IT advancements. One such mechanism is online training via FEMA's Education and Training Center, a virtual campus where more than fifty independent study courses are offered, all of which come with official certificates upon completion. Local HAZUS user groups are another venue, especially for those using this loss estimation software. Another way to keep abreast of the current trends in IT is to subscribe to a research magazine in the field of hazards or disasters, or in a more specialized field such as geographic information science or RS. *The Natural Hazards Observer,* a free publication from the Hazards Research and Application Center at the University of Colorado at Boulder, is an invaluable resource for those interested in the latest research publications, grant opportunities, upcoming training, conferences, and relevant, real-world hazards information. Short courses offered by commercial vendors such as ESRI are another way to maintain and improve skill sets, especially in the GIS arena. And major trade shows and annual meetings of professional societies, such as the American Society for Photogrammetry and Remote Sensing and the Annual Hazards Research and Applications Workshop, offer a venue to discuss relevant issues, unveil new products and tools applicable to disaster management, and promote networking among users.

Issues in human-technology interactions

As computer processing and data manipulation capabilities continue to increase, so too do the difficulties in designing and creating easy-to-use computer programs (often called "graphical user interfaces"). One of the main reasons for this difficulty is a general disconnect between users of technology and designers. Essentially, user interfaces must be designed with the end user in mind from the very beginning, but it is far too costly to customize them for all types of users in all types of situations. Therefore, most interfaces specifically support one function at a specific time and place. The human-technology interfaces become even more of a problem when completely different groups of people are responding to a disaster, and each group has different setup problems and issues that need to be addressed. The problems of interoperability or the ability of different software and hardware systems to "talk" to each other hampers vendors who try to assist in the response efforts. Even during pre-event planning, issues of system compatibility between government offices in the same jurisdiction pose significant barriers.

Availability and affordability

To facilitate decision making, emergency managers need the right information, at the right time, and in the right form. Many of the decisions they make are contingent on many other sources of information being compiled quickly and properly. Without the availability of accurate and timely information, the chain of response breaks down, often causing a logjam in the disaster cycle.[23] In the case of wildfire protection, for example, up-to-date aerial photography or satellite imagery is needed multiple times a day in order to spatially comprehend the fire area and plan containment measures that can prevent further damage to lives and property. The whole operation faces potentially devastating delays if there is cloud cover when the images are to be taken or if satellites are not positioned accurately when needed. Other impediments to the use of satellite imagery include acquisition (technical and cost considerations), spatial resolution, and processing time.

The key reasons that all emergency managers do not use the same set of tools and technological expertise come down to a few simple truths. First, the products, tools, data, experience, and knowledge needed at a particular disaster may not be immediately available. It may be that the available computers are not of high enough quality to run the software needed for the operation, or that RS equipment cannot be flown because of issues with contractors or the placement of satellites, or that personnel simply lack the necessary expertise. In an effort to address some of these limitations, which can hinder local communities in times of crisis, ICMA and the Public Entity Risk Institute (PERI) created the National Emergency Management Network (NEMN) in 2006. A membership organization of local government entities that work in partnership to better prepare for local emergencies and larger-scale disasters, NEMN provides an inventory of mutual aid resources and training in addition to databases of human and physical assets for preparedness, response, and recovery. Members also gain access to a software system designed to share, deploy, and manage response and recovery resources. One unique aspect of the NEMN software tool is its geospatial module, which enables users to map evacuation routes, locate critical infrastructure, and import high-resolution imagery as an aid for situational awareness.[24]

Second, the necessary disaster management tools might very well be out of the economic reach of those people and organizations that need them. While many small communities are expanding their capabilities and potential, many others still lack the technological tools needed for emergency management. One significant factor in making GIS more useful in the emergency management operations of small communities was the development of software that can be used on personal computers.

Challenges

With all technological advancements come certain limitations and potential pitfalls. As discussed above, many of the barriers to use can be overcome, yet certain challenges have always been and will continue to be inherent in the use of IT in emergency management. The critical importance of obtaining accurate data throughout the disaster management cycle to facilitate planning, improve response, increase relief, and accelerate recovery so that losses are minimized translates into two broad issues: first, the need to coordinate what data are collected and how, and second, the need to transform that data into useful and timely information, which users can access easily.[25] And these issues lead to two major concerns: maintaining data quality and quantity, and protecting privacy while ensuring freedom of information.

Data quality and quantity

Advancements in IT have made data-sharing issues essentially a thing of the past. One must only open up a favorite search engine to find vast amounts of information on almost any subject. Spatial data are no exception. In fact, a simple search of the Internet will located thousands of sites from which spatial data can be downloaded for immediate use. The explosion of data on GoogleEarth is one such example. The problem is that much of what is on the Internet does not conform to standardized levels of quality control and quality assurance.

To assist in the formulation of data that are acceptable for use at the point of download, the Federal Geographic Data Committee (FGDC) develops geospatial data standards for implementing the National Spatial Data Infrastructure, in consultation and cooperation with state, local, and tribal governments; the private sector and academic community; and, to the extent feasible, the international community.[26] Ensuring that data used in emergency management follow FGDC standards enables decision makers to have full confidence that the choices made at critical junctures in the response cycle are sound and that the spatial data associated with those decisions are of the utmost quality. It must always be remembered that the need for quality control extends beyond the production lines in manufacturing plants and into the offices, classrooms, and EOCs.

Open access versus privacy protection

Data availability and data sharing pose the biggest constraints to the use of sophisticated information management systems in disasters. This is even truer now in the post-9/11 world, as data security has taken on a new urgency at all governmental levels. The collapse of New York City's EOC when the World Trade Center went down on 9/11 required that emergency information and infrastructure data be reconstructed from other public and private sources; in the immediate aftermath of the attacks, however, the federal government restricted many spatial datasets from public access.[27] State and local governments are also restricting the use of some data for homeland security issues.

It is not only the sensitivity of the data (such as locations of pipelines or water intakes) that poses problems, but also the willingness to share. In many cases, data that are collected (at considerable cost to the local municipal government) for tax or planning purposes have application for emergency management at county or state levels. But there has been considerable reluctance to share data across jurisdictions when the data are paid for by one entity and received free by another. There are also issues of power and control. And even where there is a willingness to share, there is the problem of the technological capacity to share—or interoperability. Every state and local government has its own unique hardware and software systems, and often these systems are incompatible with those of other jurisdictions, especially for data sharing and applications.

The other side of the debate is the necessity for providing privacy protection, especially with RS technologies and GPS receivers. The high resolution of commercial satellite images allows us to see Earth's features in extraordinary detail and to pinpoint locations of such features with increasing precision. Wireless technologies for tracking, such as GPS units, are now the norm rather than the exception and are imbedded in automobiles, PDAs, and cell phones. Reverse 911 systems permit emergency response personnel to contact residents in specific locations about impending dangers such as chemical spills or wildfires. An individual's exact whereabouts (geographic coordinates) can even be tracked through his or her cell phone. In times of emergency, these advanced technologies provide enormous benefits and help to save lives. During nonemergencies, however, they trigger concerns about constitutional freedoms, including the right to privacy, and about the impact on homeland security.

Conclusions

This chapter provides an overview of current information technologies and their effective use in emergency management. While there are many other technologies that might be appropriate for disaster operations (especially preparedness and response), their utility has not been proven as of yet.

The application of technology to emergency management depends on the expertise and capacity of state and local governments to adopt it. Expense can also be an issue, although many organizations have data and software agreements with the major providers. For many of the geospatial technologies, considerable time and effort are required upfront (pre-event) to compile and store data—data that provide the foundation for planning in general but become critical to emergency planning and response. These data include population characteristics, parcel-level information about structures (residential, commercial, and industrial), infrastructure data (transportation routes, pipelines, power lines), and emergency services (hospitals, shelters, and police and fire departments). The primary obstacles to the effective use of technology in emergencies are the lack of training and expertise among emergency management officials, and cost of the technology (and training). However, when planning for emergencies becomes a routine function of local government, there will be greater adoption and usage of technology in disasters.

Notes

1 Graham A. Tobin and Burrell E. Montz, "Natural Hazards and Technology: Vulnerability, Risk, and Community Response in Hazardous Environments," in *Geography and Technology,* ed. Stanley D. Brunn, Susan L. Cutter, and J. W. Harrington Jr. (Dordrecht, the Netherlands: Kluwer Academic Publishers, 2004), 547–570.

2 John O'Looney, *Beyond Maps: GIS and Decision Making in Local Government* (Washington, D.C.: ICMA, 1997), 4.

3 Susan L. Cutter, "GI Science, Disasters, and Emergency Management," *Transactions in GIS* 7, no. 4 (2003): 439–446.

4 John R. Jensen and Michael E. Hodgson, "Remote Sensing of Natural and Man-Made Hazards and Disasters," in *Manual of Remote Sensing,* 3rd ed., vol. 5, *Remote Sensing of Human Settlements,* ed. Merrill K. Ridd and James D. Hipple (Bethesda, Md.: American Society for Photogrammetry and Remote Sensing, 2005), 401–429.

5 In this policy, which was adopted by the state to combat beach erosion, two lines of jurisdiction (a baseline and a more landward forty-year "setback line") are used to regulate the size and location of new or replacement structures located near the beach; new erosion control structures seaward of the line are banned; and new habitable structures are limited in size to 5,000 square feet of heated space, effectively eliminating new commercial hotels and condominiums. See the South Carolina Department of Health and Environmental Control, "Planning for Shoreline Change in South Carolina" (Charleston, S.C., 2007), available at scdhec.net/environment/ocrm/science/docs/2007_CSC_Shoreline.pdf (accessed July 6, 2007).

6 Mark A. Whitney, communication to Capt. Mark Lundtvedt, Center for Army Lessons Learned, regarding the "Hurricane Andrew After Action Report" (October 6, 1992), available at emergencymanagement.org/Lessons%20Learned%20Hurricane%20Andrew.doc (accessed July 2, 2007).

7 C. K. Huyck et al., *Methodologies for Post-Earthquake Building Damage Detection Using SAR and Optical Remote Sensing: Application to the August 17, 1999 Marmara, Turkey Earthquake,* Technical Report MCEER-04-0004 (Buffalo, N.Y.: Multidisciplinary Center Earthquake Engineering Research [MCEER], 2004).

8 Deborah S. K. Thomas et al., "Use of Spatial Data and Geographic Technologies in Response to the September 11th Terrorist Attack on the World Trade Center," in Natural Hazards Research and Applications Information Center, Public Entity Risk Institute, and Institute for Civil Infrastructure Systems, *Beyond September 11th: An Account of Post-Disaster Research,* Program on Environment and Behavior Special Publication no. 39 (Boulder: Natural Hazards Research and Applications Information Center, University of Colorado, 2003), 147–162.

9 Bruce Cahan and Matt Ball, "GIS at Ground Zero: Spatial Technology Bolsters World Trade Center Response and Recovery," *GEO World* (January 2002): 26–29, available at digitalgovernment.org/library/library/pdf/GeoWorldStoryJanuary20.pdf; Charles K. Huyck and Beverley J. Adams, *Emergency Response in the Wake of the World Trade Center Attack: The Remote Sensing Perspective,* MCEER Special Report Series, vol. 3 of *Engineering and Organizational Issues Related to the World Trade Center Terrorist Attack* (Buffalo, N.Y.: MCEER, 2002); Ronald J. Langhelm and Bruce A. Davis, "Remote Sensing Coordination for Improved Emergency Response," paper presented at the Pecora 15/Land Satellite Information IV/ISPRS Commission I/FIEOS 2002 Conference, November 10–15, 2002, Denver, Colorado, available at isprs.org/commission1/proceedings02/paper/00083.pdf (accessed June 25, 2007); Ray A. Williamson and John C. Baker, "Lending a Helping Hand: Using Remote Sensing to Support the Response and Recovery Operations at the World Trade Center," *Photogrammetric Engineering and Remote Sensing* 68, no. 9 (2002): 870–896; Susan L. Cutter, Douglas B. Richardson, and Thomas J. Wilbanks, eds., *The Geographical Dimensions of Terrorism* (New York: Routledge, 2003); Michael J. Kevany, "GIS in the World Trade Center Attack—Trial by Fire," *Computers, Environment and Urban Systems* 27, no. 6 (2003): 571–583.

10 Federal Emergency Management Agency (FEMA), "Need for Emergency Travel Trailer Group Sites Lessens in Florida," press release 1539-417 (March 25, 2005), available at fema.gov/news/newsrelease.fema?id = 17035.

11 Edward Tenner, *Why Things Bite Back: Technology and the Revenge of Unintended Consequences* (New York: Knopf, 1997).

12 The ShakeMap system, a product of the U.S. Geological Survey Earthquake Hazards Program, provides local

spatial ground motion information following significant earthquake events; for more information on the system, see earthquake.usgs.gov/eqcenter/shakemap/ (accessed July 6, 2007).

13 Multihazard Mitigation Council, *Natural Hazard Mitigation Saves: An Independent Study to Assess the Future Savings from Mitigation Activities,* vol. 1, *Findings, Conclusions, and Recommendations* (Washington, D.C.: National Institute for Building Sciences, 2005), 5, available at floods.org/PDF/MMC_Volume1_FindingsConclusionsRecommendations.pdf (accessed June 25, 2007).

14 Ronald T. Eguchi et al., "Real-Time Loss Estimation as an Emergency Response Decision Support System: The Early Post-Earthquake Damage Assessment Tool (EPEDAT)," *Earthquake Spectra* 13, no. 4 (1997): 815–832.

15 Susan L. Cutter, Jerry T. Mitchell, and Michael S. Scott, *Handbook for Conducting a GIS-Based Hazards Assessment at the County Level* (Columbia: South Carolina Emergency Preparedness Division, November 1997), available at cas.sc.edu/geog/hrl/Handbook.pdf (accessed June 25, 2007).

16 Ronald T. Eguchi et al., *A New Application for Remotely Sensed Data: Construction of Building Inventories Using Synthetic Aperture Radar Technology* (Buffalo, N.Y.: MCEER, 1999), available at mceer.buffalo.edu/publications/resaccom/99-sp01/ch1eguch.pdf (accessed June 25, 2007).

17 See the GEOMac Web site at geomac.gov.

18 Thomas J. Cova et al., "Setting Wildfire Evacuation Trigger Points Using Fire Spread Modeling and GIS," *Transactions in GIS* 9, no. 4 (2005): 603–617; Tae H. Kim, Thomas J. Cova, and Andrea Brunelle, "Exploratory Map Animation for Post-Event Analysis of Wildfire Protective Action Recommendations," *Natural Hazards Review* 7, no. 1 (2006): 1–11.

19 L. Chiroiu and G. André, "Damage Assessment Using High Resolution Satellite Imagery: Application to 2001 Bhuj, India Earthquake," *RiskWorld* (2001), available at riskworld.com/Nreports/2001/Bhuj,India,earthquake2001.PDF (accessed June 25, 2007); Lucian Chiroiu, "Damage Assessment of 2003 Bam, Iran, Earthquake Using Ikonos Imagery," *Earthquake Spectra* 21, no. S1 (2005): S219–S224; Fumio Yamazaki, Yoshihisa Yano, and Masashi Matsuoka, "Visual Damage Interpretation of Buildings in Bam City Using QuickBird Images Following the 2003 Bam, Iran, Earthquake," *Earthquake Spectra* 21, no. S1 (2005): S329–S336.

20 Stephanie E. Chang et al., "Coastal Ecosystems and Tsunami Protection after the December 2004 Indian Ocean Tsunami," *Earthquake Spectra* 22, no. S3 (2006): S863–S887.

21 Beverley J. Adams et al., "The Bam (Iran) Earthquake of December 26, 2003: Preliminary Reconnaissance Using Remotely Sensed Data and the VIEWS (Visualizing the Impacts of Earthquakes with Satellite Images) System" (Buffalo, N.Y.: MCEER, 2004), available at mceer.buffalo.edu/research/Reconnaissance/Bam12-26-03/bam.pdf (accessed June 25, 2007); Beverley J. Adams et al., "Collection of Satellite-Referenced Building Damage Information in the Aftermath of Hurricane Charley," MCEER Quick Response Report Series (Buffalo, N.Y.: MCEER, 2004), available at mceer.buffalo.edu/research/Reconnaissance/Charley8-13-04/04-SP04.pdf (accessed June 25, 2007); Charles Scawthorn et al., "Preliminary Observations on the Niigata Ken Chuetsu, Japan, Earthquake of October 23, 2004," *EERI Special Earthquake Report* (January 2005): 1–12, available at eeri.org/lfe/pdf/japan_niigata_eeri_preliminary_report.pdf (accessed June 25, 2007); J. Arn Womble et al., *Engineering and Organizational Issues before, during and after Hurricane Katrina,* vol. 2, *Advanced Damage Detection for Hurricane Katrina: Integrating Remote Sensing and VIEWS™ Field Reconnaissance,* MCEER Special Report Series, MCEER-06-SP02 (Buffalo, N.Y.: MCEER, March 2006), available at mceer.buffalo.edu/publications/Katrina/06SP02-web.pdf (accessed June 25, 2007).

22 A. Zerger and D. I. Smith, "Impediments to Using GIS for Real-Time Disaster Decision Support," *Computers, Environment and Urban Systems* 27, no. 2 (2003): 123–141.

23 Michael F. Goodchild, "Geospatial Data in Emergencies," in *The Geographical Dimensions of Terrorism* (see note 9), 99–104.

24 National Emergency Management Network, "About the NEMN: A National Network for Local Government Emergency Response and Recovery Initiatives," available at nemn.net/about/ (accessed June 28, 2007).

25 Deborah S. K. Thomas, "Data, Data Everywhere, but Can We Really Use Them?" in *American Hazardscapes: The Regionalization of Hazards and Disasters,* ed. Susan L. Cutter (Washington, D.C.: Joseph Henry Press, 2001), 61–76.

26 The National Spatial Data Infrastructure (NSDI) is the standard put forth by the Federal Geographic Data Committee for data infrastructure and dissemination. Datasets that conform to the standards put forth in the NSDI contain all the necessary metadata that facilitate their use without the need to ask where the data came from, when, and how they were collected.

27 Gerald E. Galloway, "Emergency Preparedness and Response—Lessons Learned from 9/11," in *The Geographical Dimensions of Terrorism* (see note 9), 27–34.

CHAPTER 15

Budgeting for emergency management

Richard T. Sylves

This chapter provides an understanding of

- The new prominence of homeland security as a local government responsibility
- Local government budgeting for emergency management, including revenue and expenditure issues, the use of taxes and fees, capital budgeting, and approaches to covering the costs of disaster
- Federal funding for emergency management
- The state context for emergency management.

Emergency managers have two principal responsibilities: one is to protect the community from the risk of disaster, and the other is to find the funding to do so. Thus, emergency managers cannot succeed in their line functions without understanding the realm of public budgeting, being skilled in grantsmanship and grant management, and learning how to effectively advocate for their programs.

Budgeting involves examining how an organization's resources have been used in the past, determining what has been accomplished and at what cost, and charting a course for the future. Budget allocations not only identify the amounts to be spent, but also set forth the intended purposes of those expenditures.[1] Emergency managers are responsible for using the allocated funds to accomplish the desired results.

Local governments have been in the business of budgeting for emergencies and disasters for a great many years, certainly well before state and federal disaster relief programs were set forth. Virtually every municipal charter requires the jurisdiction to provide for public safety, and local emergency management is an important component of public safety. Today, budgeting for local emergency management is profoundly affected by federal and state laws and regulations, many of which are embedded within intergovernmental grants.

This chapter examines the realm of local budgeting for emergency management. The first section describes the shift in emphasis from civil defense to homeland security; this shift is important because it creates the backdrop against which all emergency management activities are currently undertaken. The chapter then considers the principal constraints on local government budgets, and describes several approaches to obtaining funding for emergency management given those constraints. Finally, it discusses the complexities and contradictions of federal funding for emergency management, and the state context within which emergency management operates.

From civil defense to civil security

During the summer of 2004, Thomas Kean, co-chairman of the 9/11 Commission and former governor of New Jersey, explained expensive improvements in aviation security that the commission had recommended to prevent terrorist attacks, noting that the commission was "mindful" that the improvements were "an enormous investment." But, Kean said, "we have seen the devastating cost in human life and in economic destruction that result from a successful attack. We believe, therefore, it is a worthwhile investment and one necessary to fulfill the government's constitutional duty to provide for the common defense of its citizens."[2]

Even before the terrorist attacks of September 11, 2001, emergency management in the United States was dauntingly difficult. Yet in the years since September 11, emergency management has been redefined, as Kean's statement attests, to include provision for "the common defense." Although it could be argued that federal, state, and local governments have been in the business of civil defense at least since the passage of the Civil Defense Act of 1950, the new emphasis on homeland security—which is heavily focused on preventing and protecting against terrorism, including attacks with weapons of mass destruction—is having dramatic effects on state and local emergency management.[3]

From both a policy and budgetary vantage point, local emergency management today is largely a component of homeland security.

Local governments have not been explicitly involved in national defense since the cold war era, when they were expected to engage in public education, evacuation planning, and population-sheltering programs in preparation for a Soviet nuclear attack.[4] According to classical economic reasoning, defense provides a national public good; hence, the federal government should have exclusive responsibility for protecting the nation.[5] As part of the federal government's war on terrorism, however, state and local governments are expected to participate in efforts to protect national security. From both a policy and budgetary vantage point, local emergency management today is largely a component of homeland security.

Public goods and positive externalities

How should emergency management be viewed from the perspective of public policy? Specifically, is it a public good or a positive externality?

Public goods are indivisible (i.e., they cannot be broken up into smaller units and sold), are characterized by non-excludability (i.e., once a public good has been provided, no one can be excluded from enjoying it), and yield benefits that are of value to society as a whole. Because they cannot charge for them, for-profit firms have little or no economic incentive to produce public goods.[1] Fire protection is in some respects a public good, in that it protects entire communities, rather than being available only to those who pay for fire services. Measures that help to prevent a terrorist attack can also be considered a type of public good.

Nevertheless, emergency management is perhaps more appropriately considered a positive externality. A positive externality results when an action that is undertaken for one purpose has broader benefits that "spill over."[2] For example, funding for public education benefits children directly, but the benefits–in terms of economic competitiveness, for example–spill over, creating larger effects on society as a whole. Similarly, emergency management benefits individuals directly–either by lowering the risk of injury and property damage, or by helping them to cope with damage that has already occurred–but it also benefits society at large, by minimizing both the potential and the reality of social and economic disruption. If emergency management were left exclusively to the private sector, it is likely that it would be woefully underfunded–and that vulnerabilities would be greater.

[1]Robert D. Lee Jr., Ronald W. Johnson, and Philip G. Joyce, *Public Budgeting Systems*, 7th ed. (Boston: Jones and Bartlett Publishers, 2004), 5.

[2]Ibid., 6.

Among the new demands imposed on state and local governments are the requirements incorporated into the National Response Plan (NRP) and the National Incident Management System (NIMS); conditions included within major grant initiatives (such as the Urban Area Security Initiative, the State Emergency Responder Program, Emergency Management Performance Grants, and the Firefighter Assistance Program); the provision of port (and, in some cases, border) security; responsibilities in the areas of public health and bio-defense; and, in some cases, greater responsibility for aviation security. State and local governments have also been asked to identify new vulnerabilities, especially those related to terrorism. In order to avoid losing out in the competition to secure homeland security funding, many emergency managers and local elected officials are scrambling to find new ways to recast their emergency management activities. Given the dramatic changes in emergency management at the federal level—in particular, the emphasis on homeland security, which sometimes leads other important concerns to be overlooked—local emergency managers have no choice but to meet new demands and conform to new requirements—or risk putting in jeopardy future opportunities for federal homeland security funding.

Although the federal government did establish several new and relatively well-funded programs designed to strengthen the capacity of state and local governments to prevent, respond to, and recover from terrorist attacks,[6] the infusion of federal funding was not an unmitigated blessing. For example, most of the funds are dispensed as categorical grants (although they often come to states wrapped up in block grants) that are accompanied by numerous bureaucratic requirements: state and local emergency managers must prepare and submit an application; prove that they deserve the funds; meet ever-changing conditions (even after the grants are awarded); demonstrate how the money is being spent; document (often in painstakingly detail) how the funding has enhanced emergency management or homeland security; and obey time limits that stipulate when the federal funds will be made available, when they may be obligated, and when they may be spent.

Moreover, several new homeland security programs are not available to states and are only indirectly available to local government. Instead, they are directed to local law enforcement agencies,[7] public health departments, private utilities, or special districts.[8] For example, a variety of homeland security grants are available to two types of local authorities: those that operate and manage light-rail systems, regional commuter rail systems, subways, and bus systems, and those

Types of grants

Federal grants come in two types: categorical grants and block grants. Both block and categorical grants may be dispensed on the basis of a formula, which means that funds are apportioned according to mathematical rules. The formulas and their constituent variables are determined by Congress or by authorization committees in the House, the Senate, or both.

Categorical grants may be formula-based grants or direct payments for a specified use. They tend to have a narrow focus and are generally designed to address specific problems. They are also structured in such a way as to control the behavior of the recipient.[1] In any given year, federal categorical grants generally number in the hundreds.

Block grants, which are used commonly in federal programs that support local emergency management, generally allow the recipient government more flexibility in allocating funding, as long as the monies are applied only to the specified policy domain.[2] Many block grants were formed by consolidating various categorical grants. Emergency Management Performance Grants are block grants.

Block grants and categorical grants may be either conditional or unconditional. A *conditional grant* is designed to elicit a particular behavior from the recipient, and must be used in a specified way. A *matching grant* is a type of conditional grant, under which the grantor agrees to pay some fraction of the cost paid by the recipient. Under matching fund arrangements, a local government can increase its grant award by increasing its own spending.[3] *Unconditional grants* are known as lump-sum grants. *Revenue-sharing grants* are a type of unconditional grant. Certain types of formula-dispensed state aid—for example, a state property tax rebate paid out to local governments in lump sums—are also unconditional. The vast majority of emergency management block grants and categorical grants are conditional.

Most grant programs involve *project grants,* which may be block or categorical. Fire Management Assistance grants, for example, are project grants. Project grants may be awarded on a competitive or a noncompetitive basis. In either case, however, an application is required.

Although most federal grants subsidize equipment purchases, training, and payment for contract services and products, grants seldom underwrite the addition of emergency management staff or other personnel expenses.

[1]Robert D. Lee Jr., Ronald W. Johnson, and Philip G. Joyce, *Public Budgeting Systems,* 7th ed. (Boston: Jones and Bartlett Publishers, 2004), 537–538.

[2]Ibid., 538.

[3]Wallace E. Oates, "Fiscal Structure in the Federal System," in *Management Policies in Local Government Finance,* 5th ed., ed. J. Richard Aronson and Eli Schwartz (Washington, D.C.: ICMA, 2004), 51–52.

that operate and manage elements of the civil infrastructure such as ports and bridges. Hospitals, airports, water systems, and power systems are also eligible for homeland security grants.

Finally, the funding programs have come under fire for being wasteful and rigid. A 2005 report, for example, describes the inflexible set of formulas that Congress used to distribute some $10 billion in federal grant money to fire, police, and other first responders.[9] The formulas required a minimum amount of money to be spent in each state, regardless of risk. As a result, the state with the highest per capita homeland security spending in 2003 was Wyoming, and those with the lowest per capita spending were California, New York, and Texas. To worsen the problem, many states used similarly rigid formulas to pass the money on to local governments. California, for example, simply gave $5,000 to each county, an amount that was often too small to make a difference. In response to criticism about the distribution of funding, Congress agreed, by fiscal year (FY) 2006, to implement a more risk-based formula for Homeland Security First Responder grants; nevertheless, in that year no state or territory was denied a share of funding under the State Homeland Security Grant Program, the Law Enforcement Terrorism Prevention Program, or the Citizen Corps.

Local government budgeting for emergency management

Despite their new responsibilities for homeland security, local governments must continue to address both natural disasters and accidental technological events. Greater professionalization, higher public expectations, and the costs of advanced equipment and information technologies are also driving up the costs of local emergency management. Because there is never enough funding for every initiative that might be desirable, emergency management, like every other local government function, involves trade-offs. Which is more important: spending on terrorism prevention or spending on flood mitigation? Constructing a tornado shelter or entering into pre-disaster contracts with bus companies that will assist with evacuations? Purchasing a geographic information system (GIS) or subsidizing Red Cross efforts to house those who have been displaced from their homes by disaster?

Local emergency managers and elected officials need to identify and assign priorities to spending needs.[10] They also need to fully understand the environment in which their governmental units operate: they cannot ignore where the money comes from that pays for what they desire. For example, state and local governments are extremely sensitive to cuts in federal domestic discretionary spending.[11] Since 1980, federal aid to state and local governments has shrunk as a share of the total federal budget, and growth in federal aid to lower levels of government has often been outpaced by inflation. Consequently, any new forms of federal aid are highly coveted by state and local officials. And when these officials experience reductions in the federal subsidies that have historically supported various projects—such as highway construction—it is understandable that they would want to make up for these reductions by seeking emergency management funding to cover related, if not identical, infrastructure improvements. Nevertheless, long-term investments that promise multiple benefits—such as economic development, improved housing and transportation, and enhanced resiliency—may well be preferable to chasing after potentially ephemeral federal and state funding.

Revenue constraints

Virtually all general-purpose local governments in the United States operate under balanced-budget requirements, which means that the operating budget must be managed so that expenditures do not exceed revenues. For this reason, local government budgeting is generally referred to as "revenue driven" or "revenue constrained."[12] At certain points in the fiscal year, municipal officials must make an accounting of revenues and expenditures. Temporary imbalances are common, and local governments address them through cash management, short-term loans, and other instruments.

Local governments have the authority to levy taxes and fees in order to raise revenue, but most local governments are limited to user fees (such as permit, licensing, inspection, and document fees); a few broad-gauged taxes (such as property, sales, and income taxes); and targeted taxes (such hotel occupancy, street vendor, and gaming and gambling taxes). Moreover, there are limits on what local governments can do with these revenues. For example, revenue from user fees is often earmarked for specific purposes. Although revenues from broad-gauged and targeted taxes usually go into the general fund, a portion of the money generated by these taxes may be earmarked as well. On average, about two-thirds of local government revenue come from taxes or fees: local property taxes yield an estimated 24 percent, and another 30 percent is collected from charges, utility fees, and miscellaneous revenue sources.[13]

Only about one-third of local government revenue comes from other levels of government. Although direct federal aid represents just a small fraction of that revenue, a sizable share of state aid to local governments is made up of federal funding that is "passed through" the states. Thus, when local governments apply for federal grants, they must generally work through state channels to do so.[14] State legislatures, concerned that they may gradually lose control of programs that are heavily subsidized by federal grants, routinely incorporate federal funding into the state budget. This tactic allows them to better monitor and evaluate the effect of federal funding on budgets at both the state and local levels; it may also prevent duplication of federal and state program funding. Of course, both state aid and the federal funds passed through the state to local governments affect the ability of local governments to balance their budgets.

Expenditure issues

Because government in the United States is based on representative democracy, local emergency management is one of many public functions that is carried out in a political environment—and, as political scientist Harold Lasswell once said, politics is "who gets what, when, and how."[15] By definition, public budgetary decisions require choices among competing programs—choices that are based on many factors, including public preferences, as articulated by local elected officials; legal requirements, including those contained in entitlements; and intergovernmental obligations, such as grant conditions.

Public preferences When it comes to the local budget, taxpayers and their elected representatives demand value for money; thus, programs that provide tangible outputs and easily measurable benefits often have an advantage in budget decisions. Unfortunately, emergency management is among those governmental functions whose value is not readily translated into dollar terms.[16] Although experience may have provided the local government with a good idea of the costs and benefits associated with fire, police, and emergency services, officials may find it more difficult to pledge funds to subsidize initiatives predicated on the—apparently remote—possibility of a terrorist attack or natural disaster. In the wake of a disaster, of course, the support of both elected officials and the public is likely to favor emergency management initiatives, which is why emergency managers are advised to have on hand ready-made ordinances and regulations that can be brought before elected officials when those officials are most likely to be receptive.

Barring a disaster, however, emergency managers are often faced with the task of educating citizens and elected officials about the hazards to which the community is subject. Without such efforts, public perceptions may be a stronger determinant of spending than objective measures of risk: if public opinion gives priority to concerns other than emergency management, local officials will have difficulty sustaining funding of emergency management even if such efforts are heavily subsidized by higher levels of government. For example, since 2002, several major cities have chosen not to mobilize public safety officers in response to increased threat levels posted by the U.S. Department of Homeland Security (DHS) in its Homeland Security Advisory System.[17]

Entitlements Entitlements, a form of mandatory spending, are one of the least controllable categories of public spending. This is because laws, contracts, and regulations allow all those who meet the conditions of eligibility to receive benefits, regardless of whether a government has budgeted enough for the purpose. Local governments fund and manage a much smaller share of entitlement spending programs than do the federal and state governments. Nonetheless, municipal officials are compelled to fund local entitlement programs first.

Local emergency management itself embodies a few forms of mandatory, entitlement-type spending: employee pension payments, payments on outstanding debt, and payments on existing contracts, to name a few. Funding an emergency management activity as an entitlement program locks in local spending until such time as the locality revises or eliminates the entitlement. For example, if a local government outsources a particular category of emergency management–related work to a private corporation under a multiyear contract, this establishes a mandatory local funding commitment that must be honored until the contract is fulfilled. Moreover, local emergency management can be affected by entitlements imposed by other levels of government. Rights-based entitlements, for example, sometimes emerge from unfunded federal mandates. Under the Americans with Disabilities Act of 1990 (ADA), local governments are obligated to provide certain types of assistance, such as aids for the visually handicapped and wheelchair access in or around public buildings. Hurricane Katrina highlighted the problems that can arise during efforts to evacuate people with disabilities. Consequently, it is reasonable to expect that local governments will make satisfactory provision for such evacuations or risk being sued for discrimination under the ADA.

Because entitlement spending has first claim on local government budgets, the pool of discretionary funds available for other purposes, among them emergency management, is limited. Moreover, when local governments participate in federal and state grant programs that require matching funds, they are obligated to move these spending obligations ahead of discretionary funding demands when formulating their annual budgets. Although local decision makers do have some discretion in determining spending on emergency management, their decisions inevitably involve trade-offs, in terms of both financial costs and benefits and broader political considerations.

Intergovernmental obligations The federal government regularly uses its superior fiscal capacity—that is, its ability to raise revenue through borrowing and broad-gauged taxation—to persuade state and local governments to provide particular services.[18] The construction of schools and state highways and the provision of safe drinking water are just a few examples of services that receive generous federal subsidies. The federal government's array of grant programs for state and local emergency management and homeland security is very much in keeping with historical patterns.

Federal aid, however, does not come without strings attached. Some studies have demonstrated that substantial federal and state aid to local government can undermine local control.[19] In the case of emergency management, the greater the share of federal and state funding, the more likely it is that the local emergency management function will become an arm of state and federal emergency management. To ensure that emergency management remains a locally controlled function, it is essential for the emergency manager to work closely with local elected officials. These officials, and the people they represent, are both superiors and stakeholders, who must decide local emergency management policy and allocate funding that is commensurate with that policy.

Many federal grants for emergency management require "matching" funds. For example, federal post-disaster assistance typically covers 75 percent of costs; state and local governments are expected to cover the rest. Federal matching provisions provide strong incentives for states and localities to heavily subsidize the spending purpose. In fact, research has demonstrated that generous federal matching incentives sometimes encourage states and localities to favor federally subsidized programs over those that might actually offer greater benefits to the locality but that receive little or no federal subsidies.[20] Nevertheless, when properly designed, matching grant programs may enable a locality to achieve optimum benefits for the amount of money spent.[21]

Taxes and fees

Most of the capital and labor costs associated with emergency management are supported through taxes and fees. Recently, emergency management funding needs have been rendered more urgent by the combination of vacillating federal support and the imposition of extensive new requirements under the NRP and NIMS. These added burdens have compelled local officials to make exhaustive efforts to seek federal funding or risk having to cover costly mandates solely through local funds. Moreover, even if a local government succeeds in obtaining federal support, that support will cover only some of the new obligations—not all of them. Consequently, local government must rely on new or existing sources of revenue, mostly from taxes and fees, to cover the remaining costs of these obligations, as well as to support the day-to-day demands of emergency management.

Historically, the property tax has been the single largest source of local government revenue. Because of problems associated with this tax, however, and because of the ever-expanding need for local revenue, local governments are always searching for ways to enlarge their revenue sources and protect their tax base. The local tax base is influenced by many factors, including residents' demographic characteristics and the inward and outward migrations of people and businesses. A good mix of residential and commercial interests helps maintain the community's economic health.

The instruments that local officials choose to raise revenue may affect the local government's bond ratings and fiscal health. For example, in a jurisdiction that relies too heavily on property tax income, a significant decline in the market value of homes and businesses will affect the local creditworthiness and thereby depress the local government's bond rating. By the same token, local governments that offer superior fire protection services and active fire prevention programs put fewer assets at risk—which appeals to both insurers and bond raters. Disaster mitigation efforts, initiatives to increase security, and improvements in disaster resilience all support fiscal health and may therefore improve local government bond ratings.[22]

Local taxes, such as the property tax and local-option sales taxes, are mandatory.[23] Revenues from such taxes are used to support general local government services, including emergency management. Other approaches to raising revenue involve voluntary payments.

For example, a surcharge on the admissions fees for attendance at a publicly sponsored event may help defray the costs of providing police and fire protection for such events—costs that would have otherwise been shouldered by local taxpayers.

Special assessments, which are fees designed to support particular public improvements, can also be used to fund emergency management initiatives. In at-risk portions of a downtown, for example, a special assessment, augmented by state and federal funding, could help fund the cost of installing security cameras, biosensors, or other surveillance systems. Similarly, local governments may impose special assessments, in the form of fees on water use, to fund reservoir construction and flood abatement measures.

Disaster mitigation efforts, initiatives to increase security, and improvements in disaster resilience all support fiscal health and may therefore improve local government bond ratings.

Franchise fees are another option. Franchise fees are paid by utilities, including power companies, cable companies, and water companies; nationwide, such fees generate almost one-fifth of municipal operating budgets.[24] In the realm of mitigation, franchise fees can be used to subsidize police patrols of reservoirs, water treatment facilities, and pump stations, or to help emergency responders learn how to deal with downed power lines or natural-gas leaks.

Most local governments provide services to entities—such as nonprofit organizations, educational or religious institutions, and state or federal facilities—that are exempt from local property taxes. Payments in lieu of taxes (PILOTs) are voluntary payments meant to cover some fraction of the public services that these institutions enjoy free of charge. For example, the Port Authority of New York and New Jersey, a bi-state special district government that owns the World Trade Center, routinely makes PILOTs to New York City. The state of Connecticut operates and funds a PILOT program that reimburses local governments for a share of the property taxes that would have been paid by property tax–exempt colleges, universities, and hospitals.[25] Local emergency managers need to understand that PILOTs are a revenue-producing tool that can be used to subsidize emergency management. These managers would be wise, for example, to make local officials aware of the costs associated with providing emergency management services to nonprofit organizations, and to propose that PILOTs be used to compensate the emergency management department for some of these costs.

Finally, local governments may impose taxes or development fees to fund emergency management—particularly mitigation measures. The sidebar on page 307 describes a sales tax–funded mitigation effort undertaken in Tulsa, Oklahoma.

Linking emergency management to other community objectives

To emergency managers, it is axiomatic that emergency management initiatives are good for the community. But persuading citizens and elected officials of the value of emergency management is often another matter. One way of opening up funding for emergency management is by combining it with other community objectives. For example, property owners who might otherwise balk at the imposition of new building codes may feel differently when they discover that the construction of a safe room may increase the market value of their property and reduce their insurance rates. Similarly, when local governments (often in partnership with the state or federal government) offer property owners assistance with relocation or retrofitting, residents and businesses are more likely to undertake such mitigation measures.

Because economic development is among the categories of spending that local elected officials view as worthwhile, emergency management expenditures that mitigate disaster while simultaneously fostering economic development are likely to garner support. By the same token, emergency management initiatives that displace "resources from the private sector toward the production of security measures that, while necessary, are not terribly productive in terms of economic growth" are likely to be challenged or opposed by local authorities.[26]

Tulsa's Mooser Creek project

Since the mid-1990s, Tulsa, Oklahoma, has garnered considerable national publicity for its efforts to combine civic improvements with flood mitigation measures.

A sales tax approved by citizens in 1996 provided $2.9 million to improve portions of Mooser Creek, in the southwest portion of Tulsa. In an effort to preserve the creek's natural beauty and create recreational and educational opportunities, the city entered into a cooperative agreement with the National Park Service (NPS) to plan and construct a greenway. Flood mitigation, recreation, the preservation of natural resources, and sustainable development were all part of the community vision developed for the Mooser Creek watershed.

To ensure that the greenway plan would be compatible with the flood control needs that had been identified in the city's master drainage plan, Tulsa authorities impaneled a citizens' committee, which was asked to partner with the government agencies involved in the effort. The Tulsa-NPS technical committee included experts in engineering, stormwater drainage, trail construction and management, watershed management, parks, education, safety, environmental issues, and archeology.

In presenting the plans for the Mooser Creek greenway at the Mayors' Institute on City Design, Tulsa's mayor, M. Susan Savage, said:

> We chose Mooser Creek as a candidate for this intensive community involvement because Southwest Tulsa is prime for development. The citizens and businesses are pulling together to create new economic opportunities while residents in nearby housing associations have shown enthusiasm in working together toward common objectives.... As we complete these greenways, we will also improve our community's air quality, civic health, livability and quality of life.

Source: Smart Communities Network, "Fact Sheet: Community Involvement Shapes Vision for Mooser Creek" (Tulsa, Okla.: Department of Public Works, January 28, 1997), available at smartcommunities.ncat.org/articles/tulsafact.shtml (accessed September 12, 2007).

Local governments can make use of several tools to leverage private funding, enhance quality of life, and support emergency management—all at the same time. For example, many jurisdictions have the authority to create public improvement districts. Residents and businesses in these districts may willingly pay special fees or taxes in exchange for a desired civic improvement, such as a park, a river walk, or enhanced green space. When such improvements are designed to incorporate mitigation efforts, their value to the community is increased. Greenway projects, for example, by removing built structures from floodplains, support flood mitigation efforts. Boulder Creek Path in Boulder, Colorado, is an example of a public improvement district. In this project, the city of Boulder applied its land use powers to remove or relocate privately owned structures that were situated in a floodplain. In addition to being vulnerable to flood damage themselves, the structures created impediments that diverted floodwaters into nearby neighborhoods, creating a threat to the properties adjacent to the floodplain. The seven-mile-long Boulder Creek Path reduced the flood hazard and produced an environmental and civic amenity that features walkways, bike paths, and recreational facilities.[27]

One way of opening up funding for emergency management is by combining it with other community objectives.

Business improvement districts (BIDs), which are public-private partnerships designed to enhance, maintain, and in some cases market a defined area, are another option. The more than one thousand BIDs in the United States support a diverse array of programs and services, including policing, infrastructure improvements, and business recruitment and retention.[28] A BID may serve mitigation goals by funding the construction of stormwater retention ponds that are fed by parking-lot runoff. Empowerment zones, which provide targeted tax reductions and subsidize infrastructure improvements in order to attract businesses and residents, may also enable local governments to leverage private funds to complement mitigation efforts.

Public-private partnerships can often be employed to leverage modest local expenditures, producing high-value results. Tax exemptions and abatements are frequently used to encourage property owners, including businesses, to undertake mitigation measures. Local governments may also agree to facilitate permit processing for developers who incorporate mitigation efforts into their projects. Other instruments worthy of consideration include property tax reductions for certain types of property improvements, loan guarantees, and regulatory exceptions. Local governments may also use the power of eminent domain to acquire real property and structures located in hazardous areas.[29]

Public improvements often increase the value of private property. For example, a municipal security system that reduces crime and discourages terrorism makes commercial areas safer. Under tax-increment financing (TIF), the value of such improvements can be captured and used to support a public sector initiative: the local government issues bonds to finance improvements and then repays the bonds through the property tax "increment" realized as a consequence of those improvements.[30] TIF is yet another tool that may help fund local emergency management initiatives.

When a private project requires the construction of public improvements or infrastructure, an arrangement is sometimes made in which a portion of the after-tax profits on the private project reverts to the locality. The reasoning behind such an arrangement is that local taxpayers should not pay the full cost of the public improvements when only the developer will reap the economic benefit. Thus, in the interest of fairness, the developer indirectly contributes to the cost of the public infrastructure. Although private business interests own some 85 percent of the nation's most important infrastructure (e.g., telephone and cable communications companies, electric and natural gas utilities, and water purveyors), local government may use its land use authority to foster arrangements in which project developers share the cost of public improvements; revenues thus obtained, in turn, may be used to support local emergency management.[31]

Capital budgeting

As discussed in Chapter 6, mitigation often involves structural measures, such as the elevation of bridges and other structures, the construction of levees, seismic retrofits of buildings and infrastructure, the installation of surveillance cameras, and the creation of firebreaks. When the costs of structural mitigation are too great to be covered by the local government's operating budget, these projects may be funded through the capital budget. Most capital projects are subjected to a cost-benefit analysis before they are undertaken. Among the benefits to be considered are reductions in disaster vulnerability.

Capital budgeting involves borrowing in amounts that are usually outside the limits of balanced budget requirements. Investors lend the local government money, usually by purchasing some type of bond. In 1992, for example, the citizens of Berkeley, California, approved two general-obligation bond measures for mitigation efforts, including seismic upgrades of every major public building and the installation of a secondary water supply system to fight fires that might occur in the aftermath of an earthquake.[32]

When local governments issue bonds, the investors are repaid from a share of local revenue or through the income stream produced by the capital project. In the latter case, the income stream must be sufficient to repay bondholders' principal and interest. A municipality's bond rating gives investors some idea of the risk of possible default.[33] Most of a local government's outlays for capital improvements are for utilities, housing, and schools.[34] Capital outlays associated with emergency management include antiterrorism barriers, flood control works, seismic retrofitting of infrastructure and government buildings, tornado or hurricane shelters, and high-water rescue vehicles.

Because federal and state post-disaster assistance to local government is usually insufficient to cover full replacement or repair of damaged local infrastructure, local governments often need to pay infrastructure expenses through capital expenditures (in which rents, fees, or tolls from the capital project provide income) or through long-term debt management (structured borrowing with a planned repayment scheme). A local government has three sources of financing: capital assets (the land, property, and structures that it owns and which it may sell), debt (borrowing), and retained earnings (income from stocks, revenue-generating bonds, and other investment instruments that is not reinvested). Because infrastructure promotes economic development, it is of special significance in capital budgeting. Infrastructure embodies three possible types of investment: physical assets, economic development, and intangibles. Infrastructure that advances the goals of emergency management, for example, may be thought of as an intangible investment in quality of life.

The Federal Emergency Management Agency (FEMA) has long used counties as the unit of political geography covered by a disaster declaration. A municipality that is within a county that has received a disaster declaration is entitled to receive the same forms of federal assistance as the county, and will receive such assistance directly from the state or federal government. A county that is included in a Presidential Disaster Declaration may be eligible, along with its municipalities, to receive FEMA Public Assistance grants, which provide support for the repair, replacement, or restoration of public facilities and the facilities of certain nonprofit organizations. The federal share of such assistance is 75 percent; the funds are funneled through the state, and virtually all states have laws that set forth the respective state and local shares of the nonfederal match.[35]

One disaster loss that is often overlooked is damage to the local property tax base. Some property owners, faced with the prospect of additional borrowing to repair a heavily damaged (and heavily mortgaged) home, abandon the property or go into bankruptcy. Both of these outcomes tend to shrink the local property tax base, further complicating the community's economic recovery. A little-known federal program—FEMA's Community Disaster Loan Program—helps local governments that have experienced revenue losses, increased operating expenses, or both, as the result of a disaster.[36] Local governments are eligible only in the case of a Presidential Disaster Declaration and may apply only at the end of the tax year. Documentation is required to demonstrate that the losses stemmed directly from disaster impact.

Other funding options: Privatization and coproduction

In almost all localities, cost control is essential. One way to exercise cost control is to outsource certain duties or functions. Another way is to privatize an entire program or public purpose, transferring the job to a for-profit or nonprofit firm. Both approaches are usually undertaken through a fixed-term contract. Some components of emergency management may be candidates for outsourcing. For example, many local governments have outsourced GIS research and applications work that was previously conducted in-house. Municipal governments sometimes enter into outsourcing contract arrangements with private firms to obtain specialized training for emergency responders. Services that have sometimes been privatized include sanitation, corrections, public parking, education, road repair, parks and recreation, flood control, and water and sewer services.

Coproduction, a form of civic collaboration, is an important concept in local emergency management budgeting. By relying on volunteer organizations, nonprofits, and other nongovernmental organizations, emergency managers can often accomplish more with fewer paid workers and at minimal budgetary cost. (Chapter 5 discusses collaborative emergency management in detail.)

The Citizen Corps, for example, is a federally supported program that brings together local leaders, citizen volunteers, and a network of first-responder organizations, all of whom engage in local preparedness and response.[37] The Bush administration's FY 2006 budget request called for $50 million in grants to state Citizen Corps activities.[38] The "safe construction" initiatives promoted by the Institute for Building & Home Safety are another example of coproduction; these initiatives foster improved land use practices, better building standards and codes, and "fortified" (disaster-resistant) homes and businesses.[39] The Blue Sky Foundation in North Carolina also encourages developers and builders to adopt disaster-safe construction practices, as does the Red Cross, through its "Building Disaster Resistant Neighborhoods" program.[40] Such efforts complement local government initiatives to promote safe building practices.

Coproduction leverages local emergency management efforts so that the departmental budget accomplishes more than it otherwise would have. Moreover, the fact that those who are involved in coproduction arrangements—citizen volunteers and the staff of nonprofit organizations—enjoy the work they do and the services they provide helps garner public support for emergency management, which works to its advantage in the competition for local funding.

Covering the costs of emergencies and disasters

Many local governments self-insure against disaster by maintaining "rainy day" funds—emergency reserve accounts, contingency funds, or trust accounts—that elected officials believe will be sufficient to cover the costs of disaster.[41] Some local governments have the option of paying for the costs of disaster by imposing special sales taxes or surcharges and by billing special-district governments. A portion of post-disaster tax burdens may be shifted to nonresidents through wage tax increases. Another option—raising local sales taxes to generate post-incident revenue—is often chosen in response to pressure from the state government. Finally, some local governments attempt to cover disaster-related expenses by drawing on current-year allocations—by far the most risky option. Of course, local governments also count on mutual aid from adjacent jurisdictions, and on disaster relief from the state and federal government.

Many local governments participate in FEMA's National Flood Insurance Program (NFIP), which provides low-cost flood insurance to people residing in jurisdictions that are part of the program.[42] Local governments that agree to abide by NFIP rules must refrain from building in floodplains and shoreline areas. Under the NFIP's Community Rating System, localities that undertake specified mitigation measures enjoy lower premiums on their residents' flood insurance policies. Thus, although mitigation may cost communities development opportunities in the short run, it results in lower flood risks, lower NFIP premiums, and higher quality of life in the long run.

Local emergency management is also indirectly subsidized by private insurers. Private insurance companies, sometimes aided by state insurance commissioners,[43] provide a variety of incentives for policyholders, whether residential or commercial, to reduce risks from hazards.[44] Commercial terrorism insurers may find it beneficial to work closely with local emergency management agencies.[45]

Federal funding for emergency management

There are two principal types of federal funding for emergency management: pre-disaster funding and post-disaster funding. Pre-disaster funding is in the form of grants to support ongoing activities such as flood control, seismic retrofitting of public structures, and the elevation of structures that are subject to storm surges. Post-disaster funding supports, among other things, the repair or replacement of public infrastructure, the reimbursement of nonprofit organizations for relief to areas subject to federal disaster declarations, and direct aid to individuals and families. FEMA's three main assistance programs under a Presidential Disaster Declaration are (1) the Individual and Family Grant Program, (2) the Public Assistance Grant Program (which assists governments and some nonprofit organizations), and (3) Mitigation Grant programs.[46] In 2006, FEMA responded to fifty-two major disaster declarations and five emergencies.[47]

Before the attacks of September 11, 2001, most federal funding for state and local government emergency management took the form of relief after a federal disaster declaration: between disasters, federal funding was often paltry. As noted earlier, however, the years since 9/11 have been marked by a profusion of new and generously funded federal grant programs, most of which are focused on preventing terrorism.

The president's authority to declare a major disaster (or to issue such a declaration at the request of a governor) stems from the Federal Disaster Act of 1950. From May 1953 through July 2007, presidents from Dwight Eisenhower to George W. Bush issued more than 1,700 declarations of major disaster. The Federal Disaster Act of 1974 gave the president the authority to declare an emergency,[48] which allows the provision of immediate federal assistance for an ongoing disaster. A declaration of emergency also allows federal resources to be mobilized in the face of an impending disaster, and provides immediate financial assistance for state and local governments during the response phase. From 1974 to 2007, presidents from Richard Nixon to G. W. Bush issued more than 275 emergency declarations.[49]

Local emergency managers need to be aware that legislative (and some managerial) decisions about the design, funding levels, allocation systems, and administrative rules governing federal grant programs are made in a highly political environment. Federal legislators have a strong interest in ensuring that decisions about federal funding benefit their home districts and states; thus, they engage in an annual struggle over how much money will be dispensed and under what allocation schemes.[50] When funds are targeted on the basis of risk analysis, some areas win and some lose. How law enforcement and national security agencies define terrorism, assess the risk, and measure vulnerability certainly affects which districts receive funds. The same dynamic applies in the case of other programs, such as federal funding for flood control projects and other public works.

The effects of Hurricane Katrina

In late August 2005, Hurricane Katrina devastated the Gulf Coast areas of Alabama, Louisiana, and Mississippi and triggered the failure of the levees surrounding New Orleans. This destructive and widespread hurricane was blamed for more than a thousand deaths and displaced more than half a million people for periods ranging from weeks to months. In response to Katrina, Michael Chertoff, secretary of the U.S. Department of Homeland Security, activated the National Response Plan (NRP) by declaring an "incident of national significance"—the first-ever use of this designation.

Federal disaster declarations for Katrina covered portions of four states and 90,000 square miles of impact area.[1] In an extraordinary move, President George W. Bush allowed the governor of any state that hosted Katrina evacuees to submit a request for a presidential declaration of major disaster, which would allow the state to obtain reimbursement for the costs of hosting Katrina victims. President Bush also authorized adjustments to state and local cost sharing, under which the federal government took on 100 percent of Public Assistance costs for Alabama, Louisiana, and Mississippi, instead of the usual federal share of 75 percent.

As of May 24, 2006, FEMA's allocations and obligations for Hurricane Katrina totaled about $26 billion.[2] The accompanying bar chart vividly illustrates the difference in federal allocations, obligations, and expenditures for Hurricanes Rita, Wilma, and Katrina. What is remarkable is the huge amounts of federal spending for Katrina. As of January 31, 2007, FEMA's National Flood Insurance Program had paid out more than $15.6 billion in claims for Katrina damage to policyholders in Alabama, Florida, Louisiana, and Mississippi.[3]

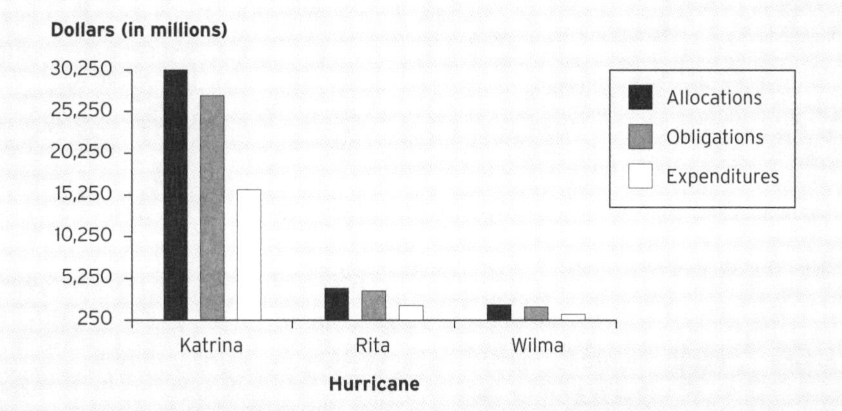

Katrina was an overwhelming challenge for emergency managers at all levels of government. At the federal level, after-action reports revealed aspects of the NRP and the National Incident Management System that need improvement. The social, political, and economic reverberations of this catastrophe will be played out for years to come. Decisions about who will cover the costs of recovery; how the cost burdens will be distributed among governments, the private sector, and disaster victims; and how all levels of government can prepare to meet the demands of such catastrophes will affect public budgeting in fundamental ways.

[1]See Federal Emergency Management Agency (FEMA), "2005 Federal Disaster Declarations, Declared Disasters by Year and State," available at fema.gov/news/disasters.fema?year=2005 (accessed September 18, 2007).

[2]Allocations represent federal spending authority made available to pay for relief and recovery costs of all types. Obligations represent commitments of federal agencies that require later payment.

[3]See FEMA, "Current Gulf Statistics: Hurricane Katrina National Flood Insurance Program (NFIP) (actual figures as of 01/31/07)," available at fema.gov/hazard/hurricane/2005katrina/statistics.shtm (accessed September 18, 2007).

Because super-majority coalitions of legislators are needed to pass bills, the distribution schemes are often designed to ensure that almost every state and local jurisdiction gets something out of a program. Naturally, such broad funding schemes have been criticized as pork barreling (the allocation of resources in excess of need) and for their failure to allocate funds in accordance with measurable risk.

Federal funding for emergency management is also affected by ongoing federal reorganizations. In particular, FEMA, the chief federal benefactor of state and local emergency management from April 1979 to March 2003, has been through a number of changes over the years. In 2003, it was folded into DHS, a new department comprising an amalgamation of twenty-two agencies and employing about 180,000 people. FEMA has struggled to survive as a cohesive organization while simultaneously trying to find its proper place within the federal government's newest and third-largest department—a department that is itself still evolving. Owing to FEMA's much-publicized problems in responding to Hurricane Katrina, in 2005 and 2006 several House and Senate committees considered replacing the agency with a revamped and renamed emergency management organization. Some proposals called for

FEMA grants and special funding sources

Assistance to Firefighters Grant Program

Provides assistance to local fire departments to protect citizens and firefighters against the effects of fire and fire-related incidents. Available from the U.S. Fire Administration (USFA) to fire departments and other first responders.

Chemical Stockpile Emergency Preparedness Program

Improves preparedness to protect the people of certain communities in the unlikely event of an accident involving this country's stockpiles of obsolete chemical munitions. Available to states, localities, and tribal governments.

Community Assistance Program, State Support Services Element

Provides funding to states to provide technical assistance to communities in the National Flood Insurance Program (NFIP) and to evaluate community performance in implementing NFIP floodplain management activities.

Comprehensive Environmental Response, Compensation, and Liability Act

Supports programs designed to improve capabilities associated with oil and hazardous materials emergency planning and exercising. Available to states, localities, and tribal governments; U.S. territories; state emergency response committees; and local emergency planning committees.

Cooperating Technical Partners

Provides technical assistance, training, and/or data to support flood hazard data development activities. Available to states, localities, and tribal governments.

Emergency Management Institute

Provides training and education to the fire service, its allied professions, emergency management officials, and the general public. Available to fire departments, other first responders, emergency management officials, and individuals.

Fire Management Assistance Grant Program

Provides assistance for the mitigation, management, and control of fires on publicly or privately owned forests or grasslands, which threaten such destruction as would constitute a major disaster. Available to state, local, and tribal governments.

Flood Mitigation Assistance Program

Provides funding to assist states and localities in implementing measures to reduce or eliminate the long-term risk of flood damage to buildings, manufactured homes, and other structures insurable under the NFIP.

Hazard Mitigation Grant Program

Provides grants to states and local governments to implement long-term hazard mitigation measures after a major disaster declaration. Also available to tribal governments, certain private-nonprofit organizations or institutions, authorized tribal organizations, and Alaska Native villages or organizations via states.

moving FEMA out of DHS and granting it the independence it had between 1979 and 2003. Even before Katrina, however, FEMA had already been diminished by the loss of some of its grant-issuing authority and mitigation and preparedness responsibilities: in summer 2005, DHS secretary Michael Chertoff explained that "to ensure that our preparedness efforts do have a focused direction, we intend to consolidate all the Department's existing preparedness efforts—including planning, training, exercising and funding—into a single directorate led by an under secretary for preparedness. Going forward, FEMA will... [directly] report to the Secretary, and it will focus on its historic and vital mission of response and recovery."[51] Chertoff's 2005 change stripped FEMA of its preparedness functions by moving those duties to the DHS Preparedness Directorate.

The six principal sources of federal grant funding for emergency management are the State Homeland Security Grant Program, the Urban Area Security Initiative (UASI), the Law Enforcement Terrorism Prevention Program (LETPP), the Emergency Management Performance Grant (EMPG) program, the Assistance to Firefighters Grant program, and the Metropolitan Medical Response System (MMRS). (Other grant programs are described in the sidebar below and in the

Map Modernization Management Support

Provides funding to supplement, not supplant, ongoing flood hazard mapping management efforts by the local, regional, or state agencies. Available to states and localities.

National Dam Safety Program

Provides financial assistance to the states for strengthening their dam safety programs.

National Earthquake Hazards Reduction Program

Provides financial assistance to the states for strengthening their earthquake hazard reduction programs.

National Fire Academy Education and Training

Provides training to increase the professional level of the fire service and others responsible for fire prevention and control. Available from USFA to fire departments and firefighting personnel.

National Flood Insurance Program

Enables property owners in participating communities to purchase insurance as a protection against flood losses in exchange for state and community floodplain management regulations that reduce future flood damages. Available to states, localities, and individuals.

National Urban Search and Rescue (US&R) Response System

Provides funding for the acquisition, maintenance, and storage of equipment, training, exercises, and training facilities to meet task force position criteria, and conduct and participate in meetings within the National US&R Response System. Available to US&R task forces.

Pre-Disaster Mitigation Program

Provides funds for hazard mitigation planning and the implementation of mitigation projects prior to a disaster event. Available to states, localities, and tribal governments.

Repetitive Flood Claims Program

Provides funding to states and localities to reduce or eliminate the long-term risk of flood damage to structures insured under the NFIP that have had one or more claims for flood damages, and that cannot meet the requirements of the Flood Mitigation Assistance program for either cost share or capacity to manage the activities.

State Fire Training System Grants

Provide financial assistance to state fire training systems for the delivery of a variety of National Fire Academy courses/programs. Available from USFA.

Superfund Amendments and Reauthorization Act

Provides funding for training in emergency planning, preparedness, mitigation, response, and recovery capabilities associated with hazardous chemicals. Available to public officials, fire and police personnel, medical personnel, first responders, and other tribal response and planning personnel.

Source: "FEMA Grants and Assistance Programs," available at fema.gov/government/grant/index.shtm (accessed September 18, 2007).

one on page 315.) It is important to note that federal agencies besides DHS and FEMA provide considerable grant assistance to state and local emergency management. In February 2005, the U.S. Office of Management and Budget reported that $3.2 billion in non-FEMA, non-DHS funding had been awarded to state and local governments.[52]

State Homeland Security Grant Program

The State Homeland Security Grant Program (known until FY 2003 as the State Domestic Preparedness Program) provides funding to states, which may then allocate these monies to their respective local governments. The Bush administration budget request to Congress for FY 2006 advocated "$1 billion for discretionary grants to States and territories" to assist states and localities in meeting the National Preparedness Goal (Homeland Security Presidential Directive 8 [HSPD 8]), which includes addressing the risks identified in state homeland security plans.[53]

Urban Area Security Initiative

Under UASI, local governments in certain metropolitan areas receive funding to support planning, training, exercising, equipment, and administrative costs. UASI requires local governments to plan, train, and conduct exercises for fifteen scenarios, twelve of which are terrorist attacks of various types and the other three are for different types of natural disasters; thus, UASI money is primarily intended to fund activities and equipment for preventing, preparing for, responding to, and recovering from terror events. The FY 2006 Bush administration budget request to Congress called for "$1 billion in discretionary grants to urban areas and regions."[54] In FY 2006, DHS identified thirty-five areas—including ninety-five cities with populations of 100,000 or more—that were eligible to apply for funding. In addition, DHS identified eleven areas that had received UASI funding in FY 2005 and were eligible to apply for sustenance funding in FY 2006.

All eligible applicants must submit an investment justification that explains how the proposed security enhancements will help their jurisdictions meet the target capabilities outlined in the National Preparedness Goal.[55] To determine which investments will be funded, DHS reviews and scores the investment justification according to a formula that takes into account three primary variables: consequence, vulnerability, and threat. Consideration is also given to such factors as population size, population density, location of critical infrastructure, existence of formal mutual aid agreements, ongoing law enforcement investigations, and the presence of international borders.

Because DHS is promoting the creation of "super" UASIs to strengthen regional capabilities, UASI jurisdictions that share boundaries were combined into a single entity for the purposes of the FY 2006 risk analysis, and urbanized areas outside city limits were included within the geographic area designated for risk analysis. Another expansion of UASI in FY 2006 was the inclusion, in the risk analysis, of information from the intelligence community.[56]

The Law Enforcement Terrorism Prevention Program

The LETPP supports law enforcement organizations in their efforts to prevent terrorism. As is the case with other categories of federal funding, support is disbursed first to the state government, which coordinates with the state's lead law enforcement agency.[57] Within sixty days of their receipt of funds, states are required to pass on at least 80 percent of the total grant amount to local governments. The LETPP has no matching grant provisions. The program supports the following activities:

- Intelligence gathering and information sharing through the creation or enhancement of terrorism intelligence/information fusion centers[58]
- Strategic planning
- Projects designed to harden high-value targets
- Interoperable communications initiatives
- Collaboration with non–law enforcement partners, other government agencies, and the private sector.[59]

U.S. Department of Homeland Security grants

Emergency Management Performance Grants[1]

Emergency Management Performance Grants (EMPGs) gives states the opportunity to structure individual emergency management programs based on needs and priorities for strengthening their emergency management capabilities while addressing issues of national concern. States have the flexibility to develop systems that encourage the building of partnerships that include government, business, volunteer, and community organizations.

Homeland Security Grant Program[2]

- **State Homeland Security Program (SHSP):** The SHSP supports the implementation of the State Homeland Security Strategies to address the identified planning, equipment, training, and exercise needs for acts of terrorism. In addition, the SHSP supports the implementation of the National Preparedness Goal, the National Incident Management System, and the National Response Plan.
- **Urban Areas Security Initiative (UASI):** UASI funds are provided to address the unique planning, equipment, training, and exercise needs of high-threat, high-density urban areas, and to assist those areas in building an enhanced and sustainable capacity to prevent, protect against, respond to, and recover from acts of terrorism.
- **Law Enforcement Terrorism Prevention Program (LETPP):** The LETPP focuses on the prevention of terrorist attacks and provides law enforcement and public safety communities with funds to support intelligence gathering and information sharing through enhancing/establishing fusion centers; interoperable communications; and collaboration with non-law enforcement partners, other government agencies, and the private sector.
- **Metropolitan Medical Response System (MMRS):** MMRS funds support MMRS jurisdictions to further enhance and sustain an integrated, systematic mass casualty incident preparedness program that enables an effective response during the first crucial hours of an incident. The program prepares jurisdictions for response to the range of mass casualty incidents, including chemical, biological, radiological/nuclear, and explosive events; agricultural and epidemic outbreaks; natural disasters; and large-scale hazardous materials incidents.
- **Citizen Corps:** The mission of the Citizen Corps is to actively involve all citizens in hometown security through personal preparedness, training, exercises, and volunteer service. Citizen Corps funds support Citizen Corps Councils and their efforts to engage citizens in all-hazards prevention, protection, response, and recovery.

The Infrastructure Protection Program[3]

- **Transit Security Grant Program (TSGP):** The TSGP provides grant funding to the nation's key high-threat urban areas to enhance security measures for their critical transit infrastructure including bus, rail, and ferry systems. This year, the TSGP will also provide funding to Amtrak for continued security enhancements for their intercity rail operations between key high-risk urban areas throughout the United States.
- **Port Security Grant Program (PSGP):** The PSGP provides grant funding to port areas for the protection of critical port infrastructure from terrorism. PSGP funds are primarily intended to assist ports in enhancing risk management capabilities; domain awareness; and capabilities to prevent, detect, respond to, and recover from attacks involving improvised explosive devices (IEDs) and other nonconventional weapons, as well as training and exercises.
- **Intercity Bus Security Grant Program (IBSGP):** The IBSGP provides funding to create a sustainable program for the protection of intercity bus systems and the traveling public from terrorism. The FY 2007 IBSGP seeks to assist owners and operators of fixed-route intercity and charter bus services in obtaining the resources required to support such security measures as enhanced planning, facility security upgrades, and vehicle and driver protection.
- **Trucking Security Program (TSP):** The TSP provides funding for the Highway Watch® Program in order to continue a sustainable national program to enhance security and overall preparedness on the nation's highways. The FY 2007 TSP will provide funding in the form of a cooperative agreement directly to the American Trucking Association for the continued modernization and management of this program.
- **Buffer Zone Protection Program (BZPP):** The BZPP provides grant funding to build security and risk-management capabilities at the state and local levels to secure predesignated Tier I and Tier II critical infrastructure sites, including chemical facilities, financial institutions, nuclear and electric power plants, dams, stadiums, and other high-risk/high-consequence facilities.

[1]U.S. Department of Homeland Security (DHS), *Fiscal Year 2005 Homeland Security Grant Program: Program Guidelines and Application Kit* (Washington, D.C.: DHS, 2005), 19-20, available at ojp.usdoj.gov/odp/docs/fy05hsgp.pdf (accessed September 18, 2007).

[2]Ibid.

[3]DHS, "Overview: FY 2007 Infrastructure Protection Program" (Washington, D.C.: DHS, January 9, 2007), 3-4, available at dhs.gov/xlibrary/assets/grants-2007-infrastructure-protection.pdf (accessed September 18, 2007).

The Emergency Management Performance Grant program

The Emergency Management Performance Grant (EMPG) program is designed to help state and local emergency managers achieve measurable results in key functional areas of emergency management. EMPGs are allocated to states, which are free to decide how much grant money they will pass on to local jurisdictions.[60] Thus, states have the flexibility to allocate funds according to risk, and to address the most urgent state and local needs. DHS encourages the states, however, to use EMPGs to foster partnerships among government, business, and volunteer organizations, and to pay for joint operations, mutual aid, local and regional support, and state-to-state cooperation. All states—including the District of Columbia, tribal governments, and U.S. territories and possessions—are eligible.[61]

The Assistance to Firefighters Grant Program and the Metropolitan Medical Response System

The Assistance to Firefighters Grant Program pays for vehicles, equipment, and training that firefighters need to protect the public. The Bush administration budget request for FY 2006 asked Congress to furnish $500 million in competitive grants to fire departments and providers of emergency medical services.[62] MMRS provides funding to support planning, training, exercising, and the purchase of equipment and pharmaceuticals to cope with events involving mass casualties.[63]

The state context for emergency management

Most state constitutions empower the governor to take certain actions to ease the impact of disasters, including military or terrorist attacks. All states provide for continuity of government and allow participation in mutual aid agreements.[64] Each state has an emergency management agency or division, led by a director or an adjutant general, that manages mitigation, preparedness, response, and recovery at the state level; this includes coordinating with other state and local agencies; developing state emergency response plans and assisting local governments with theirs; administering federal emergency management grants; providing training in cooperation with other state, local, and federal agencies; and engaging in emergency response.[65]

State budgets often make provision for funding local emergency management.[66] As noted earlier, local governments are often subgrantees under federal emergency management or homeland security grant programs.

Summary

Public budgeting involves public policy, intergovernmental relations, and a variety of tax and expenditure matters. Emergency management cannot be sustained or improved without adequate funding—and, as is the case in all areas of local government, obtaining funding and deciding how to use it involves many policy choices. Ideally, those choices will be made by people who understand the world of public budgetary instruments and possibilities. This chapter is intended to provide a base for examining those possibilities.

Because of the severity of revenue constraints and balanced budget requirements, local budgeting is largely synonymous with revenue budgeting. Local governments have several options—including taxes and fees, bonds, special assessments, and intergovernmental grants—for funding emergency management. They also have tools and methods available for encouraging citizens and businesses to take actions that will help protect the community; these include public-private partnerships, tax exemptions and abatements, and business improvement districts.

Since the attacks of September 11, 2001, federal funding for emergency management has increased dramatically; in particular, localities are eligible for much more "between-disaster" support. With the increase in federal funding, however, have come significant requirements for matching funds, greater administrative burdens, and a plethora of new demands. Moreover, the new federal emphasis on preparing for, preventing, and responding to terrorism is, in some cases, leading local governments to recast their emergency management programs to preserve their eligibility for federal funding.

The programs and policies described in this chapter will no doubt change during the coming years, sometimes in response to new events. Local emergency managers and elected officials need to attend closely to these changes in order to make wise investments in emergency management initiatives.

Notes

1 Robert D. Lee Jr., Ronald W. Johnson, and Philip G. Joyce, *Public Budgeting Systems,* 7th ed. (Boston: Jones and Bartlett Publishers, 2004), 2–3.

2 Chris Strohm, "9/11 Panel Seeks Billions in Federal Spending," *Government Executive.com,* August 18, 2004, available at govexec.com/dailyfed/0804/081804c1.htm (accessed September 12, 2007).

3 Amanda J. Dory, *Civil Security: Americans and the Challenge of Homeland Security* (Washington, D.C.: CSIS Press, 2003), 1.

4 Keith Bea, "The Formative Years: 1950–1978," in *Emergency Management: The American Experience, 1900–2005,* ed. Claire B. Rubin (Fairfax, Va.: Public Entity Risk Institute [PERI], 2007), 82, available at riskinstitute.org.

5 Local civil protection—such as fire protection, law enforcement, and emergency medical services—and the deployment of the National Guard under certain circumstances are not part of national defense.

6 Donald F. Kettl, *System under Stress: Homeland Security and American Politics* (Washington, D.C.: CQ Press, 2004), 7.

7 For example, under the Buffer Zone Protection Program, the U.S. Department of Homeland Security (DHS) dispenses grants directly to local law enforcement organizations.

8 Special districts are limited-purpose governmental units (either single-function or multifunction) that have some degree of fiscal and administrative independence from general-purpose governments; examples include water districts, sanitation districts, transportation districts, and flood control districts: see Edward J. Bierhanzl and Paul B. Downing, "User Charges and Special Districts," in *Municipal Policies in Local Government Finance,* 5th ed., ed. J. Richard Aronson and Eli Schwartz (Washington, D.C.: ICMA, 2004), 339.

9 Veronica de Rugy, "What Does Homeland Security Spending Buy?" (Washington, D.C.: American Enterprise Institute for Public Policy Research, 2005), available at aei.org/docLib/20050408_wp107.pdf (accessed September 17, 2007).

10 Ibid., 2.

11 James J. Gosling, *Budgetary Politics in American Governments,* 3rd ed. (New York: Routledge, 2002), 166.

12 Ibid., 159–194.

13 Lee, Johnson, and Joyce, *Public Budgeting Systems,* 44.

14 County governments may also work with their incorporated municipalities on grants.

15 Harold Lasswell, *Politics: Who Gets What, When, How* (New York: Meridian Books, 1958).

16 Lee, Johnson, and Joyce, *Public Budgeting Systems,* 5.

17 Benigno E. Aguirre, "Homeland Security Warnings: Lessons Learned and Unlearned" (Newark: Disaster Research Center, University of Delaware, 2004), available at udel.edu/DRC/Preliminary_Papers/PP334-Homeland%20Security.pdf (accessed September 18, 207); see also Charles V. Peña, "Homeland Security Alert System: Why Bother?" (Washington, D.C.: Cato Institute, 2002), available at cato.org/pub_display.php?pub_id=4205 (accessed September 18, 2007). The warning system itself has been strongly criticized.

18 Lee, Johnson, and Joyce, *Public Budgeting Systems,* 508.

19 Allen K. Settle, "Disaster Assistance: Securing Presidential Declarations," in *Cities and Disaster: North American Studies in Emergency Management,* ed. Richard T. Sylves and William L. Waugh Jr. (Springfield, Ill.: Charles C. Thomas Publishers, 1990), 33–57; see also Jonathan Rodden, *Hamilton's Paradox: The Promise and Peril of Fiscal Federalism* (New York: Cambridge University Press, 2006).

20 Lee, Johnson, and Joyce, *Public Budgeting Systems,* 536.

21 John Fitzgerald Due and Ann F. Friedlander, *Government Finance: Economics of the Public Sector,* 5th ed. (Homewood, Ill.: Richard D. Irwin, 1973), 491.

22 Robert Litan, "Sharing and Reducing the Financial Risks of Future 'Mega-Catastrophes,'" Brookings Institution Economic Studies Working Paper (Washington, D.C.: Brookings Institution, 2005) available at brookings.edu/~/media/Files/rc/papers/2006/03business_litan02/200603_iiep_litan.pdf (accessed September 18, 2007); see also Howard Kunreuther, "Has the Time Come for Comprehensive Natural Disaster Insurance?" in *On Risk and Disaster: Lessons from Hurricane Katrina,* ed. Ronald J. Daniels, Donald F. Kettl, and Howard Kunreuther (Philadelphia: University of Pennsylvania Press, 2006).

23 Some taxes—such as motor fuel taxes—are actually user fees. Revenues from motor fuel taxes are generally earmarked for road construction and repair.

24 Gosling, *Budgetary Politics in American Governments,* 177.

25 Judith B. Greiman, "Connecticut's Payment-in-Lieu-of-Taxes Program: A Model for the Nation," *The Independent Voice* (Connecticut Conference of Independent Colleges Quarterly Newsletter), vol. 2 (March 2004): 1, available at theccic.org/pdf/news_0304.pdf (accessed September 18, 2007).

26 De Rugy, "What Does Homeland Security Spending Buy?" 5.

27 See Total Boulder, "Boulder Creek Path," at totalboulder.com/resources/49.html (accessed September 12, 2007).

28 Richard Briffault, "A Government for Our Time? Business Improvement Districts and Urban Governance," *Columbia Law Review* 99, no. 2 (1999): 365.

29 Eminent domain refers to the government's authority to acquire property when it can be shown that the property is to be used for a public purpose and that the owner has received just compensation.

30 Gosling, *Budgetary Politics in American Governments,* 175–177.

31 In the case of utilities subject to government regulation, a local government may need to petition the state public utility commission to seek approval for its cost-recovery proposal.

32 Western States Seismic Policy Council, "WSSPC Awards in Excellence 1998 Award Recipients," available at wsspc.org/Awards/1998/award981.htm.

33 Ibid.

34 Lee, Johnson, and Joyce, *Public Budgeting Systems,* 432.

35 Federal Emergency Management Agency (FEMA), "Public Assistance Grant Program," available at fema.gov/government/grant/pa/index.shtm (accessed September 18, 2007).

36 Nonna A. Noto and Steven Maguire, *FEMA's Community Disaster Loan Program,* CRS Report for Congress RL 33174 (Washington, D.C.: Congressional Research Service, Library of Congress, 2005), available at cnie.org/NLE/CRSreports/06Aug/RL33174.pdf; for more on the program itself, go to fema.gov/government/grant/fs_cdl.shtm (both sites accessed September 18, 2007).

37 U.S. Department of Homeland Security (DHS), *Fiscal Year 2005 Homeland Security Grant Program: Program Guidelines and Application Kit* (Washington, D.C.: DHS, 2005), 79, available at ojp.usdoj.gov/odp/docs/fy05hsgp.pdf; for information on the Citizen Corps, go to citizencorps.gov/ (both sites accessed September 18, 2007).

38 U.S. Office of Management and Budget (OMB), "FY 2006 Budget Priorities: Department of Homeland Security," available at whitehouse.gov/omb/budget/fy2006/dhs.html (accessed September 17, 2007).

39 William L. Waugh Jr., "Terrorism, Homeland Security and the National Emergency Management Network," *Public Organization Review* 3, no. 4 (December 2003): 382.

40 Ibid.; see also the description of the program at the Red Cross's Web site, redcross.tallytown.com/drn.html (accessed September 18, 2007).

41 In many jurisdictions, however, the creation of rainy day funds is prohibited by laws that compel tax rebates or tax cuts if the government is holding a sizable surplus.

42 For more on the National Flood Insurance Program, go to fema.gov/business/nfip/ (accessed September 18, 2007).

43 Under the McCarran-Ferguson Act of 1945, state insurance commissioners are empowered to regulate insurance firms operating within their respective states for the purpose of preventing the formation of trusts or other arrangements that may restrain trade. The commissioners collect information from insurers in order to monitor rates, learn about the availability of coverage and policy servicing, and other issues. Insurance commissioners may be able to persuade insurers of the need to incorporate disaster mitigation into the marketing of various lines of property insurance.

44 Howard Kunreuther, "Insurability Conditions and the Supply of Coverage," in *Paying the Price: The Status and Role of Insurance against Natural Disasters in the United States,* ed. Howard Kunreuther and Richard J. Roth Sr. (Washington, D.C.: Joseph Henry Press, 1998), 17–50.

45 Howard C. Kunreuther and Erwann Michel-Kerjan, "Policy Watch: Challenges for Terrorism Risk Insurance in the United States," NBER Working Paper 10870 (Cambridge, Mass.: National Bureau of Economic Research, November 2004), available at nber.org/papers/w10870 (accessed September 18, 2007).

46 Federal support for mitigation aid may be issued either as the result of, or independently of, a Presidential Disaster Declaration.

47 FEMA, "2006 Federal Disaster Declarations, Declared Disasters by Year and State," available at fema.gov/news/disaster_totals_annual.fema (accessed September 18, 2007).

48 Richard T. Sylves, "The Politics and Budgeting of Federal Emergency Management," in *Disaster Management in the U.S. and Canada: The Politics, Policymaking, Administration and Analysis of Emergency Management,* ed. Richard T. Sylves and William L. Waugh Jr. (Springfield, Ill.: Charles C. Thomas Publishers, 1996), 26–45.

49 For current figures on major disaster declarations and emergency declarations, see FEMA, "2006 Federal Disaster Declarations, Declared Disasters by Year and State" (see note 47).

50 When it comes to allocation schemes, the behavior of state legislators often parallels that of federal legislators.

51 "Secretary Michael Chertoff U.S. Department of Homeland Security Second Stage Review Remarks," July 13, 2005, available at dhs.gov/xnews/speeches/speech_0255.shtm (accessed September 18, 2007). Chertoff's July 2005 reorganization created an undersecretary for preparedness who sits outside FEMA.

52 OMB, "FY 2006 Budget Priorities: Department of Homeland Security."

53 Ibid.

54 Ibid.

55 See DHS, "Interim National Preparedness Goal" (Washington, D.C.: DHS, March 31, 2005), available at ojp.usdoj.gov/odp/docs/InterimNationalPreparednessGoal_03-31-05_1.pdf (accessed September 18, 2007).

56 DHS, Office of the Press Secretary, "DHS Introduces Risk-Based Formula for Urban Areas Security Initiative Grants," press release, January 3, 2006, available at dhs.gov/xnews/releases/press_release_0824.shtm (accessed September 18, 2007).

57 DHS, *Fiscal Year 2005 Homeland Security Grant Program,* 74–78.

58 According to the U.S. Department of Justice, "A fusion center is an effective and efficient mechanism to exchange information and intelligence, maximize resources, streamline operations, and improve the ability to fight crime and terrorism by merging data from a variety of sources. In addition, fusion centers are a conduit for implementing portions of the National Criminal Intelligence Sharing Plan (NCISP)": see U.S. Department of Justice, Office of Justice Programs, Information Technology Initiatives, "Fusion Center Guidelines," available at it.ojp.gov/topic.jsp?topic_id=209 (accessed September 18, 2007).

59 DHS, "State Contacts and Grant Award Information," available at dhs.gov/xgovt/grants/index.shtm (accessed September 18, 2007).

60 DHS, *Fiscal Year 2005 Homeland Security Grant Program,* 85–90.

61 DHS, Preparedness Directorate, Office of Training and Grants, *FY 2007 Emergency Management Performance Grants: Program Guidance and Application Kit* (Washington, D.C.: DHS, November, 2006), available at wyohomelandsecurity.state.wy.us/pubs/info_225_FY07_EMPG_Guidance_AppKit.pdf (accessed September 18, 2007).

62 OMB, "FY 2006 Budget Priorities: Department of Homeland Security."

63 DHS, *Fiscal Year 2005 Homeland Security Grant Program,* 90–100.

64 Keith Bea, L. Cheryl Runyon, and Kae M. Warnock, *Emergency Management and Homeland Security Statutory Authorities in the States, District of Columbia, and Insular Areas: A Summary,* CRS Report for Congress RL32287 (Washington, D.C. Congressional Research Service, Library of Congress, March 2004), 4.

65 Ibid., 4–5.

66 Ibid.

CHAPTER 16

Future directions in emergency management

William L. Waugh Jr. and Kathleen Tierney

This chapter provides an understanding of

- The emergency management profession
- What Hurricane Katrina revealed
- Myths and lessons in emergency management
- The future of emergency management.

In "Future Directions," the concluding chapter of the 1991 edition of *Emergency Management: Principles and Practice for Local Government,* William Anderson and Shirley Mattingly noted that in vulnerable areas, both population density and the concentration of capital were increasing; as a result, more lives and property were being put at risk. Moreover, Americans were continuing to migrate to hazardous areas, particularly along the Gulf Coast and the West Coast.[1] In these two respects, little has changed in the nearly twenty years since that chapter was written. Anderson and Mattingly also addressed social vulnerability, noting that the elderly, people with disabilities, those living in poverty, members of minorities, and non-English-speaking groups were at particular risk from disasters. Here again, little has changed; indeed, current population trends signal increases in the size of vulnerable populations. The aftermath of Katrina provided all-too-clear evidence that the most vulnerable segments of the population continue to face inordinate risk in the event of disaster.[2] And all too often, these same groups have the greatest difficulty putting their lives back together after disaster strikes.

Anderson and Mattingly also observed that emergency management had become increasingly focused on mitigation and pre-disaster recovery planning, rather than being limited to response. Since the terrorist attacks of September 11, 2001, however, the focus on mitigation has been supplanted by a focus on terrorism prevention, at least at the national level. Other issues touched on included the difficulty of finding legal support for hazard mitigation when private property rights are involved, the growing professionalization of the field, the role of the private sector, links between disaster research and emergency management practice, and important areas of research. The authors further pointed out that two positive developments—the increasing ability to predict disasters, and the growing understanding of the social and psychological effects they produce—were likely to improve the nation's capacity to manage hazards and deal with disasters. They concluded their chapter by enjoining emergency management practitioners and students to work actively to educate the public about risk and to promote effective disaster policies. Many of the concerns they identified have not yet been resolved, and their warnings should still be heeded.

Predictions of catastrophe came true along the Gulf Coast in August of 2005, and settlement patterns continue to put lives and property at risk (see Figure 16–1). Coastal development is intensifying, and people are still migrating to earthquake country, building on floodplains and in the urban-woodland interface, and creating dense urban developments that are vulnerable to all manner of disaster. Katrina's lessons do not seem to have taken hold. Surveys indicate, for example, that people at risk may be unwilling to take protective action, even when public officials issue warnings and evacuation orders: in a 2007 survey conducted among residents of Gulf Coast and East Coast locations that are vulnerable to hurricanes, 31 percent of respondents said that they would remain in their homes even if a mandatory evacuation order were given. Respondents gave the following reasons for refusing to evacuate:

- Their homes were well built and would be safe (75 percent).
- The roads would be too crowded (56 percent).
- Evacuation would be dangerous (6 percent).
- Their possessions would be stolen or damaged (33 percent).
- They would not want to leave their pets (27 percent).[3]

Given public perceptions of the Katrina and Rita evacuations, these findings are not surprising. Although the Katrina evacuation was remarkable for the sheer number of people moved out of harm's way, the debacle at the Louisiana Superdome would certainly dissuade others from using public shelters, and the problems associated with the transfer of evacuees to almost every state in the Union cannot have failed to create uncertainty about the fate of those who evacuate.

A number of strategies might address the problems revealed by Katrina: one is to aim for sustainable development, which involves making disaster resilience the focus of both emergency management and economic development. When development is sustainable, it is intrinsically less vulnerable to disaster; moreover, where sustainability has taken hold, land use professionals and others engage in ongoing efforts to build support for better land use planning in hazardous areas.[4] Another approach, which has been suggested by sociologist Charles Perrow, is to reduce disaster losses by "shrinking the targets." In Perrow's view, population growth, high population density, and the concentration of infrastructure and industrial facilities in vulnerable areas are

Figure 16-1 Aerial view of destroyed coastal neighborhoods in the Biloxi/Gulfport area after Hurricane Katrina

Photo courtesy of FEMA/Andrea Booher

among the sources of increased risk.[5] Areas that are less densely populated, and industrial facilities that are smaller and more widely distributed, would be less attractive as terrorist targets and less vulnerable to natural and technological disasters. However, finding ways to shrink urban centers, to disperse the populations of coastal zones, and to persuade industry to forgo economies of scale in favor of distributed operations would be monumental tasks.

The emergency management profession

In 2005, Richard Sylves pointed out that "emergency management continues to have such a weakly defined self-identity that it continues to be poorly understood by policy makers and the public."[6] The review of the profession initiated in 2006 by Michael Selves, president of the International Association of Emergency Managers (IAEM), was designed to address head-on the problems that had been identified by Sylves and others. The Emergency Management Roundtable, a group of practitioners and academics that met at the National Emergency Training Center in Emmitsburg, Maryland, in 2007, took the process a step further by developing a set of core principles that are designed to be the foundation for action. Although much remains to be done, the profession of emergency management is beginning to find its center—and, increasingly, is finding its voice when it comes to dealing with government officials, the media, and the general public. The first chapter of this volume touched on the profession's emerging sense of itself: it defined the core principles of the profession, addressed the distinction between emergency management and emergency response, and discussed the difference between comprehensive emergency management and the national security functions that are the focus of homeland security programs.

One thing that has *not* changed is the foundation of the national emergency management system: it is still local government. Local, state, and regional capacity building is the among the principal goals of a number of organizations that represent emergency managers, including the IAEM, the National Emergency Management Association, the National Association of Counties, the National Governors Association, and ICMA. NFPA 1600, the current standard of the National Fire Protection Association, and the standards identified by the Emergency Management Accreditation Program (EMAP) provide benchmarks against which emergency managers can measure their programs. NFPA 1600 has a broad focus on emergency management and business continuity planning, and its 2007 standards integrate the U.S. Department of Homeland Security's "prevention" function into the emergency management model.[7] The EMAP standards focus on public emergency management programs and address the traditional roles and functions of emergency management agencies—including the need to prevent, prepare for, and respond to terrorism.[8] At the local level, the role of emergency management in relation to terrorism generally falls into one of two categories: antiterrorism (reducing vulnerabilities and potential losses) and dealing with the consequences of terrorist attacks.[9]

In short, the roles and functions of emergency management are becoming increasingly clear. While some issues—such as the content of emergency management education programs and the various approaches to ensuring quality in practice—remain subject to debate, the boundaries of the profession—including a common body of knowledge, a code of conduct, and requirements for education and training—are in place. Emergency management is still an emerging profession by most measures, but it is readily distinguished from emergency response and other risk-related professions, such as risk management.

What Katrina revealed

Although the events of September 11 had a profound impact on the practice of emergency management at all levels, it was Hurricane Katrina that raised fundamental questions about the nation's capacity to deal with disaster, and awakened government officials, the media, and the public to the crucial need to shore up the foundations of the nation's emergency management system.[10] Some of the issues highlighted by Katrina were those with which emergency managers have struggled for decades: for example, how can truly interoperable communications be achieved in order to ensure that the emergency manager can coordinate disaster operations effectively? How can emergency management agencies (and their communities) make the transition from routine to disaster operations and back again?

Additionally, Katrina raised questions about who has legal authority when catastrophic disasters cross multiple jurisdictional boundaries. During the response to Katrina, it was clear that local, state, and federal officials often failed to understand their own responsibilities and authority; as a result, some agencies overstepped their legal authority, whereas in other cases, tasks were left undone because no one assumed responsibility. The federal system may be cumbersome and frustrating, but it is the legal framework within which public officials work, and clarity about jurisdiction and authority is essential.

Precisely because they call for creativity, flexibility, and local knowledge, disaster operations cannot be managed from Washington or the state capital.

Katrina also highlighted the enormous challenges associated with managing catastrophic events, as opposed to more typical disasters. Creating large-scale devastation in an area roughly the size of Great Britain, resulting in more than 1,800 deaths, generating losses in excess of $120 billion, and virtually paralyzing response systems in the affected region, Katrina vividly revealed what many disaster experts had long pointed out: that catastrophes are qualitatively different from the "garden variety" disasters with which most emergency managers are familiar, and that strategies for dealing with disasters cannot readily be scaled up for catastrophes. Yet planning for genuinely catastrophic events is still in its infancy.

Another problem dramatized by Katrina was the need for structures and processes to support decision making under crisis conditions. For emergency management officials at all levels, the missing ingredient was situational awareness—information from officials within the disaster zone. However, lack of information was exacerbated by the fact that decision making about resource allocation was centralized: by its very nature, centralized decision making is slower and less responsive than decentralized processes. Under extreme circumstances, how can operations on the ground best be supported? How can communications problems be overcome? How can information be shared within and among agencies, including nongovernmental organizations and volunteers? How can strategies be developed to encourage and facilitate adaptation and improvisation? Finally, how can decision makers ensure that information flows from the disaster area to the officials who are responsible for allocating resources? Precisely because they call for creativity, flexibility, and local knowledge, disaster operations cannot be managed from Washington or the state capital. This is doubly true for truly catastrophic events.

Decentralized decision-making processes that incorporate well-designed decision-support systems can go a long way toward overcoming some of these problems. States such as California and Florida, for example, have invested in local and regional programs to ensure that local officials have the information they need to make operational and strategic decisions during emergencies and disasters. Statewide geographic information systems are a good example of the kinds of support that state officials can provide.

Disasters are dynamic events with political, social, and economic dimensions, and in which public and private interests sometimes collide. In the case of Katrina, the response was further complicated by a number of factors, including the sheer scale of the event, the fact that many National Guard and reserve units in Louisiana and Mississippi had been deployed to Iraq and Afghanistan, the poverty of many of the victims, the level of racial distrust in the region, poor planning, insufficient investment in mitigation, and the fact that even under the best of circumstances, local governments along the coast had limited resources with which to address social and infrastructure needs.[11] The poor and those with medical conditions or disabilities had little support before Katrina made landfall, and even less support after. The transportation and communications infrastructure was frayed and vulnerable (see Figure 16–2). The storm revealed deeper problems that need to be addressed in the redevelopment effort.

The U.S. military has been expanding its capacity to work with civilian authorities and nongovernmental organizations, particularly in humanitarian relief, international development, and complex emergencies where security is a central issue. The key to making such relationships work is the development of common objectives.[12] Greater involvement of the military in disaster operations in the United States is a certainty: the Northern Command has already been tasked with supporting civilian authorities in major disasters. Nevertheless, the use of troops to secure New Orleans was a highly unusual step in disaster response.

Although the military's logistical capabilities are not in question, the military's capacity to develop close, long-term working relationships with state and local governments is another matter. The ability to create such relationships is important because the links built during less severe emergencies provide the foundation for collaboration in major disasters. Trust is built at a personal level, and a "cavalry" approach simply does not yield the social and political connections that are needed for effective collaboration.

Figure 16-2 Highway 90 bridge from Biloxi to Ocean Springs, Mississippi, after Katrina's catastrophic winds and storm surge

Photo courtesy of FEMA/George Armstrong

Some critics of greater military involvement in disaster argue that it could be harmful in three ways: first, by diluting the military's capacity to fight wars; second, by creating confusion about whether military or civil authority is "in charge" during disasters; and third, by creating a "Samaritan's dilemma," in which the expectation of military assistance would lead subnational levels of government to decrease their own emergency management efforts. An overly militarized response to disasters could also lead to tensions with the community groups and spontaneous volunteers that respond to disasters.[13]

The logistical capacities that the military brings to relief operations are critical, and state and local officials need to be able to depend on those capacities in the event of disaster. Nevertheless, governors are reassessing their reliance on the National Guard and military reserve because units that are needed at home may be deployed overseas, and because Air Guard base closures have reduced the availability of aircraft to support disaster operations. Governors are also wary of efforts to federalize state assets and of even appearing to yield their decision-making authority to the military. The new emphasis on state-to-state, statewide, and community-to-community mutual aid is a reflection, in part, of the fact that the National Guard and the military reserve may not be available when needed.

Myths and lessons

Within the disaster research community, the lack of attention to studies that have debunked common disaster myths is a constant source of frustration. Some myths have profoundly negative effects on disaster policy and emergency management practice. For example, when government officials withhold information about a disaster out of fear that citizens will panic, they actually *increase* the likelihood of nonadaptive responses: research indicates that people respond best when they know that authorities are being honest and forthcoming with useful information.[14] Other myths have an element of truth in them but need to be qualified: for example, although it is not true that looting is typical, it can occur under certain circumstances, as was demonstrated on St. Croix after Hurricane Hugo and in New Orleans after Hurricane Katrina. (It is important to note, however, that reports of looting after Katrina were exaggerated and sensationalized by the media.) Despite widely held beliefs that public officials and emergency responders will abandon their roles during crises, such fears are largely unfounded. It is true, however, that some New Orleans police officers were unaccounted for during the disaster and its immediate aftermath; in addition, some research suggests that health care providers may choose not to go to work during a pandemic.[15]

Research indicates that people respond best when they know that authorities are being honest and forthcoming with useful information.

What is more likely than role abandonment is the large-scale convergence of first responders, volunteers, and donations. And, as Katrina demonstrated again and again, activities (such as price gouging) that take advantage of the vulnerable are unusual. On the contrary, altruistic behavior is common: the overwhelming majority of disaster victims are rescued by neighbors, family members, co-workers, and total strangers, rather than by professional search and rescue teams.[16] Finally, panic after disaster impact is very uncommon, and there is little evidence to suggest that biological or radiological threats increase the likelihood of panic.[17] Research shows that panic and social breakdown are so rare because social bonds remain strong even under extremely stressful and dangerous conditions.[18] Since the potential for confusion, anxiety, and rumors does increase under disaster conditions, especially if the public believes that information is being withheld, the best strategy is to provide as much information as possible and in usable forms (see Figure 16–3). Some people watch television, some listen to the radio, some rely on text messages and mobile telephones, and some go to the Internet for information.[19]

The profession of emergency management invests significant time and energy in the identification of "lessons learned." Some of the focus on lessons may stem from the requirement

Figure 16-3 Local fire commander instructing residents of communities in Colorado about what to do if wildfires necessitate evacuation

Photo courtesy of FEMA/Michael Rieger

to develop mitigation strategies after each disaster, to ensure that the same kinds of losses are not visited upon the community in the future. Unfortunately, however, identifying lessons and actually learning from them are two different things. In his research on disaster events and policy shifts, Thomas Birkland has identified a number of reasons that learning does not occur or is not translated into public policy. In some cases, lessons are lost as time passes; in other cases, another event prompts changes in policy that are at odds with earlier lessons.[20]

Birkland's perspective is in keeping with the common wisdom in emergency management, which holds that support for disaster programs dissipates as the memory of an event fades. Policies addressing natural disasters and terrorism alike have tended to follow this pattern.[21] For example, despite a history of effective disaster responses during the 1990s, the federal government was unable to mount a competent response to Hurricane Katrina. Evidently, the lessons of Hurricanes Hugo, Andrew, and Iniki—which had led, in the early 1990s, to the reinvention of the Federal Emergency Management Agency (FEMA) and the national emergency management system—were unlearned. In fact, the poor response to Katrina was eerily similar to the poor response to Andrew in 1992. During the Hurricane Andrew disaster, FEMA was criticized for being slow to respond, the FEMA director was replaced by Secretary of Transportation Andrew Card as the lead federal official in the response, and the after-action reports concluded that the agency included too many unqualified political appointees and had been drained of the resources needed to deal with major disasters.

Lessons should be guiding the development of emergency management programs. Nevertheless, it seems that lessons must be learned, and learned, and learned again. Because the same problems arise in one disaster after another, after-action reports are often repetitious. Reports from the ongoing series of Top Official (TOPOFF) exercises, for example, repeatedly mention problems with interorganizational and intergovernmental coordination. Reports from the Katrina disaster were filled with lessons learned that are thoroughly familiar to professional emergency managers—including, for example, the fact that approximately 20 percent of the population will not evacuate. The after-action report on the 1995 bombing of the Alfred P. Murrah Federal Building in Oklahoma City cited problems implementing the Incident Command System. Yet the same problem is mentioned in the report of the 9/11 Commission on the World Trade Center and Pentagon attacks, and in the White House report on the Katrina response.[22] One pattern that remains consistent is that genuine learning and change are difficult, even after very severe disasters.

Figure 16-4 The Lamar-Dixon Expo Center in Gonzales, Louisiana, where hundreds of lost animals were sheltered and cared for in the wake of Hurricane Katrina

Photo courtesy of FEMA/Liz Roll

Disaster policy tends to focus on the most recent catastrophic event: hence, the ongoing preoccupation with 9/11 and Katrina, and the particularly careful efforts to identify lessons learned in these two disasters. Although one of the supposed lessons of Katrina is that people are reluctant to evacuate if it means leaving their pets behind, this did not come as a surprise to professional emergency managers. Common wisdom in the field has long held that many residents will refuse to evacuate if they have to leave their pets—and, as noted earlier in this chapter, survey data support this belief. But while the television images of dogs and cats in flooded homes were profoundly disturbing to many Americans, the fact is that enlightened emergency shelter programs already provide animal shelters to accommodate pets (see Figure 16–4). The Safety of Animals in Disasters Act—sometimes referred to as the "no pets left behind" act—requires animal facilities in or near emergency shelters and is changing the design of sheltering programs. Apparently, the lesson was learned, although providing shelter for animals will not be an easy task, given the number of pet owners in the United States.

Katrina also drew attention to the impact of disaster on the most vulnerable members of society, including those with chronic illnesses. Shelters were filled with people who had left their homes without their medications; some died as a result. Work is under way to ensure that such tragedies are not repeated. For example, programs are being developed to make contingency contracting easier so that local governments can ensure continuation of essential services.

The future of emergency management

A number of issues will be crucial to emergency management in the coming years; the next five sections highlight some of the most important: the need for comprehensiveness and balance in loss-reduction programs; the expansion of local capacity; response frameworks; the evolving profession; and managing in a networked world.

Comprehensiveness and balance

Comprehensive emergency management emphasizes both all-hazards management and the reduction of losses during all four phases of the disaster cycle: mitigation, preparedness, response, and recovery. As noted earlier, during the 1980s and 1990s the field of emergency management made significant progress toward addressing all four phases in a balanced fashion. Since the attacks of September 11, 2001, however, mitigation and recovery have taken a back seat to programs that emphasize preparedness and response. Despite the passage of the

Disaster Mitigation Act of 2000, the best way for state and local governments to receive mitigation funding is to experience a disaster and then apply for post-disaster mitigation funds. Investments in mitigation are being neglected even in the aftermath of Hurricane Katrina and despite a congressionally mandated five-year-long study demonstrating that federal investment in mitigation reduces disaster losses, both for the federal treasury and for the nation at large.[23] Just as preserving health is a more balanced and effective strategy than curing disease, avoiding and reducing disaster losses is better than continually paying for them. Unless the nation and the emergency management profession commit to and invest in strong pre-event loss-reduction programs, losses will continue to escalate.

Similarly, the nation and the profession must focus on and invest in short- and long-term disaster recovery strategies for communities, households, and businesses. The post-Katrina recovery in the Gulf Coast region reveals the complexity of disaster recovery processes, the almost insurmountable challenges associated with recovering from catastrophic events, and the nation's glaring lack of attention to post-event recovery policies and strategies. Plans and programs are needed to facilitate disaster recovery, disaster resistance, and disaster resilience simultaneously. And the time for developing such strategies is not when disaster strikes; pre-disaster planning for post-disaster recovery is an essential element in a comprehensive, balanced approach to emergency management.

Expanding local capacity

One of the lessons of Hurricane Katrina, and of the 2004 hurricane season in Florida, is the value of building local and regional capacity through mutual aid agreements. Almost 66,000 people from forty-eight states, the District of Columbia, Puerto Rico, and the U.S. Virgin Islands were deployed to support response and recovery efforts in Louisiana, Mississippi, and Alabama. The mutual aid process worked best in states that had a statewide mutual aid compact, some experience with the Emergency Management Assistance Compact (EMAC), or both, and was a success in all three of the recipient states (although Alabama needed far less assistance than Louisiana and Mississippi). Statewide mutual aid compacts, EMAC, and programs such as the National Emergency Management Network (NEMN) can greatly increase the capacities of communities to deal with disasters.[24] The ability to share resources among municipalities and counties nationally greatly expands local and state surge capacities.[25] The keys to the success of EMAC and similar mutual aid programs are short request forms, clear resource categories and definitions, quick bidding and negotiation processes, and legal protections for those who are deployed. Precertified emergency response teams, support units, and other resources can speed the process of requesting assistance, negotiating terms, and deploying personnel and equipment.[26] Mutual aid arrangements can also reduce dependence on federal resources, although federal monies are generally required to reimburse donors. NEMN operates in much the same way as EMAC, and can be linked to the EMAC request procedure.

Mutual aid has long been common practice among local governments in such areas as law enforcement and fire services, but it is becoming more common in emergency management and other areas. During the Katrina disaster, for example, Florida communities provided assistance to communities in Mississippi without formal mutual aid agreements. The use of recovery strike teams in the wake of Katrina is another example (see sidebar on pages 328–329). As states, and organizations such as ICMA, step in to help address the legal and financial issues that arise with mutual aid, ad hoc arrangements are giving way to more formal agreements.

Response frameworks

The Incident Command System (ICS), unified command, multiagency coordination system, and the National Incident Management System (NIMS) provide a common terminology and set of organizational structures to deal with disaster. The fire service has almost four decades of experience with ICS—and, because NIMS is based on ICS, emergency management, disaster relief, and emergency response agencies at all levels of government are required to be proficient in the use of ICS. Standardizing the language used in an incident, and clearly delineating the roles and authority of the personnel who are involved in disaster response, greatly simplifies communication and ensures unity of command—assuming that all participants under-

Recovery strike teams

For William Whitson, who at that time was assistant city manager in Port Orange, Florida, and now is city manager of Cairo, Georgia, the feelings stirred by news coverage of Katrina were personal. He remembers seeing a picture in the *Pensacola News Journal* of a woman standing in what used to be downtown Long Beach, Mississippi, the city where he'd lived about a decade earlier. "I recognized, oddly enough, where she was," he said. "She was crying and there was just debris everywhere around her, and I said, 'This can't be.' It looked like something you'd see in Bangladesh."

"It really hit home," he added. "I said, 'I'm a professional. We've got to get in this fight. We can't just sit back and let this happen.'"

Whitson had a powerful ally in Port Orange city manager Ken Parker, who'd spent years developing an informal network for disaster recovery assistance within Florida. After Whitson showed Parker the newspaper photo, the two immediately got to work building support from their mayor and city council for plans to help Long Beach. But they also looked beyond Port Orange.

They knew the job would require far more than one city's assistance, so they started talking to Lee Feldman, city manager of Palm Bay; Frank Roberts, the former city manager of New Smyrna Beach; and other Florida managers they'd worked with on disaster aid. And by September 10, less than two weeks after Katrina made landfall on the Gulf Coast, the first Florida team was on the ground in Long Beach, assessing the city's needs. With strong support from their governing bodies, this same coalition of Florida cities ended up sending dozens of staff members and some elected officials to help their counterparts in Long Beach and two other Mississippi cities.

Soon after they started working in Long Beach, they heard about Pass Christian, a nearby Mississippi town that had been completely wiped out by Katrina, and they launched a parallel effort there. Then, in the fall of 2006, the group responded to a call for help from ICMA and the city of Pascagoula, Mississippi, which was struggling with longer-term recovery challenges related to everything from inspections and code enforcement to public relations.

Fannie Mae officials had been working with Pascagoula and, after discussions with city officials, realized the need for additional staff, so they contacted ICMA to consider solutions. The two organizations quickly made plans, with Fannie Mae offering to cover the cost of sending teams of Florida professionals to Pascagoula in the fall and ICMA agreeing to organize the effort by working with the Florida Municipal League and the Florida City and County Management Association.

The Florida teams ended up spending a combined total of four weeks helping their peers in Pascagoula catch up on a huge backlog of work. The list included building permits and insurance paperwork, a survey of residents about their needs, grant writing, and many other critical tasks. After the hurricane, Pascagoula issued more than 7,500 building permits to repair and rebuild damaged homes and other structures.

The Florida response in Mississippi marked the beginning of what is fast evolving into a concerted effort to formalize the concept of coordinated disaster recovery assistance by local governments, a concept the Florida managers call "recovery strike teams."

stand, and are willing to follow, the incident management guidelines. Nevertheless, questions remain about whether ICS is scalable—that is, whether it can work in very large, multiorganizational, intergovernmental, and intersector operations—and whether it is flexible enough to deal with rapidly changing situations.[27]

Among the other issues associated with ICS are a lack of compatibility between ICS and the decision-making processes of emergency response and disaster-relief organizations. For example, U.S. Army after-action reports on wildfire operations have criticized ICS because it does not fit the military's "battle rhythm."[28] Public health officials have complained about ICS because their decision processes tend to be more collegial—and, therefore, much slower—than those that are called for in ICS, and speeding up decision making can have significant negative consequences in the public health arena. Finally, many of the organizations that respond to disasters have nonhierarchical structures and rely on informal and consensual decision making; they are simply not interested in bureaucratic processes. Thus, emergency managers need tools to facilitate coordination and collaboration with organizations and entities whose structures and decision-making processes differ from their own. One of the

Parker, Feldman, and Roberts, who has since retired, had developed their own networks of support in the years since Hurricane Andrew devastated South Florida. They had forged strong relationships with peers in cities and counties across the state, and had done the planning to be sure that each would be prepared to respond with assistance if any of them was devastated by a hurricane or other disaster.

Whitson described it as similar to the way things have worked for years with police and fire rescue services: local governments lend personnel and equipment to help an affected community during an emergency. Typically, they send people in rotations to help with various aspects of disaster response, and they adjust plans and assignments as needed. "We're using the same template, the same organizational approach and just applying it to the long-term recovery," he said. The help of other municipal and county staff can be vital, he added, once a community gets to long-term recovery tasks.

"You're talking about a massive, massive effort. The recovery strike teams are meant to customize or mirror the delivery of [local government] services in any community," Whitson said. "They pick up garbage and send out water bills in Iowa. Whether you do it in Mississippi or do it in Iowa doesn't make much difference. That's the beauty of the model."

In the summer of 2006, Whitson and Parker took steps toward formalizing a local government role in recovery assistance when they briefed former governor Jeb Bush on their work in Long Beach and Pass Christian. Bush was impressed enough with the idea that he encouraged them to move forward in establishing teams across the state and finding a way to incorporate them into Florida's emergency management system. Last fall, Whitson, who became city manager of Cairo, Georgia, in January 2007, worked with Parker, Feldman, and others to set up the first teams in some of Florida's emergency response districts. In addition to designating team leaders and coordinators for each of four district strike teams, they identified a state coordinator who would work out of the state's emergency operations center.

Along with ensuring that the teams are trained and ready to deploy, their goal is to establish teams in the remaining emergency districts as soon as possible and ensure that all of the necessary information is available. "We want the system to be so transparent that people at the state level will know that these resources are there," Parker said of the strike team concept and any effort to build a national model. "They could be moving within hours, instead of days."

Building on what they learned through the hurricane season of 2004 as well as during Katrina, these local government managers now are working with the state to formally establish a network of recovery strike teams in different regions of Florida. And they're exploring the prospects for starting something similar on a national level. "This is where we should be going," Parker said of the strike team concept, adding that it could be expanded into a national model. The most important thing the Florida teams learned from Katrina, he noted, was the urgency of creating such networks. "It has moved us from just talk to action," he said. "We were doing it in an informal manner here in Florida, but it was largely based on personal relationships. What we've found is that it's time to go beyond that."

Source: Adapted from Christine Shenot, "Lessons Learned in Mississippi," *Public Management* (April 2007): 6-8.

critical roles of emergency management is getting the pieces to fit together—ensuring that conflicts arising from different organizational cultures do not interfere with cooperation and collaboration.

The national emergency management system includes complex networks of public, private, and nonprofit organizations; nongovernmental organizations; and volunteers. The networks are diverse, and communication—let alone collaboration—can be very difficult. Integrating volunteer organizations, faith-based organizations, for-profit organizations, and others into one unified effort can be a monumental task. Poor cultural interoperability complicates multiorganizational, intergovernmental, and intersector operations. At the extreme, clashes between centralized military organizations and decentralized civilian organizations are common, despite efforts on the part of the U.S. Department of Defense to build partnerships and find common purpose. When the warm and fuzzy meets the lean and mean, cultures and personality types clash, and differences can be hard to overcome. The attempt to impose control can be counterproductive in a system in which resources are dispersed, authority is shared, and responsibility needs to be shared.[29]

Equally important is the fact that ICS and NIMS are directed more toward solving tactical, operational, and management challenges than toward strategic and policy concerns. As in other domains, such as war and international politics, the mastery of the nuts and bolts of crisis management is no substitute for wise strategic decision making and good policy. Indeed, the most critical decisions made during disasters typically take place outside (and often far above) NIMS structures. Examples of high-level decisions that ICS and NIMS do not address include whether to order the mandatory evacuation of a major American city, whether to begin pre-positioning resources in advance of an impending disaster, whether to declare a particular disaster an incident of national significance or catastrophe, whether syndromic surveillance data warrant informing the public about a potential pandemic or bioterrorist attack, and how to balance conflicting priorities in the midst of a crisis. These are not the kinds of dilemmas that NIMS can solve. No matter how much NIMS training first responders receive, efforts at responding to large-scale and catastrophic disasters will fail unless the nation's elected and appointed leaders have the knowledge and experience—and the courage—to make well-informed decisions at policy and strategic levels before, during, and after disasters.

The evolving profession

A number of chapters in this volume address the growing professionalization of the emergency management field: among the examples of this shift are the Certified Emergency Manager program, the many state certification programs, and the increasing number of other programs awarding certificates or academic degrees. Such programs are all the more important because of the transition that is going on in the public service: at all levels of government, a large—and growing—percentage of employees is eligible for retirement.[30] The two likely effects are loss of institutional memory and a shortage of experienced personnel. Although entry-level emergency management staff are more and more likely to have education and training in the field, their experience will be minimal.

Today's emergency manager needs an understanding of disaster research and of emergency response, and should be able to draw on a knowledge base that includes sociology, political science, geography, economics, public health, and other disciplines. The emergency manager's task environment also requires familiarity with politics, public policy, economics, environmental science, and, particularly with reference to terrorism, international affairs. Despite the broad familiarity with other fields that is required, however, emergency management is fundamentally public administration: it involves providing leadership for a public, nonprofit, or private agency; managing human and financial resources; engaging in strategic planning; organizing offices and emergency operations centers; evaluating programs; negotiating contracts and labor agreements, and other tasks. Just as education and training have become essential in other local government fields, more and more emergency managers are specialists with certificates, degrees, or both.

In a small community, the emergency manager may be a volunteer, a part-time employee, or an employee—such as a firefighter, a public safety officer, or a public works manager—who has other responsibilities. However, this situation is now changing. Particularly as the members of the baby boom generation retire, even smaller local governments are finding that they need professional emergency managers to deal with the risks that their communities face.

According to Wayne Blanchard, director of FEMA's Higher Education Project, the emergency manager of the past

- Spent his or her entire career in one jurisdiction
- Came to emergency management as a second or third career
- Was often a part-time employee, a volunteer, or an employee who "wore other hats"
- Used a hazard-based approach and was oriented primarily to disaster response
- Had minimal access to policy makers and was unlikely to belong to professional associations
- Was not well paid and did not work in a well-funded agency or office.

By contrast, today's emergency managers

- Tend to be younger, college educated (increasingly with degrees in emergency management), and from more diverse career backgrounds
- Choose emergency management as their first career
- Are geographically mobile
- Have a broader knowledge base, and better analytical and communications skills
- Are more adept with technology
- Are lifelong learners
- Are more oriented toward comprehensive emergency management
- Are better paid and work in better-funded agencies
- Have a broad range of working contacts, including elected and appointed officials, developers and others in the business community, academics and researchers, professional organizations, and community-based organizations.[31]

In short, the new manager comes to the position with a broader range of skills, a wider perspective, and higher expectations of his or her role in, and value to, the community. The emergence of the "new" emergency manager explains the popularity of emergency management academic programs, and the pressure to clarify just what it means to be a professional in the field.

It is important to note that whatever their education and experience, emergency managers still have to deal with the nitty-gritty of disaster management. For example, pragmatic solutions may raise ethical issues. Following recent hurricanes, it was suggested that by facilitating the reopening of big-box retail stores, communities could provide a major distribution point for food, water, ice, and other essential resources. But restoring utilities and clearing debris in order to enable a big-box retailer to reopen means delaying addressing the needs of other area businesses and homeowners. Moreover, assisting a large retail facility may indirectly harm smaller, family-owned businesses in the area. Such actions raise questions that would perhaps best be answered in a community meeting or before the local governing body, as part of the pre-disaster emergency planning process. There will not be time to debate such issues in the middle of the disaster response.

Conclusion: Local emergency management in a networked world

One of the themes running throughout this volume is that the social, political, and administrative world in which emergency managers operate is made up of networks. Public, private, and nonprofit organizations are the most visible elements in the political and administrative environment; volunteer organizations and individual volunteers, although less visible, are nevertheless critical to increasing the surge capacity of local, state, and national emergency management systems. Working within a networked environment requires new approaches to management and leadership. Collaboration is essential. Flexibility is essential. Classic command-and-control structures and processes are often ineffective. As Admiral Thad Allen, of the U.S. Coast Guard, observed, dealing with the effects of Hurricane Katrina called for "unity of effort," not "unity of command."[32]

Emergency management, like any complex undertaking, requires leaders who are "bridge builders"—who can bring together disparate organizations and help them find a common purpose.[33] The critical skill for an emergency manager is the ability to collaborate effectively.[34] The critical task is to sell emergency management to the community: to help elected officials and the public understand the importance of emergency management and the value that it adds. The critical objective is to increase resilience so that the community can deal with disaster and recovery quickly. To accomplish the critical task and to achieve the critical objective, emergency managers must cultivate respect and attention. They must work with other officials to promote mitigation and preparedness, and to organize response and recovery. It is essential for policy makers to understand the role and function of emergency management, and to respect emergency managers' expertise.

Figure 16-5 The center of Greensburg, Kansas, twelve days after being hit by an F5 tornado with 200-mph winds on May 4, 2007

Photo courtesy of FEMA/Greg Hanshall

One of the goals of this volume is to serve as a resource for local officials, so that they can better understand emergency management and can make the best use of the resources it offers.[35] All public officials are responsible for protecting life and property. A little education can generate support and facilitate interaction. Although being involved in politics makes some emergency managers uneasy, it is essential if communities are to effectively mitigate, prepare for, respond to, and recover from disaster.

We live in an era of government accountability, and local emergency managers will be held responsible for protecting the residents of their communities. The standards against which their work will be evaluated are based on effectiveness and efficiency. Thus, local emergency managers must focus their attention on the hazards they see, and on the disasters that are most likely to occur in their communities (see Figure 16–5). For managers to do otherwise would be a failure of responsibility and a breach of ethics.

Notes

1 William A. Anderson and Shirley Mattingly, "Future Directions," in *Emergency Management: Principles and Practice for Local Government* (Washington, D.C.: ICMA, 1991), 331–335

2 Ironically, a report on the increasing poverty in the United States was issued on the day that Katrina drew attention to the problem in New Orleans; see Carmen DeNavas-Walt, Bernadette D. Proctor, and Cheryl Lee Hill, "Income, Poverty, and Health Insurance Coverage in the United States: 2004," *Current Population Reports* P60-229 (Washington, D.C.: Bureau of the Census, U.S. Department of Commerce, August 2005), available at census.gov/prod/2005pubs/p60-229.pdf (accessed August 16, 2007). The Census Bureau's press release on the report was issued on August 30, 2005, the day after Katrina made landfall and as the media began reporting conditions in New Orleans.

3 Robert J. Blendon et al., "Hurricane Preparedness in High-Risk Areas, June 18–July 10, 2007" (Cambridge, Mass.: Project on the Public and Biological Security, Harvard School of Public Health, 2007).

4 See, for example, Raymond J. Burby, "Hurricane Katrina and the Paradoxes of Government Disaster Policy: Bringing about Wise Governmental Decisions for Hazardous Areas," *Annals of the American Academy of Political and Social Science* 604, no. 1 (2006): 171–191.

5 Charles Perrow, *The Next Catastrophe: Reducing Our Vulnerabilities to Natural, Industrial, and Terrorist Disasters* (Princeton, N.J.: Princeton University Press, 2007).

6 Richard T. Sylves, "Why Revolutionary Change Is Needed in Emergency Management" (paper presented at the Eighth Annual Emergency Management Higher Education Conference, Emmitsburg, Md., June 7–9, 2005).

7 NFPA 1600: Standard on Disaster/Emergency Management and Business Continuity Programs, 2007 Edition (Quincy, Mass.: National Fire Protection Association, 2007). Emergency managers have generally considered disaster prevention a part of mitigation; in homeland security parlance, however, prevention refers to counterterrorism: anticipating and precluding attacks.

8 Emergency Management Accreditation Program (EMAP), *Emergency Management Standards by EMAP* (Lexington, Ky.: EMAP, August 2007), avail-

able at emaponline.org/?342 (accessed September 23, 2007).

9 William L. Waugh Jr., "Terrorism as Disaster," in *Handbook of Disaster Research,* ed. Havidán Rodríguez, E. L. Quarantelli, and Russell R. Dynes (New York: Springer, 2007), 388–404.

10 William L. Waugh Jr., "The Political Costs of Failure in the Responses to Hurricanes Katrina and Rita," *Annals of the American Academy of Political and Social Science* 604, no. 1 2006): 10–25.

11 Ibid.

12 See, for example, Linton Wells and Charles Hauss, "Odd Couples: The DoD and NGOs," *PS: Political Science and Politics* 40, no. 3 (2007): 485–487.

13 Kathleen Tierney and Christine Bevc, "Disaster as War: Militarism and the Social Construction of Disaster in New Orleans," in *The Sociology of Katrina: Perspectives on a Modern Catastrophe,* ed. David L. Brunsma, David Overfelt, and J. Steven Picou (Lanham, Md.: Rowman and Littlefield, 2007).

14 See, for example, Waugh, "Terrorism as Disaster."

15 K. Qureshi et al., "Health Care Workers' Ability and Willingness to Report to Duty during Catastrophic Disasters," *Journal of Urban Health* 82, no. 3 (2005): 378–388.

16 See, for example, Kathleen Tierney, "Metaphors Matter: Disaster Myths, Media Frames, and Their Consequences in Hurricane Katrina," in *Annals of the American Academy of Political and Social Science* 604, no. 1 (2006): 57–81; and David A. McEntire, *Disaster Response and Recovery* (Hoboken, N.J.: John Wiley & Sons, 2007), 65–74.

17 Ibid.

18 Norris R. Johnson, "Panic and the Breakdown of Social Order: Popular Myth, Social Theory, Empirical Evidence," *Sociological Focus* 20, no. 8 (1987): 171–183; William E. Feinberg and Norris R. Johnson, "The Ties That Bind: A Macro-Level Approach to Panic," *International Journal of Mass Emergencies and Disasters* 19, no. 3 (2001): 269–295; and Lee Clarke, "Panic: Myth or Reality?" *Contexts* 1, no. 3 (2002): 21–26, available at contextsmagazine.org/content_sample_v1-3.php (accessed September 23, 2007).

19 Some "high-end" users want information in every conceivable form, from statistics to satellite images to estimations of probabilities.

20 Thomas A. Birkland, *Lessons of Disaster: Policy Change after Catastrophic Events* (Washington, D.C.: Georgetown University Press, 2006).

21 William L. Waugh Jr., *Terrorism and Emergency Management* (New York: Marcel Dekker, 1990).

22 Amy K. Donahue and Robert V. Tuohy, "Lessons We Don't Learn: A Study of the Lessons of Disasters, Why We Repeat Them, and How We Can Learn Them," *Homeland Security Affairs* 2, no. 2 (2006), available at hsaj.org/pages/volume2/issue2/pdfs/2.2.4.pdf (accessed September 23, 2007).

23 Multihazard Mitigation Council, *Natural Hazard Mitigation Saves: An Independent Study to Assess the Future Savings from Mitigation Activities,* vol. 1, *Findings, Conclusions, and Recommendations* (Washington, D.C.: National Institute of Building Sciences, 2005), available at floods.org/PDF/MMC_Volume1_FindingsConclusionsRecommendations.pdf (accessed September 23, 2007).

24 For more information on the National Emergency Management Network, see its Web site at nemn.net (accessed September 23, 2007).

25 William L. Waugh Jr., "EMAC, Katrina, and the Governors of Louisiana and Mississippi," *Public Administration Review,* Special Issue on Katrina (forthcoming, December 2007); William L. Waugh Jr., "Mechanisms for Collaboration in Emergency Management: ICS, NIMS, and the Problem of Command and Control," *The Collaborative Manager,* ed. Rosemary O'Leary (Washington, D.C.: Georgetown University Press, forthcoming 2008).

26 Robert J. O'Neill Jr., "A New Model for Disaster Response," *Management Insights,* August 16, 2006, available at governing.com/mgmt_insight.aspx?id=3234 (accessed September 23, 2007).

27 See Waugh, "Mechanisms for Collaboration."

28 See, for example, 20th Engineer Battalion (Mechanized), 1st Cavalry Division, "2000 Wildland Fire Support After Action Review" (2000).

29 William L. Waugh Jr., "Hurricane Katrina and Cultural Interoperability" (paper presented at the American Society for Public Administration National Conference, Washington, D.C., March 23–27, 2007); and William L. Waugh Jr., "Organizational Culture, Communication, and Decision-Making: Making Multi-Organizational, Inter-Sector and Intergovernmental Operations Work" (paper presented at the National Conference on Catastrophic Care for the Nation, National Disaster Medical System, Atlanta, Ga., April 13–17, 2002). See also O'Neill, "A New Model for Disaster Response."

30 The impact of retirements is already being felt in the federal government. At the time of this writing, the U.S. Department of Homeland Security, including the Federal Emergency Management Agency (FEMA), was significantly understaffed to the point that Congress was investigating to find out why senior positions had not been filled. Similarly, half or more of the employees in many local and state civil service systems were eligible for—or very close to—retirement. Federal agencies had noted "brain-drains" due to retirements.

31 B. Wayne Blanchard, "The Emergency Management Higher Education Project" (Emmitsburg, Md.: Emergency Management Institute, FEMA), available at training.fema.gov/emiweb/edu (accessed September 23, 2007).

32 James Kitfield, "New Coast Guard Chief Discusses Lessons Learned from Katrina," *GovExec.com,* June 2, 2006, available at govexec.com/story_page.cfm?articleid=34234&dcn=todaysnews (accessed September 23, 2007).

33 Donald F. Kettl, *System under Stress: Homeland Security and American Politics,* 2nd ed. (Washington, D.C.: CQ Press, 2007), 143.

34 William L. Waugh Jr. and Gregory Streib, "Collaboration and Leadership for Effective Emergency Management," *Public Administration Review* 66, Special Issue on Collaborative Management (December 2006): 131–140.

35 Guides are provided to other public officials, from governors to mayors, as well. See, for example, Ann Beauchesne, *A Governor's Guide to Emergency Management,* vol. 1, *Natural Disasters* (Washington, D.C.: National Governors Association, 2001), available at nga.org/cda/files/REPORTEMERGUIDE2001.pdf (accessed September 23, 2007); and National Emergency Management Association (NEMA), *If Disaster Strikes Today. Are You Ready to Lead? A Governor's Primer on All-Hazards Emergency Management* (Lexington, Ky: NEMA, n.d.), available at nemaweb.org/docs/Gov_Primer.pdf (accessed September 23, 2007).

FOR FURTHER REFERENCE

Chapter 1 The post-9/11 world

Agranoff, Robert. *Managing within Networks: Adding Value to Public Organizations.* Washington, D.C.: Georgetown University Press, 2007.

Birkland, Thomas E. *Lessons from Disaster: Policy Change after Catastrophic Events.* Washington, D.C.: Georgetown University Press, 2006.

Boin, Arjen; Paul 'tHart, Eric Stern, and Bengt Sundelius. *The Politics of Crisis Management: Public Leadership under Pressure.* Cambridge: Cambridge University Press, 2005.

Burby, Raymond J., ed. *Cooperating with Nature: Confronting Natural Hazards with Land-Use Planning and Sustainable Communities.* Washington, D.C.: Joseph Henry Press, 1998.

Canton, Lucien G. *Emergency Management: Concepts and Strategies for Effective Programs.* Hoboken, N.J.: Wiley-Interscience, 2006.

Cutter, Susan L., ed. *American Hazardscapes: The Regionalization of Hazards and Disasters.* Washington, D.C.: Joseph Henry Press, 2001.

Drabek, Thomas E. *The Professional Emergency Manager.* Boulder: Institute for Behavioral Science, University of Colorado, 1987.

Drabek, Thomas E., and Gerald J. Hoetmer, eds. *Emergency Management: Principles and Practice for Local Government.* Washington, D.C.: ICMA, 1991.

Emergency Management Roundtable. *The Principles of Emergency Management.* 2007.

Kelman, Steven. "The Transformation of Government in the Decade Ahead." In *Reflections on 21st Century Government Management*, edited by Donald F. Kettl and Steven Kelman, 33–61. Washington, D.C.: IBM Center for the Business of Government, 2007. businessofgovernment.org/pdfs/KettlKelmanReport.pdf (accessed October 15, 2007).

Kemp, Roger L. *Emergency Management and Homeland Security: An Overview.* Washington, D.C.: ICMA, August 2006.

Kettl, Donald F. *System under Stress: Homeland Security and American Politics.* 2nd ed. Washington, D.C.: CQ Press, 2007.

Kunreuther, Howard, and Richard J. Roth Sr., eds. *Paying the Price: The Status and Role of Insurance against Natural Disasters in the United States.* Washington, D.C.: Joseph Henry Press, 1998.

Mileti, Dennis S. *Disasters by Design: A Reassessment of Natural Hazards in the United States.* Washington, D.C.: Joseph Henry Press, 1999.

O'Neill, Bob. "Local Government's Role in Natural Disaster." Keynote address presented at the National Congress, Local Government Manager Australia, 2006. lgma.org.au/national/2006NationalCongress/papers/Bobs%20speech.pdf (accessed August 1, 2007).

Perry, Ronald W., and E. L. Quarantelli. *What Is a Disaster? New Answers to Old Questions.* Philadelphia: Xlibris, 2005.

Rubin, Claire B., ed. *Emergency Management: The American Experience 1900–2005.* Fairfax, Va.: Public Entity Risk Institute (PERI), 2007.

Tierney, Kathleen J., Michael K. Lindell, and Ronald W. Perry. *Facing the Unexpected: Disaster Preparedness and Response in the United States.* Washington, D.C.: Joseph Henry Press, 2001.

Waugh, William L., Jr. *Living with Hazards, Dealing with Disasters.* Armonk, N.Y.: M. E. Sharpe, 2000.

———, ed. *Shelter from the Storm: Repairing the National Emergency Management System after Hurricane Katrina.* Thousand Oaks, Calif.: Sage, 2006.

Chapter 2 Origins and evolution

Drabek, Thomas E., and Gerard J. Hoetmer, eds. *Emergency Management: Principles and Practice for Local Government.* Washington, D.C.: ICMA, 1991.

Hogue, Henry B., and Keith Bea. *Federal Emergency Management and Homeland Security Organization: Historic Developments and Legislative Options.* CRS RL-33369. Washington, D.C.: Congressional Research Service, Library of Congress, June 1, 2006. fas.org/sgp/crs/homesec/RL33369.pdf (accessed October 15, 2007).

Lindell, Michael K., and Ronald W. Perry. *Behavioral Foundations of Community Emergency Planning.* Washington, D.C.: Hemisphere Publishing, 1992.

Mileti, Dennis S. *Disasters by Design: A Reassessment of Natural Hazards in the United States.* Washington, D.C.: Joseph Henry Press, 1999.

National Research Council. Committee on Disaster Research in the Social Sciences: Future Challenges and Opportunities. *Facing Hazards and Disasters: Understanding Human Dimensions.* Washington, D.C.: National Academies Press, 2006.

Perry, Ronald W., and Michael K. Lindell. *Emergency Planning.* New York: John Wiley & Sons, 2007.

Perry, Ronald W., and E. L. Quarantelli. *What Is a Disaster? New Answers to Old Questions.* Philadelphia: Xlibris, 2005.

Platt, Rutherford H. *Disasters and Democracy: The Politics of Extreme Natural Events.* Washington, D.C.: Island Press, 1999.

Rubin, Claire B., ed. *Emergency Management: The American Experience, 1900–2005.* Fairfax, Va.: Public Entity Risk Institute (PERI), 2007. Available at riskinstitute.org.

Rubin, Claire B., Irmak Renda-Tanali, and William Cumming. *Disaster Time Line: Major Focusing Events and U.S. Outcomes, 1978–2006.* Arlington, Va.: Claire B. Rubin & Associates, April 2007. disaster-timeline.com.

Tobin, Graham A., and Burrell E. Montz. *Natural Hazards: Explanation and Integration.* New York: Guilford Press, 1997.

U.S. Department of Homeland Security. Federal Emergency Management Agency (FEMA) Higher Education Program. *Instructor's Guide for Hazards Risk Management Course.* 2004. training.fema.gov/EMIWeb/edu/hazardmgt.asp.

Chapter 3 Organizing for emergency management

Grant, Nancy. "Emergency Management Training and Education for Public Administration." In *Disaster Management in the U.S. and Canada: The Politics, Policymaking, Administration and Analysis of Emergency Management.* 2nd ed., edited by Richard T. Sylves and William L. Waugh Jr. Springfield, Ill.: Charles C Thomas Publishers, 1996.

Jenkins, Brian Michael, and Frances Edwards-Winslow. *Saving City Lifelines: Lessons Learned in the 9-11 Terror Attacks.* San Jose, Calif.: Mineta Transportation Institute, September 2003.

Kellman, Barry. *Managing Terrorism's Consequences: Legal Issues.* Oklahoma City, Okla.: Memorial Institute for the Prevention of Terrorism, 2002.

Maniscalco, Paul M., and Hank T. Christen. "The Basics of the Incident Management System."

Chap. 2 in *Understanding Terrorism and Managing the Consequences.* Upper Saddle River, N.J.: Prentice Hall, 2002.

Monday, Jacquelyn L. *Beyond September 11th: An Account of Post-Disaster Research.* Special Publication #39. Boulder: Natural Hazards Center, Institute of Behavioral Science, University of Colorado, 2003. colorado.edu/hazards/publications/sp/sp39/ (accessed October 15, 2007).

Waugh, William L., Jr. *Living with Hazards, Dealing with Disasters.* Armonk, N.Y.: M. E. Sharpe, 2000.

Winslow, Frances E. "Planning for Weapons of Mass Destruction/Nuclear, Biological, and Chemical Agents: A Local/Federal Partnership." In *Handbook of Crisis and Emergency Management,* edited by Ali Farazmand, 677–692. New York: Marcel Dekker, Inc., 2001.

———. "The Role of the Emergency Manager." In *Cities and Disasters: North American Studies in Emergency Management,* edited by Richard T. Sylves and William L. Waugh Jr. Springfield, Ill.: Charles C Thomas Publishers, 1990.

Chapter 4 The intergovernmental context

Agranoff, Robert, and Michael McGuire. *Collaborative Public Management: New Strategies for Local Governments.* Washington, D.C.: Georgetown University Press, 2003.

Bardach, Eugene. *Getting Agencies to Work Together: The Practice and Theory of Managerial Craftsmanship.* Washington, D.C.: Brookings Institution Press, 1998.

Birkland, Thomas A. *After Disaster: Agenda Setting, Public Policy, and Focusing Events.* Washington, D.C.: Georgetown University Press, 1997.

Drabek, Thomas E. *Strategies for Coordinating Disaster Responses.* Monograph #61. Boulder: Program on Environment and Behavior, Natural Hazards Center, University of Colorado, 2003.

Drabek, Thomas E., Harriet L. Tamminga, Thomas S. Kilijanek, and Christopher R. Adams. *Managing Multiorganizational Emergency Responses: Emergent Search and Rescue Networks in Natural Disaster and Remote Area Settings.* Monograph #33. Boulder: Program on Technology, Environment and Man, Natural Hazards Center, University of Colorado, 1981.

Majone, Giandomenico. *Evidence, Argument, and Persuasion in the Policy Process.* New Haven, Conn.: Yale University Press, 1989.

May, Peter J., and Walter Williams. *Disaster Policy Implementation: Managing Programs under Shared Governance.* New York: Plenum Press, 1986.

Platt, Rutherford H. *Disasters and Democracy: The Politics of Extreme Natural Events.* Washington, D.C.: Island Press, 1999.

Schneider, Saundra K. *Flirting with Disaster: Public Management in Crisis Situations.* Armonk, N.Y.: M. E. Sharpe, 1995.

Sylves, Richard T., and William L. Waugh Jr., eds. *Disaster Management in the U.S. and Canada: The Politics, Policymaking, Administration and Analysis of Emergency Management.* Springfield, Ill.: Charles C Thomas Publishers, 1996.

Walker, David Bradstreet. *The Rebirth of Federalism: Slouching toward Washington.* 2nd ed. Washington, D.C.: Chatham House, a division of CQ Press, 2000.

Chapter 5 Collaborative emergency management

Armstrong, Michael J. "Back to the Future: Charting the Course for Project Impact." *Natural Hazards Review* 1 (August 2000): 138–144.

"Building Successful Partnerships with Faith-Based Organizations." In *Church Mentoring Network: A Program Manual for Linking and Supporting Mentoring Ministries.* Alexandria, Va.: National Mentoring Partnership, 1999.

Cooper, Christopher, and Robert Block. *Disaster: Hurricane Katrina and the Failure of Homeland Security.* New York: Times Books, 2006.

Geis, Donald E. "By Design: The Disaster Resistant and Quality-of-Life Community." *Natural Hazards Review* 1 (August 2000): 151–160.

Godschalk, David R., Timothy Beatley, Philip Berke, David J. Brower, Edward J. Kaiser, Charles C. Bohl, and Matthew R. Goebel. "Natural Hazard Mitigation: Planning for Sustainable Communities." Chap. 13 in *Natural Hazard Mitigation: Recasting Disaster Policy and Planning.* Washington, D.C., and Covelo, Calif.: Island Press, 1999.

Hinshaw, Robert E. "Floodplain Management at the Local Level: Tulsa, Oklahoma, and Boulder, Colorado." Chap. 10 in *Living with Nature's Extremes: The Life of Gilbert Fowler White.* Boulder, Colo.: Johnson Books, 2006.

Kendra, James M., and Tricia Wachtendorf. "Creativity in Emergency Response after the World Trade Center Attack." In *Beyond September 11th: An Account of Post-Disaster Research,* Special Publication #39. Boulder: Natural Hazards Center, Institute of Behavioral Science, University of Colorado, 2003. colorado.edu/hazards/publications/sp/sp39/ (accessed October 15, 2007).

Meo, Mark, Becky Ziebro, and Ann Patton. "Tulsa Turnaround: From Disaster to Sustainability." *Natural Hazards Review* 5 (February 2004): 1–9.

Mileti, Dennis S. "Influences on the Adoption and Implementation of Mitigation." Chap. 5 in *Disasters by Design: A Reassessment of Natural Hazards in the United States.* Washington, D.C.: Joseph Henry Press, 1999.

Moynihan, Donald P. "Leveraging Collaborative Networks in Infrequent Emergency Situations." Washington, D.C.: IBM Center for the Business of Government, 2005.

Multihazard Mitigation Council. "Community Studies Results." Chap. 5 in *Natural Hazard Mitigation Saves: An Independent Study to Assess the Future Savings from Mitigation Activities.* Vol. 1, *Findings, Conclusions, and Recommendations.* Washington, D.C.: National Institute of Building Sciences, 2004.

Nathe, Sarah K. "Public Education for Earthquake Hazards," *Natural Hazards Review* 1 (November 2000): 191–196.

Natural Hazards Research and Applications Information Center. *Holistic Disaster Recovery: Ideas for Building Local Sustainability after a Natural Disaster.* Boulder: Natural Hazards Center, University of Colorado, 2006.

Passerini, Eve. "Disasters as Agents of Social Change in Recovery and Reconstruction." *Natural Hazards Review* 1 (May 2000): 67–72.

Schoch-Spana, Monica, Crystal Franco, Jennifer B. Nuzzo, and Christiana Usenza, on behalf of the Working Group on Community Engagement in Health Emergency Planning. "Community Engagement: Leadership Tool for Catastrophic Health Events." *Biosecurity and Bioterrorism* 5, no. 1 (2007): 8–25.

Schwab, Jim. "Nature Bats Last: The Politics of Floodplain Management." *Environment Development,* newsletter of the American Planning Association (January–February 1996).

Wachtendorf, Tricia, Rory Connell, Brian Monahan, and Kathleen Tierney. "Disaster Resistant Communities Initiative Assessment." Newark: Disaster Research Center, University of Delaware, 2002. udel.edu/DRC/projectreport48.pdf (accessed October 15, 2007).

Witt, James Lee. *Stronger in the Broken Places: Nine Lessons for Turning Crisis into Triumph.* New York: Times Books, 2002.

Chapter 6 Mitigation

Alesch, Daniel, Peter May, Robert Olshansky, William Petak, and Kathleen Tierney. *Promoting Seismic Safety: Guidance for Advocates.* Buffalo, N.Y.: Multidisciplinary Center for Earthquake Engineering Research, 2004.

Birkland, Thomas A. *After Disaster: Agenda Setting, Public Policy, and Focusing Events.* Washington, D.C.: Georgetown University Press, 1997.

Bolin, Robert C., and Lois Stanford. *The Northridge Earthquake: Vulnerability and Disaster.* London: Routledge, 1998.

Burby, Raymond J. *Cooperating with Nature: Confronting Natural Hazards with Land Use Planning for Sustainable Communities.* Washington, D.C.: Joseph Henry Press, 1998.

Clarke, Lee, ed. *Terrorism and Disaster: New Threats, New Ideas.* Amsterdam and Boston: JAI Press, 2003.

Godschalk, David R. "Urban Hazard Mitigation: Creating Resilient Cities." *Natural Hazards Review* 4, no. 3 (August 2003): 136–143.

Godschalk, David R., Timothy Beatley, Philip Berke, David J. Brower, Edward J. Kaiser, Charles C. Bohl, and Matthew R. Goebel. *Natural Hazard Mitigation: Recasting Disaster Policy and Planning.* Washington, D.C., and Covelo, Calif.: Island Press, 1999.

Godschalk, David R., David J. Brower, and Timothy Beatley. *Catastrophic Coastal Storms: Hazard Mitigation and Development Management.* Durham, N.C.: Duke University Press, 1989.

McDonald, Roxanna. *Introduction to Natural and Man-Made Disasters and Their Effects on Buildings.* Oxford: Architectural Press, 2003.

Mileti, Dennis S. *Disasters by Design: A Reassessment of Natural Hazards in the United States.* Washington, D.C.: Joseph Henry Press, 1999.

Simpson, Robert H., Richard A. Anthes, and Michael Garstang. *Hurricane! Coping with Disaster: Progress and Challenges since Galveston, 1900.* Washington, D.C.: American Geophysical Union, 2003.

Tierney, Kathleen J., Michael K. Lindell, and Ronald W. Perry. *Facing the Unexpected: Disaster Preparedness and Response in the United States.* Washington, D.C.: Joseph Henry Press, 2001.

Chapter 7 Planning and preparedness

Auf der Heide, Erik. *Disaster Response: Principles of Preparation and Coordination.* St. Louis, Mo.: C. V. Mosby, 1989. orgmail2.coe-dmha.org/dr/flash.htm.

Erickson, Paul A. *Emergency Response Planning for Corporate and Municipal Managers.* New York: Butterworth-Heinemann, 2007.

Lesak, David M. *Hazardous Materials: Strategies and Tactics.* Upper Saddle River, N.J.: Prentice Hall, 1999.

Lindell, Michael K., and Ronald W. Perry. *Communicating Environmental Risk in Multiethnic Communities.* Thousand Oaks, Calif.: Sage, 2004.

Perry, Ronald W., and Michael K. Lindell. *Emergency Planning.* Hoboken, N.J.: John Wiley & Sons, 2006.

Walsh, Donald W. *National Incident Management System: Principles and Practice.* Boston: Jones and Bartlett, 2005.

Chapter 8 Applied response strategies

Alexander, David E. *Principles of Emergency Planning and Management.* New York: Oxford University Press, 2002.

Auf der Heide, Erik. *Disaster Response: Principles in Preparation and Coordination.* St. Louis, Mo.: C. V. Mosby, 1989.

Boin, Arjen, and Paul 't Hart. "Public Leadership in Times of Crisis." *Public Administration Review* 63 (September/October 2003): 544–552.

Canton, Lucien G. *Emergency Management: Concepts and Strategies for Effective Programs.* Hoboken, N.J.: Wiley-Interscience, 2007.

Dombrowsky, Wolf R. "Again and Again: Is a Disaster What We Call a 'Disaster.'" In *What Is a Disaster? Perspectives on the Question*, ed. E. L. Quarantelli, 19–30. London: Routledge, 1998.

Drabek, Thomas E. *The Social Dimensions of Disaster.* 2nd ed. Emmitsburg, Md.: Emergency Management Institute, Federal Emergency Management Agency, U.S. Department of Homeland Security, 2004. training.fema.gov/EMIWeb/edu/sdd.asp (accessed October 15, 2007).

Drabek, Thomas E., and Gerard J. Hoetmer, eds. *Emergency Management: Principles and Practice for Local Government.* Washington, D.C.: ICMA, 1991.

Dynes, Russell R., E. L. Quarantelli, and Gary A. Kreps. *A Perspective on Disaster Planning.* 3rd ed. Research Notes/Report 11. Newark: Disaster Research Center, University of Delaware, 1981.

Perry, Ronald W., and E. L. Quarantelli. *What Is a Disaster? New Answers to Old Questions.* Philadelphia: Xlibris, 2005.

Rosenthal, Uriel, and Charles M. Hart. *Coping with Crisis: The Management of Disasters, Riots, and Terrorism.* Springfield, Ill.: Charles C Thomas Publishers, 1989.

Wachtendorf, Tricia. *Improvising 9/11: Organizational Improvisation following the World Trade Center Disaster.* Newark: Disaster Research Center, University of Delaware, 2004.

Weick, Karl E. "The Collapse of Sense-Making in Organizations: The Mann Gulch Disaster." *Administrative Science Quarterly* 38, no. 4 (1993): 628–652.

Chapter 9 Disaster response

Boisvert, Andrew. *Understanding Hazardous Materials, Operations, and Emergency Response.* New York: AuthorHouse, 2007.

Brunacini, Alan V. *Fire Command.* 2nd ed. Quincy, Mass.: National Fire Protection Association, 2002.

Canton, Lucien G. *Emergency Management.* New York: Wiley-Interscience, 2007.

Erickson, Paul A. *Emergency Response Planning for Corporate and Municipal Managers.* New York: Butterworth-Heinemann, 2007.

Lindell, Michael K., and Ronald W. Perry. *Communicating Environmental Risk in Multiethnic Communities.* Thousand Oaks, Calif.: Sage, 2004.

Lindell, Michael K., Carla A. Prater, and Ronald W. Perry. *Introduction to Emergency Management.* New York: John Wiley & Sons, 2007.

McEntire, David A. *Emergency Response and Recovery.* New York: John Wiley & Sons, 2007.

Nicholson, William C. *Emergency Response and Emergency Management Law.* Springfield, Ill.: Charles C Thomas Publishers, 2003.

Perry, Ronald W., and Michael K. Lindell. *Emergency Planning.* New York: John Wiley & Sons, 2007.

Pine, John C. *Technology in Emergency Management.* New York: John Wiley & Sons, 2007.

Rodriguez, Havidán, Enrico L. Quarantelli, and Russell R. Dynes, eds. *Handbook of Disaster Research.* New York: Springer, 2006.

Sachs, Gordon M. *Terrorism Emergency Response: A Workbook for Responders.* Upper Saddle River, N.J.: Prentice Hall, 2002.

Tierney, Kathleen, Michael K. Lindell, and Ronald W. Perry. *Facing the Unexpected: Disaster Preparedness and Response in the United States.* Washington, D.C.: John Henry Press, 2001.

Chapter 10 The role of the health sector

Agency for Toxic Substances and Disease Registry (ATSDR). *Managing Hazardous Materials Incidents: A Planning Guide for the Management of Contaminated Patients.* 3 Vols. Atlanta, Ga.: ATSDR, Public Health Service, U.S. Department of Health and Human Services, 2003. atsdr.cdc.gov/MHMI/ (accessed October 15, 2007).

Auf der Heide, Erik. *Community Medical Disaster Planning and Evaluation Guide.* Dallas: American College of Emergency Physicians, 1995.

———. *Disaster Response: Principles of Preparation and Coordination.* St. Louis, Mo.: C. V. Mosby, 1989. orgmail2.coe-dmha.org/dr/index.htm (accessed October 15, 2007).

———. "The Importance of Evidence-Based Disaster Planning." *Annals of Emergency Medicine* 47, no. 1 (2006): 34–49. atsdr.cdc.gov/emergency_response/importance_disaster_planning.pdf (accessed October 15, 2007).

———. "Principles of Hospital Disaster Planning." In *Disaster Medicine,* edited by David E. Hogan and Jonathan L. Burstein, 57–89. Philadelphia: Lippincott Williams & Wilkins, 2002.

Barbera, Joseph A., and Anthony G. Macintyre. *Medical and Health Incident Management (MaHIM) System: A Comprehensive Functional System Description for Mass Casualty Medical and Health Incident Management.* Washington, D.C.: Institute for Crisis, Disaster, and Risk Management, The George Washington University, October 2002.

Brown, Sonia. *Clinic Disaster Plan Guidance.* Los Alamos, Calif.: Governor's Office of Emergency Services, June 2002. oes.ca.gov/Operational/OESHome.nsf/PDF/ClinicDisasGuide?$file/ClinicDisasGuide.pdf (accessed October 15, 2007).

California Office of Emergency Services. *Hospital Earthquake Preparedness Guidelines.* Oakland: Earthquake Program, California Office of Emergency Services, 1997.

Ciottone, Gregory R., Philip D. Anderson, Erik Auf der Heide, Robert G. Darling, Irving Jacoby, Eric K. Noji, and Selim Suner, eds. *Disaster Medicine.* Philadelphia, Pa.: Mosby Elsevier, 2006.

Glass, Thomas A., and Monica Schoch-Spana. "Bioterrorism and the People: How to Vaccinate a City against Panic." *Clinical Infectious Diseases* 34, no. 2 (2002): 217–223. journals.uchicago.edu/CID/journal/issues/v34n2/011333/011333.web.pdf (accessed October 15, 2007).

Hogan, David E., and Jonathan L. Burstein, eds. *Disaster Medicine.* 2nd ed. Philadelphia: Lippincott Williams & Wilkins, 2007.

Joint Commission on Accreditation of Healthcare Organizations (JCAHO). *Guide to Emergency Management Planning in Health Care.* Oakbrook Terrace, Ill.: Joint Commission Resources, Inc., 2002.

———. *Health Care at the Crossroads: Strategies for Creating and Sustaining Community-wide Emergency Preparedness Systems.* Oakbrook Terrace, Ill.: JCAHO, 2003.

Moores, Clark A., Frederick M. Burkle Jr., and Scott R. Lillibridge, eds. "Disaster Medicine." *Emergency Medicine Clinics of North America* 14, no. 2 (May 1996).

Noji, Eric K., ed. *The Public Health Consequences of Disasters.* New York: Oxford University Press, 1997.

Occupational Safety and Health Administration. *OSHA Best Practices for Hospital-Based First Receivers of Victims from Mass Casualty Incidents Involving the Release of Hazardous Substances.* Washington, D.C.: Occupational Safety and Health Administration, 2005. osha.gov/dts/osta/bestpractices/html/hospital_firstreceivers.html (accessed October 15, 2007).

Reitherman, Robert. "How to Prepare a Hospital for an Earthquake." *Journal of Emergency Medicine* 4, no. 2 (1986): 119–131.

Sidell, Frederick R., Ernest Takafuji, and David R. Franz. *Medical Aspects of Chemical and Biological Warfare.* In Part 1 of *Textbook of Military Medicine,* edited by Russ Zajtchuk and Ronald F. Bellamy. Washington, D.C.: Office of the Surgeon General, Borden Institute, Walter Reed Army Medical Center, 1997. globalsecurity.org/wmd/library/report/1997/cwbw/ (accessed October 15, 2007).

U.S. Department of Justice, Federal Bureau of Investigation, and the U.S. Army Soldier Biological Chemical Command. *Criminal and Epidemiological Investigation Handbook.* 2003. ecbc.army.mil/downloads/mirp/ECBC_ceih.pdf (accessed October 15, 2007).

Chapter 11 Recovery

Chrislip, David. *The Collaborative Leadership Fieldbook: A Guide for Citizens and Civic Leaders.* San Francisco: Jossey Bass, 2002.

Comerio, Mary. *Disaster Hits Home: New Policy for Urban Housing Recovery.* Berkeley: University of California Press, 1998.

Emergency Operations Organization. *Recovery and Reconstruction Plan.* Los Angeles, Calif., City of Los Angeles, 1994. lacity.org/epd/pdf/mpa/r&r%20annex%20plan.pdf (accessed October 15, 2007).

Federal Emergency Management Agency (FEMA). *The Great Midwest Flood: Voices 10 Years Later.* Washington, D.C.: FEMA, U.S. Department of Homeland Security, 2003. fema.gov/library/viewRecord.do?id=1789 (accessed October 15, 2007).

Laye, John. *Avoiding Disaster: How to Keep Your Business Going When Catastrophe Strikes.* Hoboken, N.J.: John Wiley & Sons, 2002.

National Research Council. *Practical Lessons from the Loma Prieta Earthquake.* Washington, D.C.: National Academies Press, 1994.

Natural Hazards Research and Applications Information Center. *Holistic Disaster Recovery: Ideas for Building Local Sustainability after a Natural Disaster.* Boulder: Natural Hazards Center, University of Colorado, 2006.

Schwab, Jim, Kenneth C. Topping, Charles C. Eadie, Robert E. Deyle, and Richard A. Smith. *Planning for Post-Disaster Recovery and Reconstruction.* PAS Report #483/484. Washington, D.C.: Federal Emergency Management Agency and American Planning Association, 1998.

Wilson, Richard C. *The Loma Prieta Earthquake: What One City Learned.* Washington, D.C.: ICMA, 1991.

Chapter 12 Legal issues

Baltic, Scott. "ICS for Everyone." *Homeland Preparedness Professional* 3, no. 1 (January/February 2004).

Bea, Keith. *Emergency Management Preparedness Standards: Overview and Options for Congress.* CRS RL-32520. Washington, D.C.: Congressional Research Service, Library of Congress, February 4, 2005. fas.org/sgp/crs/homesec/RL32520.pdf (accessed October 15, 2007).

Bea, Keith, L. Cheryl Runyon, and Kae M. Warnock. *Emergency Management and Homeland Security Statutory Authorities in the States, District of Columbia, and Insular Areas: A Summary.* CRS RL-32287. Washington, D.C.: Congressional Research Service, Library of Congress, March 17, 2004. digital.library.unt.edu/govdocs/crs//data/2004/upl-meta-crs-6566/RL32287_2004Mar17.pdf (accessed October 15, 2007).

Comfort, Louise K., ed. *Managing Disasters: Strategies and Policy Perspectives.* Durham, N.C.: Duke University Press, 1988.

Haddow, George D., Jane A. Bullock, and Damon P. Coppola. *Introduction to Emergency Management.* 3rd ed. Burlington, Mass.: Butterworth-Heinemann, 2008.

Hogue, Henry B., and Keith Bea. *Federal Emergency Management and Homeland Security Organization: Historical Developments and Legislative Options.* CRS RL-33369. Washington, D.C.: Congressional Research Service, Library of Congress, June 1, 2006. fas.org/sgp/crs/homesec/RL33369.pdf (accessed October 15, 2007).

Kamien, David G., ed. *The McGraw-Hill Homeland Security Handbook.* New York: The McGraw-Hill Companies, 2005.

National Commission on Terrorist Attacks upon the United States. *The 9/11 Commission Report.* New York: W. W. Norton, July 2004.

Nicholson, William C. *Emergency Response and Emergency Management Law: Cases and Materials.* Springfield, Ill.: Charles C Thomas Publishers, 2003.

———. *Homeland Security Law and Policy.* Springfield, Ill.: Charles C Thomas Publishers, 2005.

———. "Legal Issues in Emergency Response to Terrorism Incidents Involving Hazardous Materials: The Hazardous Waste Operations and Emergency Response ('HAZWOPER') Standard, Standard Operating Procedures, Mutual Aid and the Incident Management System." *Widener Symposium Law Journal* 9, no. 2 (2003): 295, 298–300.

———. "National Incident Management System (NIMS) Basics: What Managers Need to Know." *Best Practices in Emergency Services* 7, no. 7 (July 2004): 76.

———. "Seeking Consensus on Homeland Security Standards: Adopting the National Response Plan and the National Incident Management System." *Widener Law Review* 11, no. 2 (2006).

Platt, Rutherford H. *Disasters and Democracy: The Politics of Extreme Natural Events.* Washington, D.C.: Island Press, 1999.

Rubin, Claire B., William Cumming, and Irmak Renda-Tanali. *Terrorism Time Line: Major Focusing Events and Their U.S. Outcomes (1993–2006).* Arlington, Va.: Claire B. Rubin & Associates, May 2005. disaster-timeline.com (accessed October 15, 2007).

Sylves, Richard T., and William L. Waugh, Jr., eds. *Disaster Management in the U.S. and Canada.* Springfield, Ill.: Charles C Thomas Publishers, 1996.

Waugh, William L., Jr. "The 'All-Hazards' Approach Must Be Continued." *Journal of Emergency Management* 2, no. 1 (2004): 11–12.

Chapter 13 Social vulnerabilities

Bender, Stephen O., Lisa Flax, Russell W. Jackson, and David N. Stein, eds. "Special Issue: Application of Vulnerability Assessment Techniques in the Americas." *Natural Hazards Review* (November 2002).

Bolin, Robert, with Lois Stanford. *The Northridge Earthquake: Vulnerability and Disaster.* London: Routledge, 1998.

Cutter, Susan L., ed. *American Hazardscapes: The Regionalization of Hazards and Disasters.* Washington, D.C.: Joseph Henry Press, 2001.

———. *Hazards, Vulnerability and Environmental Justice.* London: Earthscan, 2006.

Enarson, Elaine, Cheryl Childers, Deborah Thomas, and Ben Wisner. "A Social Vulnerability Approach to Disasters." FEMA Higher Education Project. May 2003. training.fema.gov/emiweb/edu/sovul.asp (accessed October 15, 2007).

Enarson, Elaine, and Betty Hearn Morrow, eds. *The Gendered Terrain of Disaster: Through Women's Eyes.* Westport, Conn.: Greenwood, 1998.

Fothergill, Alice. "Gender, Risk, and Disaster." *International Journal of Mass Emergencies and Disasters* 14 (1996): 33–56.

Fothergill, Alice, Jo Anne Darlington, and Enrique Maestas. "Race, Ethnicity and Disasters in the United States: A Review of the Literature." *Disasters* 23, no. 2 (1999): 156–173.

Hewitt, Kenneth. *Regions of Risk: A Geographical Introduction to Disasters.* London: Longman, 1997.

Mileti, Dennis. *Disasters by Design: A Reassessment of Natural Hazards in the United States.* Washington, D.C.: Joseph Henry Press, 1999.

Morrow, Betty Hearn. "Identifying and Mapping Community Vulnerability." *Disasters* 23, no. 1 (1999): 1–18. sciencepolicy.colorado.edu/about_us/meet_us/roger_pielke/envs_5120/week_12/Morrow.pdf (accessed October 15, 2007).

Natural Hazards Research and Applications Information Center. *Holistic Disaster Recovery: Ideas for Building Local Sustainability after a Natural Disaster.* Boulder: Natural Hazards Center, University of Colorado, 2001.

Parr, Arnold. "Disasters and Disabled Persons: An Examination of the Safety Needs of a Neglected Minority." *Disasters* 11, no. 2 (1987): 148–159.

Peacock, Walter Gillis, Betty Hearn Morrow, and Hugh Gladwin, eds. *Hurricane Andrew: Ethnicity, Gender and the Sociology of Disasters.* New York: Routledge, 1997.

Perry, Ronald W., and Alvin H. Mushkatel. *Minority Citizens in Disasters.* Athens: University of Georgia Press, 1986.

Phillips, Brenda D. "Cultural Diversity in Disasters: Sheltering, Housing, and Long-Term Recovery." *International Journal of Mass Emergencies and Disasters* 11, no. 1 (1993): 99–110.

Wisner, Benjamin, Piers Blaikie, Terry Cannon, and Ian Davis. *At Risk: Natural Hazards, People's Vulnerability and Disasters.* 2nd ed. London: Routledge, 2004.

Chapter 14 New information technologies

Chang, N.-B., W. L. Wei, C. C. Tseng, and C.-Y.J. Kao. "The Design of a GIS-based Decision Support System for Chemical Emergency Preparedness and Response in an Urban Environment." *Computer, Environment and Urban Systems* 21, no. 1 (1997): 67–94.

Cutter, Susan L. "GI Science, Disasters, and Emergency Management." *Transactions in GIS* 7, no. 4 (2003): 439–446.

"GIS for Emergency Management." ESRI White Paper J-8283. Redlands, Calif.: ESRI, July 1999. esri.com/library/whitepapers/pdfs/emermgmt.pdf (accessed October 15, 2007).

Gunes, A. Ertug, and Jacob P. Kovel. "Using GIS in Emergency Management Operations." *Journal of Urban Planning and Development* 126, no. 3 (2000): 136–149.

Johnson, Russ. "GIS Technology for Disasters and Emergency Management." ESRI White Paper J-8474. Redlands, Calif.: ESRI, May 2000. esri.com/library/whitepapers/pdfs/disastermgmt.pdf (accessed October 15, 2007).

National Research Council, Committee on Planning for Catastrophe: A Blueprint for Improving Geospatial Data, Tools, and Infrastructure. *Successful Response Starts with a Map: Improving Geospatial Support for Disaster Management.* Washington, D.C.: National Academies Press, 2007.

Oosterom, Peter van, Siyka Zlatanova, and Elfriede M. Fendel, eds. *Geo-Information for Disaster Management.* New York: Springer, 2005.

Parkinson, Bradford W., James J. Spilker Jr., Penina Axelrad, and Per Enge, eds. *Global Positioning System: Theory and Applications.* Vol. 1, *Progress in Astronautics and Aeronautics.* Reston, Va.: American Institute of Aeronautics and Astronautics, 1996.

Perry, Ronald W., and Michael K. Lindell. "Preparedness for Emergency Response: Guidelines for the Emergency Planning Process." *Disasters* 27, no. 4 (2003): 336–350.

Zerger, Andre, and David I. Smith. "Impediments to Using GIS for Real-Time Disaster Decision Support." *Computers, Environment and Urban Systems* 27, no. 2 (2003): 123–141.

Chapter 15 Budgeting

Bullock, Jane A., George D. Haddow, Dammon P. Coppola, Erdem Ergin, Lissa Westerman, and Sarp Yeletaysi. *Introduction to Homeland Security.* 2nd ed. Burlington, Mass.: Butterworth-Heinemann, 2006.

Canton, Lucien G. *Emergency Management: Concepts and Strategies for Effective Programs.* Hoboken, N.J.: Wiley-Interscience, 2006.

Daniels, Ronald J., Donald F. Kettl, and Howard Kunreuther, eds. *One Risk and Disaster: Lessons from Hurricane Katrina.* Philadelphia: University of Pennsylvania Press, 2006.

Kunreuther, Howard, and Richard J. Roth Sr., eds. *Paying the Price: The Status and Role of Insurance against Natural Disasters in the United States.* Washington, D.C.: Joseph Henry Press, 1998.

Lee, Robert D., Jr., Ronald W. Johnson, and Philip G. Joyce. *Public Budgeting Systems.* 8th ed. Sudbury, Mass.: Jones and Bartlett, 2008.

Chapter 16 Future directions

Anderson, William A., and Shirley Mattingly. "Future Directions." In *Emergency Management: Principles and Practice for Local Government,* edited by Thomas E. Drabek and Gerard J. Hoetmer, 311–335. Washington, D.C.: ICMA, 1991.

Beauchesne, Ann M. *A Governor's Guide to Emergency Management.* Vol. 1, *Natural Disasters.* Washington, D.C.: National Governors Association, 2001. nga.org/cda/files/REPORTEMERGUIDE2001.pdf (accessed October 15, 2007).

Birkland, Thomas E. *Lessons from Disaster: Policy Change after Catastrophic Events.* Washington, D.C.: Georgetown University Press, 2006.

Boin, Arjen; Paul 'tHart, Eric Stern, and Bengt Sundelius. *The Politics of Crisis Management: Public Leadership under Pressure.* Cambridge: Cambridge University Press, 2005.

Burby, Raymond J., ed. *Cooperating with Nature: Confronting Natural Hazards with Land-Use Planning and Sustainable Communities.* Washington, D.C.: Joseph Henry Press, 1998.

Canton, Lucien G. *Emergency Management: Concepts and Strategies for Effective Programs.* Hoboken, N.J.: Wiley-Interscience, 2006.

Comfort, Louise K. *Shared Risk: Complex Systems in Seismic Response.* Oxford: Pergamon/Elsevier Science Ltd., 1999.

Cooper, Christopher, and Robert Block. *Disaster: Hurricane Katrina and the Failure of Homeland Security.* New York: Times Books, 2006.

Cutter, Susan L., ed. *American Hazardscapes: The Regionalization of Hazards and Disasters.* Washington, D.C.: Joseph Henry Press, 2001.

Kettl, Donald F. *System under Stress: Homeland Security and American Politics.* 2nd ed. Washington, D.C.: CQ Press, 2007.

Kunreuther, Howard, and Richard J. Roth Sr., eds. *Paying the Price: The Status and Role of Insurance against Natural Disasters in the United States.* Washington, D.C.: Joseph Henry Press, 1998.

National Emergency Management Association (NEMA). *If Disaster Strikes Today, Are You Ready to Lead? A Governor's Primer on All-Hazards Emergency Management.* Lexington, Ky.: NEMA, 2003. nemaweb.org/docs/Gov_Primer.pdf (accessed October 15, 2007).

Perrow, Charles. *The Next Catastrophe: Reducing Our Vulnerabilities to Natural, Industrial, and Terrorist Disasters.* Princeton, N.J.: Princeton University Press, 2007.

Perry, Ronald W., and E. L. Quarantelli. *What Is a Disaster? New Answers to Old Questions.* Philadelphia: Xlibris, 2005.

Rodriguez, Havidán, Enrico L. Quarantelli, and Russell R. Dynes, eds. *Handbook of Disaster Research.* New York: Springer, 2007.

Tierney, Kathleen J., Michael K. Lindell, and Ronald W. Perry. *Facing the Unexpected: Disaster Preparedness and Response in the United States.* Washington, D.C.: Joseph Henry Press, 2001.

Waugh, William L., Jr. "Assessing Quality in Emergency Management." In *Performance and Quality Measurement in Government: Issues and Experiences*, edited by Arie Halachmi, 665–682. Burke, Va.: Chatelaine Press, 1999.

———. "EMAC, Katrina, and the Governors of Louisiana and Mississippi," *Public Administration Review*, Special Issue on the Katrina Disaster (December 2007).

———. "Mechanisms for Collaboration in Emergency Management: ICS, NIMS, and the Problem of Command and Control." *The Collaborative Manager*, ed. Rosemary O'Leary. Washington, D.C.: Georgetown University Press, forthcoming 2008.

———, ed. *Shelter from the Storm: Repairing the National Emergency Management System after Hurricane Katrina.* Thousand Oaks, Calif.: Sage, 2006.

Waugh, William L., Jr., and Gregory Streib. "Collaboration and Leadership for Effective Emergency Management," *Public Administration Review* 66 (Suppl. 1) (December 2006): 131–140.

Waugh, William L., Jr., and Richard T. Sylves. "Organizing the War on Terrorism." *Public Administration Review* 62 (Suppl. 1) (September 2002): 145–153.

CONTRIBUTORS

William L. Waugh Jr. (Editor and Chapters 1 and 16) is professor of public administration in the Andrew Young School of Policy Studies at Georgia State University in Atlanta. He also teaches in the graduate program in crisis and emergency management at the University of Nevada, Las Vegas. He serves on the Emergency Management Accreditation Program (EMAP) Commission and has also served on the Certified Emergency Manager (CEM) Commission. Recognized nationally and internationally for his scholarship and contributions to the emergency management profession, Dr. Waugh is the editor-in-chief of the *Journal of Emergency Management* and is on the editorial board of *Public Administration Review.* He attended the University of Maryland in Munich, Germany, and College Park and earned his bachelor's degree from the University of North Alabama, his master's degree from Auburn University, and his doctorate in political science from the University of Mississippi.

Kathleen Tierney (Editor and Chapter 16) is professor of sociology and director of the Natural Hazards Center at the University of Colorado at Boulder. The Hazards Center is housed in the Institute of Behavioral Science, where Dr. Tierney holds a joint appointment. Prior to her arrival at the University of Colorado in 2003, she was the director of the Disaster Research Center at the University of Delaware. Dr. Tierney is also a member of the executive committee of the National Consortium for the Study of Terrorism and Responses to Terrorism, a Department of Homeland Security social science academic center of excellence headquartered at the University of Maryland. Her research interests include the study of crisis-related collective behavior; extreme event preparedness among governmental, for-profit, and nonprofit organizations; the use of new information technologies in emergency management; and disaster and homeland security policy and institutions. Her publications include numerous articles and book chapters, among which are chapters on business disaster preparedness and on homeland security policy. Dr. Tierney teaches courses on qualitative research methods and on societal dimensions of hazards and disasters at the University of Colorado.

Beverley J. Adams (Chapter 14) is technical director of ImageCat's Remote Sensing operations and managing director of the company's European division in London. She is an expert in the application of remote sensing (RS) and geographic information system (GIS) technology for disaster management and risk assessment, and her current focus is on commercially implementing ImageCat's suite of RS-based risk/damage assessment products and services. After obtaining a PhD from University College London (UCL), where she was awarded the prestigious College and Broderick-Parry prizes for academic achievement, Dr. Adams spent four years as a member and then leader of the RS team at ImageCat headquarters in Los Angeles. During this time, she served as the principal investigator (PI) or co-PI on research grants from the National Science Foundation, the U.S. Department of Transportation, and the National Aeronautics and Space Administration, and was invited to brief U.S. White House Homeland Security Council members on emerging advanced technology applications for disaster management. Dr Adams and her Los Angeles- and London-based RS teams currently participate in research collaborations with U.S. and international organizations, including the Multidisciplinary Center for Earthquake Engineering Research (MCEER); Stanford University; Louisiana State University (LSU); University of California-Irvine; University of British Columbia; Cambridge University; and the University of Bologna. In addition to these activities, Dr. Adams has managed large international projects using airborne, satellite, and UAV (unmanned aerial vehicle) data; LIDAR (light detection and ranging) imagery; stereoscopic photography; and SAR (synthetic aperture radar) data to characterize, assess, and model natural and man-made disasters; in so doing, she has pioneered the application of object-based processing and spatially tiered analytical methodologies for multihazard damage assessment. Dr. Adams has also directed field reconnaissance missions following the Bam earthquake; the Indian Ocean tsunami; and Hurricanes Katrina, Rita, Charley, and Ivan. Recent publications include milestone reviews on the use of RS technology following both Hurricane Katrina and the World Trade Center attack, as well as journal articles, book chapters, and industry monographs. She currently serves as a reviewer for the journal *Earthquake Spectra* and sits on the panel of graduate students at LSU and UCL.

Erik Auf der Heide, MD, MPH, FACEP (Chapter 10), is a medical officer with the Agency for Toxic Substances and Disease Registry (ATSDR), U.S. Department of Health and Human Services. The agency is administered by and collocated with the Centers for Disease Control and Prevention (CDC) in Atlanta. His undergraduate training was at the University of California at the Riverside and Berkeley campuses. He received his medical degree from Baylor College of Medicine in Houston; he then completed his specialty training in emergency medicine at Akron City Hospital in Akron, Ohio, and a fellowship in disaster medicine cosponsored by the CDC and the School of Medicine and Rollins School of Public Health at Emory University. As a part of his fellowship, he earned a master's degree in public health with a focus on epidemiology. He is board certified in emergency medicine and is a Fellow of the American College of Emergency Physicians. He practiced clinical emergency medicine for eighteen years before coming to government service. Dr. Auf der Heide has published numerous works and lectured widely, both nationally and internationally, on the topics of disaster management and disaster research.

Susan L. Cutter (Chapter 14) is a Carolina Distinguished Professor of Geography at the University of South Carolina and director of the university's Hazards and Vulnerability Research Institute. She received her BA from California State University-Hayward and her MA and PhD (1976) from the University of Chicago. She has authored or edited twelve books, and more than a hundred peer-reviewed articles and book chapters. Dr. Cutter is a member of the Executive Committee of the U.S. Department of Homeland Security's National Consortium for the Study of Terrorism and Responses to Terrorism (START), and she serves on many national advisory boards and committees, including those of National Research Council, the American Association for the Advancement of Science (AAAS), the National Science Foundation, the Natural Hazards Center, the H. John Heinz III Center, and the U.S. Army Corps of Engineers's Interagency Performance Evaluation Task Force (IPET) team studying the impacts of Hurricane Katrina. She is an elected Fellow of the AAAS (1999), past president of the Association of American Geographers (2000), and current president of the Consortium of Social Science Associations (COSSA). In 2006, Dr. Cutter received the Decade of Behavior Research Award, given by a multidisciplinary consortium of more than fifty national and international scientific organizations in the social and behavioral sciences.

Gregg Dawson (Chapter 4) is director of the Emergency Preparedness and Homeland Security Programs for the North Central Texas Council of Governments. Previously he served as emergency management coordinator and emergency operations center manager for Fort Worth–Tarrant County, Texas. He serves on the Texas governor's Homeland Security Advisory Board and on a number of other state and national performance standards and advisory boards. He has also served as a curriculum adviser and adjunct instructor for the University of North

Texas, the University of Texas, community colleges, and state agencies. The recipient of local and national awards for leadership in emergency services, he has published numerous articles in national and international publications and is proactive in promoting innovative emergency management practices. He earned his bachelor's degree from the University of North Texas, has an associate's degree in nursing, completed additional studies in health and society at Regents College in London, and is a certified Alternative Dispute Mediation specialist.

Frances L. Edwards, PhD, CEM (Chapter 3), is a research associate with the Mineta Transportation Institute and associate professor at San José State University in California, which she is also director of the Master of Public Administration program. She served as director of the Office of Emergency Services in San José for fourteen years, during which time she also served as director of San José's Metropolitan Medical Task Force (MMTF) and head of the city's Urban Area Security Initiative, a U.S. Department of Homeland Security grant program. Her current research interests include emergency management and homeland security, and immigration issues. She has published several books, more than thirty professional articles, numerous chapters for textbooks, and three op-ed columns for the San José *Mercury News*. She was a commissioner for the California Seismic Safety Commission from 1991 to 1995.

Ronald T. Eguchi (Chapter 14) is president and chief executive officer of ImageCat, Inc., an advanced technology company specializing in innovative solutions to risk assessment and management. He has more than thirty years of experience in risk analysis and risk management studies, and has directed major research and application studies in these areas for government agencies and private industry. He serves or has served on editorial boards for the *Natural Hazards Review,* published by the American Society of Civil Engineers (ASCE); the Natural Hazards Research and Applications Information Center at the University of Colorado; the *Journal on Uncertainties in Engineering Mechanics,* published by Resonance Publications, Inc.; and *SPECTRA,* the journal of the Earthquake Engineering Research Institute (EERI). He was appointed to a National Research Council (NRC) panel that prepared a report titled "Assessing the Costs of Natural Disasters," and he currently serves on the NRC's Disaster Roundtable. He has been an active member of the ASCE Technical Council on Lifeline Earthquake Engineering, having chaired the council's Executive Committee in 1991, and in 1997, he was awarded the council's C. Martin Duke Award for his contributions to the area of lifeline earthquake engineering. He was recently awarded the 2008 EERI Distinguished Lecturer Award. He has authored more than two hundred publications, many dealing with the seismic risk of utility lifeline systems and the application of remote sensing technologies for disaster response.

Christopher T. Emrich (Chapter 14) received his PhD in geography from the University of South Carolina, where he specialized in geographic information science and hazards. His primary research interests are in the application of geospatial analysis techniques and applications to real-world problems, including emergency management, social vulnerability, environmental issues, and planning. In 2004, he was involved in FEMA's Florida hurricane response and recovery, where he conducted geographic information system (GIS)-based analysis of recovery operations and mitigation programs. Recently, he directed all FEMA Florida GIS activities as the GIS unit leader of the Florida Long Term Recovery Office. He is currently a visiting assistant professor in the Department of Geography at the University of South Florida, where he teaches undergraduate- and graduate-level GIS courses, and continues his research into the use and implementation of geospatial techniques in emergency management and general planning activities.

Elaine Enarson (Chapter 13) is a disaster sociologist and an international consultant on gender equity and disaster risk reduction, currently teaching for Brandon University's Department of Applied Disaster and Emergency Studies Department in Brandon, Manitoba. After earning her bachelor's degree from the University of California at Santa Cruz, she earned her master's and doctorate in sociology from the University of Oregon; since then, she has worked as an independent scholar in the United States, Canada, and Australia. Dr. Enarson served as lead course developer of the college course "A Social Vulnerability Approach to Disaster" for FEMA's Higher Education Project, helped develop risk assessment guidelines for grassroots women in the Caribbean, and initiated the online Gender and Disaster Sourcebook project. Her field studies have focused on social vulnerability to natural disasters in the United States, Canada, and India.

David R. Godschalk (Chapter 6) is Stephen Baxter Professor Emeritus in the Department of City and Regional Planning and is an adjunct professor in the Kenan-Flagler Business School at the University of North Carolina (UNC) at Chapel Hill. A Fellow of the American Institute of Certified Planners, he has been a city planning director and a planning consultant in Florida. As a consultant, he worked with local governments in North Carolina and Florida on smart growth, hazard mitigation, coastal management, and conflict resolution; he advised Portugal's Natural Park Service on coastal protection and Saudi Arabia's minister of Municipal and Rural Affairs on growth management. Dr. Godschalk has done extensive research on natural hazard mitigation and serves on the Multihazard Mitigation Council of the National Institute of Building Sciences, where he oversaw the preparation of a report to Congress on the benefits of natural hazard mitigation. A former member of the North Carolina Smart Growth Commission and a former member of the Chapel Hill Town Council, he is a registered architect (inactive) in the state of Florida. He holds a bachelor's degree in art from Dartmouth College, a bachelor's degree in architecture from the University of Florida, and a master's degree and a doctorate in city planning from UNC.

Daniel C. Goodrich, MPA, CEM (Chapter 3) is an instructor and research associate for the Mineta Transportation Institute at the San José State University College of Business. He was a 2006 Fellow of the Foundation for Defence of Democracy. He has been the director for eight exercises for the San José Metropolitan Medical Task Force, where he created facilitated exercises from which Harvard University's Kennedy School of Government created a case study. He has delivered eight professional papers, including "Fourth Generation Warfare" at the 2006 NATO Security through Science—Conference on National Armaments Directors (STS-CNAD) meeting in Portugal, and in 2004 he chaired a session on "First Responders" at the NATO Advanced Research Workshop in Germany. He served in the U.S. Marine Corps for ten years, during which time he held leadership positions in Security Forces. He is a consultant to the California Department of Transportation and has trained National Aeronautics and Space Administration/Ames Research Center staff in emergency management. He has three entries on nuclear topics in the *WMD Encyclopedia.* Mr. Goodrich has a master's degree in public administration.

Charles K. Huyck (Chapter 14) is a geographer specializing in the integration of advanced geospatial technologies and emergency management. As executive vice president of ImageCat, Inc., he oversees a team of engineers, scientists, and programmers developing software tools and data-processing algorithms for loss estimation and risk reduction. He has more than fifteen years of experience in geographic information system (GIS) analysis and application development. He introduced GIS and remote sensing to EQE International, Ltd., where he served as GIS programmer analyst on several loss estimation and research projects. At the California Governor's Office of Emergency Services, he was responsible for geospatial analysis, database development, and mapping disaster information for the Northridge earthquake, California winter storms, and California firestorms. He has authored more than fifty papers, many pertaining to damage detection and inventory development with remotely sensed data.

Michael K. Lindell (Chapters 7 and 9) is a senior scholar in the Hazard Reduction and Recovery Center, professor of landscape architecture and urban planning, and adjunct professor of psychology at Texas A&M University. Since receiving his PhD in social and quantitative psychology from the University of Colorado in 1975, he has conducted more than forty major research projects, many funded by the National Science Foundation; he has made more than

170 presentations before scientific societies and short courses for emergency planners; and he has been an invited participant in workshops on risk communication and emergency management in the United States and abroad. He has also served as a consultant to nuclear utilities and chemical companies, as well as to local, state, and national emergency management agencies. He recently served as consultant to the International Atomic Energy Agency and as a member of two National Academy of Sciences panels-"Disasters Research in Social Sciences" and "Assessing Vulnerabilities Related to the Nation's Chemical Infrastructure." He has written extensively on emergency management and has authored seventy technical reports, ninety journal articles and book chapters, and ten books. Professor Lindell is the current editor of the *International Journal of Mass Emergencies and Disasters.*

David A. McEntire (Chapter 4) is an associate professor in the Emergency Administration and Planning Program at the University of North Texas (UNT), where he also serves as the PhD coordinator in the Department of Public Administration. His research interests include emergency management theory, vulnerability reduction, community preparedness, response coordination, and homeland security. He serves on the editorial board of the *Journal of Emergency Management* and is a member of the FEMA Region VI Advisory Board. His work has received awards for outstanding articles in *Disaster Prevention and Management,* and he has published books with John Wiley and Charles C Thomas. Prior to his career in academia, he worked for the American Red Cross in Denver, Colorado. Dr. McEntire received his bachelor's degree from Brigham Young University, and his master's and doctorate from the University of Denver.

David M. Neal (Chapter 11) is the director of the Center for the Study of Disasters and Extreme Events and professor of political science at Oklahoma State University. He is also a member of the Fire and Emergency Management Program. He has published on a wide range of topics related to disaster, and has been actively engaged in promoting higher education for professionals in disaster management. His current work focuses on the National Incident Management System and effective disaster response. He received his bachelor's degree in education with a major in political science and minor in sociology from Bowling Green State University, where he also earned his master's degree in sociology. He obtained his PhD in sociology from the Ohio State University, where he also served as a research assistant at the Disaster Research Center.

William C. Nicholson, Esq. (Chapter 12) is an assistant professor in North Carolina Central University's Criminal Justice Department, where he is also affiliated with the Institute for Homeland Security and Workforce Development. He previously served as general counsel to the Indiana State Emergency Management Agency. Recent notable publications include two books focusing on emergency management and homeland security law, as well as two major law review articles addressing legal aspects of incident management, the National Incident Management System, and the National Response Plan. He has received a number of awards for his scholarly work in the area of emergency management and homeland security, and he serves as a member of editorial boards for emergency management publications. Dr. Nicholson earned his bachelor's degree from Reed College in Portland, Oregon, and a Juris Doctor from Washington and Lee University's School of Law in Lexington, Virginia.

Ann Patton (Chapter 5) is a writer, consultant, and grassroots advocate for disaster-resistant, sustainable communities. She has more than thirty five years' experience in Tulsa, Oklahoma, as a citizen activist, newspaper reporter, city programs manager, and writer/editor. She was the founding director of three award-winning local programs-Tulsa Partners Inc., Project Impact, and Citizen Corps-all working through partnerships to create safe, sustainable families and communities. She is also a charter member of the team that built Tulsa's flood control and hazard mitigation programs. Among her commendations are a first-place national award from the American Society of Planning Officials and two local journalism awards from the National Council of Christians and Jews (1970s); FEMA's national public services award (1998); Oklahoma's Ben Frizzell Award for leadership (2003); Oklahoma Floodplain Managers

Association lifetime achievement award (2004); Tulsa Partners' J. D. Metcalfe Building Bridges Award for community service (2004); and a Goodwill Appreciation Award from the Islamic Society of Tulsa and the Council of American Islamic Relations (2007). In 2004, the city of Tulsa named an open-space floodplain park "Ann Patton Commons."

Ronald W. Perry (Chapters 7 and 9) is professor of public affairs at Arizona State University. His principal research interests are in disaster preparedness, public hazard education, and disaster operations. He currently serves on the steering committees for the Phoenix Urban Area Security Initiative and the Phoenix Metropolitan Medical Response System, as well as on the Arizona Council for Earthquake Safety. He holds the "Award for Excellence in Emergency Management" from the Arizona Emergency Services Association and the "Award for Outstanding Environmental Achievement" from the U.S. Environmental Protection Agency, and the Phoenix fire department selected him for the Pearce Memorial Award for Contributions to the Hazardous Incident Response Team.

Brenda D. Phillips (Chapter 11) is a professor in the Fire and Emergency Management Program at Oklahoma State University and a senior researcher with the Center for the Study of Disasters and Extreme Events. She has provided extensive consulting and voluntary work on disaster recovery and special-needs populations to organizations across the United States. Dr. Phillips has also assisted with the development of emergency management programs in the United States, Mexico, Costa Rica, and Canada. She earned her bachelor's degree from Bluffton College in Ohio and both a master's and a doctorate in sociology from the Ohio State University.

Richard (Richie) A. Rotanz (Chapter 8) is special adviser to the provost for emergency management studies at Adelphi University in New York, where he has contributed to the development of the university's graduate certificate program for emergency management and to Adelphi's currently developing master's degree program. He is also a chief adviser for the development of a new Applied Research Center for Homeland Security on Long Island. Previously, he was the first commissioner of emergency management for Nassau County, New York, in which capacity he directly supervised the creation of that agency, its first emergency operation plan, its emergency operation center (EOC), and many other related functions. Prior to that, he was deputy commissioner of New York City's Office of Emergency Management (a detailed officer from the New York fire department), responsible for all citywide disaster planning and for managing the city's EOC, specifically during the 9/11 attacks. Mr. Rotanz has been involved in various articles and texts in the field of emergency management as well as the fire service. He earned his bachelor's degree in fire science and a master's degree in protection management, and he is currently working on his doctorate in public policy administration.

Claire B. Rubin, MA (Chapter 2), is a social scientist with almost thirty years of experience as a researcher, practitioner, and academic in the fields of emergency management and homeland security. She is president of Claire B. Rubin & Associates (www.clairerubin.com), a small business specializing in disaster research and consulting located in Arlington, Virginia, and is also a visiting research scholar at the George Washington University's Institute for Crisis, Disaster, and Risk Management. She is a cofounder and currently the managing editor of the *Journal of Homeland Security and Emergency Management.* She has written more than sixty-five publications on hazards and disasters, and has delivered numerous lectures and presentations on emergency management and homeland security topics. Currently, Ms. Rubin is working on the development of several educational products, including the *Time Line Series* (www.disaster-timeline.com).

Joseph Scanlon (Chapter 10) is professor emeritus and director of the Emergency Communications Research Unit at Carleton University in Ottawa, Canada. He has been doing disaster research for thirty-seven years and has published scores of book chapters, monographs, and articles in academic and professional journals. In 2002, he received the Charles Fritz award for his lifetime contribution to the area of the sociology of disaster. His most recent research has involved a study of the overseas response to the handling of the tsunami dead. His work on the problems of dealing with mass casualties began with a study of the Woodstock and Edmonton tornadoes in Canada, and it continued when he had an Oak Ridge fellowship at the Agency for Toxic Substances and Disease Registry, a CDC agency in Atlanta, where he worked with his co-author, Dr. Erik Auf der Heide. Before joining the faculty at Carleton, he was executive assistant to the Honourable Judy LaMarsh, minister of health in the Canadian federal cabinet.

Richard T. Sylves (Chapter 15) is a professor of political science and international relations and a senior policy fellow of the Center on Energy and Environmental Policy at University of Delaware, where he previously directed the Environmental and Energy Policy graduate program. He recently completed a three-year term as an appointed member of the National Academy of Science Disaster Roundtable Executive Committee. He has held two postdoctorates, one as a producer with WHYY-TV 12 (Philadelphia) public television news and special productions. He has done extensive research and writing on presidential declarations of disaster (see www.peripresdecusa.org, and he won research grant projects from FEMA, the U.S. Department of Homeland Security-FEMA, National Oceanic and Atmospheric Administration SeaGrant, and the Public Entity Risk Institute. He earned his bachelor's and master's degrees from the State University of New York and his doctorate in political science from University of Illinois at Urbana-Champaign.

INDEX

Emergency Management:
Principles and Practice for Local Government
Second Edition

Text type
Interstate, ITC Slimbach

Composition
Circle Graphics, Columbia, Maryland

Printing and binding
United Book Press, Inc., Baltimore, Maryland

Design
Will Kemp, ICMA, Washington, D.C.

07-302